CONTEMPORARY ABSTRACT ALGEBRA

CONTEMPORARY ABSTRACT ALGEBRA

SECOND EDITION

Joseph A. Gallian

University of Minnesota, Duluth

D. C. HEATH AND COMPANY

Lexington, Massachusetts   Toronto

In memory of my father, Joseph Gallian (1912–1975)

Cover: Photograph by Scott A. Burns. Design by Dustin Graphics.

International Standard Book Number: 0-669-19496-4

Library of Congress Catalog Number: 89-80258

10 9 8 7 6 5 4 3 2 1

Acknowledgments

Page xi. Line from "Paperback Writer" by John Lennon and Paul McCartney, © 1965 Northern Songs Limited. All rights for the United States, Canada, and Mexico controlled and administered by SBK Blackwood Music, Inc. Under license from ATV Music (MacLen). All rights reserved. International copyright secured. Used by permission.

Page 12. "Brain Boggler" by Maxwell Carver, © 1988 by Discover Publications, Inc.

Page 29. Photograph from *Snow Crystals* by W. A. Bentley and W. J. Humphreys, 1962, Dover Publications, N.Y.

Page 31. Cross section of AIDS virus © 1987 by Scientific American, Inc. All rights reserved.

Page 64. Poem from *The Compleat Computer* by Dennie L. Van Tassel and Cynthia Van Tassel. Copyright © 1983, 1976, Science Research Associates, Inc. Reprinted by permission of the publisher.

Page 94. Poem "T.T.T." from *Grooks,* 1966 by Piet Hein. Used with permission of Piet Hein.

Page 215. Poem "Problems" from *Grooks,* 1966 by Piet Hein. Used with permission of Piet Hein.

Page 264. Newspaper article in collage copyright © 1988 by the New York Times Company. Reprinted by permission.

Page 285. Photograph courtesy of Michael Artin.

Pages 317, 355, 356. Photographs courtesy of the American Mathematical Society.

Page 357. Photograph by the author.

Page 372. Photograph reproduced by permission of the Masters and Fellows of Pembroke College, Cambridge, England.

Pages 375, 383, 394, 399, 400. Symmetry patterns from *Symmetry in Science and Art* by A. V. Shubnikov and V. A. Kopstik, 1974. Reprinted by permission of Plenum Publishing Corporation.

Page 378. Figure as reprinted in *Excursions into Mathematics* by A. Beck, M. Blecher, and D. Crowe, Worth Publications, 1969. Reprinted by permission of the publisher.

Pages 389, 391, 397. Symmetry drawings of M. C. Escher from the Collection Haags Gemeentemuseum—The Hague.

Pages 390, 397. Escher-like drawings from *Creating Escher-Type Drawings,* © 1977. Creative Publications, Palo Alto, California.

Pages 392, 393, 395. Designs reprinted from the *American Mathematical Monthly,* Vol. 85 (June 1978) with permission of the publisher.

Page 417, 419. Computer duplications of Escher patterns reproduced here by permission of Douglas Dunham.

Page 426. Photograph by Seymour Schuster.

Page 442. Line from "The Ballad of John and Yoko" by John Lennon and Paul McCartney, © 1969 Northern Songs Limited. All rights for the United States, Canada, and Mexico controlled and administered by SBK Blackwood Music, Inc. Under license from ATV Music (MacLen). All rights reserved. International copyright secured. Used by permission.

Page 471. Poem "The Road to Wisdom" from *Grooks,* 1966 by Piet Hein. Used with permission of Piet Hein.

Page A3. Line from "All You Need Is Love" by John Lennon and Paul Mc-Cartney, © 1967 Northern Songs Limited. All rights for the United States, Canada, and Mexico controlled and administered by SBK Blackwood Music, Inc. Under license from ATV Music (MacLen). All rights reserved. International copyright secured. Used by permission.

Page A5. Line from "Let It Be" by John Lennon and Paul McCartney, © 1970 Northern Songs Limited. All rights for the United States, Canada, and Mexico controlled and administered by SBK Blackwood Music, Inc. Under license from ATV Music (MacLen). All rights reserved. International copyright secured. Used by permission.

Page A9. Line from "While My Guitar Weeps" by George Harrison. Copyright 1968, Harrisongs Limited. Used by permission. All rights reserved.

Page A14. Line from "Day Tripper" by John Lennon and Paul McCartney, © 1965 Northern Songs Limited. All rights for the United States, Canada, and Mexico controlled and administered by SBK Blackwood Music, Inc. Under license from ATV Music (MacLen). All rights reserved. International copyright secured. Used by permission.

Page A22. Line from "With a Little Help from My Friends" by John Lennon and Paul McCartney, © 1967 Northern Songs Limited. All rights for the United States, Canada, and Mexico controlled and administered by SBK Blackwood Music, Inc. Under license from ATV Music (MacLen). All rights reserved. International copyright secured. Used by permission.

Preface to the Second Edition

I began collecting material for the second edition of this book the day after I sent the final version of the first edition to the publisher. This second edition includes hundreds of new exercises, scores of new illustrations, and dozens of new examples. It contains additional applications, computer exercises, historical notes, and biographies, as well as updated references and suggested readings. Several proofs that were omitted in the first edition are presented in this one.

With few exceptions, the organization and content of the first edition remain intact. In response to the feedback I received, nearly all changes for the second edition are additions to the text. These include proofs of the fundamental theorem of finite Abelian groups, the uniqueness of splitting fields, the primitive element theorem for field extensions, and the Sylow theorems. The sections on preliminaries, symmetry groups, divisibility in integral domains, geometric constructions, and electric circuits have been expanded. The chapter in the first edition on designing a ZIP Code reader has been deleted. Tomorrow I begin collecting material for the third edition.

Answers or key steps for virtually all of the odd-numbered exercises are given at the end of the text. Answers or solutions for all of the even-numbered exercises are available in the *Answer Key*.

I am grateful to the following people for serving as reviewers for this edition: Ronald Bercov, University of Alberta; William C. Fox, State University of New York at Stony Brook; Larry C. Grove, University of Arizona; Marshall Hall, Jr., Emory University; Robin Hartshorne, University of California, Berkeley; Loren C. Larson, St. Olaf College; Thomas Q. Sibley, Saint John's University (Minnesota); and Mark L. Teply, University of Wisconsin at Milwaukee.

I'm indebted to the following people who have kindly contributed to this edition in one way or another: Duane Anderson, University of Minnesota, Duluth; Louis Friedler, University of Bridgeport; Anthony Gaglione, United States Naval Academy; Steve Galovich, Carleton College; Ladnor Geissinger, University of North Carolina at Chapel Hill; Branko Grünbaum, University of Washington: Henry Heatherly, University of Southwestern Louisiana; Victor Katz, University of the District of Columbia; James Loats, Metropolitan State College; Monty Strauss, Texas Tech University; and Robin J. Wilson, The Open University.

Finally, I wish to express my appreciation to the typist Daniel Ellison and to Cathy Cantin, Ann Marie Jones, and Karen Wise of D. C. Heath for their assistance in the preparation of this edition.

Joseph A. Gallian

Preface to the First Edition

Dear Sir or Madam will you read my book, it took me years to write, will you take a look.

John Lennon and Paul McCartney, Paperback Writer, single

The academic goals of students in the undergraduate abstract algebra course have changed dramatically in the past decade. Far fewer students plan for advanced degrees in mathematics or intend to enter the teaching profession. Many students now seek degrees in computer science, and most students enter business or industry upon graduation. Despite this trend, abstract algebra is important in the education of a mathematically trained person. The terminology and methodology of algebra are used ever more widely in computer science, physics, chemistry, and data communications, and of course, algebra still has a central role in advanced mathematics itself. What I have attempted to do here is to capture the traditional spirit of abstract algebra while giving it a concrete computational foundation and including applications. I believe that students will best appreciate the abstract theory when they have a firm grasp of just what it is that is being abstracted.

Nearly every student who enrolls in an abstract algebra course brings to it good intuition for numbers and space. I have tried to capitalize on this by including many examples and exercises with a number theoretic or geometric basis.

I agree with George Polya that guessing and conjecturing are important in mathematics and should be taught and encouraged whenever possible. Here the computer is the ideal tool, for it can provide the students with long lists of data from which patterns can be gleaned and conjectures made. I have included computer exercises that ask the students to produce such lists for this purpose. Most of these can be done on an inexpensive programmable calculator.

In my opinion, every undergraduate mathematics course should have a liberal arts character. I have tried to achieve this with comments, historical notes, quotations, biographies and photographs, and in general, by my approach to the entire subject. The lines from popular songs, humorous quotations, and to some extent, the selection of topics and biographies are intended to make the book enjoyable for students to read.

In addition to topics such as symmetry, crystallographic groups, algebraic coding, and Boolean algebras, which are available in a few of the popular texts on abstract algebra, I have included features not found in other texts. Among these are chapters on finite simple groups and digraphs of groups, Escher and Escher-like graphics, annotated lists of films, and an application of the dihedral group of order 8 to the design of a letter-facing machine. An application of cyclotomic polynomials to the labeling of dice to yield the standard probability distribution for the sum of the faces and an application of conjugacy classes to determine the probability that two elements from a group commute add insight to these topics.

Several theorems are presented without proof. In each instance I feel that it is the understanding of the statement of the theorem, not its proof, that is the important issue.

Because of the importance of matrix groups in both theory and applications, I introduce 2×2 matrix addition and multiplication and the determinant function early in the text and use these examples throughout. The field theory and algebraic coding theory portions of the book are preceded by a chapter on the rudiments of vector space theory: basis, linear independence, and dimension. Knowledge of the material covered in a basic linear algebra course would be an advantage but is not prerequisite.

For the most part, the special topics are independent of each other. However, the chapter on Sylow theorems is a prerequisite for the chapter on finite simple groups, and the chapters on symmetry and Cayley digraphs require an intuitive understanding of group generators and relations. Chapter 28 gives a formal treatment of these ideas. I, myself, avoid this formalism and present generators and relations by way of examples.

This book has hundreds of examples, 116 figures, 32 tables, 24 photographs, and over 1200 exercises, including 46 computer exercises. Many of the exercises are computational and concrete. Some extend the topics discussed in the text proper; others lay the groundwork and provide motivation for the ideas presented in subsequent chapters. Answers or hints are provided for nearly all of the odd-numbered exercises. Answers for the even-numbered exercises are available in a supplement.

I would like to acknowledge the thoughtful comments received from the reviewers: Hubert S. Butts, Louisiana State University; Herbert E. Kasube, Bradley University; Hiram Paley, University of Illinois at Urbana-Champaign; Philip Quartararo, Jr., Southern University. Sabra Anderson and Louis Friedler used the manuscript in their classes at the University of Minnesota and the College of Saint Scholastica and made helpful suggestions.

I owe an enormous debt to my friend, David Witte, who meticulously read the entire manuscript. His insightful comments and incisive criticisms have immeasurably improved both the style and content of the text.

I wish to express my appreciation to Jane Lounsberry, Sonja Rasmussen, and Carol Stockman who cheerfully typed and retyped the manuscript.

Finally, I'm grateful to Mary LeQuesne, Margaret Roll, Antoinette Tingley Schleyer, and Mary Lu Walsh of D. C. Heath for their kind cooperation and assistance in the preparation of this text. It has been a pleasure to work with them.

Joseph A. Gallian

Contents

6

Isomorphisms 99

7

External Direct Products 113

8

Internal Direct Products 119

Supplementary Exercises for Chapters 5–8

9

Cosets and Lagrange's Theorem 130

CONTEMPORARY ABSTRACT ALGEBRA

INTEGERS AND EQUIVALENCE RELATIONS

PART 1

Preliminaries

0

The whole of science is nothing more than a refinement of everyday thinking.

Albert Einstein, *Physics and Reality*

Properties of Integers

Much of abstract algebra involves properties of integers and sets. In this chapter we collect the ones we need for future reference.

An important property of the integers, which we will often use, is the so-called Well Ordering Principle. Since this property cannot be proved from the usual properties of arithmetic, we will take it as an axiom.

Well Ordering Principle
Every nonempty set of positive integers contains a smallest member.

The concept of divisibility plays a fundamental role in the theory of numbers. We say a nonzero integer t is a *divisor* of an integer s if there is an integer u such that $s = tu$. In this case, we write $t \mid s$ (read "t divides s"). When t is not a divisor of s, we write $t \nmid s$. A *prime* is a positive integer greater than 1 whose only positive divisors are 1 and itself.

As our first application of the Well Ordering Principle, we establish a fundamental property of integers that we will use often.

Division Algorithm
Let a and b be integers with $b > 0$. Then there exist unique integers q and r with the property that $a = bq + r$ where $0 \leqslant r < b$.

Proof. We begin with the existence portion of the theorem. Consider the set $S = \{a - bk \mid k \text{ is an integer and } a - bk \geq 0\}$. If $0 \in S$, then b divides a and we may obtain the desired result with $q = a/b$ and $r = 0$. Now assume $0 \notin S$. Since S is nonempty (if $a > 0$, $a - b \cdot 0 \in S$; if $a < 0$, $a - b(2a) = a(1 - 2b) \in S$), we may apply the Well Ordering Principle to conclude that S has a smallest member, say $r = a - bq$. Then $a = bq + r$ and $r \geq 0$, so all that remains to be proved is that $r < b$.

On the contrary, if $r > b$, then $a - b(q + 1) = a - bq - b = r - b > 0$ so that $a - b(q + 1) \in S$. But $a - b(q + 1) < a - bq$, and $a - bq$ is the *smallest* member of S. Thus $r \leq b$. If $r = b$, then $0 \in S$ and this case is already taken care of. So, $r < b$.

To establish the uniqueness of q and r, let us suppose that there are integers of q, q', r and r' such that

$$a = bq + r, \quad 0 \leq r < b \qquad \text{and} \qquad a = bq' + r', \quad 0 \leq r' < b.$$

For convenience, we may also suppose that $r' \geq r$. Then $bq + r = bq' + r'$ and $b(q - q') = r' - r$. So, b divides $r' - r$ and $0 \leq r' - r \leq r' < b$. It follows that $r' - r = 0$ and therefore $r' = r$ and $q = q'$. ∎

The integer q in the division algorithm is called the *quotient* upon dividing a by b; the integer r is called the *remainder* upon dividing a by b.

Example 1 For $a = 17$ and $b = 5$, the division algorithm gives $17 = 5 \cdot 3 + 2$; for $a = -23$ and $b = 6$, the division algorithm gives $-23 = -4 \cdot 6 + 1$. □

Several states use linear functions to encode the month and date of birth into a three digit number that is incorporated into driver's license numbers. If the encoding function is known, the division algorithm can be used to recapture the month and date of birth from the three digit number. For instance, the last three digits of a Florida male driver's license number are those given by the formula $40(m - 1) + b$ where m is the number of the month of birth and b is the day of birth. Thus, since $177 = 40 \cdot 4 + 17$, a person with these last three digits was born on May 17. In Missouri, the final three digits of a male's driver's license number are $63m + 2b$, where m denotes the number of the month of birth and b is the day of birth. So, a license with the last three digits of $701 = 63 \cdot 11 + 2 \cdot 4$ indicates that the holder is a male born on November 4. Incidentally, Wisconsin uses the same method as Florida to encode birth information, but the numbers immediately precede the last pair of digits.

DEFINITIONS Greatest Common Divisor, Relatively Prime Integers
The *greatest common divisor* of two nonzero integers a and b is the largest of all common divisors of a and b. We denote this integer by $\gcd(a, b)$. When $\gcd(a, b) = 1$, we say a and b are *relatively prime*.

We leave it as an exercise (exercise 10) to prove that every common divisor of a and b divides gcd(a, b).

The following property of the greatest common divisor of two integers plays a critical role in abstract algebra. The proof provides an application of the division algorithm and our second application of the Well Ordering Principle. This result is Proposition 1 in Book Seven of Euclid's *Elements*, written about 300 B.C.

GCD *Is a Linear Combination*

For any nonzero integers a and b, there exists integers s and t such that gcd(a, b) = $as + bt$.

Proof. Consider the set $S = \{am + bn \mid m, n$ are integers and $am + bn > 0\}$. Since S is obviously nonempty (if some choice of m and n makes $am + bn < 0$, then replace m and n by $-m$ and $-n$), the Well Ordering Principle asserts that S has a smallest member, say, $d = as + bt$. We claim that $d = $ gcd(a, b). To verify this claim, use the division algorithm to write $a = dq + r$ where $0 \le r < d$. If $r > 0$, then $r = a - dq = a - (as + bt)q = a - asq - btq = a(1 - sq) + b(-tq) \in S$, contradicting the fact that d is the smallest member of S. So, $r = 0$ and d divides a. Analogously (or better yet, by symmetry), d divides b as well. This proves that d is a common divisor of a and b. Now suppose d' is another common divisor of a and b and write $a = d'h$ and $b = d'k$. Then $d = as + bt = (d'h)s + (d'k)t = d'(hs + kt)$ so that d' is a divisor of d. Thus, among all common divisors of a and b, d is the greatest. ∎

Example 2 gcd(4, 15) = 1; gcd(4, 10) = 2; gcd($2^2 \cdot 3^2 \cdot 5, 2 \cdot 3^3 \cdot 7^2$) = $2 \cdot 3^2$. Note that 4 and 15 are relatively prime while 4 and 10 are not. Also, $4 \cdot 4 + 15(-1) = 1$ and $(-2)4 + 1 \cdot 10 = 2$. □

Although the above result is a powerful theoretical tool, it does not provide a practical method for calculating the gcd(a, b) or specific values for s and t. The next example shows how this can be done.

Example 3 Euclidean Algorithm (Proposition 2 of Book Seven of the *Elements*). For any pair of positive integers a and b we may find the gcd(a, b) by repeated use of division to produce a decreasing sequence of integers: $r_1 > r_2 > \cdots$ as follows.

$$a = bq_1 + r_1 \qquad 0 < r_1 < b,$$
$$b = r_1q_2 + r_2 \qquad 0 < r_2 < r_1,$$
$$r_1 = r_2q_3 + r_3 \qquad 0 < r_3 < r_2,$$
$$\vdots$$
$$r_{k-3} = r_{k-2}q_{k-1} + r_{k-1} \qquad 0 < r_{k-1} < r_{k-2},$$

$$r_{k-2} = r_{k-1}q_k + r_k \qquad 0 < r_k < r_{k-1},$$
$$r_{k-1} = r_k q_{k+1} + 0.$$

(It is a consequence of the Well Ordering Principle that this process must eventually result in a remainder of 0.) Then r_k, the last nonzero remainder, is the gcd (a, b).

To see that $r_k = \gcd(a, b)$, observe that since $r_k \mid r_{k-1}$, the next to the last equation implies that $r_k \mid r_{k-2}$. This, in turn, implies $r_k \mid r_{k-3}$. Continuing to work backwards in this fashion we see that $r_k \mid b$ and $r_k \mid a$. This proves that r_k is a common divisor of a and b. Now, if r is another common divisor of a and b, the first equation above shows that $r \mid r_1$. The second equation then shows that $r \mid r_2$. Continuing, we see that $r \mid r_k$ and indeed r_k is the greatest of all common divisors of a and b.

Let's apply this process to compute the gcd(2520, 154).

$$2520 = 154 \cdot 16 + 56,$$
$$154 = 56 \cdot 2 + 42,$$
$$56 = 42 \cdot 1 + 14,$$
$$42 = 14 \cdot 3.$$

Since 14 is the last nonzero remainder, gcd(2502, 154) = 14.

The preceding equations also provide a method to express the gcd(a, b) as a linear combination of a and b. For instance, we may use the equations to write the remainders in terms of 2520 and 154. Namely,

$$56 = 2520 + 154(-16),$$
$$42 = 154 + 56(-2),$$
$$14 = 56 + 42(-1)$$
$$= 56 + [154 + 56(-2)](-1)$$
$$= 56 \cdot 3 + 154(-1)$$
$$= [2520 + 154(-16)]3 + 154(-1)$$
$$= 2520 \cdot 3 + 154(-49). \qquad \square$$

The next lemma is frequently used.

Euclid's Lemma* $p \mid ab$ *implies* $p \mid a$ *or* $p \mid b$
If p is a prime that divides ab, then p divides a or p divides b.

Proof. Suppose p is a prime that divides ab but does not divide a. We must show p divides b. Since p does not divide a, there are integers s and t such that $1 = as + pt$. Then $b = abs + ptb$, and since p divides the right-hand side of this equation, p also divides b. ∎

---

*This result is Proposition 14 in Book Nine of Euclid's *Elements*.

Note that Euclid's Lemma may fail when p is not a prime, since $6 \mid 4 \cdot 3$ but $6 \nmid 4$ and $6 \nmid 3$.

Our next property shows that the primes are the building blocks for all integers. We will often use this property without explicitly saying so.

Theorem 0.1   *Fundamental Theorem of Arithmetic*
Every integer greater than 1 is a prime or a product of primes. This product is unique, except for the order in which the factors appear. Thus, if $n = p_1 p_2 \ldots p_r$ and $n = q_1 q_2 \ldots q_s$ where the p's and q's are primes, then $r = s$ and after renumbering the q's, we have $p_i = q_i$ for all i.

We will prove the existence portion of Theorem 0.1 later in the chapter. The uniqueness portion is a consequence of Euclid's Lemma.

Another concept that frequently arises is that of the least common multiple of two integers.

DEFINITION   Least Common Multiple
The least common multiple of two nonzero integers a and b is the smallest positive integer that is a multiple of both a and b. We will denote this intege by $\text{lcm}(a, b)$.

We leave it as an exercise (exercise 10) to prove that every common m of a and b is a multiple of $\text{lcm}(a, b)$.

Example 4 $\text{lcm}(4, 6) = 12; \text{lcm}(4, 8) = 8; \text{lcm}(10, 12) = 60; \text{lc}$
$\text{lcm}(2^2 \cdot 3^2 \cdot 5, 2 \cdot 3^3 \cdot 7^2) = 2^2 \cdot 3^3 \cdot 5 \cdot 7^2$.

Modular Arithmetic

Another application of the division algorithm that modular arithmetic.

DEFINITION   Arithmetic Modulo n
Let n be a fixed positive integer. For
n (read "a plus b modulo n") is th
$(a \cdot b) \bmod n$ (read "a times b r
$a \cdot b$ by n.

Example 5

$(7 + 4) \bmod 3 = 2$
$(10 + 5) \bmod 6 =$
$(6 + 5) \bmod 7 = 4;$

Throughout this text, we will make frequent use of modular equations.

DEFINITION   Modular Equations
If a and b are integers and n is a positive integer, we write $a = b \bmod n$ when n divides $a - b$.

Example 6

$$13 = 3 \bmod 5 \qquad 22 = 10 \bmod 6 \qquad -10 = 4 \bmod 7. \qquad \square$$

Example 7   Consider the equations

1. $x = 1 \bmod 5$.
2. $x = 3 \bmod 5$.
3. $2x = 5 \bmod 6$.
4. $ax = b \bmod n$.

Note that x is a solution of equation 1 if and only if $x - 1 = 5k$ for some integer k. That is, $x - 1 = 0, \pm 5, \pm 10, \ldots$ or x belongs to $\{\ldots, -9, -4, 1, 6, 11, \ldots\}$. Similarly, x is a solution of equation 2 if and only if x belongs to $\{\ldots, -7, -2, 3, 8, 13, \ldots\}$. Equation 3 has no solution, since the left side of the equation $2x - 5 = 6k$ is always odd, while the right side is always even. In view of equation 3, we see that equation 4 need not have a solution, but one important case in which it does is whenever a and n are relatively prime. For then there are integers s and t such that $as + nt = 1$. So, $asb + ntb = b$. Thus, $asb - b = -ntb$ is divisible by n and sb is a solution to the equation $ax = b \bmod n$. $\qquad \square$

Modular arithmetic is an indispensable tool in computer science. One application concerns the generation of "random" numbers: a sequence of integers that is easily produced and passes a number of statistical tests for randomness. A common method for generating such numbers is to use a recursion formula of the form

$$x_{n+1} = (ax_n + c) \bmod m.$$

To use such a formula one chooses the modulus m, the multiplier a, and the increment c, as well as an initial value for x_0 (often called the *seed*). Over the years guidelines have evolved for selecting a, c, and m. Typically, m is a large power of 2 (e.g., 2^{32}) and $a = 5$. Of course, numbers generated in this way completely determined by the choice of these parameters and the initial value, so are not random at all. Nevertheless, these numbers do possess many rties that we expect of random numbers and so are useful for simulation. odular arithmetic is often used in assigning an extra digit to identification s for the purpose of detecting forgery or errors. We present two such ons.

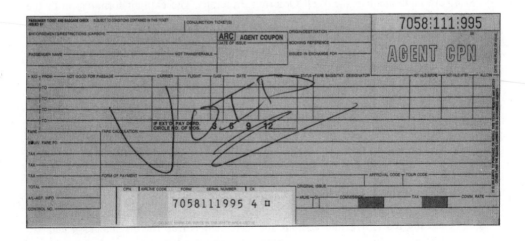

Figure 0.1

Example 8 The United States postal service money-order shown in Figure 0.1 has an identification number consisting of ten digits together with an extra digit called a *check*. The check digit is the 10-digit number modulo 9. Thus, the number 3953988164 has the check digit 2 since 3953988164 = 2 mod 9. If the number 39539881642 were incorrectly entered into a computer (programmed to calculate the check digit) as, say, 39559881642, the machine would calculate the check as 4 while the entered check digit is 2. Thus the error is detected.

Example 9 Airline companies and United Parcel Service use the modulo 7 value of identification numbers to assign check digits. Thus, the identification number 7058111995 (see Figure 0.2) has appended the check digit 4 because

Figure 0.2

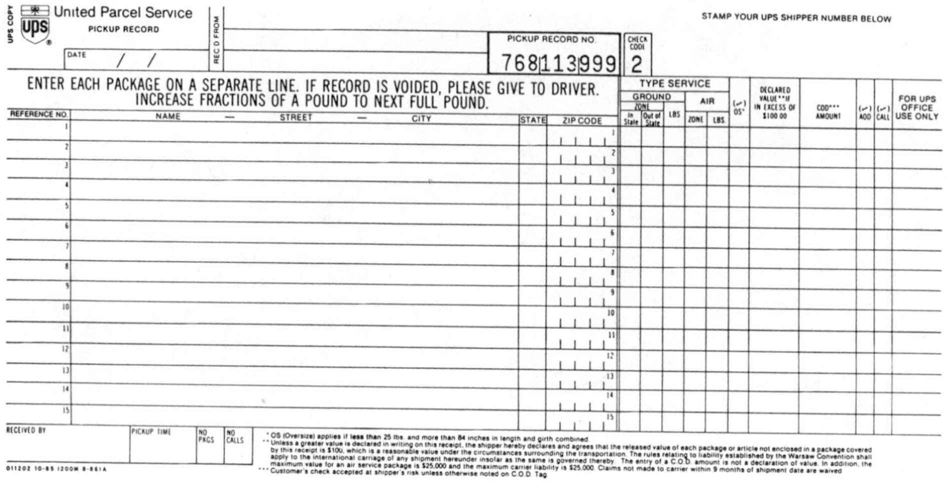

Figure 0.3

$7058111995 = 4 \bmod 7$. Similarly, the UPS pickup record number 768113999, shown in Figure 0.3, has appended a check digit 2.

In Chapters 2 and 5 we will examine more sophisticated means of assigning check digits to numbers.

Mathematical Induction

There are two forms of proof by mathematical induction that we will use. Both are equivalent to the Well Ordering Principle. The explicit formulation of the method of mathematical induction came in the sixteenth century. Francisco Maurolycus (1494–1575), a teacher of Galileo, used it in 1575 to prove $1 + 3 + 5 + \cdots + (2n - 1) = n^2$ and Blaise Pascal (1623–1662) used it when he presented what we now call Pascal's triangle for the coefficients of the binomial expansion. The term was coined by Augustus DeMorgan.

First Principle of Mathematical Induction
Let S be a set of integers containing a. Suppose S has the property that whenever some integer $n \geq a$ belongs to S, then the integer $n + 1$ also belongs to S. Then, S contains every integer greater than or equal to a.

So, to use induction to prove that a statement involving positive integers is true for every positive integer, we must first verify that the statement is true for the integer 1. We then *assume* the statement is true for the integer n and use this assumption to prove the statement is true for the integer $n + 1$.

Example 10   We will use the First Principle of Mathematical Induction to prove that for every positive integer n, the expression $2^{2n} - 1$ is divisible by 3. Let S be the set of all positive integers k for which $2^{2k} - 1$ is divisible by 3. Clearly, $1 \in S$. Now assume that some integer $n \in S$. We must show that $2^{2(n+1)} - 1$ is divisible by 3, so that $n + 1 \in S$. Since $2^{2(n+1)} - 1 = 2^{2n+2} - 1 = 4 \cdot 2^{2n} - 1 = 4 \cdot 2^{2n} - 4 + 3 = 4(2^{2n} - 1) + 3$ and $n \in S$, it follows that 3 divides this last expression. Thus, by induction, $2^{2n} - 1$ is divisible by 3 for all positive integers n. $\square$

Example 11   DeMoivre's Theorem
We use induction to prove that for every positive integer n and every real number θ, $(\cos \theta + i \sin \theta)^n = \cos n\theta + i \sin n\theta$, where i is the complex number $\sqrt{-1}$. Obviously, the statement is true for $n = 1$. Now assume it is true for n. We must prove $(\cos \theta + i \sin \theta)^{n+1} = \cos(n + 1)\theta + i \sin(n + 1)\theta$. Observe that

$$
\begin{aligned}
(\cos \theta + i \sin \theta)^{n+1} &= (\cos \theta + i \sin \theta)^n(\cos \theta + i \sin \theta) \\
&= (\cos n\theta + i \sin n\theta)(\cos \theta + i \sin \theta) \\
&= \cos n\theta \cos \theta + i(\sin n\theta \cos \theta + \sin \theta \cos n\theta) \\
&\quad - \sin n\theta \sin \theta.
\end{aligned}
$$

Now, using trigonometric identities for $\cos(\alpha + \beta)$ and $\sin(\alpha + \beta)$, we see this last term is $\cos(n + 1)\theta + i \sin(n + 1)\theta$. So, by induction, the statement is true for all positive integers. $\square$

In many instances, the assumption that a statement is true for an integer n does not readily lend itself to a proof that the statement is true for the integer $n + 1$. In such cases, the following equivalent form of induction may be more convenient.

> ### Second Principle of Mathematical Induction
> *Let S be a set of integers containing a. Suppose S has the property that n belongs to S whenever every integer less than n and greater than or equal to a belongs to S. Then, S contains every integer greater than or equal to a.*

To use this form of induction, we first show the statement is true for the integer a. We then *assume* the statement is true for *all* integers that are greater than or equal to a and less than n, and use this assumption to prove the statement is true for all integers greater than or equal to a.

Example 12   We will use the Second Principle of Mathematical Induction with $a = 2$ to prove the existence portion of the Fundamental Theorem of Arithmetic. Let S be the set of integers greater than 1 that are primes or a product of positive primes. Clearly, $2 \in S$. Now we assume that for some integer n, S contains all integers k with $2 \leq k < n$. We must show that $n \in S$. If n is a prime, then $n \in S$ by definition. If n is not a prime, then n can be written in the form ab where $1 < a < n$ and $1 < b < n$. Since we are assuming that both a and b belong to S, we know that each of them is prime or a product of positive primes. Thus, n is also a product of positive primes. This completes the proof. $\qquad\square$

Notice that it is more natural to prove the Fundamental Theorem of Arithmetic with the Second Principle of Mathematical Induction than the First Principle. Knowing that a particular integer factors as a product of primes does not tell you something about factoring the next larger integer. (Does knowing 5280 is a product of primes help you to factor 5281 as a product of primes?)

The following problem appeared in the "Brain Boggler" section of the January 1988 issue of the science magazine *Discover*.

Example 13   The Quakertown Poker Club plays with blue chips worth $5.00 and red chips worth $8.00. What is the largest bet that cannot be made?

To gain insight into this problem we try various combinations of blue and red chips to obtain: 5, 8, 10, 13, 15, 16, 18, 20, 21, 23, 24, 26, 28, 29, 30, 31, 32, 33, 34, 35, 36, 37, 38, 39, 40. It appears that the answer is 27. But how can we be sure? Well, we need only prove that every integer greater than 27 can be written in the form $a \cdot 5 + b \cdot 8$, where a and b are nonnegative integers. This will solve the problem since a represents the number of blue chips and b the number of red chips needed to make a bet of $a \cdot 5 + b \cdot 8$. For the purpose of contrast, we will give two proofs—one using the First Principle of Mathematical Induction and one using the Second Principle.

Let S be the set of all integers of the form $a \cdot 5 + b \cdot 8$, where a and b are nonnegative. Obviously, $28 \in S$. Now assume that some integer $n \in S$, say, $n = a \cdot 5 + b \cdot 8$. We must show that $n + 1 \in S$. First, note that since $n \geq 28$, we cannot have both a and b less than 3. If $a \geq 3$, then

$$
\begin{aligned}
n + 1 &= (a \cdot 5 + b \cdot 8) + (-3 \cdot 5 + 2 \cdot 8) \\
&= (a - 3) \cdot 5 + (b + 2) \cdot 8.
\end{aligned}
$$

(Regarding chips, this last equation says we may increase a bet from n to $n + 1$ by removing 3 blue chips from the pot and adding 2 red chips.) If $b \geq 3$, then

$$
n + 1 = (a \cdot 5 + b \cdot 8) + (5 \cdot 5 - 3 \cdot 8) = (a + 5) \cdot 5 + (b - 3) \cdot 8.
$$

(The bet can be increased by 1 by removing 3 red chips and adding 5 blue chips.) This completes the proof.

To prove the same statement by the Second Principle, we note that each of 28, 29, 30, 31, and 32 are in S. Now assume that for some integer $n > 32$, S contains all integers k with $28 \leq k < n$. We must show $n \in S$. Since $n - 5 \in S$, there are nonnegative integers a and b such that $n - 5 = a \cdot 5 + b \cdot 8$. But then $n = (a + 1) \cdot 5 + b \cdot 8$. Thus n is in S. $\qquad\square$

In the remainder of the text we shall be less formal about induction and dispense with the phrase "Let S be the set of integers. . . ."

Equivalence Relations

In mathematics, things that are considered different in one context may be viewed as equivalent in another context. We have already seen one such example. Indeed, the sums $2 + 1$ and $4 + 4$ are certainly different in ordinary arithmetic, but are the same under modulo 5 arithmetic. Congruent triangles that are situated differently in the plane are not the same, but they are considered as the same in plane geometry. In physics, vectors of the same magnitude and direction can produce different effects—a 10-pound weight placed 2 feet from a fulcrum produces a different effect than a 10-pound weight placed 1 foot from a fulcrum. But in linear algebra, vectors of the same magnitude and direction are considered as the same. What is needed to make these distinctions precise is an appropriate generalization of the notion of equality; that is, we need a formal mechanism for specifying whether or not two quantities are the same in a given setting. This mechanism is an equivalence relation.

DEFINITION Equivalence Relation
An *equivalence relation* on a set S is a set R of ordered pairs of elements of S such that

1. $(a, a) \in R$ for all $a \in S$ (reflexive property).
2. $(a, b) \in R$ implies $(b, a) \in R$ (symmetric property).
3. $(a, b) \in R$ and $(b, c) \in R$ imply $(a, c) \in R$ (transitive property).

When R is an equivalence relation on a set S, it is customary to write aRb instead of $(a,b) \in R$. Also, since an equivalence relation is just a generalization of equality, a suggestive symbol such as $\approx$, $\equiv$, or $\sim$ is usually used to denote the relation. If $\sim$ is an equivalence relation on a set S and $a \in S$, then the set $[a] = \{x \in S \mid x \sim a\}$ is called the *equivalence class of S containing a*.

Example 14 Let S be the set of all triangles in a plane. If $a, b \in S$, define $a \sim b$ if a and b are similar, that is, if a and b have corresponding angles the same. Then, $\sim$ is an equivalence relation on S. □

Example 15 Let S be the set of all polynomials with real coefficients. If f, $g \in S$, define $f \sim g$ if $f' = g'$, where f' is the derivative of f. Then, $\sim$ is an equivalence relation on S. Since two functions with equal derivatives differ by a constant, we see that for any f in S, $[f] = \{f + c \mid c$ is real$\}$. □

Example 16 Let S be the set of integers and n a positive integer. If $a, b \in S$, define $a \equiv b$, if $a = b$ modulo n (that is, if $a - b$ is divisible by n). Then, $\equiv$

is an equivalence relation on S and $[a] = \{a + kn \mid k \in S\}$. Since this particular relation is important in abstract algebra, we will take the trouble to verify that it is indeed an equivalence relation. Certainly, $a - a$ is divisible by n so that $a \equiv a$ for a in S. Next, assume that $a \equiv b$, say, $a - b = rn$. Then, $b - a = (-r)n$ and therefore $b \equiv a$. Finally, assume $a \equiv b$ and $b \equiv c$, say, $a - b = rn$ and $b - c = sn$. Then, we have $a - c = (a - b) + (b - c) = rn + sn = (r + s)n$, so that $a \equiv c$. $\square$

Example 17 Let $\equiv$ be as in Example 16 and let $n = 7$. Then we have $16 \equiv 2$; $9 \equiv -5$; $24 \equiv 3$. Also, $[1] = \{\ldots -20, -13, -6, 1, 8, 15, \ldots\}$ and $[4] = \{\ldots -17, -10, -3, 4, 11, 18, \ldots\}$. $\square$

Example 18 Let $S = \{(a, b) \mid a, b \text{ are integers}, b \neq 0\}$. If $(a, b), (c, d) \in S$, define $(a, b) \approx (c, d)$ if $ad = cb$. Then $\approx$ is an equivalence relation on S. (The motivation for this example comes from fractions. In fact, the pairs (a, b) and (c, b) are equivalent if the fractions a/b and c/d are equal.) $\square$

DEFINITION Partition

A *partition* of a set S is a collection of nonempty disjoint subsets of S whose union is S. Figure 0.4 illustrates a partition of a set into four subsets.

Example 19 The sets $\{0\}, \{1, 2, 3, \ldots\}$, and $\{\ldots, -3, -2, -1\}$ constitute a partition of the set of integers. $\square$

Example 20 The set of nonnegative integers and the set of nonpositive integers do not partition the integers since both contain 0. $\square$

The next theorem reveals that equivalence relations and partitions are intimately intertwined.

Theorem 0.2 *Equivalence Classes Partition*

The equivalence classes of an equivalence relation on a set S constitute a partition of S. Conversely, for any partition P of S, there is an equivalence relation on S whose equivalence classes are the elements of P.

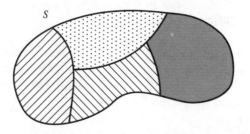

Figure 0.4 Partition of S into 4 subsets.

Proof. Let ~ be an equivalence relation on a set S. For any $a \in S$, the reflexive property shows $a \in [a]$. So, $[a]$ is nonempty. Also, the union of all equivalence classes is S. Now, suppose that $[a]$ and $[b]$ are distinct equivalence classes. We must show $[a] \cap [b] = \emptyset$. On the contrary, assume $c \in [a] \cap [b]$. We will show $[a] \subseteq [b]$. To this end, let $x \in [a]$. We then have $c \sim a$, $c \sim b$, and $x \sim a$. By the symmetric property, we also have $a \sim c$. Thus, by transitivity, $x \sim c$, and by transitivity again, $x \sim b$. This proves $[a] \subseteq [b]$. Analogously, $[b] \subseteq [a]$. Thus, $[a] = [b]$, in contradiction to our assumption that $[a]$ and $[b]$ were distinct equivalence classes.

We leave the proof of the converse portion of the theorem as an exercise. ■

Functions (Mappings)

Although the concept of a function plays a central role in nearly every branch of mathematics, the terminology and notation associated with functions vary quite a bit. In this section, we fix ours.

DEFINITION   Function (Mapping)
A *function* ϕ (or *mapping*) *from a set A to a set B* is a rule that assigns to each element a of A exactly one element b of B. The set A is called the *domain of* ϕ and B is called the *codomain of* ϕ. If ϕ assigns b to a, then b is called the *image of a under* ϕ. The subset of B comprised of all the images of elements of A is called the *image of A under* ϕ.

We use the shorthand $\phi: A \rightarrow B$ to mean that ϕ is a mapping from A to B. We will usually write $a\phi = b$ to indicate that ϕ carries a to b. However, when dealing with polynomials and number-theoretic functions, we use the more familiar notation $f(a)$ for the functional value of a under f and $f(x)$ for formulas. Readers who have not previously seen functions written to the right of the domain values are no doubt puzzled by our choice. Our motivation comes from the next definition.

DEFINITION   Composition of Functions
Let $\phi: A \rightarrow B$ and $\psi: B \rightarrow C$. The *composition* $\phi\psi$ is the mapping from A to C defined by $a(\phi\psi) = (a\phi)\psi$ for all a in A. The composition function $\phi\psi$ can be visualized as in Figure 0.5.

In calculus courses the composition of f with g is written $(f \circ g)(x)$ and is defined as $f(g(x))$; that is, in calculus, composition is done from right to left. When we compose functions we omit the "circle" and we work from left to right, the way one ordinarily reads and does arithmetic.

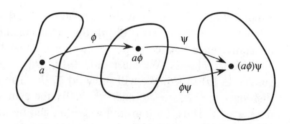

Figure 0.5 Composition of functions ϕ and ψ.

There are several kinds of functions that occur often enough to be given names.

DEFINITION One-to-One Function
A function ϕ from a set A to a set B is called *one-to-one* if $a\phi = b\phi$ implies $a = b$.

The term *one-to-one* is suggestive since the definition ensures that one element of B can be the image of only one element of A. Alternately, ϕ is one-to-one if $a_1 \neq a_2$ implies $a_1\phi \neq a_2\phi$. That is, different elements of A map to different elements of B. See Figure 0.6.

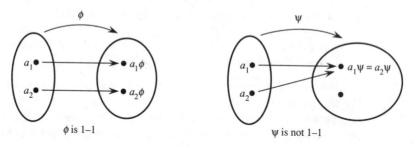

Figure 0.6

DEFINITION Function from A onto B
A function ϕ from a set A to a set B is said to be *onto B* if each element of B is the image of at least one element of A. In symbols, $\phi: A \rightarrow B$ is onto if for each b in B there is at least one a in A such that $a\phi = b$. See Figure 0.7.

The next theorem summarizes the facts about functions we will need.

Theorem 0.3 *Properties of Functions*
Given functions $\alpha: A \rightarrow B$, $\beta: B \rightarrow C$ *and* $\gamma: C \rightarrow D$. *Then*

1. $(\alpha\beta)\gamma = \alpha(\beta\gamma)$ *(associativity)*
2. *If* α *and* β *are one-to-one, then* $\alpha\beta$ *is one-to-one.*

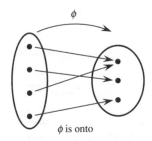
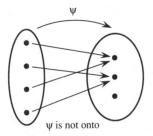

Figure 0.7

3. *If α and β are onto, then αβ is onto.*
4. *If α is one-to-one and onto, then there is a function α^{-1} from B onto A such that $a(\alpha\alpha^{-1}) = a$ for all a in A and $b(\alpha^{-1}\alpha) = b$ for all b in B.*

Proof. We prove only part 1. The remaining parts are left as exercises.

Let $a \in A$. Then $a(\alpha\beta)\gamma = (a(\alpha\beta))\gamma = ((a\alpha)\beta)\gamma$. On the other hand, $a\alpha(\beta\gamma) = (a\alpha)(\beta\gamma) = ((a\alpha)\beta)\gamma$.   ∎

Example 21   Let **Z** denote the set of integers, **R** the set of real numbers, and **N** the set of nonnegative integers. The following table illustrates the properties of one-to-one and onto.

Domain	Codomain	Rule	1–1	Onto
Z	**Z**	$x \to x^3$	yes	no
R	**R**	$x \to x^3$	yes	yes
Z	**N**	$x \to \lvert x \rvert$	no	yes
Z	**Z**	$x \to x^2$	no	no

To verify that $x \to x^3$ is one-to-one in the first two cases, notice that if $x^3 = y^3$, we may take the cube root of both sides of the equation to obtain $x = y$. Clearly, the mapping from **Z** to **Z** given by $x \to x^3$ is not onto since 2 is the cube of no integer. However, $x \to x^3$ defines an onto function from **R** to **R** since every real number is the cube of its cube root (that is, $\sqrt[3]{b} \to b$). The remaining verifications are left as exercises.

EXERCISES

Failure is instructive. The person who really thinks learns quite as much from his failures as from his successes.

John Dewey

1. For each value of n listed, find all integers less than n and relatively prime to n. $n = 8, 12, 20, 25$.

2. Determine $\gcd(2^4 \cdot 3^2 \cdot 5 \cdot 7^2,\ 2 \cdot 3^3 \cdot 7 \cdot 11)$ and $\operatorname{lcm}(2^3 \cdot 3^2 \cdot 5,\ 2 \cdot 3^3 \cdot 7 \cdot 11)$.

3. Calculate $(7 \cdot 3) \bmod 5$, $(7 + 3) \bmod 5$, $(15 \cdot 4) \bmod 7$, $(15 + 4) \bmod 7$.

4. Find integers s and t so that $1 = 7 \cdot s + 11 \cdot t$. Show that s and t are not unique.

5. In Florida, the last three digits of the driver's license number of a female with birth month m and birth date b are represented by $40(m - 1) + b + 500$. For both males and females, the fourth and fifth digits from the end give the year of birth. Determine the dates of birth of people with the numbers whose last five digits are 42218 and 53953.

6. In Missouri, the last three digits of the driver's license number of a female with birth month m and birth date b are represented by $63m + 2b + 1$. Determine the birth months, birth dates, and sex corresponding to the numbers 248 and 601.

7. Show that if a and b are positive integers, then $ab = \operatorname{lcm}(a, b) \cdot \gcd(a, b)$. (This exercise is referred to in Chapter 7.)

8. Suppose a and b are integers that divide the integer c. If a and b are relatively prime, show ab divides c. Show, by example, that if a and b are not relatively prime, then ab need not divide c.

9. Show that $5n + 3$ and $7n + 4$ are relatively prime for all n.

10. Let a and b be positive integers and let $d = \gcd(a, b)$ and $m = \operatorname{lcm}(a, b)$. If t divides both a and b, prove that t divides d. If s is a multiple of both a and b, prove that s is a multiple of m.

11. Let n and a be positive integers and let $d = \gcd(a, n)$. Show that the equation $ax = 1 \bmod n$ has a solution if and only if $d = 1$. (This exercise is referred to in Chapter 2.)

12. Let n be a fixed positive integer greater than 1. For any integers i and j, prove that $i \bmod n + j \bmod n = (i + j) \bmod n$. This exercise is referred to in Chapter 6.

13. Does the set of positive rational numbers satisfy the Well Ordering Principle?

14. Show that $\gcd(a, bc) = 1$ if and only if $\gcd(a, b) = 1$ and $\gcd(a, c) = 1$.

15. If there are integers a, b, s, and t with the property that $at + bs = 1$, show that $\gcd(a, b) = 1$.

16. Let $d = \gcd(a, b)$. If $a = da'$ and $b = db'$, show that $\gcd(a', b') = 1$.

17. Use the Euclidean algorithm to find the $\gcd(34, 126)$ and write it as a linear combination of 34 and 126.

18. Let $p_1, p_2, \ldots, p_n$ be distinct positive primes. Show that $p_1 p_2 \cdots p_n + 1$ is divisible by none of these primes.

19. Prove that there are infinitely many primes. (Hint: Use the previous exercise.)

20. For every positive integer n, prove $1 + 2 + \cdots + n = n(n + 1)/2$.

21. For every positive integer n, prove that a set with exactly n elements has exactly 2^n subsets (counting the empty set and the entire set).

22. (Generalized Euclid's Lemma) If p is a prime and p divides $a_1a_2 \cdots a_n$, prove p divides a_i for some i.

23. Use the Generalized Euclid's Lemma (see exercise 22) to establish the uniqueness portion of the Fundamental Theorem of Arithmetic.

24. What is the largest bet that cannot be made with chips worth $7.00 and $9.00? Verify that your answer is correct with both forms of induction.

25. The Fibonacci numbers are: 1, 1, 2, 3, 5, 8, 13, 21, 34, ... In general, the Fibonacci numbers are defined by $f_1 = 1$, $f_2 = 1$, and $f_{n+2} = f_{n+1} + f_n$ for $n = 1, 2, 3, \ldots$. Prove the nth Fibonacci number f_n satisfies $f_n < 2^n$.

26. Prove that the First Principle of Mathematical Induction is a consequence of the Well Ordering Principle.

27. In the cut "As" from *Songs in the Key of Life,* Stevie Wonder mentions the equation $8 \times 8 \times 8 \times 8 = 4$. Find all integers n for which this statement is true, modulo n.

28. Determine the check digit for a money order with identification number 7234541780.

29. Suppose that in one of the noncheck positions of a money order number, the digit 0 is substituted for the digit 9 or vice versa. Prove that this error will not be detected by the check digit. Prove that all other errors involving a single position are detected.

30. Suppose a money order with the identification number and check digit 21720421168 is erroneously copied as 27750421168. Will the check digit detect the error?

31. A transposition error involving distinct adjacent digits is one of the form $\ldots ab \ldots \rightarrow \ldots ba \ldots$ with $a \neq b$. Prove that the money order check digit scheme will not detect such errors unless the check digit itself is transposed.

32. Determine the check digit for the United Parcel Service (U.P.S.) identification number 873345672.

33. Show that a substitution of a digit a_i' for the digit a_i ($a_i' \neq a_i$) in a noncheck position of a U.P.S. number is detected if and only if $|a_i - a_i'| \neq 7$.

34. Determine which transposition errors involving adjacent digits are detected by the U.P.S. check digit.

35. Prove that for every integer n, $n^3 = n \bmod 6$.

36. If it were 2:00 A.M. now, what time would it be 3736 hours from now?

37. If n is an odd integer, prove $n^2 = 1 \bmod 8$.

38. If the odometer of an automobile read 97,000 now, what would it read 12,000 miles later?

39. Let $S = \{(x, y) \mid x, y \text{ are real}\}$. If (a, b) and (c, d) belong to S, define

$(a, b)R(c, d)$ if $a^2 + b^2 = c^2 + d^2$. Prove that R is an equivalence relation on S. Give a geometrical description of the equivalence classes of S.

40. Let S be the set of real numbers. If $a, b \in S$, define $a \sim b$ if $a - b$ is an integer. Show $\sim$ is an equivalence relation on S. Describe the equivalence classes of S.

41. Let S be the set of integers. If $a, b \in S$, define aRb if $ab \geq 0$. Is R an equivalence relation on S?

42. Let S be the set of integers. If $a, b \in S$, define aRb if $a + b$ is even. Prove that R is an equivalence relation and determine the equivalence classes of S.

43. A *relation* on a set S is a set of ordered pairs of elements of S. Find an example of a relation that is reflexive and symmetric, but not transitive.

44. Find an example of a relation that is reflexive and transitive, but not symmetric.

45. Find an example of a relation that is symmetric and transitive, but not reflexive.

SUGGESTED READINGS

Mary Joan Collison, "The Unique Factorization Theorem: From Euclid to Gauss," *Mathematics Magazine* 53 (1980): 96–100.

This article examines the history of the Fundamental Theorem of Arithmetic.

Linda Dineen, "Secret Encryption with Public Keys," *The UMAP Journal* 8(1987): 9–29.

This well-written article describes several ways in which modular arithmetic can be used to code secret messages. They range from a simple scheme used by Julius Caesar, to a highly sophisticated scheme invented in 1978 and based on modular n arithmetic, where n has more than 200 digits.

Ian Richards, "The Invisible Prime Factor," *American Scientist* 70(1982): 176–179.

The author explains how elementary number theory such as Euclid's Lemma and modular arithmetic can be used to test whether an integer is prime. He then discusses how prime numbers can be used to create secret codes that are extremely difficult to break.

GROUPS

PART 2

Introduction to Groups

Symmetry is a vast subject, significant in art and nature. Mathematics lies at its root, and it would be hard to find a better one on which to demonstrate the working of the mathematical intellect.

Hermann Weyl, *Symmetry*

Symmetries of a Square

Suppose we remove a square from a plane, move it in some way, then put the square back into the space it originally occupied. Our goal in this chapter is to describe in some reasonable fashion all possible ways this can be done. More specifically, we want to describe the possible relationships between the starting position of the square and its final position in terms of motions. However, we are interested in the net effect of a motion, rather than the motion itself. Thus, for example, we consider a 90° rotation and a 450° rotation as equal, since the relationship between the starting position and the final position of the square is the same for each of these motions. With this simplifying convention, it is an easy matter to achieve our goal.

To begin, we can think of the square as being transparent (glass, say), with the corners marked on one side with the colors blue, white, pink, and green. This makes it easy to distinguish between motions that have different effects. With this marking scheme, we are now in a position to describe, in simple fashion, all possible ways a square object can be repositioned. See Figure 1.1. We now claim that any motion—no matter how complicated—is equivalent to one of these eight. To verify this claim, observe that the final position of the square is completely determined by the location and orientation (that is, face up or face down) of any particular corner. But, clearly, there are only four locations and two orientations for a given corner, so there are exactly eight distinct final positions for the corner.

R_0 = Rotation of 0° (no change in position)

P	W
G	B

$\xrightarrow{R_0}$

P	W
G	B

R_{90} = Rotation of 90° (counterclockwise)

P	W
G	B

$\xrightarrow{R_{90}}$

W	B
P	G

R_{180} = Rotation of 180°

P	W
G	B

$\xrightarrow{R_{180}}$

B	G
W	P

R_{270} = Rotation of 270°

P	W
G	B

$\xrightarrow{R_{270}}$

G	P
B	W

H = Rotation of 180° about a horizontal axis

P	W
G	B

$\xrightarrow{H}$

G	B
P	W

V = Rotation of 180° about a vertical axis

P	W
G	B

$\xrightarrow{V}$

W	P
B	G

D = Rotation of 180° about the main diagonal

P	W
G	B

$\xrightarrow{D}$

P	G
W	B

D' = Rotation of 180° about the other diagonal

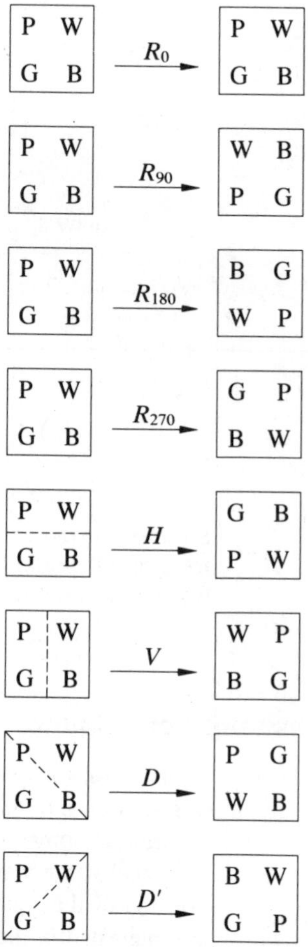

P	W
G	B

$\xrightarrow{D'}$

B	W
G	P

Figure 1.1

Let's investigate some consequences of the fact that every motion is equal to one of the eight listed in Figure 1.1. Suppose a square is repositioned by a rotation of 90° followed by a rotation of 180° about the horizontal axis of symmetry. In pictures,

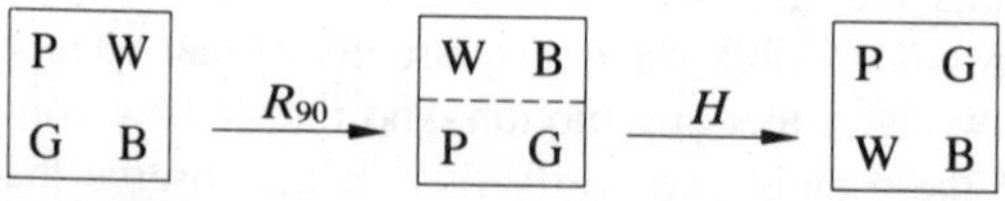

P	W
G	B

$\xrightarrow{R_{90}}$

W	B
P	G

$\xrightarrow{H}$

P	G
W	B

Thus, we see that this pair of motions—taken together—is equal to the single motion D. This observation motivates us to define an algebraic operation "followed by" on the set of motions.

DEFINITION A Followed by B

We say a motion A *followed by* a motion B (written AB) is equal to the motion C if, upon performing the motion A and then the motion B, the net effect is the same as that produced by performing C alone.

With this definition, we may now write $R_{90}H = D$. The eight motions R_0, R_{90}, R_{180}, R_{270}, H, V, D, and D', together with the operation "followed by," form a mathematical system called the *dihedral group of order 8* (the order of a group is the number of elements it contains). It is denoted by D_4. Rather than introduce the formal definition of a group here, let's look at some properties of groups by way of the example D_4.

To facilitate future computations, we construct an *operation table* or *Cayley table* (so named in honor of the prolific English mathematician Arthur Cayley, who first introduced them in 1854) for D_4 below. The circled entry represents the fact that $D = R_{90}H$.

	R_0	R_{90}	R_{180}	R_{270}	H	V	D	D'
R_0	R_0	R_{90}	R_{180}	R_{270}	H	V	D	D'
R_{90}	R_{90}	R_{180}	R_{270}	R_0	$\circled{D}$	D'	V	H
R_{180}	R_{180}	R_{270}	R_0	R_{90}	V	H	D'	D
R_{270}	R_{270}	R_0	R_{90}	R_{180}	D'	D	H	V
H	H	D'	V	D	R_0	R_{180}	R_{270}	R_{90}
V	V	D	H	D'	R_{180}	R_0	R_{90}	R_{270}
D	D	H	D'	V	R_{90}	R_{270}	R_0	R_{180}
D'	D'	V	D	H	R_{270}	R_{90}	R_{180}	R_0

Notice how beautiful this table looks! This is no accident! Perhaps the most important feature of this table is that it is completely filled in without introducing any new motions. Of course, this is because, as we have already pointed out, any sequence of motions turns out to be the same as one of these eight. Algebraically, this says that if A and B are in D_4, then so is AB. This property is called *closure* and it is one of the requirements for a mathematical system to be a group. Next, notice that if A is any element of D_4, then $AR_0 = R_0A = A$. Thus, combining any element A on either side with R_0 yields A back again. An element R_0 with this property is called an *identity*, and every group must have one. Another striking feature of the table is that every element of D_4 appears exactly once in each row and column. This feature is something that all groups must have, and, indeed, it is quite useful to keep this fact in mind when constructing the table in the first place. As a consequence of this condition, we see that for each element A in D_4, there is exactly one element B in D_4 so that $AB = BA = R_0$. In this case, B is said to be the *inverse* of A and vice versa. For example, R_{90} and R_{270} are inverses of each other, and H is its own inverse. The term *inverse* is a descriptive one, for if A and B are inverses of each other, the B "undoes" whatever A "does," in the sense that A and B taken together in either order produce R_0 representing no change.

Another property of D_4 deserves special comment. Observe that $HD \neq DH$ but $R_{90}R_{180} = R_{180}R_{90}$. Thus, in a group, AB may or may not be the same as BA. If it happens that $AB = BA$ for *all* choices of group elements A and B, we say the group is *commutative* or—better yet—*Abelian* (in honor of the great Norwegian mathematician Niels Abel). Otherwise, we say the group is *non-Abelian*.

Thus far, we have illustrated, by way of D_4, three of the four conditions that define a group; namely, closure, existence of an identity, and existence of inverses. The remaining condition required for a group is *associativity*; that is, $(ab)c = a(bc)$ for all a, b, c in the set. To be sure that D_4 is indeed a group, we should check this equation for each of the $8^3 = 512$ possible choices of a, b, and c in D_4. In practice, however, this is rarely done! Here, for example, we simply observe instead that each of the eight motions defines a function from the plane to itself, and the operation "followed by" is nothing other than function composition. Then, since function composition is associative, so is "followed by."

The Dihedral Groups

The analysis carried out above for a square can similarly be done for an equilateral triangle or regular pentagon or, indeed, any regular n-gon ($n \geq 3$). The corresponding group is denoted by D_n and is called the *dihedral group of order 2n*.

The dihedral groups arise frequently in art and nature. Many of the decorative designs used on floor coverings, pottery, and buildings have one of the dihedral groups as a group of symmetry. Corporation logos are rich sources for dihedral symmetry [1]. Chrysler's logo has D_5 as a symmetry group, and Mercedes-Benz has D_3. The ubiquitous five-pointed star has symmetry group D_5. This symbol appears on the flags of 37 nations, as well as the highest military medals of both the United States and the Soviet Union. The family of sea animals that includes the starfish, sea cucumbers, feather stars, and sand dollars exhibit patterns with D_5 symmetry group.

Perhaps the most spectacular manifestation of the dihedral groups are the huge circular stained glass rose windows in the Gothic cathedrals of Europe. In the beautiful book *Rose Windows* [2], one can find magnificent representations of D_{12} and D_{16}, D_{24}, D_{30}, and many others. These windows exhibit an intricate blend of religion, art, geometry, architecture, and science. The groups D_{12} and D_{24} are the most common ones associated with rose windows.*

---

*The number 12 had special significance to medieval Christians. It was considered the number of perfection, of the universe, of the Logos of Time, the Apostles, the zodiac, the tribes of Israel, and the precious stones in the foundation of New Jerusalem. There are 24 elders of the Apocalypse.

Figure 1.2

The dihedral group of order $2n$ is often called the *group of symmetries of a regular n-gon*. A *plane symmetry* of a figure F in a plane is a function from the plane to itself that carries F onto F and preserves distances; that is, for any points p and q in the plane, the distance from the image of p to the image of q is the same as the distance from p to q. (The term symmetry is from the Greek word *symmetron* meaning "well-ordered.") The *symmetry group* of a plane figure is the set of all symmetries of the figure. Symmetries in three dimensions are defined analogously. Obviously, a rotation of a plane about a point in the plane is a symmetry of the plane and a rotation about a line in three dimensions is a symmetry in three-dimensional space. Similarly, any translation of a plane or of three-dimensional space is a symmetry. A *reflection across a line L* is that function that leaves every point of L fixed and takes any point q, not on L, to the point q' so that L is the perpendicular bisector of the line segment joining q and q' (see Figure 1.2). A reflection across a plane in three dimensions is defined analogously. Notice that the restriction of a 180° rotation about a line L in three dimensions to a plane containing L is a reflection across L in the plane. Thus, in the dihedral groups, the motions that we described as 180° rotations about axes of symmetry in three dimensions (for example, H, V, D, D') are reflections across lines in two dimensions. Just as a reflection across a line is a plane symmetry that cannot be achieved by a physical motion of the plane in

Figure 1.3 Logos with Cyclic Rotation Symmetry Groups.

two dimensions, a reflection across a plane is a three-dimensional symmetry that cannot be achieved by a physical motion of three-dimensional space. A cup, for instance, has reflective symmetry across the plane bisecting the cup, but this symmetry cannot be duplicated with a physical motion in three dimensions.

Many objects and figures have rotational symmetry but not reflective symmetry. A symmetry group consisting of the rotational symmetries of $0°$, $360°/n$, $2(360°)/n$, . . . , $(n - 1)360°/n$ and no other symmetries is called a *cyclic rotation group of order n* and is denoted by $\langle R_{360/n} \rangle$. Cyclic rotation groups along with dihedral groups are favorites of artists, designers, and nature. Figure 1.3 (page 27) illustrates with corporate logos the cyclic rotation groups of orders 2, 3, 4, 5, 6, 8, 16, and 20.

Further examples of the occurrence of dihedral groups and cyclic groups in art and nature can be found in the references. A study of symmetry in greater depth is done in Chapters 29 and 30.

EXERCISES

If you think you can or can't, you are right.

Henry Ford

1. With pictures and words, describe each symmetry in D_3 (the set of symmetries of an equilateral triangle).

2. Write out a complete multiplication table for D_3.

3. Is D_3 Abelian?

4. Describe in pictures or words the elements of D_5 (symmetries of a regular pentagon).

5. For $n \geq 3$, describe the elements of D_n. (Hint: You will need to consider two cases—n even and n odd.) How many elements does D_n have?

6. In D_n, explain geometrically why a reflection followed by a reflection must be a rotation.

7. In D_n, explain geometrically why a rotation followed by a rotation must be a rotation.

8. In D_n, explain geometrically why a rotation and a reflection taken together in either order must be a reflection.

9. Associate the number $+1$ with a rotation and the number -1 with a reflection. Describe an analogy between multiplying these two numbers and multiplying elements of D_n.

10. If r_1, r_2, and r_3 represent rotations from D_n and f_1, f_2, f_3 represent reflections from D_n, determine whether $r_1 \, r_2 f_1 \, r_3 f_2 f_3 \, r_3$ is a rotation or reflection.

11. Find elements A, B, and C in D_4 such that $AB = BC$ but $A \neq C$. (Thus, "cross" cancellation is not valid.)

12. Explain what the following diagram proves about the group D_n.

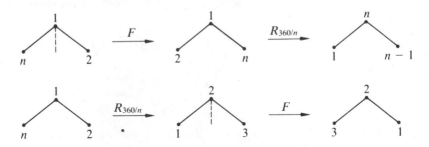

13. Describe the symmetries of a nonsquare rectangle. Construct the corresponding Cayley table.

14. Describe the symmetries of a parallelogram that is neither a rectangle nor a rhombus. Describe the symmetries of a rhombus that is not a rectangle.

15. Describe the symmetries of a noncircular ellipse. Do the same for a hyperbola.

16. Consider an infinitely long strip of equally spaced H's:

$$\cdots \text{H H H H} \cdots$$

Describe the symmetries of this strip. Is the group of symmetries of the strip Abelian?

17. For each of the snowflakes in Figure 1.4, find the symmetry group and locate the axes of reflective symmetry.

Figure 1.4 Photographs of Snowflakes from the Bentley and Humphrey atlas.

18. Find the symmetry group of a Bic pen. (Disregard the cap.)

19. Describe the symmetries of a cigarette. Describe the symmetries of a cigar.

20. Bottle caps that are pried off typically have 22 ridges around the rim. Find the symmetry group of such a cap.

21. Find the symmetry group for the Canadian hendecagonal one dollar coin in Figure 1.5. Disregard the printing and scene on the coin.

Figure 1.5

22. For each design below, determine the symmetry group.

23. Determine the symmetry group of the outer shell of the cross-section of the AIDS virus shown below.

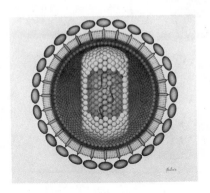

REFERENCES

1. B. B. Capitman, *American Trademark Designs,* New York: Dover, 1976.
2. P. Cowen, *Rose Windows,* San Francisco: Chronicle Books, 1979.
3. I. Hargittai and M. Hargittai, *Symmetry Through the Eyes of a Chemist,* New York: VCH, 1986.
4. Caroline MacGillavry, *Fantasy and Symmetry: The Periodic Drawings of M. C. Escher,* New York: Harry N. Abrams, 1976.
5. A. V. Shubnikov and V. A. Koptsik, *Symmetry in Science and Art,* New York: Plenum Press, 1974.
6. P. S. Stevens, *Handbook of Regular Patterns,* Cambridge, Mass: MIT Press, 1980.
7. H. Weyl, *Symmetry,* Princeton: Princeton University Press, 1952.

Niels Abel

He [Abel] has left mathematicians something to keep them busy for five hundred years.

Charles Hermite

Norway has issued five stamps to honor Abel. This one was issued in 1929 to commemorate the 100th anniversary of his death.

NIELS HENRIK ABEL, one of the foremost mathematicians of the nineteenth century, was born in Norway on August 5, 1802. At the age of sixteen, he began reading the classic mathematical works of Newton, Euler, Lagrange, and Gauss. When Abel was eighteen years old, his father died and the burden of supporting the family fell upon him. He took in private pupils and did odd jobs, while continuing to do mathematical research. At the age of nineteen, Abel solved a problem that had vexed leading mathematicians for hundreds of years. He proved that, unlike the situation for equations of degree 4 or less, there is no finite (closed) formula for the solution of the general fifth degree equation.

Although Abel died long before the advent of the subjects that now comprise abstract algebra, his solution to the quintic problem laid groundwork for many of these subjects. In addition to his work in the theory of equations, Abel made outstanding contributions to the theory of elliptic functions, elliptic integrals, Abelian integrals, and infinite series. Just when his work was beginning to receive the attention it deserved, Abel contracted tuberculosis. He died on April 6, 1829, at the age of twenty-six. In 1870, Camille Jordan introduced the term *Abelian group* to honor Abel. A statue of Abel now stands in the Royal Park in Oslo.

Groups

2

A good stock of examples, as large as possible, is indispensable for a thorough understanding of any concept, and when I want to learn something new, I make it my first job to build one.

Paul R. Halmos

Facts about groups of symmetries have been known for hundreds of years. In the thirteenth century, the Moors demonstrated a sophisticated understanding of symmetry with their decorative wall patterns at the Alhambra, a fortress-palace in Southern Spain. The Alhambra is the crowning glory of Moorish architecture (see [2, pp. 169–176]). In his study of buildings, Leonardo da Vinci (1452–1519) determined all possible symmetry groups of planar objects.

Definition and Examples of Groups

The term *group* was coined by Galois about 150 years ago to describe sets of one-to-one functions on finite sets that could be grouped together to form a closed set. As is the case with most fundamental concepts in mathematics, the modern definition of a group that follows is the result of a long evolutionary process. Although this definition was given by both Heinrich Weber and Walther von Dyck in 1882, it did not gain universal acceptance until this century.

DEFINITION Binary Operation

Let G be a set. A *binary operation* on G is a function that assigns each ordered pair of elements of G an element of G.

A binary operation on a set G, then, is simply a method (or formula) by which an ordered pair from G combine to yield a new member of G. The most familiar binary operations are ordinary addition, subtraction, and multiplication of integers. Division of integers is not a binary operation on the integers (because an integer divided by an integer need not be an integer).

The binary operations addition modulo n and multiplication modulo n on the set $\{0, 1, 2, \ldots, n - 1\}$, which we denote by Z_n, play an extremely important role in abstract algebra. In certain situations we will want to combine the elements of Z_n by addition modulo n only; in other situations we will want to use both addition modulo n and multiplication modulo n to combine the elements. It will be clear from the context whether we are using addition only or addition and multiplication. For example, when multiplying matrices with entries from Z_n, we will need both addition modulo n and multiplication modulo n.

> *DEFINITION* Group
>
> Let G be a nonempty set together with a binary operation (usually called multiplication) that assigns to each ordered pair (a, b) of elements of G an element in G denoted by ab. We say G is a *group* under this operation if the following three properties are satisfied.
>
> 1. *Associativity.* The operation is associative; that is, $(ab)c = a(bc)$ for all a, b, c in G.
> 2. *Identity.* There is an element e (called the *identity*) in G, such that $ae = ea = a$ for all a in G.
> 3. *Inverses.* For each element a in G, there is an element b in G (called an inverse of a) such that $ab = ba = e$.

In words, then, a group is a set together with an associative operation such that every element has an inverse and any pair of elements can be combined without going outside the set. This latter condition is called *closure*. Be sure to verify closure when testing for a group. (See Example 5.)

If a group has the property that $ab = ba$ for every pair of elements a and b, we say the group is *Abelian*. A group is *non-Abelian* if there is some pair of elements a and b for which $ab \neq ba$. When encountering a particular group for the first time, one should determine whether or not it is Abelian.

Now that we have the formal definition of a group, our first job is to build a good stock of examples. These examples will be used throughout the text to illustrate the theorems. (The best way to grasp the meat of a theorem is to see what it says in specific cases.) As we progress, the reader is bound to have hunches and conjectures that can be tested against the stock of examples. To develop a complete understanding of these examples, the reader should supply the missing details.

Example 1 The set of integers Z (so denoted because the German word for integers is *Zahlen*), the set of rational numbers Q (for quotient), and the set of

real numbers **R** are all groups under ordinary addition. In each case the identity
is 0 and the inverse of a is $-a$. □

Example 2 The set of integers under ordinary multiplication *is not* a group.
Property (3) fails. For example, there is no *integer b* such that $5b = 1$. □

Example 3 The subset $\{1, -1, i, -i\}$ of the complex numbers is a group
under complex multiplication. Note that -1 is its own inverse, while the inverse
of i is $-i$. □

Example 4 The set Q^+ of positive rationals *is* a group under ordinary multiplication. The inverse of any a is $1/a = a^{-1}$. □

Example 5 The set S of positive irrational numbers together with 1 under
multiplication satisfies the three properties given in the definition of a group
but is *not* a group. Indeed, $\sqrt{2} \cdot \sqrt{2} = 2$, so S is not closed under multiplication. □

Example 6 A rectangular array of the form $\begin{bmatrix} a & b \\ c & d \end{bmatrix}$ is called a *2 × 2 matrix*.
The set of all 2 × 2 matrices with real entries is a group under componentwise
addition. That is,

$$\begin{bmatrix} a_1 & b_1 \\ c_1 & d_1 \end{bmatrix} + \begin{bmatrix} a_2 & b_2 \\ c_2 & d_2 \end{bmatrix} = \begin{bmatrix} a_1 + a_2 & b_1 + b_2 \\ c_1 + c_2 & d_1 + d_2 \end{bmatrix}$$

The identity is $\begin{bmatrix} 0 & 0 \\ 0 & 0 \end{bmatrix}$ and the inverse of $\begin{bmatrix} a & b \\ c & d \end{bmatrix}$ is $\begin{bmatrix} -a & -b \\ -c & -d \end{bmatrix}$. □

Example 7 The set $Z_n = \{0, 1, \ldots, n - 1\}$ for $n \geq 1$ *is* a group under
addition modulo n. For any j in Z_n, the inverse of j is $n - j$. This group is
usually referred to as the *group of integers modulo n*. □

As we have seen, the real numbers, the 2 × 2 matrices with real entries,
and the integers modulo n are all groups under the appropriate addition. But
what about multiplication? In each case the existence of some elements that do
not have inverses prevents the set from being a group under the usual multiplication. However, we can form a group in each case by simply throwing out the
rascals.

Example 8 The set $\mathbf{R}^\#$ of nonzero real numbers is a group under ordinary
multiplication. The identity is 1. The inverse of a is $1/a$. □

Example 9* The *determinant* of 2×2 matrix $\begin{bmatrix} a & b \\ c & d \end{bmatrix}$ is the number $ad - bc$. If A is a 2×2 matrix, *det A* denotes the determinant of A.

The set

$$GL(2, \mathbf{R}) = \left\{ \begin{bmatrix} a & b \\ c & d \end{bmatrix} \middle| a, b, c, d \in \mathbf{R}, ad - bc \neq 0 \right\}$$

of 2×2 matrices with real entries and nonzero determinant is a non-Abelian group under the operation

$$\begin{bmatrix} a_1 & b_1 \\ c_1 & d_1 \end{bmatrix} \begin{bmatrix} a_2 & b_2 \\ c_2 & d_2 \end{bmatrix} = \begin{bmatrix} a_1a_2 + b_1c_2 & a_1b_2 + b_1d_2 \\ c_1a_2 + d_1c_2 & c_1b_2 + d_1d_2 \end{bmatrix}.$$

The first step in verifying that this set is a group is to show that the product of two matrices with nonzero determinant also has nonzero determinant. This follows from the fact that for any pair of 2×2 matrices A and B, det $(AB) =$ (det A)(det B). (See exercise 5.)

Associativity can be verified by direct (but cumbersome) calculations. The identity is $\begin{bmatrix} 1 & 0 \\ 0 & 1 \end{bmatrix}$; the inverse of $\begin{bmatrix} a & b \\ c & d \end{bmatrix}$ is

$$\begin{bmatrix} \dfrac{d}{ad - bc} & \dfrac{-b}{ad - bc} \\ \dfrac{-c}{ad - bc} & \dfrac{a}{ad - bc} \end{bmatrix}$$

(explaining the requirement that $ad - bc \neq 0$). This very important group is called the *general linear group* of 2×2 matrices over **R**. □

Example 10 The set of 2×2 matrices with real number entries is *not* a group under the operation defined in Example 9. Inverses do not exist when the determinant is zero. □

Now that we have shown how to make subsets of the real numbers and subsets of the set of 2×2 matrices into multiplicative groups, we next consider the integers under multiplication modulo n.

Example 11 (L. Euler, 1761)
By exercise 11 of Chapter 0 an integer a has a multiplicative inverse modulo n if and only if a and n are relatively prime. So, for each $n > 1$, we define $U(n)$ to be the set of all positive integers less than n and relatively prime to n. Then

---

*For simplicity we have restricted our matrix examples to the 2×2 case. However, readers who have had linear algebra can readily generalize to $n \times n$ matrices.

$U(n)$ is a group under multiplication modulo n. (We leave it as an exercise that this set is closed under this operation.)

For $n = 10$, we have $U(10) = \{1, 3, 7, 9\}$. The Cayley table for $U(10)$ is

mod 10	1	3	7	9
1	1	3	7	9
3	3	9	1	7
7	7	1	9	3
9	9	7	3	1

(Recall that ab mod 10 is defined as the unique integer r with the property $a \cdot b = nq + r$, where $0 \leqslant r < n$ and $a \cdot b$ is ordinary multiplication.) □

Example 12 The set $\{0, 1, 2, 3\}$ is *not* a group under multiplication modulo 4. Although 1 and 3 have inverses, the elements 0 and 2 do not. □

Example 13 The set of integers under subtraction *is not* a group, since the operation is not associative. □

With the examples given this far as a guide, it is wise for the reader to pause here and think of his or her own examples. Study actively! Don't just read along and be spoon-fed by the book.

Example 14 For all integers $n \geqslant 1$, the set of complex roots of unity

$$\left\{ \cos \frac{k \cdot 360}{n} + i \sin \frac{k \cdot 360}{n} \ \middle| \ k = 0, 1, 2, \ldots, n - 1 \right\}$$

(i.e., complex zeros of $x^n - 1$) is a group under multiplication. (See De Moivre's Theorem—Example 11 of Chapter 0.) Compare this group with the one of Example 3. □

The complex number $a + bi$ can be represented geometrically as the point (a, b) in a plane coordinatized by a horizontal real axis and a vertical i (or imaginary) axis. The distance from the point $a + bi$ to the origin is $\sqrt{a^2 + b^2}$ and is often denoted by $|a + bi|$. For any angle θ, the line segment joining the complex number $\cos \theta + i \sin \theta$ and the origin forms an angle of θ with the positive real axis. Thus, the six complex zeros of $x^6 = 1$ are located at points around the circle of radius 1, 60° apart, as shown in Figure 2.1.

Example 15 The set $\mathbf{R}^n = \{(a_1, a_2, \ldots, a_n) \mid a_1, a_2, \ldots, a_n \in \mathbf{R}\}$ is a group under componentwise addition [i.e., $(a_1, a_2, \ldots, a_n) + (b_1, b_2 \ldots, b_n) = (a_1 + b_1, a_2 + b_2, \ldots, a_n + b_n)$]. □

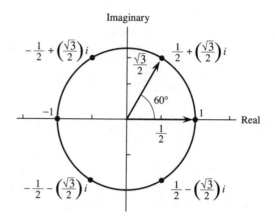

Figure 2.1

Example 16 For a fixed point (a, b) in $\mathbf{R}^2$, define $T_{a,b}: \mathbf{R}^2 \to \mathbf{R}^2$ by $(x, y) \to (x + a, y + b)$. Then $T(\mathbf{R}^2) = \{T_{a,b} \mid a, b \in \mathbf{R}\}$ is a group under function composition. Straightforward calculations show that $T_{a,b}T_{c,d} = T_{a+c,b+d}$. From this formula we may observe that $T(\mathbf{R}^2)$ is closed, $T_{0,0}$ is the identity, the inverse of $T_{a,b}$ is $T_{-a,-b}$, and $T(\mathbf{R}^2)$ is Abelian. Function composition is always associative. The elements of $T(\mathbf{R}^2)$ are called *translations*.   □

Example 17 The set of all 2×2 matrices with nonzero determinant and entries from Z_p (p a prime), Q (rational numbers), or $\mathbf{C}$ (complex numbers) *is* a group under matrix multiplication. The addition and multiplication of the entries are to be done with Z_p, Q, or $\mathbf{C}$, respectively. If you consider, for instance, the entries of the matrices below as belonging to Z_5 and combine them under addition and multiplication modulo 5, we have

$$\begin{bmatrix} 3 & 2 \\ 2 & 4 \end{bmatrix} \begin{bmatrix} 3 & 1 \\ 1 & 1 \end{bmatrix} = \begin{bmatrix} 1 & 0 \\ 0 & 1 \end{bmatrix}.$$

Thus, the two matrices on the left are inverses of each other.   □

Example 18 The set of all 2×2 matrices with determinant 1 with entries form Q, $\mathbf{R}$, $\mathbf{C}$, or Z_p (p a prime) *is* a non-Abelian group under matrix multiplication. This group is called the *special linear group* of 2×2 matrices over Q, $\mathbf{R}$, $\mathbf{C}$, or Z_p, respectively. If the entries are from F, where F is any of the above, we denote this group by $SL(2,F)$.   □

Example 19 The set $\{1, \ldots, n - 1\}$ *is* a group under multiplication modulo n if and only if n is prime.   □

Example 20 The set of all symmetries of the infinite ornamental pattern in which the arrowheads are spaced uniformly a unit apart along a line *is* an Abelian

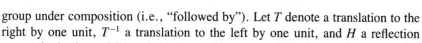

group under composition (i.e., "followed by"). Let T denote a translation to the right by one unit, T^{-1} a translation to the left by one unit, and H a reflection across the horizontal line of the figure. Then, every member of the group is of the form $x_1 x_2 \cdots x_n$, where each $x_i \in \{T, T^{-1}, H\}$. In this case, we say T, T^{-1}, and H *generate* the group. □

Table 2.1 summarizes many of the specific groups that we have presented thus far.

Table 2.1 Summary of Group Examples

Group	Operation	Identity	Form of element	Inverse	Abelian
Z	addition	0	k	$-k$	yes
Q^+	multiplication	1	m/n $m, n > 0$	n/m	yes
Z_n	addition mod n	0	k	$n - k$	yes
$\mathbf{R}^{\#}$	multiplication	1	x	$1/x$	yes
$GL(2, \mathbf{R})$	matrix multiplication	$\begin{bmatrix} 1 & 0 \\ 0 & 1 \end{bmatrix}$	$\begin{bmatrix} a & b \\ c & d \end{bmatrix}$ $ad - bc \neq 0$	$\begin{bmatrix} \dfrac{d}{ad-bc} & \dfrac{-b}{ad-bc} \\ \dfrac{-c}{ad-bc} & \dfrac{a}{ad-bc} \end{bmatrix}$	no
$U(n)$	multiplication mod n	1	k $\gcd(k, n) = 1$	solution to $kx = 1 \bmod n$	yes
$\mathbf{R}^n$	componentwise addition	$(0, 0, \ldots, 0)$	$(a_1, a_2, \ldots, a_n)$	$(-a_1, -a_2, \ldots, -a_n)$	yes
$SL(2, \mathbf{R})$	matrix multiplication	$\begin{bmatrix} 1 & 0 \\ 0 & 1 \end{bmatrix}$	$\begin{bmatrix} a & b \\ c & d \end{bmatrix}$ $ad - bc = 1$	$\begin{bmatrix} d & -b \\ -c & a \end{bmatrix}$	no
D_n	composition	R_0	R_α, F	$R_{360-\alpha}, F$	no

As the above examples demonstrate, the notion of a group is a very broad one indeed. The goal of the axiomatic approach is to find a set of properties general enough to permit many diverse examples that possess these properties and are specific enough to allow one to deduce many interesting consequences of these properties.

The goal of abstract algebra is to discover truths about algebraic systems (that is, sets with one or more binary operations) that are independent of the specific nature of the operations. All one knows or needs to know is that these operations, whatever they may be, have certain properties. We then seek to deduce consequences of these properties. This is why this branch of mathematics is called *abstract* algebra. It must be remembered, however, that when a specific group is being discussed, a specific operation must be given (at least implicitly).

Elementary Properties of Groups

Now that we have seen many diverse examples of groups, we wish to deduce some properties they share. The definition itself raises some fundamental questions. Every group has *an* identity. Could a group have more than one? Every group element has *an* inverse. Could an element have more than one? The examples suggest not.

Theorem 2.1 *Uniqueness of the Identity*
In a group G, there is only one identity element.

Proof. Suppose both e and e' are identities of G. Then,

1. $ae = a$ for all a in G, and
2. $e'a = a$ for all a in G.

The choice of $a = e'$ in (1) and $a = e$ in (2) yields $e'e = e'$ and $e'e = e$. Thus, e and e' are both equal to $e'e$ and so are equal to each other. ∎

Because of this theorem, we may unambiguously speak of "the identity" of a group and denote it by "e" (because the German word for identity is *Einheit*).

Theorem 2.2 *Cancellation*
In a group G, the right and left cancellation laws hold; that is, $ba = ca$ implies $b = c$, and $ab = ac$ implies $b = c$.

Proof. Suppose $ba = ca$. Let a' be an inverse of a. Then, multiplying on the right by a' gives $(ba)a' = (ca)a'$. Associativity yields $b(aa') = c(aa')$. Then, $be = ce$ and, therefore, $b = c$ as desired. Similarly, one can prove $ab = ac$ implies $b = c$ by multiplying by a' on the left. ∎

A consequence of the cancellation property is the fact that in a Cayley table for a group, each group element occurs exactly once in each row and column. (See exercise 28.) Another consequence of the cancellation property is the uniqueness of inverses.

Theorem 2.3 *Uniqueness of Inverses*
For each element a in a group G, there is a unique element b in G such that $ab = ba = e$.

Proof.   Suppose b and c are both inverses of a. Then $ab = e$ and $ac = e$, so that $ab = ac$. Now cancel a.   ■

As was the case with the identity element, it is reasonable, in view of Theorem 2.3, to speak of "the inverse" of an element g of a group; and, in fact, we may unambiguously denote it by g^{-1}. This notation is suggested by that used for ordinary real numbers under multiplication. Similarly, when n is a positive integer, g^n is used to denote the product $\underbrace{gg \cdots g}_{n \text{ factors}}$; we define $g^0 = e$; when n is negative, we define $g^n = (g^{-1})^{-n}$. With this notation, the familiar laws of exponents hold for groups; that is, for all integers m and n and any group element g, we have $g^m g^n = g^{m+n}$ and $(g^m)^n = g^{mn}$. Although the way one manipulates the group expressions $g^m g^n$ and $(g^m)^n$ coincides with the laws of exponents for real numbers, the law of exponents fail to hold for expressions involving two group elements. Thus, for groups in general, $(ab)^n \neq a^n b^n$ (see exercise 19).

Also, one must be careful with this notation when the group operation is addition and is denoted by $+$. In this case, the inverse of g is written as $-g$. Likewise, for example, g^3 means $g + g + g$ and is usually written as $3g$, while g^{-3} means $(-g) + (-g) + (-g)$ and is written as $-3g$. When this notation is used, do not interpret "ng" as the group operation; n may not even be an element of the group! Unlike the case for real numbers, in an abstract group, we do not permit nonintegers exponents such as $g^{1/2}$. Table 2.2 shows the common notation and corresponding terminology for groups under multiplication and groups under addition. As is the case for real numbers, we use $a - b$ as an abbreviation for $a + (-b)$.

Because of the associative property, we may unambiguously write the expression abc, for this can be reasonably interpreted as only $(ab)c$ or $a(bc)$, which are equal. In fact, by using induction and repeated application of the associative property, one can prove a general associative property that essentially means parentheses can be inserted or deleted at will without affecting the value of a product involving any number of group elements. Thus,

$$a^2(bcdb^2) = a^2 b(cd)b^2 = (a^2 b)(cd)b^2 = a(abcdb)b,$$

and so on.

Table 2.2

Multiplicative Group		Additive Group	
$a \cdot b$ or ab	Multiplication	$a + b$	Addition
e or 1	Identity or one	0	Zero
a^{-1}	Multiplicative inverse of a	$-a$	Additive inverse of a
a^n	Power of a	na	Multiple of a
ab^{-1}	Quotient	$a - b$	Difference

Applications of Modular Arithmetic

With the terminology and notation introduced thus far, we can describe some interesting applications of modular arithmetic. In Chapter 0 we discussed the methods used by the postal service and United Parcel Service to append check digits to identification numbers. We observed that neither method was able to detect all possible single digit errors. Detection of all single digit errors as well as nearly all errors involving the transposition of two adjacent digits is easily achieved, however. In this section we present several such methods that are in use.

Most products sold in supermarkets have an identification number coded with bars that are read by optical scanners. See Figure 2.2. This code is called the Universal Product Code (UPC). Each coded item is assigned a 12-digit number. The first six digits identify the manufacturer, the next five the product, the last is a check. (For many items, the twelfth digit is not printed but it is always bar coded.) In Figure 2.2, the check digit is 8.

To explain how the check digit is calculated it is convenient to introduce the dot product notation for two k-tuples:

$$(a_1, a_2, \ldots, a_k) \cdot (w_1, w_2, \ldots, w_k)$$
$$= a_1 w_1 + a_2 w_2 + \cdots + a_k w_k.$$

An item with the UPC identification number $a_1 a_2 \ldots a_{12}$ satisfies the condition

$$(a_1, a_2, \ldots, a_{12}) \cdot (3, 1, 3, 1, \ldots, 3, 1) = 0 \bmod 10.$$

In particular, the check digit is

$$-(a_1, a_2, \ldots, a_{11}) \cdot (3, 1, 3, 1, \ldots, 3) \bmod 10.$$

To verify that the number in Figure 2.2 satisfies the above condition we calculate

$$0 \cdot 3 + 2 \cdot 1 + 1 \cdot 3 + 0 \cdot 1 + 0 \cdot 3 + 0 \cdot 1 + 6 \cdot 3$$
$$+ 5 \cdot 1 + 8 \cdot 3 + 9 \cdot 1 + 7 \cdot 3 + 8 \cdot 1 = 90 = 0 \bmod 10.$$

Figure 2.2

The fixed k-tuple used in the calculation of check digits is called the *weighting vector*.

Now, suppose a single error is made in entering the number in Figure 2.2 into a computer. Say, for instance, that 021000958978 is entered (notice that the seventh digit is incorrect). Then the computer calculates

$$0 \cdot 3 + 2 \cdot 1 + 1 \cdot 3 + 0 \cdot 1 + 0 \cdot 3 + 0 \cdot 1 + 9 \cdot 3$$
$$+ 5 \cdot 1 + 8 \cdot 3 + 9 \cdot 1 + 7 \cdot 3 + 8 \cdot 1 = 99.$$

Since $99 \neq 0$ mod 10, the entered number cannot be correct. In general, as we will see in Theorem 2.4, when the numbers used in the weighting vector are relatively prime to n, any error involving a single digit will be detected by a check digit computed modulo n.

But why should a weighting vector be used at all? One could detect all single digit errors in a string $a_1 a_2 \ldots a_k$ by simply choosing a check digit a_{k+1} so that $a_1 + a_2 + \cdots + a_k + a_{k+1} = 0$ mod 10. For surely any error in exactly one position will result in a sum that is not 0 modulo 10. (In fact, this method is used on U.S. mail with bar coded zip codes—see [3]). The advantage of the UPC scheme is that it will detect nearly all errors involving the transposition of two adjacent digits as well as errors involving one digit.*
For doubters, let us say that the identification number given in Figure 2.2 is entered as 021000658798. Notice that the last two digits preceding the check digit have been transposed. But by calculating the dot product, we obtain $94 \neq 0$ mod 10 so we have detected an error. In fact, the only undetected transposition errors of adjacent digits a and b are those where $|a - b| = 5$. To verify this we observe that a transposition error of the form

$$a_1 a_2 \ldots a_i a_{i+1} \ldots a_{12} \rightarrow a_1 a_2 \ldots a_{i+1} a_i \ldots a_{12}$$

is undetected if and only if

$$(a_1, a_2, \ldots, a_{i+1}, a_i, \ldots, a_{12}) \cdot (3, 1, 3, 1, \ldots, 3, 1) = 0 \text{ mod } 10.$$

That is, if and only if

$$(a_1, a_2, \ldots, a_{i+1}, a_i, \ldots, a_{12}) \cdot (3, 1, 3, 1, \ldots, 3, 1) =$$
$$(a_1, a_2, \ldots, a_i, a_{i+1}, \ldots, a_{12}) \cdot (3, 1, 3, 1, \ldots, 3, 1)$$

This equality simplifies to either

$$3a_{i+1} + a_i = 3a_i + a_{i+1} \quad \text{or}$$
$$a_{i+1} + 3a_i = a_i + 3a_{i+1}$$

---

*A highly publicized error of this type recently occurred when Lt. Col. Oliver North gave United States Assistant Secretary of State Elliott Abrams an incorrect Swiss bank account number for depositing \$10 million for the contras. The correct account number began with "386"; the number North gave Abrams began with "368."

depending upon whether i is even or odd. Both cases reduce to $2(a_{i+1} - a_i) = 0 \bmod 10$.

Identification numbers printed on bank checks consist of an eight digit number $a_1 a_2 \ldots a_8$ and a check digit a_9 so that

$$(a_1, a_2, \ldots, a_9) \cdot (7, 3, 9, 7, 3, 9, 7, 3, 9) = 0 \bmod 10.$$

As was the case for the UPC scheme, this method detects all single digit errors and all errors involving the transposition of adjacent digits a and b except when $|a - b| = 5$. But it also detects most errors of the form $\ldots abc \ldots \rightarrow \ldots cba \ldots$, while the UPC method detects no errors of this form.

The next theorem reveals the relationship between the weighting vector and its ability to detect errors. In particular, it shows why it is not possible to detect all single digit errors and all transposition errors with a dot product modulo 10 scheme.

Theorem 2.4 *Error-Detecting Capability*
Suppose an identification number $a_1 a_2 \ldots a_k$ satisfies

$$(a_1, a_2, \ldots, a_k) \cdot (w_1, w_2, \ldots, w_k) = 0 \bmod n$$

where $0 \leq a_i < n$ for each i. Then all single digit errors in the ith position are detected if and only if w_i is relatively prime to n and all errors of the form

$$\ldots a_i a_{i+1} \ldots a_j a_{j+1} \ldots \rightarrow \ldots a_j a_{i+1} \ldots a_i a_{j+1} \ldots$$

(that is, the digits in the ith and jth position are interchanged) are detected if and only if $w_i - w_j$ is relatively prime to n.

Proof.   Consider a single error in the ith position, say, a_i' is substituted for a_i. Then the dot product of the correct number and the incorrect number differ by $(a_i - a_i')w_i$. Thus, the error is undetected if and only if $(a_i - a_i')w_i = 0 \bmod n$. If w_i is relatively prime to n, then w_i belongs to $U(n)$ and therefore, $w_i^{-1} \bmod n$ exists. So $(a_i - a_i')w_i w_i^{-1} = 0 \bmod n$ and $a_i' = a_i$. If w_i is not relatively prime to n, then an error a_i' with $|a_i - a_i'|$ divisible by $n/\gcd(w_i, n)$ will not be detected.

Now consider an error of the form

$$\ldots a_{ij} a_{i+1} \ldots a_j a_{j+1} \ldots \rightarrow \ldots a_j a_{i+1} \ldots a_i a_{i+1} \ldots$$

Then the dot products of the correct number and the incorrect number differ by

$$(a_i w_i + a_j w_j) - (a_j w_i + a_i w_j) = (a_i - a_j)(w_i - w_j)$$

Thus, the error is undetected if and only if

$$(a_i - a_j)(w_i - w_j) = 0 \bmod n$$

The conclusion now follows as before.   ■

In light of Theorem 2.4 we see that to detect all single digit errors and all transposition errors, the modulus n must be a prime. However, since the standard number system is base 10, a penalty must be paid for using a prime modulus. For instance, if we select $n = 7$, we must restrict our identification numbers to at most five digits, excluding the check digit itself, and each digit must be between 0 and 6. This gives at most $7^5 = 16,807$ identification numbers. Another possibility is to choose $n = 11$. In this case all the digits from 0 to 9 are available and we can have up to nine digits in length, excluding the check digit. Thus there are $10^9 = 1,000,000,000$ possible identification numbers. The drawback here is that the check digit may turn out to be 10, which is actually *two* digits. In practice, this is handled by ad hoc methods. For example, the ten digit International Standard Book Number (ISBN) used throughout the world has the property $(a_1, a_2, \ldots, a_{10}) \cdot (10, 9, 8, 7, 6, 5, 4, 3, 2, 1) = 0 \bmod 11$. The first nine digits identifies the language of the publishing company's country, the publishers, and the title, while the tenth digit is the check. When a_{10} is required to be 10 to make the dot product 0, the character X is used as the check digit.

In Chapter 5 we will present a check digit scheme, based on the dihedral group of order 10, that detects all single digit errors and all transposition errors without introducing any new characters or restricting the number of digits involved.

Historical Note

We conclude this chapter with a bit of history concerning the noncommutativity of matrix multiplication. In 1925, quantum theory was replete with annoying and puzzling ambiguities. It was Werner Heisenberg who recognized the cause. He observed that the product of the quantum-theoretical analogs of the classical Fourier series did not necessarily commute. For all his boldness, this shook Heisenberg. As he later recalled [4, p. 94]:

> In my paper the fact that XY was not equal to YX was very disagreeable to me. I felt this was the only point of difficulty in the whole scheme, otherwise I would be perfectly happy. But this difficulty had worried me and I was not able to solve it.

Heisenberg asked his teacher, Max Born, if his ideas were worth publishing. Born was fascinated and deeply impressed by Heisenberg's new approach. Born wrote [1, p. 217]:

> After having sent off Heisenberg's paper to the *Zeitschrift für Physik* for publication, I began to ponder over his symbolic multiplication, and was soon so involved in it that I thought about it for the whole day and could hardly sleep at night. For I felt there was something fundamental behind it, the consummation of our endeavors of many years. And one morning, about the 10 July 1925, I suddenly saw light: Heisenberg's symbolic multiplication was nothing but the matrix calculus, well-known to me since my student days from Rosanes' lectures is Breslau.

Born and his student, Pascual Jordan, reformulated Heisenberg's ideas in terms of matrices, but it was Heisenberg who was credited with the formulation. In his autobiography, Born laments [1, p. 219]:

Nowadays the textbooks speak without exception of Heisenberg's matrices, Heisenberg's commutation law, and Dirac's field quantization.

In fact, Heisenberg knew at that time very little of matrices and had to study them.

Upon learning in 1933 that he was to receive the Nobel Prize with Dirac and Schrödinger for this work, Heisenberg wrote to Born [1, p. 220]:

If I have not written to you for such a long time, and have not thanked you for your congratulations, it was partly because of my rather bad conscience with respect to you. The fact that I am to receive the Nobel Prize alone, for work done in Göttingen in collaboration—you, Jordan, and I—this fact depresses me and I hardly know what to write to you. I am, of course, glad that our common efforts are now appreciated, and I enjoy the recollection of the beautiful time of collaboration. I also believe that all good physicists know how great was your and Jordan's contribution to the structure of quantum mechanics—and this remains unchanged by a wrong decision from outside. Yet I myself can do nothing but thank you again for all the fine collaboration, and feel a little ashamed.

The story has a happy ending, however, because Born received the Nobel Prize in 1954 for his fundamental work in quantum mechanics.

EXERCISES

I went this far with him: "Sir, allow me to ask you one question. If the Church should say to you, 'two and three make ten,' what would you do?" "Sir," said he, "I should believe it, and I should count like this: one, two, three, four, ten." I was now fully satisfied.

Boswell's Journal for 31st May, 1764

1. Give two reasons why the set of odd integers under addition is not a group.
2. Referring to Example 13, verify the assertion that subtraction is not associative.
3. Show that $\begin{bmatrix} 2 & 2 \\ 1 & 1 \end{bmatrix}$ does not have a multiplicative inverse in $GL(2, \mathbf{R})$.
4. Show that the group $GL(2, \mathbf{R})$ of Example 9 is non-Abelian, by exhibiting a pair of matrices A and B in $GL(2, \mathbf{R})$ such that $AB \neq BA$.
5. If A and B are 2×2 matrices with real entries, prove that $\det (AB) = (\det A)(\det B)$.
6. Give an example of group elements a and b with the property that $a^{-1}ba \neq b$.

7. Translate each of the following multiplicative expressions to its additive counterpart.
   a. a^2b^3
   b. $a^{-2}(b^{-1}c)^2$
   c. $(ab^2)^{-3}c^2 = e$.

8. For any elements a and b from a group and any integer n, prove that $(a^{-1}ba)^n = a^{-1}b^na$.

9. Write out a complete operation table for Z_4 (see Example 7).

10. Is the binary operation defined by the following table associative? Is it commutative?

	a	b	c	d
a	a	b	c	d
b	b	a	d	c
c	c	d	a	b
d	a	d	b	c

11. Show $\{1, 2, 3\}$ under multiplication modulo 4 is not a group but $\{1, 2, 3, 4\}$ under multiplication modulo 5 is a group.

12. Referring to Example 20, describe in words the symmetry corresponding to $T^{-4}H$.

13. Referring to Example 18, show that in the group $SL(2, Z_3)$, the element $\begin{bmatrix} 2 & 1 \\ 1 & 1 \end{bmatrix}$ is the inverse of $\begin{bmatrix} 1 & 2 \\ 2 & 2 \end{bmatrix}$.

14. In the notation of Example 16, verify that $T_{a,b}T_{c,d} = T_{a+c,b+d}$.

15. Prove that the set of all 2×2 matrices with entries from **R** and determinant $+1$ is a group under matrix multiplication.

16. For any integer $n > 2$, show that there are at least two elements in $U(n)$ that satisfy $x^2 = 1$.

17. An abstract algebra teacher intended to give a typist a list of nine integers that form a group under multiplication modulo 91. Instead, one of the nine integers was inadvertently left out so that the list appeared as 1, 9, 16, 22, 53, 74, 79, 81. Which integer was left out? (This really happened!)

18. Suppose G is a group with the property that whenever a, b and c belong to G and $ab = ca$ then $b = c$. Prove that G is Abelian.

19. (Law of Exponents for Abelian groups) Let a and b be elements of an Abelian group and let n be any integer. Show $(ab)^n = a^nb^n$. Is this also true for non-Abelian groups?

20. (Socks-Shoes Property) In a group, prove that $(ab)^{-1} = b^{-1}a^{-1}$. Find an example that shows it is possible to have $(ab)^{-2} \neq b^{-2}a^{-2}$. Find a non-Abelian example that shows it is possible to have $(ab)^{-1} = a^{-1}b^{-1}$ for some

distinct nonidentity elements a and b. Draw an analogy between the statement $(ab)^{-1} = b^{-1}a^{-1}$ and the act of putting on and taking off your socks and shoes.

21. Prove that a group G is Abelian if and only if $(ab)^{-1} = a^{-1}b^{-1}$ for all a and b in G.

22. In a group, prove $(a^{-1})^{-1} = a$ for all a.

23. Show that the set $\{5, 15, 25, 35\}$ is a group under multiplication modulo 40. What is the identity element of this group? Can you see any relationship between this group and $U(8)$?

24. What is the inverse of $a_1a_2 \ldots a_n$?

25. The integers 5 and 15 are among a collection of twelve integers that form a group under multiplication modulo 56. List all twelve.

26. Give an example of a group with 105 elements. Give two examples of groups with 42 elements.

27. Construct a Cayley table for $U(12)$.

28. Prove that every group table is a *Latin square**; that is, each element of the group appears exactly once in each row and each column. (This exercise is referred to in this chapter.)

29. Suppose the table below is a group table. Fill in the blank entries.

	e	a	b	c	d
e	e	—	—	—	—
a	—	b	—	—	e
b	—	c	d	e	—
c	—	d	—	a	b
d	—	—	—	—	—

30. Prove that if $(ab)^2 = a^2b^2$ in a group G, then $ab = ba$.

31. Let a, b, and c be elements of a group. Solve the equation $axb = c$ for x. Solve $a^{-1}xa = c$ for x.

32. Prove that the set of all rational numbers of the form 3^m6^n, where m and n are integers, is a group under multiplication.

33. Let G be a finite group. Show that there is an odd number of elements x of G such that $x^3 = e$. Show that there is an even number of elements x of G such that $x^2 \neq e$.

34. Let G be a group and let $g \in G$. Define a function ϕ_g from G to G by $x\phi_g = g^{-1}xg$ for all x in G. Show that ϕ_g is one-to-one and onto. (That is,

---

*Latin squares are useful in designing statistical experiments. There is also a close connection between Latin squares and finite geometries.

$x\phi_g = y\phi_g$ implies $x = y$ and, for each y in G, there is an x in G such that $x\phi_g = y$.)

35. Let G be a group and g, $h \in G$. Define ϕ_g, ϕ_h, ϕ_{gh} as in the previous problem (that is, $x\phi_h = h^{-1}xh$ and $x\phi_{gh} = (gh)^{-1}x(gh)$). Show $\phi_g \circ \phi_h = \phi_{gh}$.

36. Explain why division is not a binary operation on the set of real numbers.

37. Consider the set $G = \{0, 1, 2, 3, 4, 5, 6, 7\}$. Suppose there is a group operation $*$ on G that satisfies the following two conditions:
   a. $a * b \le a + b$ for all a, b in G,
   b. $a * a = 0$ for all a in G.
   Construct the multiplication table for G. (This group is sometimes called the *Nim group*.)

38. Prove that if G is a group with the property that the square of every element is the identity, then G is Abelian. (This exercise is referred to in Chapters 9, 24, and 28.)

39. Let F denote a reflection in D_{10}. If R_α denotes a rotation of α degrees, express the element $(R_{36} F)^{-1}$ as a product without using negative exponents.

40. Prove that the set of all 3×3 matrices with real entries of the form

$$\begin{bmatrix} 1 & a & b \\ 0 & 1 & c \\ 0 & 0 & 1 \end{bmatrix}$$

is a group. (Multiplication is defined by

$$\begin{bmatrix} 1 & a & b \\ 0 & 1 & c \\ 0 & 0 & 1 \end{bmatrix}\begin{bmatrix} 1 & a' & b' \\ 0 & 1 & c' \\ 0 & 0 & 1 \end{bmatrix} = \begin{bmatrix} 1 & a + a' & b' + ac' + b \\ 0 & 1 & c' + c \\ 0 & 0 & 1 \end{bmatrix}.$$

This group, sometimes called the *Heisenberg group* after the Nobel Prize–winning physicist Werner Heisenberg, is intimately related to the Heisenberg Uncertainty Principle of Quantum Physics.)

41. Prove the assertion made in Example 19 that the set $\{1, 2, \ldots, n - 1\}$ is a group under multiplication modulo n if and only if n is prime.

42. In a finite group, show that the number of nonidentity elements that satisfy the equation $x^5 = e$ is a multiple of 4.

43. Use the UPC scheme to determine the check digit for the number 07312400508.

44. Use the bank scheme to determine the check digit for the number 09190204.

45. Verify the check digit for the ISBN assigned to this book.

46. The invalid ISBN 0–669–03925-4 is the result of a transposition of two adjacent digits not involving the check digit. Determine the correct ISBN.

47. The State of Utah appends a ninth digit a_9 to their eight-digit driver's license number $a_1a_2 \ldots a_8$ so that $(a_1, a_2, \ldots, a_8, a_9) \cdot (9, 8, 7, 6, 5, 4, 3, 2, 1) = 0$ mod 10.

   a. If the first eight digits of a Utah license number are 14910573, what is the ninth digit?
   b. Suppose a legitimate Utah license number 149105767 is miscopied as 149105267. How would you know a mistake was made? Is there any way you could determine the correct number? Suppose you know the error was in the seventh position, could you correct the mistake?
   c. If a legitimate Utah number 149105767 were miscopied as 199105767, would you be able to tell a mistake was made? Explain.
   d. Explain why any transposition error involving adjacent digits of a Utah number would be detected.

48. The Canadian province of Quebec uses the weighting vector (12, 11, 10, . . . , 2, 1) and modulo 10 arithmetic to append a check digit to the driver's license numbers. Criticize this method. Describe all single digit errors that are undetected by this scheme. How does the transposition of two adjacent digits of a number affect the check digit of a number?

49. (IBM check digit method) Major credit cards, banks in West Germany, many libraries in the United States, and South Dakota driver's license numbers employ the following check digit method. To a number $a_1 a_2 \ldots a_k$, where k is odd, append

$$-((a_1, a_2, \ldots, a_k) \cdot (2, 1, 2, 1, \ldots) + r) \bmod 10$$

where r is the number of summands in the dot product ≥ 10. [If k is even, the weighting vector (1, 2, 1, 2, . . .) is used instead.] Calculate the check digit for the number 3125600196431. Prove that this method detects all single digit errors. Why does this not contradict Theorem 2.4? Determine which transposition errors involving adjacent digits go undetected by this method.

PROGRAMMING EXERCISES

Almost immediately after the war, Johnny [Von Neumann] and I also began to discuss the possibilities of using computers heuristically to try to obtain insights into questions of pure mathematics. By producing examples and by observing the properties of special mathematical objects, one could hope to obtain clues as to the behavior of general statements which have been tested on examples.

S. M. Ulam, *Adventures of a Mathematician*

1. Write a program to print out the following information about $U(n)$ (see Example 11). Assume $n < 100$.
   a. The elements of $U(n)$.
   b. The inverse of each member of $U(n)$.
   Run your program for $n = 12, 15, 30, 36, 63$.

2. Determine the size of $U(k)$ for $k = 9, 27, 81, 243, 25, 125, 49, 121$. On the basis of this information, try to guess a formula for the size of $U(p^n)$ as a function of the prime p and the integer n.

3. Determine the size of $U(k)$ for $k = 18, 54, 162, 486, 50, 250, 98, 242$. Make a conjecture about the relationship between the size of $U(2p^n)$ and the size of $U(p^n)$ where p is a prime other than 2.

4. Write a program that calculates the UPC check digit. Check your program with the numbers 011300101373 and 038000001277.

5. Write a program that calculates the bank number check digit.

6. Write a program that calculates the ISBN check digit.

7. Write a program that calculates the IBM check digit.

REFERENCES

1. Max Born, *My Life: Recollections of a Nobel Laureate,* New York: Charles Scribner's Sons, 1978.

2. J. Bronowski, *The Ascent of Man,* Boston: Little, Brown and Company, 1973.

3. J. A. Gallian, "The Zip Code Bar Code," *UMAP Journal,* 7 (1986): 191–195.

4. J. Mehra and H. Rechenberg, *The Historical Development of Quantum Theory,* Vol. 3, New York: Springer-Verlag, 1982.

SUGGESTED READINGS

P. J. Denning, "Computer Viruses," *American Scientist,* 76(1988): 236–238.

The author discusses computer viruses (programs that attack other programs) and possible immunization schemes. One of these employs check digits.

J. A. Gallian and S. Winters, "Modular Arithmetic in the Marketplace," *The American Mathematical Monthly,* 95(1988): 548–551.

This article provides a more detailed analysis of the check digit schemes presented in this chapter and in Chapter 0. In particular, the error detection rates for the various schemes are given.

J. E. White, "Introduction to Group Theory for Chemists," *Journal of Chemical Education* 44(1967): 128–135.

Students interested in the physical sciences may find this article worthwhile. It begins with easy examples of groups and builds up to applications of group theory concepts and terminology to chemistry.

SUGGESTED FILM

Dihedral Kaleidoscopes with H. S. M. Coxeter, International Film Bureau, $13\frac{1}{2}$ minutes, in color.

A dihedral kaleidoscope is formed by two plane mirrors intersecting at an angle π/n. When an object is placed before the mirrors of such a kaleidoscope, $2n$ objects are seen: the original and $2n - 1$ reflections. This illustrates the fact that a dihedral kaleidoscope generates, by two reflections, the dihedral group of order $2n$. The film discusses the special cases for $n = 1, 2, 3, 4$, makes some observations in the general case, and proceeds to the limiting case in which the two mirrors are parallel. The dihedral kaleidoscopes are then used to show some interesting regular figures, their stellations and tessellations of the plane. This film was awarded first prize for mathematics at the International Festival of Scientific and Technical Films.

Finite Groups; Subgroups

3

In our own time, in the period 1960–1980, we have seen particle physics emerge as the playground of group theory.

Freeman Dyson

Terminology and Notation

As we will soon discover, finite groups—that is, groups with finitely many elements—have interesting arithmetical properties. To facilitate the study of finite groups, it is convenient to introduce some terminology and notation.

DEFINITION Order of a Group
The number of elements of a group (finite or infinite) is called its *order*. We will use $|G|$ to denote the order of G.

Thus, the group Z of integers under addition has infinite order, while the group $U(10) = \{1, 3, 7, 9\}$ under multiplication modulo 10 has order 4.

DEFINITION Order of an Element
The *order* of an element g in a group G is the smallest positive integer n such that $g^n = e$. (In additive notation this would be $ng = 0$.) If no such integer exists, we say g has *infinite order*. The order of an element g is denoted by $|g|$.

So, to find the order of a group element g you need only compute the sequence of products $g, g^2, g^3, \ldots$, until you reach the identity for the first time. The exponent of this product (or coefficient if the operation is addition) is the order of g. If the identity never appears in the sequence, then g has infinite order.

Example 1   Consider $U(15) = \{1, 2, 4, 7, 8, 11, 13, 14\}$ under multiplication modulo 15. To find the order of 7, say, we compute the sequence $7^1 = 7$, $7^2 = 4$, $7^3 = 13$, $7^4 = 1$, so $|7| = 4$. To find the order of 11, we compute $11^1 = 11$, $11^2 = 1$, so $|11| = 2$. Similar computations show $|1| = 1$, $|2| = 4$, $|4| = 2$, $|8| = 4$, $|13| = 4$, $|14| = 2$. (Here is a trick that makes these calculations easier. Rather than compute the sequence $13^1, 13^2, 13^3, 13^4$, we may observe that $13 = -2$ modulo 15 [since $13 + 2 = 0$ mod 15] so that $13^2 = (-2)^2 = 4$, $13^3 = -2 \cdot 4 = -8$, $13^4 = (-2)(-8) = 1$.) $\square$

Example 2   Consider Z_{10}. Since $1 \cdot 2 = 2, 2 \cdot 2 = 4, 3 \cdot 2 = 6, 4 \cdot 2 = 8$, $5 \cdot 2 = 0$ we know $|2| = 5$. Similar computations show $|0| = 1$, $|7| = 10$, $|5| = 2$ and $|6| = 5$. $\square$

Example 3   Consider Z. Here every nonzero element has infinite order since the sequence $a, 2a, 3a, \ldots$ never includes 0 when $a \neq 0$. $\square$

The perceptive reader may have noticed among our examples of groups in Chapter 2 that some are subsets of others with the same binary operation. The group in Example 18 with real entries, for instance, is a subset of the group in Example 9. Similarly, the group of complex numbers $\{1, -1, i, -i\}$ is a subset of the group described in Example 14 for n equal to any multiple of 4. This situation arises so often that we introduce a special term to describe it.

> *DEFINITION*   Subgroup
> If a subset H of a group G is itself a group under the operation of G, we say H is a *subgroup* of G.

We use the notation $H \leqslant G$ to mean H is a subgroup of G. If we want to indicate that H is a subgroup of G, but not equal to G itself, we write $H < G$. Such a subgroup is called a *proper subgroup*. The subgroup $\{e\}$ is called the *trivial subgroup* of G; a subgroup that is not $\{e\}$ is called a *nontrivial subgroup* of G.

Notice that Z_n under addition modulo n is *not* a subgroup of Z under addition, since addition modulo n is not the operation of Z.

Subgroup Tests

When determining whether or not a subset H of a group G is a subgroup of G, one need not directly verify the group axioms. The next three results provide simple tests that suffice to show a subset of a group is a subgroup.

Theorem 3.1 *One-Step Subgroup Test*
*Let G be a group and H a nonempty subset of G. Then, H is a subgroup
of G if H is closed under division; that is, if ab^{-1} is in H whenever a and
b are in H.*

Proof. Since the operation of H is the same as that of G, it is clear that
this operation is associative. Next, we show that e is in H. Since H is
nonempty, we may pick some x in H. Then, letting $a = x$ and $b = x$ in
the hypothesis, we have $e = xx^{-1} = ab^{-1}$ is in H. To verify that x^{-1} is
in H whenever x is in H, all we need to do is choose $a = e$ and $b = x$ in
the statement of the theorem. Finally, the proof will be complete when we
show that H is closed; that is, if x, y belong to H, we must show xy is in
H also. Well, we have already shown that y^{-1} is in H whenever y is; so
letting $a = x$ and $b = y^{-1}$, we have $xy = x(y^{-1})^{-1} = ab^{-1}$ is in H. ■

Although we have dubbed Theorem 3.1 the "One-Step Test," there are
actually four steps involved in applying the theorem. (After you gain some
experience, the first three steps are routine.) Notice the similarity between the
last three steps listed below and the three steps involved in the Principle of
Mathematical Induction.

1. Identify the property P that distinguishes the elements of H; that is, identify
   a defining condition.
2. Prove the identity has property P. (This verifies that H is nonempty.)
3. *Assume* two elements a and b have property P.
4. Use the assumption about a and b to show ab^{-1} has property P.

Of course, when the operation of G is addition, step 4 becomes $a - b$. The
procedure is illustrated in Examples 4 and 5.

Example 4 Let G be an Abelian group with identity e. Then $H = \{x \in
G \mid x^2 = e\}$ is a subgroup of G. Here, the defining property of H is the condition
$x^2 = e$. So, we first note that $e^2 = e$ so that H is nonempty. Now we assume
that a and b belong to H. This means $a^2 = e$ and $b^2 = e$. Finally, we must
show that $(ab^{-1})^2 = e$. Since G is Abelian, $(ab^{-1})^2 = ab^{-1}ab^{-1} = a^2(b^{-1})^2 =
a^2(b^2)^{-1} = ee^{-1} = e$. Therefore, ab^{-1} belongs to H and, by the One-Step
Subgroup Test, H is a subgroup of G. □

In many instances, a subgroup will consist of all elements that have a
particular form. Here, the property P is the particular form.

Example 5 Let G be an Abelian group under multiplication with identity e.
Then $H = \{x^2 \mid x \in G\}$ is a subgroup of G. (In words, H is the set of all
"squares.") Since $e^2 = e$, the identity has the correct form. Next we write two
elements of H in the correct form, say, a^2 and b^2. We must show $a^2(b^2)^{-1}$ also

has the correct form. Since G is Abelian, we may write $a^2(b^2)^{-1}$ as $(ab^{-1})^2$, which is the correct form. Thus, H is a subgroup of G.   ☐

How do you prove a subset of a group is *not* a subgroup? Here are three possible ways:

1. Show that the identity is not in the set.
2. Exhibit an element of the set whose inverse is not in the set.
3. Exhibit two elements of the set whose product is not in the set.

Example 6   Let G be the group of nonzero real numbers under multiplication, $H = \{x \in G \mid x = 1 \text{ or } x \text{ is irrational}\}$ and $K = \{x \in G \mid x \geq 1\}$. Then H is not a subgroup of G since $\sqrt{2} \in H$ but $\sqrt{2} \cdot \sqrt{2} = 2 \notin H$. Also, K is not a subgroup since $2 \in K$ but $2^{-1} \notin K$.   ☐

Beginning students often prefer to use the next theorem instead of Theorem 3.1.

Theorem 3.2   *Two-Step Subgroup Test*
Let G be a group and H a nonempty subset of G. Then, H is a subgroup of G if $ab \in H$ whenever a, $b \in H$ (closed under multiplication), and $a^{-1} \in H$ whenever $a \in H$ (closed under inverse).

Proof.   By Theorem 3.1, it suffices to show that a, $b \in H$ implies $ab^{-1} \in H$. So, we suppose that a, $b \in H$. Since H is closed under inverse, we also have $b^{-1} \in H$. Thus, $ab^{-1} \in H$ by closure under multiplication.   ∎

When dealing with finite groups, it is easier to use the following subgroup test.

Theorem 3.3   *Finite Subgroup Test*
Let H be a nonempty finite subset of a group G. Then, H is a subgroup of G if H is closed under the operation of G.

Proof.   In view of Theorem 3.2, we need only prove that $a^{-1} \in H$ whenever $a \in H$. If $a = e$, then $a^{-1} = a$ and we are done. If $a \neq e$, consider the sequence $a, a^2, a^3, \ldots$. Since H is finite and closure implies that all positive powers of a are in H, not all of these elements are distinct. Say, $a^i = a^j$ and $i > j$. Then $a^{i-j} = e$; and since $a \neq e$, $i - j > 1$. Thus, $a^{i-j} = a \cdot a^{i-j-1} = e$ and, therefore, $a^{i-j-1} = a^{-1}$. But, $i - j - 1 \geq 1$ implies $a^{i-j-1} \in H$ and we are done.   ∎

Examples of Subgroups

The proofs of the next few theorems show how the above tests work.

Theorem 3.4 *$\langle a \rangle$ Is a Subgroup*
Let G be a group, and let a be any element of G. Then

$$\langle a \rangle = \{a^n \mid n \in Z\}$$

is a subgroup of G. (a^0 is defined to be the identity.)

Proof. Let a^n, $a^m \in \langle a \rangle$. Then, $a^n \cdot (a^m)^{-1} = a^{n-m} \in \langle a \rangle$; so, by Theorem 3.1, $\langle a \rangle$ is a subgroup of G. ∎

The subgroup $\langle a \rangle$ defined in Theorem 3.4 is called the *cyclic subgroup of G generated by a*. Notice that, although the list . . . , $a^{-2}, a^{-1}, a^0, a^1, a^2$, . . . has infinitely many entries, the set $\{a^n \mid n \in Z\}$ may have only finitely many elements. Also note that, since $a^i \cdot a^j = a^{i+j} = a^{j+i} = a^j \cdot a^i$, every cyclic group is Abelian.

Example 7 In $U(10)$, $\langle 3 \rangle = \{3, 9, 7, 1\} = U(10)$, for $3^1 = 3$, $3^2 = 9$, $3^3 = 7$, $3^4 = 1$, $3^5 = 3^4 \cdot 3 = 1 \cdot 3$, $3^6 = 3^4 \cdot 3^2 = 9$, ... □

Example 8 In Z_{10}, $\langle 2 \rangle = \{2, 4, 6, 8, 0\}$. Remember, a^n means na when the operation is addition. □

Example 9 In Z, $\langle -1 \rangle = Z$. Here each entry in the list . . . , $-2(-1)$, $-1(-1)$, $0(-1)$, $1(-1)$, $2(-1)$, . . . represents a distinct group element. □

Example 10 In D_n, the dihedral group of order $2n$, let R denote a rotation of $360/n$ degrees. Then,

$$R^n = R_{360°} = e, \qquad R^{n+1} = R, \qquad R^{n+2} = R^2, \ldots .$$

Similarly, $R^{-1} = R^{n-1}$, $R^{-2} = R^{n-2}$, . . . so that $\langle R \rangle = \{e, R, \ldots, R^{n-1}\}$. We see, then, that the powers of R "cycle back" periodically with period n. Diagrammatically, raising R to successive positive powers is the same as moving clockwise around the following circle one node at a time, while raising R to successive negative powers is the same as moving around the circle counterclockwise one node at a time. □

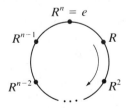

□

In Chapter 4 we will show that $|\langle a \rangle| = |a|$; that is, the order of the subgroup generated by a is the order of a itself. (Actually, the definition of $|a|$ was chosen to ensure the validity of this equation.)

We next consider one of the most important subgroups.

DEFINITION   Center of a Group

The *center* $Z(G)$ of a group G is the subset of elements in G that commute with every element of G. In symbols,

$$Z(G) = \{a \in G \mid ax = xa \text{ for all } x \text{ in } G\}.$$

(The notation $Z(G)$ comes from the fact that the German word for center is *Zentrum*. The term was coined by J. A. de Seguier in 1904.)

Theorem 3.5   *Center Is a Subgroup*
The center of a group G is a subgroup G.

Proof.   For variety, we shall use Theorem 3.2 to prove this result. Clearly, $e \in Z(G)$, so $Z(G)$ is nonempty. Now, suppose $a, b \in Z(G)$. Then $(ab)x = a(bx) = a(xb) = (ax)b = (xa)b = x(ab)$ for all x in G; and, therefore, $ab \in Z(G)$.

Next, assume that $a \in Z(G)$. Then, we have $ax = xa$ for all x in G. What we want is $a^{-1}x = xa^{-1}$ for all x in G. Informally, all we need do to obtain the second equation from the first one is to simultaneously bring the a's across the equal sign:

$$a\overset{\frown}{x} = \underset{\smile}{x}a$$

becomes $xa^{-1} = a^{-1}x$. (Be careful here; groups need not be commutative. The a on the left comes across as a^{-1} on the left and the a on the right comes across as an a^{-1} on the right.) Formally, the desired equation can be obtained from the original one by multiplying it on the left and right by a^{-1} like so:

$$a^{-1}(ax)a^{-1} = a^{-1}(xa)a^{-1},$$
$$(a^{-1}a)xa^{-1} = a^{-1}x(aa^{-1}),$$
$$exa^{-1} = a^{-1}xe,$$
$$xa^{-1} = a^{-1}x.$$

This shows that $a^{-1} \in Z(G)$ whenever a is.   ∎

For practice, let's determine the centers of the dihedral groups.

Example 11   For $n \geq 3$,

$$Z(D_n) = \begin{cases} \{R_0, R_{180}\} & \text{when } n \text{ is even,} \\ \{R_0\} & \text{when } n \text{ is odd.} \end{cases}$$

We begin by showing that $Z(D_n)$ cannot contain a reflection. If F is a reflection, there are two possible cases for the reflection axis for F. Either this axis passes through a vertex of the n-gon, or it joins the midpoints of two opposite sides of the n-gon. Let's assume first that the axis passes through a vertex. Label the n-gon as shown below.

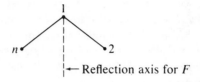

Now, $FR_{360/n}$ gives

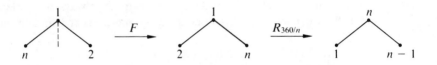

while $R_{360/n}F$ gives

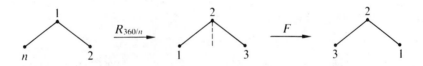

Thus, $FR_{360/n}$ sends vertex 1 to vertex n, while $R_{360/n}F$ sends vertex 1 to vertex 2. Since $n \geqslant 3$, we have $FR_{360/n} \neq R_{360/n}F$ so that F is not in the center of D_n. A similar argument on the following diagram rules out reflections that join midpoints of opposite sides (this case arises when n is even).

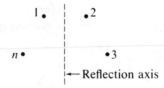

We have proved, then, that no reflection is in the center of D_n.

Next, consider a rotation $R = R_{k \cdot 360/n} \ (1 \leqslant k < n)$ in D_n. Let's assume that $0 < k \cdot 360/n < 180°$. Label the n-gon as shown in the following figure, and let F denote a reflection across the axis passing through vertex 1.

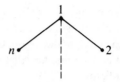

Now, RF sends vertex 1 to a vertex on the right side of the reflection axis, while FR sends vertex 1 to a vertex on the left side of the reflection axis. Thus, $FR \neq RF$. A similar argument shows that $FR \neq RF$ when $180° < k \cdot 360°/n < 360°$. This proves that R_0 and R_{180} are the only possible elements in the center of D_n. When n is odd, D_n has no 180° rotation; and we leave it to the reader to show that when n is even, R_{180} does, indeed, commute with every member of D_n. $\qquad\qquad\square$

Although an element from a non-Abelian group need not necessarily commute with every element of the group, there are always some elements with which it will commute. This observation prompts the next definition and theorem.

DEFINITION Centralizer of a in G

Let a be a fixed element of a group G. The *centralizer of a in G*, $C(a)$, is the set of all elements in G that commute with a. In symbols, $C(a) = \{g \in G \mid ga = ag\}$.

Example 12 In D_4, we have the following centralizers.

$$C(R_0) = D_4 = C(R_{180}),$$
$$C(R_{90}) = \{R_0, R_{90}, R_{180}, R_{270}\} = C(R_{270}),$$
$$C(H) = \{R_0, H, R_{180}, V\} = C(V),$$
$$C(D) = \{R_0, D, R_{180}, D'\} = C(D').\qquad\square$$

Notice that each of the centralizers in Example 12 is actually a subgroup of D_4. The next theorem shows that this was not a coincidence.

Theorem 3.6 *$C(a)$ Is a Subgroup*

For each a in a group G, the centralizer of a is a subgroup of G.

Proof. A proof similar to that of Theorem 3.5 is left to the reader to supply. $\qquad\qquad\blacksquare$

EXERCISES

If I were to prescribe one process in the training of men which is fundamental to success in any direction, it would be thoroughgoing training in the habit of

accurate observation. It is a habit which every one of us should be seeking ever more to perfect.

Eugene G. Grace

1. For each group in the following list, find the order of the group and the order of each element in the group. In each case, how are the orders of the elements of a group related to the order of the group?

$$Z_{12}, \quad U(10), \quad U(12), \quad U(20), \quad D_4$$

2. Let Q be the group of rational numbers under addition and Q^* be the group of nonzero rational numbers under multiplication. In Q, list the elements in $\langle \frac{1}{2} \rangle$. In Q^*, list the elements in $\langle \frac{1}{2} \rangle$.

3. Let Q and Q^* be as in exercise 2. Find the order of each element in Q and Q^*.

4. Prove that in any group an element and its inverse have the same order.

5. Without actually computing the orders, explain why the two elements in each of the following pairs of elements from Z_{30} must have the same order: $\{2, 28\}, \{8, 22\}$. Do the same for the pairs of elements from $U(15)$: $\{2, 8\}$, $\{7, 13\}$.

6. Let x belong to a group. If $x^2 \neq e$ while $x^6 = e$, prove that $x^4 \neq e$ and $x^5 \neq e$. What can we say about the order of x?

7. Show that $U(14) = \langle 3 \rangle = \langle 5 \rangle$. (Hence $U(14)$ is cyclic.) Is $U(14) = \langle 11 \rangle$?

8. Show that $Z_{10} = \langle 3 \rangle = \langle 7 \rangle = \langle 9 \rangle$. Is $Z_{10} = \langle 2 \rangle$?

9. Show that $U(20) \neq \langle k \rangle$ for any k in $U(20)$. (Hence $U(20)$ is not cyclic.)

10. Prove that an Abelian group with two elements of order 2 must have a subgroup of order 4.

11. Find a group that contains elements a and b such that $|a| = |b| = 2$ while
    a. $|ab| = 3$
    b. $|ab| = 4$
    c. $|ab| = 5$
    Can you see any relationship between $|a|, |b|$ and $|ab|$?

12. For each divisor k of n, let $U_k(n) = \{x \in U(n) \mid x = 1 \bmod k\}$. List the elements of $U_4(20)$, $U_5(20)$, $U_5(30)$, and $U_{10}(30)$. Prove that $U_k(n)$ is a subgroup of $U(n)$. (This exercise is referred to in Chapter 8.)

13. Suppose m is a divisor of k and k is a divisor of n. Show $U_k(n)$ is a subgroup of $U_m(n)$. (This notation is explained in exercise 12.)

14. If H and K are subgroups of G, show that $H \cap K$ is a subgroup of G. (Can you see that the same proof shows that the intersection of any number of subgroups of G, finite or infinite, is again a subgroup of G?)

15. Let G be a group. Show that $Z(G) = \bigcap_{a \in G} C(a)$. (This means the intersection of *all* subgroups of the form $C(a)$.)

16. Let G be a group, and let $a \in G$. Prove $C(a) = C(a^{-1})$.

17. Suppose G is the group defined by the following Cayley table.

	1	2	3	4	5	6	7	8
1	1	2	3	4	5	6	7	8
2	2	1	8	7	6	5	4	3
3	3	4	5	6	7	8	1	2
4	4	3	2	1	8	7	6	5
5	5	6	7	8	1	2	3	4
6	6	5	4	3	2	1	8	7
7	7	8	1	2	3	4	5	6
8	8	7	6	5	4	3	2	1

   a. Find the centralizer of each member of G.

   b. Find $Z(G)$.

   c. Find the order of each element of G. How are these orders arithmetically related to the order of the group?

18. Let G be a group, and let $a \in Z(G)$. In a Cayley table for G, how does the row headed by a compare with the column headed by a?

19. Prove Theorem 3.6.

20. If H is a subgroup of G, then by the *centralizer* $C(H)$ of H we mean the set $\{x \in G \mid xh = hx$ for all $h \in H\}$. Prove that $C(H)$ is a subgroup of G.

21. Let G be an Abelian group with identity e and let n be some integer. Prove that the set of all elements of G that satisfy the equation $x^n = e$ is a subgroup of G. Give an example of a group G in which the set of all elements of G that satisfy the equation $x^2 = e$ does not form a subgroup of G.

22. Suppose a belongs to a group and $|a| = 5$. Prove $C(a) = C(a^3)$. Find an element a from some group such that $|a| = 6$ and $C(a) \neq C(a^3)$.

23. Show that a group of order 6 cannot have a subgroup of order 4. (Hint: Suppose there is a group G of order 6 that has a subgroup H of order 4. Let x belong to G but not H. Now show $xH = \{xh \mid h \in H\}$ and H have no elements in common.) Is the proof you used also valid for a subgroup H of order 5? Generalize to the case where G has order n.

24. Suppose n is an even positive integer and H is a subgroup of Z_n. Prove that either every member of H is even or exactly half of the members of H are even.

25. Suppose a group contains elements a and b such that $|a| = 4$, $|b| = 2$, and $a^3b = ba$. Find $|ab|$.

26. Consider the elements $A = \begin{bmatrix} 0 & -1 \\ 1 & 0 \end{bmatrix}$ and $B = \begin{bmatrix} 0 & 1 \\ -1 & -1 \end{bmatrix}$ from $SL(2, \mathbf{R})$. Find $|A|$, $|B|$, and $|AB|$. Does your answer surprise you?

27. Consider the element $A = \begin{bmatrix} 1 & 1 \\ 0 & 1 \end{bmatrix}$ in $SL(2, \mathbf{R})$. What is the order of A? If we view $A = \begin{bmatrix} 1 & 1 \\ 0 & 1 \end{bmatrix}$ as a member of $SL(2, Z_n)$, what is the order of A?

28. For any positive integer n and any angle θ show that in the group $SL(2, \mathbf{R})$

$$\begin{bmatrix} \cos\theta & -\sin\theta \\ \sin\theta & \cos\theta \end{bmatrix}^n = \begin{bmatrix} \cos n\theta & -\sin n\theta \\ \sin n\theta & \cos n\theta \end{bmatrix}.$$

   Use this formula to find the order of

$$\begin{bmatrix} \cos 60° & -\sin 60° \\ \sin 60° & \cos 60° \end{bmatrix} \quad \text{and} \quad \begin{bmatrix} \cos \sqrt{2}° & -\sin \sqrt{2}° \\ \sin \sqrt{2}° & \cos \sqrt{2}° \end{bmatrix}.$$

   (Geometrically, $\begin{bmatrix} \cos\theta & -\sin\theta \\ \sin\theta & \cos\theta \end{bmatrix}$ represents a rotation of the plane θ degrees.)

29. Let G be the symmetry group of a circle. Show G has elements of every finite order as well as elements of infinite order.

30. Let $|x| = 6$. Find $|x^2|, |x^3|, |x^4|, |x^5|$. Let $|y| = 9$. Find $|y^i|$ for $i = 2, 3, \ldots, 8$. Do these examples suggest any relationship between the order of the power of an element and the order of the element?

31. D_4 has 7 cyclic subgroups. List them. Find a subgroup of D_4 of order 4 that is not cyclic.

32. $U(15)$ has 6 cyclic subgroups. List them.

33. If $|a| = n$ and k divides n, prove $|a^{n/k}| = k$.

34. Suppose G is a group that has exactly 8 elements of order 3. How many subgroups of order 3 does G have?

35. Let $H = \{x \in U(20) \mid x = 1 \bmod 3\}$. Is H a subgroup of $U(20)$?

36. Compute the orders of the following groups:
    a. $U(3), U(4), U(12)$
    b. $U(5), U(7), U(35)$
    c. $U(4), U(5), U(20)$
    d. $U(3), U(5), U(15)$

37. On the basis of your answers for exercise 36, make a conjecture about the relationship between $|U(r)|, |U(s)|$, and $|U(rs)|$.

38. Compute $|U(4)|, |U(10)|$, and $|U(40)|$. Do these groups provide a counterexample to your answer to exercise 37? If so, revise your conjecture.

39. Find a cyclic subgroup of order 4 in $U(40)$.

40. Find a noncyclic subgroup of order 4 in $U(40)$.

41. Let $G = \left\{ \begin{bmatrix} a & b \\ c & d \end{bmatrix} \,\middle|\, a, b, c, d, \in Z \right\}$ under addition.

Let $H = \left\{ \begin{bmatrix} a & b \\ c & d \end{bmatrix} \,\middle|\, a + b + c + d = 0 \right\}$. Prove that H is a subgroup of G. What if 0 is replaced by 1?

42. Let $G = GL(2, \mathbf{R})$. Let $H = \{A \in G \mid \det A$ is a power of 2$\}$. Show H is a subgroup of G.

43. Let H be a subgroup $\mathbf{R}$ under addition. Let $K = \{2^a \mid a \in H\}$. Prove that K is a subgroup of $\mathbf{R}^{\#}$ under multiplication.

44. Let G be a group of functions from $\mathbf{R}$ to $\mathbf{R}^{\#}$ under multiplication. Let $H = \{f \in G \mid f(1) = 1\}$. Prove that H is a subgroup of G.

45. Let $G = GL(2, \mathbf{R})$ and $H = \left\{ \begin{bmatrix} a & 0 \\ 0 & b \end{bmatrix} \,\middle|\, a$ and b nonzero integers $\right\}$.

Prove or disprove that H is a subgroup of G.

46. Let $H = \{a + bi \mid a, b \in \mathbf{R}, ab \geq 0\}$. Prove or disprove that H is a subgroup of $\mathbf{C}$ under addition.

47. Let $H = \{a + bi \mid a, b \in \mathbf{R}, a^2 + b^2 = 1\}$. Prove or disprove that H is a subgroup of $\mathbf{C}^{\#}$ under multiplication. Describe the elements of H geometrically.

48. Find the smallest subgroup of Z containing.
   a. 8 and 14 (the notation for this is $\langle 8, 14 \rangle$)
   b. 8 and 13
   c. 6 and 15
   d. m and n

In each part find an integer k so that the subgroup is $\langle k \rangle$.

PROGRAMMING EXERCISES

A Programmer's Lament

I really hate this damned machine;
I wish that they would sell it
It never does quite what I want
But only what I tell it.

Dennie L. Van Tassel,
The Compleat Computer

1. Write a program to print out the cyclic subgroups of $U(n)$ generated by each k in $U(n)$. Assume $n < 100$. Run the program for $n = 12, 15, 30$. Compare the order of the subgroups with the order of the group itself. What arithmetical relationship do these integers have?

2. Repeat exercise 1 for Z_n. Have the program list the elements of Z_n that generate all of Z_n, that is, those elements k, $0 \leq k \leq n - 1$, for which $Z_n = \langle k \rangle$.

How does this set compare with $U(n)$? (See programming exercise 1a of Chapter 2.) Make a conjecture.

3. Write a program that does the following. For each pair of elements a and b from $U(n)$ print out $|a|$, $|b|$, and $|ab|$ on the same line. Assume $n < 100$. Run your program for $n = 15$, 30, 42. What is the arithmetical relationship between $|ab|$ and $|a|$ and $|b|$? Make a conjecture.

4. Repeat exercise 3 for Z_n using $a + b$ in place of ab.

5. Write a program that prints out the elements of $U_k(n)$. Assume $n < 100$. (See exercise 12 for notation.) Run your program for the following choices of n and k: (75, 3), (50, 5), (40, 5), (40, 8), (44, 4), (60, 4), and (40, 4).

SUGGESTED READING

Gina Kolata, "Perfect Shuffles and Their Relation to Math," *Science* 216 (1982): 505–506.

This is a delightful nontechnical article that discusses how group theory and computers were used to solve a difficult problem about shuffling a deck of cards. Serious work on the problem was begun by an undergraduate student as part of a programming course.

SUGGESTED SOFTWARE

L. D. Geissinger, *Exploring Small Groups in Abstract Algebra*, Chicago: Harcourt Brace Jovanovich, 1988.

This award-winning program helps students visualize and understand many of the basic properties of groups that are covered in this text. Among these are subgroups, order of an element, centralizer, conjugacy classes, cosets, and factor groups. The system requirements are IBM PC, XT, or AT with 256K RAM, one double-sided disk drive, color graphics adaptor, and color monitor. The software requires DOS version 2.0 or higher.

Cyclic Groups

4

To carry out his role of abstractor, the mathematician must continually pose such questions as "What is the common aspect of diverse situations?" or "What is the heart of the matter?" He must always ask himself, "What makes such and such a thing tick?" Once he has discovered the answer to these questions and has extracted the crucial simple parts, he can examine these parts in isolation. He blinds himself, temporarily, to the whole picture, which may be confusing.

Philip Davis and William Chinn

Properties of Cyclic Groups

DEFINITION   Cyclic Group

A group G is called *cyclic* if there is an element a in G such that $G = \{a^n \mid n \in Z\}$. Such an element a is called a *generator* of G.

In view of the notation introduced in the previous section, we may indicate that G is a cyclic group generated by a by writing $G = \langle a \rangle$.

In this chapter, we examine cyclic groups in detail and determine their important characteristics. We begin with a few examples.

Example 1   The set of integers Z under ordinary addition is cyclic. Both 1 and -1 are generators. (Recall that, when the operation is addition, 1^n is interpreted as $\underbrace{1 + 1 + \cdots + 1}_{n \text{ terms}}$ when n is positive and $\underbrace{(-1) + (-1) + \cdots + (-1)}_{|n| \text{ terms}}$ when n is negative.) □

Example 2   The set $Z_n = \{0, 1, \ldots, n - 1\}$ for $n \geq 1$ is a cyclic group under addition modulo n. Again, 1 and $-1 = n - 1$ are generators. □

Unlike Z, which has only two generators, Z_n may have many generators (depending on which n we are given).

Example 3 $Z_8 = \langle 1 \rangle = \langle 3 \rangle = \langle 5 \rangle = \langle 7 \rangle$. To verify, for instance, that $Z_8 = \langle 3 \rangle$, we note that $\langle 3 \rangle = \{3, (3 + 3) \bmod 8, (3 + 3 + 3) \bmod 8, \ldots\}$ is the set $\{3, 6, 1, 4, 7, 2, 5, 0\} = Z_8$. Thus, 3 is a generator of Z_8. On the other hand, 2 is not a generator since $\langle 2 \rangle = \{0, 2, 4, 6\} \neq Z_8$.   □

Example 4   (See Example 11 of Chapter 2.)
$U(10) = \{1, 3, 7, 9\} = \{3^0, 3^1, 3^3, 3^2\} = \langle 3 \rangle$. Also, $\{1, 3, 7, 9\} = \{7^0, 7^3, 7^1, 7^2\} = \langle 7 \rangle$. So both 3 and 7 are generators for $U(10)$.   □

Quite often in mathematics, a "nonexample" is as helpful in understanding a concept as an example. With regard to cyclic groups, $U(8)$ serves this purpose; that is, $U(8)$ is not a cyclic group. How can we verify this? Well, note that $U(8) = \{1, 3, 5, 7\}$. But,

$$\langle 1 \rangle = \{1\}$$
$$\langle 3 \rangle = \{3, 1\}$$
$$\langle 5 \rangle = \{5, 1\}$$
$$\langle 7 \rangle = \{7, 1\}$$

so that $U(8) \neq \langle a \rangle$ for any a in $U(8)$.

With these examples under our belts, we should now be ready to tackle cyclic groups in an abstract way and state their key properties.

Theorem 4.1   *Criterion for $a^i = a^j$*
Let G be a group, and let a belong to G. If a has infinite order, then all distinct powers of a are distinct group elements. If a has finite order, say, n, then $\langle a \rangle = \{e, a, a^2, \ldots, a^{n-1}\}$ and $a^i = a^j$ if and only if n divides $i - j$.

Proof.   If a has infinite order, there is no nonzero n such that a^n is the identity. Since $a^i = a^j$ implies $a^{i-j} = e$, we must have $i - j = 0$ and the first statement of the theorem is proved.

Now assume $|a| = n$. We will prove $\langle a \rangle = \{e, a, \ldots, a^{n-1}\}$. Certainly, the elements $e, a, \ldots, a^{n-1}$ are distinct. For if $a^i = a^j$ with $0 \leq j < i \leq n - 1$, then $a^{i-j} = e$. But this contradicts the fact that n is the least positive integer such that a^n is the identity.

Now, suppose that a^k is an arbitrary member of $\langle a \rangle$. By the division algorithm, there exists integers q and r such that

$$k = qn + r   \text{with}   0 \leq r < n.$$

Then $a^k = a^{qn+r} = a^{qn} \cdot a^r = (a^n)^q \cdot a^r = e \cdot a^r = a^r$ so that $a^k \in \{e, a, \ldots, a^{n-1}\}$. This proves $\langle a \rangle = \{e, a, \ldots, a^{n-1}\}$.

Next, we assume that $a^i = a^j$ and prove that n divides $i - j$. We begin by observing that $a^i = a^j$ implies $a^{i-j} = e$. Again, by the division algorithm, there are integers q and r such that

$$i - j = qn + r \quad \text{with} \quad 0 \leq r < n$$

Then $a^{i-j} = a^{qn+r}$ and, therefore, $e = ea^r = a^r$. Since n is the least positive integer such that a^n is the identity, we must have $r = 0$ so that n divides $i - j$.

Conversely, if n divides $i - j$, then $a^{i-j} = a^{nq} = e^q = e$ so that $a^i = a^j$. ■

One special case of Theorem 4.1 occurs so often it deserves singling out.

Corollary $a^k = e$ *Implies* $|a|$ *Divides k*
Let G be a group and a an element of G of order n. If $a^k = e$, then n divides k.

Proof.   Since $a^k = e = a^0$, we have, by Theorem 4.1, that n divides $k - 0$. ■

Theorem 4.1 and its corollary for the case $|a| = 6$ are illustrated in Figure 4.1.

What is important about Theorem 4.1 in the finite case is that it says that multiplication in $\langle a \rangle$ is essentially done by *addition* modulo n. That is, if $(i + j)$ mod $n = k$, then $a^i \cdot a^j = a^k$. Thus, no matter what group G is, or how the element a is chosen, multiplication in $\langle a \rangle$ works the same as addition in Z_n whenever $|a| = n$. Similarly, if a has infinite order, then multiplication in $\langle a \rangle$ works the same as addition in Z, since $a^i \cdot a^j = a^{i+j}$ and no modular arithmetic is done.

For these reasons, the cyclic groups Z_n and Z serve as prototypes for all cyclic groups, and algebraists say that there is essentially only one cyclic group of each order. What is meant by this is that, although there may be many different sets of the form $\{a^n \mid n \in Z\}$, there is essentially only one way to operate on these sets, depending on the order of a. Algebraists do not really care what the

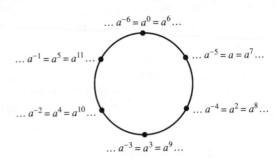

Figure 4.1

elements of a set are; they care only about the algebraic properties of the set—
that is, the ways the elements of a set can be combined. We will return to this
theme in the chapter on isomorphisms.

In Example 3 we saw that 3 was a generator for Z_8 while 2 was not. Similarly,
3 and 7 are generators for $U(10)$ while 9 is not. It would be nice to be able to
"eyeball" the generators for Z_n and for cyclic groups in general. Theorem 4.2
and its corollary give us a simple arithmetic method for identifying generators.

Theorem 4.2 Generators of Cyclic Groups
*Let $G = \langle a \rangle$ be a cyclic group of order n. Then $G = \langle a^k \rangle$ if and only if the
$\gcd(k, n) = 1$.*

Proof. If the $\gcd(k, n) = 1$ we may write $1 = ku + nv$ for some integers
u and v. Then $a = a^{ku+nv} = a^{ku} \cdot a^{nv} = a^{ku}$. Thus, a belongs to $\langle a^k \rangle$ and
therefore all powers of a belong to $\langle a^k \rangle$. So, $G = \langle a^k \rangle$ and a^k is a generator
of G.

Now suppose that the $\gcd(k, n) = d > 1$. Write $k = td$ and $n = sd$.
Then $(a^k)^s = (a^{td})^s = (a^{sd})^t = (a^n)^t = e$ so that $|a^k| \le s < n$. This shows
a^k is not a generator of G. ■

Taking $G = Z_n$ and $a = 1$ in Theorem 4.2, we have the following useful
result. (In particular, note that the generators of Z_n are precisely the elements
of $U(n)$.)

Corollary Generators of Z_n
An integer k is a generator of Z_n if and only if $\gcd(k, n) = 1$.

The value of Theorem 4.2 is that once one generator of a cyclic group has
been found, all generators of the cyclic group can easily be determined. For
example, consider the subgroup of all rotations in D_6. Clearly one generator is
R_{60}. And, since $|R_{60}| = 6$, we see by Theorem 4.2 that the only other generator
is $(R_{60})^5 = R_{300}$. Of course, we could have readily deduced this information
without the aid of Theorem 4.2 by direct calculations. So, to illustrate the real
power of Theorem 4.2, let us use it to find all generators of the cyclic group
$U(50)$. First, note that direct computations show $|U(50)| = 20$ and 3 is one of
its generators. Thus, in view of Theorem 4.2, the complete list of generators
for $U(50)$ is

$$3 \bmod 50 = 3, \qquad 3^{11} \bmod 50 = 47,$$
$$3^3 \bmod 50 = 27, \qquad 3^{13} \bmod 50 = 23,$$
$$3^7 \bmod 50 = 37, \qquad 3^{17} \bmod 50 = 13,$$
$$3^9 \bmod 50 = 33, \qquad 3^{19} \bmod 50 = 17.$$

Admittedly, we had to do some arithmetic here, but it certainly entailed much
less work than finding all the generators by simply determining the order of each
element of $U(50)$ one by one.

Classification of Subgroups of Cyclic Groups

The last theorem in this chapter tells us how many subgroups a finite cyclic group has and how to find them.

> *Theorem 4.3   Fundamental Theorem of Cyclic Groups*
> *Every subgroup of a cyclic group is cyclic. If $|\langle a \rangle| = n$, then the order of any subgroup of $\langle a \rangle$ is a divisor of n; and, for each divisor k of n, the group $\langle a \rangle$ has exactly one subgroup of order k, namely, $\langle a^{n/k} \rangle$.*

Before we prove this theorem, let's see what it means. Understanding what a theorem means is a prerequisite to understanding its proof. Suppose $G = \langle a \rangle$ and G has order 30, say. The first part of the theorem says that if H is any subgroup of G, then H has the form $\langle a^k \rangle$ for some k. The second part of the theorem says that G has one subgroup of each of the orders 1, 2, 3, 5, 6, 10, 15, or 30—and no others. The proof will also show how to find these subgroups.

Proof.   Let $G = \langle a \rangle$ and suppose H is a subgroup of G. We must show H is cyclic. If it consists of the identity alone, then clearly H is cyclic. So we may assume $H \neq \{e\}$. We now claim that H contains an element of the form a^t, where t is positive. Since $G = \langle a \rangle$, every element of H has the form a^t; and when a^t belongs to H with $t < 0$, then a^{-t} belongs to H also and $-t$ is positive. Thus, our claim is verified. Now let m be the least positive integer such that $a^m \in H$. By closure, $\langle a^m \rangle \leq H$. We next claim that $H = \langle a^m \rangle$. To prove this claim, it suffices to let b be an arbitrary member of H and show that b is in $\langle a^m \rangle$. Since $b \in G = \langle a \rangle$, we have $b = a^k$ for some k. Now, apply the division algorithm to k and m to obtain integers q and r such that $k = mq + r$ where $0 \leq r < m$. Then $a^k = a^{mq+r} = a^{mq} \cdot a^r$ so that $a^r = a^{-mq}a^k$. But, $a^k = b \in H$ and $a^{-mq} = (a^m)^{-q}$ is in H also so that $a^r \in H$. But, m is the *least* positive integer such that $a^m \in H$, and $0 \leq r < m$, so r must be 0. Thus, $a^{-mq}a^k = e$, and therefore $b = a^k = a^{mq} = (a^m)^q \in \langle a^m \rangle$. This proves the assertion of the theorem that every subgroup of a cyclic group is cyclic.

To prove the next portion of the theorem, suppose $|\langle a \rangle| = n$ and H is any subgroup of $\langle a \rangle$. We have already shown that $H = \langle a^m \rangle$ for some m. And, since $(a^m)^n = (a^n)^m = e^m = e$, we know from the corollary to Theorem 4.1 that $|a^m|$ is a divisor of n. Thus, $|H| = |a^m|$ is a divisor of n.

Finally, let k be any divisor of n, and let $t = n/k$. Clearly, $(a^t)^k = (a^{n/k})^k = e$ and $(a^t)^s = e$ for no positive $s < t$ so $\langle a^t \rangle$ has order k. We next show that $\langle a^t \rangle$ is the only subgroup of order k. To this end, let H be any subgroup of order k. We have previously shown that $H = \langle a^m \rangle$, where m is the least positive integer such that a^m is in H. Now, writing $n = mq + r$, where $0 \leq r < m$, we then have $e = a^n = a^{mq+r} = a^{mq} \cdot a^r$ so that $a^r = a^{-mq} = (a^m)^{-q} \in H$. Thus, $r = 0$ and $n = mq$. Also, $k = |H| = |\langle a^m \rangle| = n/m = q$ and, therefore, $m = n/q = n/k = t$. This proves $H = \langle a^m \rangle = \langle a^t \rangle$.   ■

Returning for a moment to our discussion of the cyclic group $\langle a \rangle$ where a has order 30, we may conclude from Theorem 4.3 that the subgroups of $\langle a \rangle$ are precisely those of the form $\langle a^m \rangle$ where m is a divisor of 30. Moreover, the subgroup of order k is $\langle a^{n/k} \rangle$. So the list of subgroups of $\langle a \rangle$ is:

$$
\begin{array}{lll}
\langle a \rangle = \{e, a, a^2, \ldots, a^{29}\} & & \text{order 30,} \\
\langle a^2 \rangle = \{e, a^2, a^4, \ldots, a^{28}\} & & \text{order 15,} \\
\langle a^3 \rangle = \{e, a^3, a^6, \ldots, a^{27}\} & & \text{order 10,} \\
\langle a^5 \rangle = \{e, a^5, a^{10}, a^{15}, a^{20}, a^{25}\} & & \text{order 6,} \\
\langle a^6 \rangle = \{e, a^6, a^{12}, a^{18}, a^{24}\} & & \text{order 5,} \\
\langle a^{10} \rangle = \{e, a^{10}, a^{20}\} & & \text{order 3,} \\
\langle a^{15} \rangle = \{e, a^{15}\} & & \text{order 2,} \\
\langle a^{30} \rangle = \{e\} & & \text{order 1.}
\end{array}
$$

In general, if $\langle a \rangle$ has order n and k divides n, then $\langle a^{n/k} \rangle$ is the unique subgroup of order k.

Taking the group in Theorem 4.3 to be Z_n and a to be 1, we obtain the following important special case.

Corollary *Subgroups of Z_n*
For each divisor k of n the set $\langle n/k \rangle$ is the unique subgroup of Z_n of order k; moreover, these are the only subgroups of Z_n.

Example 5   The list of subgroups of Z_{30} is

$$
\begin{array}{lll}
\langle 1 \rangle = \{0, 1, 2, \ldots, 29\} & & \text{order 30,} \\
\langle 2 \rangle = \{0, 2, 4, \ldots, 28\} & & \text{order 15,} \\
\langle 3 \rangle = \{0, 3, 6, \ldots, 27\} & & \text{order 10,} \\
\langle 5 \rangle = \{0, 5, 10, 15, 20, 25\} & & \text{order 6,} \\
\langle 6 \rangle = \{0, 6, 12, 18, 24\} & & \text{order 5,} \\
\langle 10 \rangle = \{0, 10, 20\} & & \text{order 3,} \\
\langle 15 \rangle = \{0, 15\} & & \text{order 2,} \\
\langle 30 \rangle = \{0\} & & \text{order 1.} \qquad \square
\end{array}
$$

By combining Theorems 4.2 and 4.3, we can easily count the number of elements of each order in a finite cyclic group. For convenience, we introduce an important number-theoretic function called the *Euler phi function.* Let $\phi(1) = 1$, and for any integer $n > 1$, let $\phi(n)$ denote the number of positive integers less than n and relatively prime to n. (We write this function on the left because that is the way it is always done in number theory and because we will never compose it with another function.) Notice that $|U(n)| = \phi(n)$.

Theorem 4.4   *Number of Elements of Each Order in a Cyclic Group*
If d is a divisor of n, the number of elements of order d in a cyclic group of order n is $\phi(d)$.

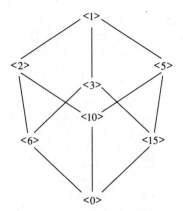

Figure 4.2

The relationship between the various subgroups of a group can be illustrated with a *subgroup lattice* of the group. This is a diagram that includes all the subgroups of the group and connects a subgroup H at one level to a subgroup K at a higher level with a sequence of line segments if and only if H is a proper subgroup of K. Although there are many ways to draw such a diagram, the connections between the subgroups must be the same. Typically one attempts to present the diagram in an eye-pleasing fashion. The lattice diagram for Z_{30} is shown in Figure 4.2.

The precision of Theorem 4.3 can be appreciated by comparing the ease with which we are able to identify the subgroups of Z_{30} with that of, say, doing the same for $U(30)$ or D_{30}. And these groups have relatively simple structures among noncyclic groups.

We will prove in Chapter 9 that a certain portion of Theorem 4.3 extends to arbitrary finite groups; namely, the order of a subgroup divides the order of the group itself. We will also see, however, that a finite group need not have exactly one subgroup corresponding to each divisor of the order of the group. For some divisors, there may be none at all, while for other divisors, there may be many.

One final remark about the importance of cyclic groups is appropriate. Although cyclic groups constitute a very narrow class of finite groups, we will see in Chapter 12 that they play the role of building blocks for all finite Abelian groups in much the same way that primes are the building blocks for the integers and the chemical elements are the building blocks for the chemical compounds.

EXERCISES

A mathematician, like a painter or a poet, is a maker of patterns. If his patterns are more permanent than theirs, it is because they are made with ideas.... The mathematician's patterns, like the painter's or poet's must be beautiful; the ideas, like the colors or the words, must fit together in a harmonious way. Beauty is the first test; there is no permanent place in the world for ugly mathematics.

G. H. Hardy

1. Find all generators of Z_6, Z_8, and Z_{20}.
2. Suppose $\langle a \rangle$, $\langle b \rangle$, and $\langle c \rangle$ are cyclic groups of orders 6, 8, and 20, respectively. Find all generators of $\langle a \rangle$, $\langle b \rangle$, and $\langle c \rangle$, respectively.
3. List the elements of the subgroups $\langle 20 \rangle$ and $\langle 10 \rangle$ in Z_{30}.
4. List the elements of the subgroups $\langle 3 \rangle$ and $\langle 15 \rangle$ in Z_{18}.
5. List the elements of the subgroups $\langle 3 \rangle$ and $\langle 7 \rangle$ in $U(20)$.
6. What do exercises 3, 4, and 5 have in common? Try to make a generalization that includes these three cases.
7. Find an example of a noncyclic group, all of whose proper subgroups are cyclic.
8. Let a be an element of a group and $|a| = 15$. Compute the orders of the following elements of G.
   a. a^3, a^6, a^9, a^{12}
   b. a^5, a^{10}
   c. a^2, a^4, a^8, a^{14}
9. How many subgroups does Z_{20} have? List a generator for each of these subgroups. Suppose $G = \langle a \rangle$ and $|a| = 20$. How many subgroups does G have? List a generator for each of these subgroups.
10. Let $G = \langle a \rangle$ and $|a| = 24$. List all generators for the subgroup of order 8.
11. Let G be a group and let $a \in G$. Prove $\langle a^{-1} \rangle = \langle a \rangle$.
12. Suppose a has infinite order. Find all generators of the subgroup $\langle a^3 \rangle$.
13. Suppose $|a| = 24$. Find a generator for $\langle a^{21} \rangle \cap \langle a^{10} \rangle$. In general, what is a generator for the subgroup $\langle a^m \rangle \cap \langle a^n \rangle$?
14. Suppose a cyclic group G has exactly three subgroups: G itself, $\{e\}$, and a subgroup of order 7. What is $|G|$?
15. Let G be an Abelian group and let $H = \{g \in G \mid |g| \text{ divides } 12\}$. Prove that H is a subgroup of G. Is there anything special about 12 here? Would your proof be valid if 12 is replaced by some other positive integer? State the general result.
16. If a cyclic group has an element of infinite order, how many elements of finite order does it have?
17. List the cyclic subgroups of $U(30)$.
18. Let G be a group and a an element of G of order n. For each integer k between 1 and n, show that $|a^k| = |a^{n-k}|$.
19. Let G be a group and a an element of G.
    a. If $a^{12} = e$, what can we say about the order of a?

b. If $a^m = e$, what can we say about the order of a?

c. Suppose $|G| = 24$ and G is cyclic. If $a^8 \neq e$ and $a^{12} \neq e$, show $\langle a \rangle = G$.

20. Prove a group of order 3 must be cyclic.

21. Let Z denote the group of integers under addition. Is every subgroup of Z cyclic? Why? Describe all the subgroups of Z.

22. For any element a in any group G, prove $\langle a \rangle$ is a subgroup of $C(a)$ (the centralizer of a).

23. If $|a| = n$, show that $|a^t| = n/\gcd(n, t)$.

24. Find all generators of Z.

25. Let a be an element of a group and suppose a has infinite order. How many generators does $\langle a \rangle$ have?

26. For each value of n listed below, determine whether or not $U(n)$ is a cyclic group. When it is cyclic, list all of the generators of $U(n)$, $n = 5, 9, 10, 14, 15, 18, 20, 22, 25$. Make a conjecture about the prime power decomposition of integers n for which $U(n)$ is cyclic. Are $n = 8$ and $n = 16$ counterexamples to your conjecture (try them)? If so, modify your conjecture.

27. List all the elements of order 8 in $Z_{8000000}$. How do you know your list is complete?

28. Let i denote the complex number $\sqrt{-1}$. Why might it make sense in certain physical situations to interpret i geometrically as a 90° rotation?

29. Let G be a finite group. Show that there exists a fixed positive integer n such that $a^n = e$ for all a in G. (Note that n is independent of a.)

30. Determine the subgroup lattice for Z_{12}.

31. Determine the subgroup lattice for Z_{p^2q} where p and q are distinct primes.

32. Determine the subgroup lattice for Z_8.

33. Determine the subgroup lattice for Z_{p^n} where p is a prime and n is some positive integer.

34. Determine the subgroup lattice for $U(12)$.

35. Show that the group of positive rational numbers under multiplication is not cyclic.

36. Consider the set $\{4, 8, 12, 16\}$. Show that this set is a group under multiplication modulo 20 by constructing its Cayley table. What is the identity element? Is the group cyclic? If so, find all of its generators.

37. Consider the set $\{7, 35, 49, 77\}$. Show that this set is a group under multiplication modulo 84 by constructing its Cayley table. What is the identity element? Is the group cyclic?

38. Let m and n be elements of the group Z. Find a generator for the group $\langle m \rangle \cap \langle n \rangle$.

39. Suppose a and b are group elements that commute and have orders m and n. If $\langle a \rangle \cap \langle b \rangle = \{e\}$, prove that the group contains an element whose order is the least common multiple of m and n. Show that this need not be true if a and b do not commute.

40. Prove that an infinite group must have an infinite number of subgroups.

41. Let p be a prime. If a group has more than $p - 1$ elements of order p, why can't the group be cyclic?

42. Suppose G is a cyclic group and 6 divides $|G|$. How many elements of order 6 does G have? If 8 divides $|G|$, how many elements of order 8 does G have? If a is one element of order 8, list the other elements of order 8.

43. List all the elements of Z_{40} that have order 10.

44. Let $|x| = 40$. List all the elements of $\langle x \rangle$ that have order 10.

45. Determine the orders of the elements of D_{33} and how many there are of each.

46. How many elements of order 4 does D_{12} have? How many elements of order 4 does D_{4n} have?

47. If G is an Abelian group and contains cyclic subgroups of order 4 and 5, what other size cyclic subgroups must G contain?

48. If G is an Abelian group and contains cyclic subgroups of orders 4 and 6, what other size cyclic subgroups must G contain?

49. If G is an Abelian group and contains a pair of cyclic subgroups of order 2, show that G must contain a subgroup of order 4. Must this subgroup be cyclic?

50. Given the fact that $U(49)$ is cyclic and has 42 elements, deduce the number of generators $U(49)$ has without actually finding any of the generators.

51. Let a and b be elements of a group. If $|a| = 10$ and $|b| = 21$, show $\langle a \rangle \cap \langle b \rangle = \{e\}$.

52. Let a and b be elements of a group. If $|a| = m$, $|b| = n$, and m and n are relatively prime, show $\langle a \rangle \cap \langle b \rangle = \{e\}$.

53. Suppose G is a group of order 16 and, by direct computation, you know that G has at least nine elements x such that $x^8 = e$. Can you conclude that G is not cyclic? What if G has at least five elements x such that $x^4 = e$? Generalize.

54. Prove that Z_n has an even number of generators if $n > 2$.

55. Bertrand's Postulate from number theory says that for any integer $N > 1$, there is always a prime between N and $2N$. Use this fact to prove that Z_n has more than two generators whenever $n > 6$.

56. Suppose $|x| = n$. Find a necessary and sufficient condition on r and s so that $\langle x^r \rangle \subseteq \langle x^s \rangle$.

57. Let $|x| = n$. Show $\langle x^r \rangle = \langle x^s \rangle$ if and only if $\gcd(n, r) = \gcd(n, s)$.

58. Prove that $H = \left\{ \begin{bmatrix} 1 & n \\ 0 & 1 \end{bmatrix} \;\middle|\; n \in Z \right\}$ is a cyclic subgroup of $GL(2, \mathbf{R})$.

59. Let G be a group, H a subgroup and a an element of G. Suppose $|a| = n$ and for each proper divisor k of n, $a^k \notin H$. Show that $\langle a \rangle \cap H = \{e\}$.

60. Suppose that G is a finite group with the property that every nonidentity element has prime order (for example, D_3 and D_5). If $Z(G)$ is not trivial, prove that every nonidentity element of G has the same order.

PROGRAMMING EXERCISES

It seems to me the impact and role of the electronic computer will significantly affect pure mathematics also, just as it has already done so in the mathematical sciences, principally physics, astronomy and chemistry.

S. M. Ulam, *Adventures of a Mathematician*

1. For all $1 < n < 100$, have the computer determine whether $U(n)$ is cyclic. When $U(n)$ is cyclic, list all of the generators and all of the subgroups. Run your program for $n = 8, 9, 18, 20, 25, 30, 49, 50$, and 60.

2. For any pair of positive integers m and n, let $Z_m \oplus Z_n = \{(a, b) \mid a \in Z_m, b \in Z_n\}$. For any pair of elements (a, b) and (c, d) in $Z_m \oplus Z_n$, define $(a, b) + (c, d) = ((a + c) \bmod m, (b + d) \bmod n)$. (For example, in $Z_3 \oplus Z_4$, we have $(1, 2) + (2, 3) = (0, 1)$. Write a program to check whether or not $Z_m \oplus Z_n$ is cyclic. Run your program for the following choices for m and n: $(2, 2)$, $(2, 3)$, $(2, 4)$, $(2, 5)$, $(3, 4)$, $(3, 5)$, $(3, 6)$, $(3, 7)$, $(3, 8)$, $(3, 9)$, and $(4, 6)$. On the basis of this output, guess how m and n must be related for $Z_m \oplus Z_n$ to be cyclic.

3. In this exercise we assume $a, b \in U(n)$. Define $\langle a, b \rangle = \{a^i b^j \mid 0 \le i < |a|, 0 \le j < |b|\}$. Write a program to compute the orders of $\langle a, b \rangle$, $\langle a \rangle$, $\langle b \rangle$, and $\langle a \rangle \cap \langle b \rangle$. Run your program for the following choices for a, b, and n: $(21, 101, 550)$, $(21, 49, 550)$, $(7, 11, 100)$, $(21, 31, 100)$, $(63, 77, 100)$. On the basis of your output, make a conjecture about the arithmetical relationship between $|\langle a, b \rangle|$, $|\langle a \rangle|$, $|\langle b \rangle|$, and $|\langle a \rangle \cap \langle b \rangle|$.

SUGGESTED READINGS

S. R. Cavior, "The Subgroups of the Dihedral Group," *Mathematics Magazine* 48 (1975): 107.

For each positive integer n, let $d(n)$ denote the number of positive divisors of n, and $\sigma(n)$ the sum of the positive divisors of n. This paper gives a short proof that the number of subgroups of the dihedral group $D_n(n \ge 3)$ is $d(n) + \sigma(n)$.

Deborah L. Massari, "The Probability of Generating a Cyclic Group," *Pi Mu Epsilon Journal* 6 (1979): 3–6.

In this easy-to-read paper, it is shown that the probability of a randomly chosen element from a cyclic group being a generator of the group depends only on the set of prime divisors of the order of the group, rather than the order itself. The article, written by an undergraduate student, received First Prize in a National Pi Mu Epsilon Paper Contest.

J. J. Sylvester

I really love my subject.

J. J. Sylvester

JAMES JOSEPH SYLVESTER was the most colorful and influential mathematician in America in the nineteenth century. Sylvester was born on September 3, 1814, in London and showed his mathematical genius early. At the age of fourteen he studied under DeMorgan and won several prizes for his mathematics, and at the unusually young age of twenty-five, he was elected a Fellow of the Royal Society.

After receiving B.A. and M.A. degrees from Trinity College in Dublin in 1841, Sylvester began a professional life that would include academics, law, and actuarial careers. In 1876, at the age of sixty-two, he was appointed to a prestigious position at the newly founded Johns Hopkins University. During his seven years at Johns Hopkins, Sylvester pursued research in pure mathematics, the first ever done in America, with tremendous vigor and enthusiasm. He also founded the *American Journal of Mathematics,* the first journal in America devoted to mathematical research. Sylvester returned to England in 1884 to a professorship at Oxford, a position he held until his death on March 15, 1897.

Sylvester's major contributions to mathematics were in the theory of equations, matrix theory, determinant theory, and invariant theory (which he founded with Cayley). His writings and lectures—flowery and eloquent, pervaded with poetic flights, emotional expressions, bizarre utterances, and paradoxes—reflected the personality of this sensitive, excitable, and enthusiastic man. We quote three of his students. E. W. Davis commented* on Sylvester's teaching methods.

> Sylvester's methods! He had none. "Three lectures will be delivered on a New Universal Algebra," he would say; then, "The course must be extended to twelve." It did last all the rest of that year. The following year the course was to be *Substitutions-Theorie,* by Netto. We all got the text. He lectured about three times, following the text closely and stopping sharp at the end of the hour. Then he began to think about matrices again. "I must give one lecture a week on those," he said. He could not confine himself to the hour, nor to the one lecture a week. Two weeks were passed, and Netto was forgotten entirely and never mentioned again. Statements like the following were not infrequent in his lectures: "I haven't proved this, but I am as sure as I can be of anything that it must be so. From

this it will follow, etc." At the next lecture it turned out that what he was so sure of was false. Never mind, he kept on forever guessing and trying, and presently a wonderful discovery followed, then another and another. Afterward he would go back and work it all over again, and surprise us with all sorts of side lights. He then made another leap in the dark, more treasures were discovered, and so on forever.

Sylvester's enthusiasm for teaching and his influence on his students is captured in the following passage written by Sylvester's first student at Johns Hopkins, G. B. Halsted.*

A short, broad man of tremendous vitality, . . . Sylvester's capacious head was ever lost in the highest cloud-lands of pure mathematics. Often in the dead of night he would get his favorite pupil, that he might communicate the very last product of his creative thought. Everything he saw suggested to him something new in the higher algebra. This transmutation of everything into new mathematics was a revelation to those who knew him intimately. They began to do it themselves.

Another characteristic of Sylvester, which is very unusual among mathematicians, was his apparent inability to remember mathematics! W. P. Durfee had the following to say*:

Sylvester had one remarkable peculiarity. He seldom remembered theorems, propositions, etc., but had always to deduce them when he wished to use them. In this he was the very antithesis of Cayley, who was thoroughly conversant with everything that had been done in every branch of mathematics.

I remember once submitting to Sylvester some investigations that I had been engaged on, and he immediately denied my first statement, saying that such a proposition had never been heard of, let alone proved. To his astonishment, I showed him a paper of his own in which he had proved the proposition; in fact, I believe the object of his paper had been the very proof which was so strange to him.

---

*(F. Cajori, *Teaching and History of Mathematics in the U.S.*, Washington, 1890, 265–266.)

SUPPLEMENTARY EXERCISES for Chapters 1–4

It is better to wear out than to rust out.

Bishop Richard Cumberland

1. Let G be a group and H a subgroup of G. For any x in G, define $x^{-1}Hx = \{x^{-1}hx \mid h \in H\}$. Prove
   a. $x^{-1}Hx$ is a subgroup of G.
   b. If H is cyclic, then $x^{-1}Hx$ is cyclic.
   c. If H is Abelian, then $x^{-1}Hx$ is Abelian.
   The group $x^{-1}Hx$ is called a *conjugate* of H. (Note that conjugation preserves structure.)

2. Let G be a group and let H be a subgroup of G. Define $N(H) = \{x \in G \mid x^{-1}Hx = H\}$. Prove that $N(H)$ (called the *normalizer* of H) is a subgroup of G.*

3. Let G be a group. For each $a \in G$, define $\text{cl}(a) = \{x^{-1}ax \mid x \in G\}$. Prove that these subsets of G partition G. ($\text{cl}(a)$ is called the *conjugacy class* of a.)

4. The group defined by the following table is called the *group of quaternions*. Use the table to determine each of the following:
   a. the center
   b. $\text{cl}(a)$
   c. $\text{cl}(b)$
   d. all cyclic subgroups

	e	a	a^2	a^3	b	ba	ba^2	ba^3
e	e	a	a^2	a^3	b	ba	ba^2	ba^3
a	a	a^2	a^3	e	ba^3	b	ba	ba^2
a^2	a^2	a^3	e	a	ba^2	ba^3	b	ba
a^3	a^3	e	a	a^2	ba	ba^2	ba^3	b
b	b	ba	ba^2	ba^3	a^2	a^3	e	a
ba	ba	ba^2	ba^3	b	a	a^2	a^3	e
ba^2	ba^2	ba^3	b	ba	e	a	a^2	a^3
ba^3	ba^3	b	ba	ba^2	a^3	e	a	a^2

5. Prove that in any group, $|ab| = |ba|$.

---

*This very important subgroup was first used by L. Sylow in 1872 to prove the existence of certain kinds of subgroups in a group. His work is discussed in Chapter 26.

6. (Conjugation preserves order.) Prove that in any group $|x^{-1}ax| = |a|$. (This exercise is referred to in Chapter 26.)

7. Prove that if a is the only element of order 2 in a group, then a lies in the center of the group.

8. Let G be the plane symmetry group of the infinite strip of equally spaced H's shown below.

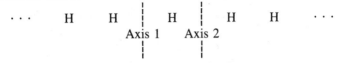

Let x be the reflection about Axis 1 and let y be the reflection about Axis 2. Calculate $|x|$, $|y|$, and $|xy|$. Must the product of elements of finite order have finite order?

9. What are the orders of the elements of D_{15}? How many elements have each of these orders?

10. Prove that a group of order 4 is Abelian.

11. Prove that a group of order 5 must be cyclic.

12. Prove that an Abelian group of order 6 must be cyclic.

13. Let G be an Abelian group and n a fixed positive integer. Let $G^n = \{g^n \mid g \in G\}$. Prove that G^n is a subgroup of G. Give an example showing that G^n need not be a subgroup of G when G is non-Abelian.

14. Let $G = \{a + b\sqrt{2}\}$, where a and b are rational numbers not both zero. Prove that G is a group under ordinary multiplication.

15. (1969 Putnam Competition) Prove that no group is the union of two proper subgroups. Does the statement remain true if "two" is replaced by "three"?

16. Prove that the subset of elements of finite order in an Abelian group forms a subgroup. (This subgroup is called the *torsion subgroup*.) Is the same thing true for non-Abelian groups?

17. Let p be a prime and G an Abelian group. Show that the set of all elements whose orders are powers of p is a subgroup of G.

18. Let x and y belong to a group G. Assume $x \neq e$, $|y| = 2$ and $yxy^{-1} = x^2$. Find $|x|$.

19. Suppose a group is generated by two elements a and b (that is, every element of the group can be expressed as some product of a's and b's). Given that $a^3 = b^2 = e$ and $ba^2 = ab$, construct the Cayley table for the group. We have already seen an example of a group that satisfies these conditions. Name it.

20. Suppose a group is generated by two elements a and b. Given that $a^4 = b^2 = e$ and $ba = a^3b$, construct the Cayley table for the group. Give an example of a group that satisfies these conditions.

21. Let x, y belong to a group G. If $xy \in Z(G)$, prove $xy = yx$.

22. Suppose H and K are nontrivial subgroups of Q under addition. Show $H \cap K$ is a nontrivial subgroup of Q. Is this true if Q is replaced by $\mathbf{R}$?

23. Let H be a subgroup of G and g an element of G. Prove $N(g^{-1}Hg) = g^{-1}N(H)g$.

24. Let H be a subgroup of a group G and $|g| = n$. If g^m belongs to H and m and n are relatively prime, prove g belongs to H.

Permutation Groups

5

Wigner's discovery about the electron permutation group was just the beginning. He and others found many similar applications and nowadays group theoretical methods—especially those involving characters and representations—pervade all branches of quantum mechanics.

George Mackey, Proceedings of the American Philosophical Society

Definition and Notation

In this chapter, we study certain groups of functions, called permutation groups, from a set A to itself. In the early and mid–nineteenth century, groups of permutations were the only groups investigated by mathematicians. It was not until around 1850 that the notion of an abstract group was introduced by Cayley, and it took another quarter century before the idea firmly took hold.

> *DEFINITIONS* Permutation of A—Permutation Group of A
> A *permutation* of a set A is a function from A to A that is both one-to-one and onto. A *permutation group of a set A* is a set of permutations of A that forms a group under function composition.

Although groups of permutations of any nonempty set A of objects exist, one is usually only interested in the case where A is finite. Furthermore, it is customary, as well as convenient, to take A to be a set of the form $\{1, 2, 3, \ldots, n\}$ for some positive integer n. Unlike in calculus, where most functions are defined on infinite sets and are given by formulas, in algebra, permutations of finite sets are usually given by an explicit listing of each element of the domain and its corresponding functional value. In practice, the sets that are permuted are fairly small, so listing the elements and the values they are assigned is reasonable. For example, we define a permutation α of the set $\{1, 2, 3, 4\}$ by specifying

$$1\alpha = 2, \quad 2\alpha = 3, \quad 3\alpha = 1, \quad 4\alpha = 4.$$

A more convenient way to express this correspondence is to write α in array form as

$$\alpha = \begin{bmatrix} 1 & 2 & 3 & 4 \\ 2 & 3 & 1 & 4 \end{bmatrix}.$$

Here $j\alpha$ is placed directly below j for each j. Similarly, the permutation β of the set $\{1, 2, 3, 4, 5, 6\}$ is given by

$$1\beta = 5, \quad 2\beta = 3, \quad 3\beta = 1, \quad 4\beta = 6, \quad 5\beta = 2, \quad 6\beta = 4$$

is expressed in array form as

$$\beta = \begin{bmatrix} 1 & 2 & 3 & 4 & 5 & 6 \\ 5 & 3 & 1 & 6 & 2 & 4 \end{bmatrix}.$$

Composition of permutations expressed in array notation is carried out from left to right by going from top to bottom, then top to bottom. For example, let

$$\sigma = \begin{bmatrix} 1 & 2 & 3 & 4 & 5 \\ 2 & 4 & 3 & 5 & 1 \end{bmatrix}$$

and

$$\gamma = \begin{bmatrix} 1 & 2 & 3 & 4 & 5 \\ 5 & 4 & 1 & 2 & 3 \end{bmatrix};$$

then

$$\sigma\gamma = \begin{bmatrix} 1 & 2 & 3 & 4 & 5 \\ 2 & 4 & 3 & 5 & 1 \end{bmatrix}\begin{bmatrix} 1 & 2 & 3 & 4 & 5 \\ 5 & 4 & 1 & 2 & 3 \end{bmatrix}$$
$$= \begin{bmatrix} 1 & 2 & 3 & 4 & 5 \\ 4 & 2 & 1 & 3 & 5 \end{bmatrix}.$$

On the right we have 4 under 1, since $1(\sigma\gamma) = (1\sigma)\gamma = 2\gamma = 4$, so $\sigma\gamma$ sends 1 to 4. The remainder of the bottom row of $\sigma\gamma$ is obtained in a similar fashion.

We are now ready to give some examples of permutation groups.

Example 1 Symmetric Group S_3

Let S_3 denote the set of all one-to-one functions from $\{1, 2, 3\}$ to itself. Then S_3, under function composition, is a group with six elements. The six elements are

$$\varepsilon = \begin{bmatrix} 1 & 2 & 3 \\ 1 & 2 & 3 \end{bmatrix}, \qquad \alpha = \begin{bmatrix} 1 & 2 & 3 \\ 2 & 3 & 1 \end{bmatrix}, \qquad \alpha^2 = \begin{bmatrix} 1 & 2 & 3 \\ 3 & 1 & 2 \end{bmatrix},$$

$$\beta = \begin{bmatrix} 1 & 2 & 3 \\ 1 & 3 & 2 \end{bmatrix}, \qquad \beta\alpha = \begin{bmatrix} 1 & 2 & 3 \\ 2 & 1 & 3 \end{bmatrix}, \qquad \beta\alpha^2 = \begin{bmatrix} 1 & 2 & 3 \\ 3 & 2 & 1 \end{bmatrix}.$$

Note that $\alpha\beta = \begin{bmatrix} 1 & 2 & 3 \\ 3 & 2 & 1 \end{bmatrix} \neq \beta\alpha$, so that S_3 is non-Abelian. $\qquad\qquad\square$

Example 1 can be generalized as follows.

Example 2 Symmetric Group S_n

Let $A = \{1, 2, \ldots, n\}$. The set of all permutations of A is called the *symmetric group of degree n* and is denoted by S_n. Elements of S_n have the form

$$\alpha = \begin{bmatrix} 1 & 2 & \cdots & n \\ 1\alpha & 2\alpha & \cdots & n\alpha \end{bmatrix}.$$

It is easy to compute the order of S_n. There are n choices of 1α. Once 1α has been determined, there are $n - 1$ possibilities for 2α (since α is one-to-one, we must have $1\alpha \neq 2\alpha$). After choosing 2α, there are exactly $n - 2$ possibilities for 3α. Continuing along in this fashion, we see that S_n must have $n(n - 1) \cdots 3 \cdot 2 \cdot 1 = n!$ elements. We leave it to the reader to prove S_n is non-Abelian when $n \geq 3$. $\qquad\qquad\square$

Example 3 Symmetries of a Square

As a third example, we associate each motion in D_4 with the permutation of the locations of each of the four corners of a square. For example, if we label the four corner positions as in the figure below and keep these labels fixed for reference, we may describe a 90° rotation by the permutation

$$\rho = \begin{bmatrix} 1 & 2 & 3 & 4 \\ 2 & 3 & 4 & 1 \end{bmatrix},$$

While a reflection across a horizontal axis yields

$$\phi = \begin{bmatrix} 1 & 2 & 3 & 4 \\ 2 & 1 & 4 & 3 \end{bmatrix}.$$

As we have seen in Chapter 1, these two elements generate the entire group. In fact, the group table is

	ε	ρ	ρ²	ρ³	φ	ρφ	ρ²φ	ρ³φ
ε	ε	ρ	ρ²	ρ³	φ	ρφ	ρ²φ	ρ³φ
ρ	ρ	ρ²	ρ³	ε	ρφ	ρ²φ	ρ³φ	φ
ρ²	ρ²	ρ³	ε	ρ	ρ²φ	ρ³φ	φ	ρφ
ρ³	ρ³	ε	ρ	ρ²	ρ³φ	φ	ρφ	ρ²φ
φ	φ	ρ³φ	ρ²φ	ρφ	ε	ρ³	ρ²	ρ
ρφ	ρφ	φ	ρ³φ	ρ²φ	ρ	ε	ρ³	ρ²
ρ²φ	ρ²φ	ρφ	φ	ρ³φ	ρ²	ρ	ε	ρ³
ρ³φ	ρ³φ	ρ²φ	ρφ	φ	ρ³	ρ²	ρ	ε

Even though we have used a different notation for the elements of the group than we did in earlier chapters, *algebraically,* we still have the same group. Our previous group table for D_4 could be obtained from this one by merely replacing ρ by R_{90}, φ by H, ρφ by $R_{90} H = D$, and so on. ☐

Cycle Notation

There is another notation commonly used to specify permutations. It is called cycle notation and was first introduced by the great French mathematician Cauchy in 1815. Cycle notation has theoretical advantages in that certain important properties of the permutation can be readily determined when cyclic notation is used.

As an illustration of cycle notation, let us consider the permutation $\alpha = \begin{bmatrix} 1 & 2 & 3 & 4 & 5 & 6 \\ 2 & 1 & 4 & 6 & 5 & 3 \end{bmatrix}$. This assignment of values could be presented schematically as follows:

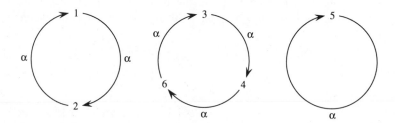

Although mathematically satisfactory, such diagrams are cumbersome. Instead, we leave out the arrows and simply write $\alpha = (1, 2)(3, 4, 6)(5)$. As a second example consider $\beta = \begin{bmatrix} 1 & 2 & 3 & 4 & 5 & 6 \\ 5 & 3 & 1 & 6 & 2 & 4 \end{bmatrix}$. In cycle notation β can be written $(2, 3, 1, 5)(6, 4)$ or $(4, 6)(3, 1, 5, 2)$, since both of these unambiguously specify the function β. An expression of the form $(a_1, a_2, \ldots, a_m)$ is called a *cycle of length m* or an *m-cycle.*

A multiplication of cycles can be introduced by thinking of a cycle as a permutation that fixes any symbol not appearing in the cycle. Thus, the cycle (4, 6) can be thought of as representing the permutation $\begin{bmatrix} 1 & 2 & 3 & 4 & 5 & 6 \\ 1 & 2 & 3 & 6 & 5 & 4 \end{bmatrix}$. In this way, we can multiply cycles by thinking of them as permutations given in array form. Consider the following example from S_8. Let $\alpha = (13)(27)(4568)$ and $\beta = (1572)(34)(6)(8)$. (When the domain consists of single digit integers, it is common practice to omit the commas between the digits.) What is the cycle form of $\alpha\beta$? Of course, one could say that $\alpha\beta = (13)(27)(4568)(1572)(34)(6)(8)$, but it is usually more desirable to express a permutation in *disjoint* cycle form (that is, the various cycles have no member in common). To this end, observe that the cycle (13) sends 1 to 3. Now, the cycle (27) does not have a "3" in it, so this cycle leaves "3" alone. Similarly, the next two cycles in the product leave "3" fixed. Finally, (34) means 3 is sent to 4. So the net effect of $\alpha\beta$ is to send 1 to 4. So we begin $\alpha\beta = (14 \cdots)$. .... Now, repeating the entire process beginning with 4, we have, cycle by cycle, $4 \to 4$, $4 \to 4$, $4 \to 5$, $5 \to 7$, $7 \to 7$, $7 \to 7$, $7 \to 7$, so that $\alpha\beta = (147 \cdots)$. .... Ultimately, we have $\alpha\beta = (147)(2)(3568)$. The important thing to bear in mind when multiplying cycles is to "keep moving" from one cycle to the next.

To be sure you understand how to switch from one notation to the other and how to multiply permutations, we will do one more example of each.

If the array notation for a permutation β is

$$\begin{bmatrix} 1 & 2 & 3 & 4 & 5 & 6 & 7 & 8 \\ 2 & 1 & 3 & 8 & 6 & 4 & 5 & 7 \end{bmatrix},$$

then in cycle notation, β is (12)(3)(48756).

One can convert this back to array form without converting each cycle into array form by simply observing that (12) means 1 goes to 2 and 2 goes to 1; (3) means 3 goes to 3; (48756) means $4 \to 8$, $8 \to 7$, $7 \to 5$, $5 \to 6$, $6 \to 4$.

One final remark about the cycle notation: Most mathematicians prefer not to write cycles that have only one entry. In that case, it is understood that any missing element is mapped to itself. With this convention, the permutation β above can be written as (12)(48756). Similarly,

$$\alpha = \begin{bmatrix} 1 & 2 & 3 & 4 & 5 \\ 3 & 2 & 4 & 1 & 5 \end{bmatrix}$$

can be written $\alpha = (134)$. Of course, the identity permutation consists only of cycles with one entry, so we cannot omit all of these! In this case, one usually writes just one cycle. For example,

$$\varepsilon = \begin{bmatrix} 1 & 2 & 3 & 4 & 5 \\ 1 & 2 & 3 & 4 & 5 \end{bmatrix}$$

could be written as $\varepsilon = (5)$ or $\varepsilon = (1)$. Just remember that missing elements are mapped to themselves.

Properties of Permutations

We are now ready to state a number of theorems about permutations and cycles. The proof of the first theorem is implicit in our discussion of writing permutations in cycle form.

Theorem 5.1 *Products of Disjoint Cycles*
Every permutation can be written as a cycle or as a product of disjoint cycles.

Proof. Let α be a permutation on $A = \{1, 2, \ldots, n\}$. To write α in cycle form, we start by choosing any member of A, say, a_1, and let

$$a_2 = a_1\alpha, \qquad a_3 = a_1\alpha^2$$

and so on, until we arrive at $a_1 = a_1\alpha^m$ for some m. We know such an m exists because the sequence $a_1, a_1\alpha, a_1\alpha^2, \ldots$ must be finite; so there must eventually be a repeat, say, $a_1\alpha^i = a_1\alpha^j$ for some i and j with $i < j$. Then $a_1 = a_1\alpha^m$ where $m = j - i$. We express this relationship between a_1, $a_2, \ldots, a_m$ as

$$\alpha = (a_1, a_2, \ldots, a_m) \ldots .$$

The three dots at the end indicate the possibility that we may not have exhausted the set A in this process. In that case, we merely choose any element b_1 of A not appearing in the first cycle and proceed to create a new cycle as before. That is, we let $b_2 = b_1\alpha$, $b_3 = b_1\alpha^2$, and so on, until we reach $b_1 = b_1\alpha^k$ for some k. This new cycle will have no elements in common with the previously constructed cycle. For, if so, then $a_1\alpha^i = b_1\alpha^j$ for some i and j. But then $b_1 = a_1\alpha^{i-j}$ and therefore $b_1 = a_t$ for some t. This contradicts the way b_1 was chosen. Continuing this process until we run out of elements of A, our permutation will appear as

$$\alpha = (a_1, a_2, \ldots, a_m)(b_1, b_2, \ldots, b_k) \cdots (c_1, c_2, \ldots, c_t).$$

In this way, we see that every permutation can be written as a product of disjoint cycles. ∎

Theorem 5.2 *Disjoint Cycles Commute*
If the pair of cycles $\alpha = (a_1, a_2, \ldots, a_m)$ and $\beta = (b_1, b_2, \ldots, b_n)$ have no entries in common, then $\alpha\beta = \beta\alpha$.

Proof. For definiteness, let us say that α and β are permutations of the set

$$S = \{a_1, a_2, \ldots, a_m, b_1, b_2, \ldots, b_n, c_1, c_2, \ldots, c_k\}$$

where the c's are the members of S left fixed by both α and β. To prove $\alpha\beta = \beta\alpha$ we must show that $x(\alpha\beta) = x(\beta\alpha)$ for all x in S. If x is one of the a elements, say, a_i, then

$$a_i(\alpha\beta) = (a_i\alpha)\beta = a_{i+1}\beta = a_{i+1}$$

since β fixes all a elements. (We interpret a_{i+1} as a_1 if $i = m$.) For the same reason,

$$a_i(\beta\alpha) = (a_i\beta)\alpha = a_i\alpha = a_{i+1}.$$

Hence, the functions $\alpha\beta$ and $\beta\alpha$ agree on a elements. A similar argument shows $\alpha\beta$ and $\beta\alpha$ agree on the b elements, as well. Finally, suppose x is a c element, say, c_i. Then, since both α and β fix c elements, we have

$$c_i(\alpha\beta) = (c_i\alpha)\beta = c_i\beta = c_i \quad \text{and}$$
$$c_i(\beta\alpha) = (c_i\beta)\alpha = c_i\alpha = c_i.$$

This completes the proof. ∎

Corollary   *(Ruffini—1799) Order of a Permutation*
The order of a permutation written in disjoint cycle form is the least common multiple of the lengths of the cycles.

Proof.   First, observe that a cycle of length n has order n. (Verify this yourself.) Next, suppose α and β are disjoint cycles of lengths m and n, and let k be the least common multiple of m and n. It follows from Theorem 4.1 that both α^k and β^k are the identity permutation ε and, since α and β commute, $(\alpha\beta)^k = \alpha^k\beta^k$ is also the identity. Thus, we know by the corollary to Theorem 4.1 ($a^k = e$ implies $|a|$ divides k) that the order of $\alpha\beta$—let us call it t—must divide k. But then $(\alpha\beta)^t = \alpha^t\beta^t = \varepsilon$ so that $\alpha^t = \beta^{-t}$. However, it is clear that if α and β are disjoint, then the same must be true of α^t and β^t, since raising a cycle to some power simply rearranges the symbols of the cycle. But, if α^t and β^{-t} are disjoint and equal, there must be no symbols at all in either one! That is, $\alpha^t = \beta^{-t} = \varepsilon$ (remember, fixed points are omitted). It follows, then, that both m and n must divide t. This means that k, the least common multiple of m and n, divides t also. This shows that $k = t$.

Thus far, we have proved the corollary is true in the cases where the permutation is a single cycle or a product of two disjoint cycles. The general case involving more than two cycles can be handled in an analogous way. ∎

As we will soon see, a particularly important kind of permutation is a cycle of length 2, that is, a permutation of the form (ab). Many authors call these permutations *transpositions*, since the effect of (ab) is to interchange or transpose a and b.

Theorem 5.3   *Product of 2-Cycles*
Every permutation in S_n, $n > 1$, is a product of 2-cycles.

Proof. First, note that the identity can be expressed as (12)(12), so it is a product of 2-cycles. By Theorem 5.1, we know that every permutation can be written in the form

$$(a_1a_2 \cdots a_k)(b_1b_2 \cdots b_t) \cdots (c_1c_2 \cdots c_s).$$

A direct computation shows that this is the same as

$$(a_1a_2)(a_1a_3) \cdots (a_1a_k)(b_1b_2)(b_1b_3) \cdots (b_1b_t) \cdots (c_1c_2)(c_1c_3) \cdots (c_1c_s).$$

This completes the proof.   ■

The example below demonstrates this technique. It also shows that the decomposition of a permutation into a product of 2-cycles is not unique.

Example 4

$$
\begin{aligned}
(12345) &= (12)(13)(14)(15) \\
&= (15)(25)(35)(45) \\
&= (23)(24)(25)(21) \\
&= (13)(23)(25)(21)(52)(54)
\end{aligned}
$$ □

Example 4 even shows that the *number* of 2-cycles may vary from one decomposition to the next. Theorem 5.4 (due to Cauchy) says, however, that there is one aspect of a decomposition that never varies.

We isolate a special case of Theorem 5.4 as a lemma.

Lemma.   *If* $\varepsilon = \beta_1\beta_2 \cdots \beta_r$ *where the* β's *are 2-cycles, then r is even.*

Proof.   Clearly $r \neq 1$, since a 2-cycle is not the identity. If $r = 2$, we are done. So, we suppose $r > 2$ and we proceed by induction. Since $(ij) = (ji)$ the product $\beta_1\beta_2$ can be expressed in one of the following forms on the left

$$
\begin{aligned}
(ab)(ab) &= \varepsilon \\
(ab)(ac) &= (bc)(ab) \\
(ab)(cd) &= (cd)(ab) \\
(ab)(bc) &= (bc)(ac).
\end{aligned}
$$

If the first case occurs, we may delete $\beta_1\beta_2$ from the original product to obtain $\varepsilon = \beta_3 \cdots \beta_r$ and therefore, by the Second Principle of Mathematical Induction, $r - 2$ is even. In the other three cases, we replace the form of $\beta_1\beta_2$ on the left by its counterpart on the right to obtain a new product of r 2-cycles that is still the identity, but where the first occurrence of the integer a is in the second 2-cycle of the product instead of the first. We now repeat the procedure just described with $\beta_2\beta_3$ and, as before, we obtain a product of $r - 2$ 2-cycles equal to the identity or a new product of r 2-cycles, where the first occurrence of a is in the third 2-cycle. Continuing this process we must obtain a product of $r - 2$ 2-cycles equal to the identity. For otherwise, we have a product equal to the identity in which first occurrence of the integer a is in the last 2-cycle, and such a product does not fix a while the identity does. Hence, by induction, $r - 2$ is even and r is even as well. ∎

Theorem 5.4 *Always Even or Always Odd*
If a permutation α can be expressed as a product of an even number of 2-cycles, then every decomposition of α into a product of 2-cycles must have an even number of 2-cycles. In symbols, if

$$\alpha = \beta_1\beta_2 \cdots \beta_r \quad and \quad \alpha = \gamma_1\gamma_2 \cdots \gamma_s$$

where the β's and the γ's are 2-cycles, then r and s are both even or both odd.

Proof. Observe that $\beta_1\beta_2 \cdots \beta_r = \gamma_1\gamma_2 \cdots \gamma_s$ implies

$$\varepsilon = \gamma_1\gamma_2 \cdots \gamma_s\beta_r^{-1} \cdots \beta_2^{-1}\beta_1^{-1}.$$

Since a 2-cycle is its own inverse, the Lemma guarantees that $s + r$ is even. Thus, r and s are both even or both odd. ∎

DEFINITION Even and Odd Permutations
A permutation that can be expressed as a product of an even number of 2-cycles is called an *even* permutation. A permutation that can be expressed as a product of an odd number of 2-cycles is called an *odd* permutation.

Theorems 5.3 and 5.4 together show that every permutation can be unambiguously classified as either even or odd, but not both. At this point, it is natural to ask what significance this observation has. The answer is given in Theorem 5.5.

Theorem 5.5 *Even Permutations Form a Group*
The set of even permutations in S_n forms a subgroup of S_n.

Proof. This proof is left to the reader. ∎

The group of even permutations of n symbols is denoted by A_n and is called the *alternating group of degree n*. The alternating groups are among the most important examples of groups. The groups A_4 and A_5 will arise on several occasions in later chapters. In particular, A_5 has great historical significance. One of the problems in the exercises asks the reader to prove that A_n has order $n!/2$ (when $n > 1$). Thus, we see that exactly half of the members of S_n are even permutations. A geometrical interpretation of A_4 is given in Example 5, and a multiplication table for A_4 is given in Table 5.1.

Table 5.1 The Alternating Group A_4 of Even Permutations of $\{1, 2, 3, 4\}$

(In this table, the permutations of A_4 are designated as $\alpha_1, \alpha_2, \ldots, \alpha_{12}$ and an entry k inside the table represents α_k. For example, $\alpha_3 \alpha_8 = \alpha_7$.)

	α_1	α_2	α_3	α_4	α_5	α_6	α_7	α_8	α_9	α_{10}	α_{11}	α_{12}
$(1) = \alpha_1$	1	2	3	4	5	6	7	8	9	10	11	12
$(12)(34) = \alpha_2$	2	1	4	3	8	7	6	5	11	12	9	10
$(13)(24) = \alpha_3$	3	4	1	2	6	5	8	7	12	11	10	9
$(14)(23) = \alpha_4$	4	3	2	1	7	8	5	6	10	9	12	11
$(123) = \alpha_5$	5	6	7	8	9	10	11	12	1	2	3	4
$(243) = \alpha_6$	6	5	8	7	12	11	10	9	3	4	1	2
$(142) = \alpha_7$	7	8	5	6	10	9	12	11	4	3	2	1
$(134) = \alpha_8$	8	7	6	5	11	12	9	10	2	1	4	3
$(132) = \alpha_9$	9	10	11	12	1	2	3	4	5	6	7	8
$(143) = \alpha_{10}$	10	9	12	11	4	3	2	1	7	8	5	6
$(234) = \alpha_{11}$	11	12	9	10	2	1	4	3	8	7	6	5
$(124) = \alpha_{12}$	12	11	10	9	3	4	1	2	6	5	8	7

Example 5 Rotations of a Tetrahedron

The twelve rotations of a regular tetrahedron can be conveniently described with the elements of A_4. The top row of Figure 5.1 illustrates the identity and three "edge" rotations about axes joining midpoints of two edges. The other two rows depict the product of the edge rotations and two "face" rotations. $\square$

A Check-Digit Scheme Based on D_5

In Chapters 0 and 4 we presented several schemes for appending a check digit to an identification number. Among these schemes, only the International Standard Book Number method was capable of detecting all single digit errors and all transposition errors involving adjacent digits. However, recall that this success was achieved by introducing the alphabetical character X to handle the case where 10 was required to make the dot product 0 modulo 11. In the late 1960s, a scheme was devised that detects all single digit errors and all transposition errors involving adjacent digits without the introduction of a new character (see [1] and [2]).

To describe this method we need the permutation $\sigma = (0)(14)(23)(58697)$ and the dihedral group of order 10 as represented in Table 5.2. (Here we use 0 through 4 for the rotations and 5 through 9 for the reflections.) To append a check digit to any string of digits, we weight the digits of the string with successive powers of σ, starting with the right digit and, using the operation ∗ defined in Table 5.2, multiply the result and take the inverse of the product. To illustrate, consider 793. Beginning with the digit 3, we apply successive powers of σ to transform each digit of 793. That is, $3\sigma = 2$; $9\sigma^2 = (9\sigma)\sigma = 7\sigma = 5$; $7\sigma^3 = ((7\sigma)\sigma)\sigma = (5\sigma)\sigma = 8\sigma = 6$. Then we compute $(6 ∗ 5 ∗ 2)^{-1}$ using

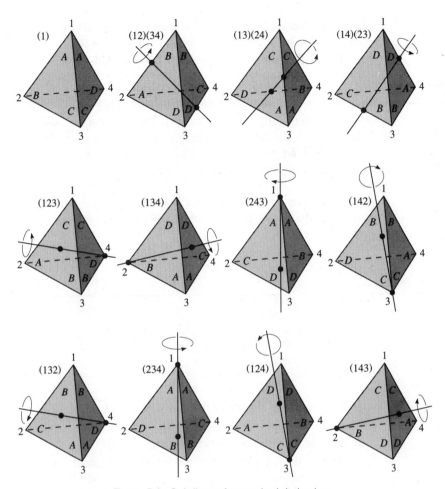

Figure 5.1 Rotations of a regular tetrahedron.

Table 5.2 The Multiplication Table of D_5

*	0	1	2	3	4	5	6	7	8	9
0	0	1	2	3	4	5	6	7	8	9
1	1	2	3	4	0	6	7	8	9	5
2	2	3	4	0	1	7	8	9	5	6
3	3	4	0	1	2	8	9	5	6	7
4	4	0	1	2	3	9	5	6	7	8
5	5	9	8	7	6	0	4	3	2	1
6	6	5	9	8	7	1	0	4	3	2
7	7	6	5	9	8	2	1	0	4	3
8	8	7	6	5	9	3	2	1	0	4
9	9	8	7	6	5	4	3	2	1	0

Table 5.2 to obtain $3^{-1} = 2$, the check digit. For the number 17326, we determine $(1\sigma^5 * 7\sigma^4 * 3\sigma^3 * 2\sigma^2 * 6\sigma)^{-1} = (4 * 9 * 2 * 2 * 9)^{-1} = 0^{-1} = 0$.

To see that this scheme detects all single digit errors, we observe that an error-free number $a_n a_{n-1} \cdots a_1 a_0$ (where a_0 is the check digit) has the property that $a_n \sigma^n * a_{n-1} \sigma^{n-1} * \cdots * a_1 \sigma * a_0 = 0$ and, therefore, any particular factor in this product is uniquely determined by all of the others. Thus, a single digit error does not result in a product of 0. The fact that all transposition errors of adjacent digits are detected can be verified by showing that for all distinct a and b, $a\sigma * b \neq b\sigma * a$. Then, for all i, $a\sigma^{i+1} * b\sigma^i \neq b\sigma^{i+1} * a\sigma^i$, and consequently a transposition of distinct adjacent digits, will not result in a product of 0.

In addition to being foolproof in detecting single digit errors and transpositions of adjacent digits, this method will detect approximately 90% of all other types of errors.

EXERCISES

When you feel how depressingly
slowly you climb,
it's well to remember that
Things Take Time.

Piet Hein, "T. T. T.," *Grooks* (1966)*

1. Find the order of each of the following permutations.
   a. (14)   b. (147)   c. (14762)
2. What is the order of a k-cycle $(a_1\ a_2\ \ldots\ a_k)$?
3. What is the order of each of the following permutations?
   a. (124)(357)   b. (124)(356)
   c. (124)(35)   d. (124)(3578)
4. What is the order of each of the following permutations?
   a. $\begin{bmatrix} 1 & 2 & 3 & 4 & 5 & 6 \\ 2 & 1 & 5 & 4 & 6 & 3 \end{bmatrix}$ b. $\begin{bmatrix} 1 & 2 & 3 & 4 & 5 & 6 & 7 \\ 7 & 6 & 1 & 2 & 3 & 4 & 5 \end{bmatrix}$
5. What is the order of the product of a pair of disjoint cycles of lengths 4 and 6?
6. What are the possible orders for the elements of S_6 and A_6? What about S_7 and A_7? (This exercise is referred to in Chapter 27.)
7. Show A_8 contains an element of order 15.
8. What is the maximum order of any element in A_{10}?

---

*Hein is a Danish engineer, poet, and inventor of the game *Hex*.

9. Determine whether the following permutations are even or odd.
   a. (135)   b. (1356)   c. (13567)
   d. (12)(134)(152)   e. (1243)(3521)

10. Show that a function from a finite set S to itself is one-to-one if and only if it is onto. Is this true when S is infinite?

11. Let n be a positive integer. If n is odd, is an n-cycle an odd or an even permutation? If n is even, is an n-cycle an odd or an even permutation?

12. If α is even, prove α^{-1} is even. If α is odd, prove α^{-1} is odd.

13. Prove that A_n is a subgroup of S_n.

14. Prove that the product of an even permutation and an odd permutation is odd.

15. Is the product of two odd permutations an even or an odd permutation?

16. Associate an even permutation with the number $+1$ and an odd permutation with the number -1. Draw an analogy between the result of multiplying two permutations and the result of multiplying their corresponding numbers $+1$ or -1.

17. Let

$$\alpha = \begin{bmatrix} 1 & 2 & 3 & 4 & 5 & 6 \\ 2 & 1 & 3 & 5 & 4 & 6 \end{bmatrix} \quad \text{and} \quad \beta = \begin{bmatrix} 1 & 2 & 3 & 4 & 5 & 6 \\ 6 & 1 & 2 & 4 & 3 & 5 \end{bmatrix}.$$

Compute each of the following.
   a. α^{-1} b. $\alpha\beta$ c. $\beta\alpha$

18. Let

$$\alpha = \begin{bmatrix} 1 & 2 & 3 & 4 & 5 & 6 & 7 & 8 \\ 2 & 1 & 3 & 5 & 4 & 7 & 6 & 8 \end{bmatrix} \text{ and } \beta = \begin{bmatrix} 1 & 2 & 3 & 4 & 5 & 6 & 7 & 8 \\ 1 & 3 & 8 & 7 & 6 & 5 & 2 & 4 \end{bmatrix}.$$

Write α and β as
   a. products of disjoint cycles
   b. products of 2-cycles

19. Show that if H is a subgroup of S_n, then either every member of H is an even permutation or exactly half of them are even. (This exercise is referred to in Chapters 27 and 32.)

20. Compute the order of each member of A_4. What arithmetical relation do these orders have with the order of A_4?

21. Do the odd permutations in S_n form a group? Why?

22. Let α and β belong to S_n. Prove that $\alpha^{-1}\beta^{-1}\alpha\beta$ is an even permutation.

23. Use Table 5.1 to compute the following.
   a. The centralizer of $\alpha_3 = (13)(24)$.
   b. The centralizer of $\alpha_{12} = (124)$.

24. Determine the subgroup lattice for S_3. (Hint: exercise 23 of Chapter 3 is relevant here.)

25. Show $(123)^{-1} = (321)$ and $(1478)^{-1} = (8741)$.

26. What cycle is $(a_1a_2 \cdots a_n)^{-1}$?

27. Let G be a group of permutations on a set X. Let $a \in X$ and define stab$(a) = \{\alpha \in G \mid a\alpha = a\}$. We call stab$(a)$ the *stabilizer of a in G* (since it consists of all members of G that leave a fixed). Prove that stab(a) is a subgroup of G. (This subgroup was introduced by Galois in 1832.) Exercise 27 is referred to in Chapter 9.

28. In S_3, find elements α and β so that $|\alpha| = 2$, $|\beta| = 2$, and $|\alpha\beta| = 3$.

29. Find group elements α and β so that $|\alpha| = 3$, $|\beta| = 3$, and $|\alpha\beta| = 5$.

30. Prove that S_n is non-Abelian for all $n \geq 3$.

31. Represent the symmetry group of an equilateral triangle as a group of permutations of its vertices (see Example 3).

32. Let α and β belong to S_n. Prove that $\beta^{-1}\alpha\beta$ and α are both even or both odd.

33. Let G be the set of all permutations of the positive integers. Let H be the subset of elements of G that can be expressed as a product of a finite number of cycles. Prove that H is a subgroup of G.

34. Show that A_5 has 24 elements of order 5, 20 elements of order 3, and 15 elements of order 2.

35. Show that every element in A_n for $n \geq 3$ can be expressed as a 3-cycle or a product of three cycles.

36. Show that for $n \geq 3$, $Z(S_n) = \{\varepsilon\}$.

37. Use the check digit scheme based on D_5 to append a check digit to 45723.

38. Verify the statement made in the discussion of the check digit scheme based on D_5 that $a\sigma * b \neq b\sigma * a$ for a distinct a and b. Use this to prove that $a\sigma^{i+1} * b\sigma^i \neq b\sigma^{i+1} * a\sigma^i$ for all i. Prove that this implies that all transposition errors involving adjacent digits are detected.

39. Let $\sigma = (124875)(36)$. For any string $a_1a_2 \cdots a_k$ (k odd), assign the check digit $- (a_1\sigma + a_2 + a_3\sigma + a_4 + a_5\sigma + \cdots + a_k\sigma)$ mod 10. Calculate the check digit for the number 3125600196431. Prove that this method detects all single digit errors. Determine which transposition errors involving adjacent digits go undetected by this method. How does this method compare with the method given in exercise 49 of Chapter 2?

PROGRAMMING EXERCISE

ASCII and ye shall receive.

Jeffrey Armstrong

1. Write a program to implement the check digit scheme discussed in this chapter.

REFERENCES

1. H. P. Gumm, "A New Class of Check-Digit Methods for Arbitrary Number Systems," *IEEE Transactions on Information Theory* 31(1985): 102–105.

2. J. Verhoeff, *Error Detecting Decimal Codes,* Amsterdam: Mathematisch Centrum, 1969.

SUGGESTED READINGS

J. Alperin, "Groups and Symmetry," *Mathematics Today,* New York: Springer-Verlag, 1978.

> This beautifully written article is intended to convey to the intelligent nonmathematician something of the nature and development of group theory. It succeeds admirably. In a manner that is accessible to all, Alperin discusses symmetry groups, Galois theory, Lie groups, and simple groups.

H. P. Gumm, "Encoding of Numbers to Detect Typing Errors," *International Journal of Applied Engineering Education* 2(1986): 61–65.

> Gumm discusses some of the standard methods for assigning check digits. He includes a program for implementing the scheme presented in this chapter.

I. N. Herstein and I. Kaplansky, *Matters Mathematical,* New York: Chelsea, 1978.

> Chapter 3 of this book discusses several interesting applications of permutations to games.

Douglas Hofstadter, "The Magic Cube's Cubies Are Twiddled by Cubists and Solved by Cubemeisters," *Scientific American* 244 (1981): 20–39.

> This article written by a Pulitzer Prize recipient, discusses the group theory involved in the solution of the Magic (Rubik's) Cube. In particular, permutation groups, subgroups, conjugates (elements of the form $x^{-1}yx$), commutators (elements of the form $x^{-1}y^{-1}xy$), and the "always-even-always-odd" theorem (Theorem 5.4) are prominently mentioned. At one point, Hofstadter says, "It is this kind of marvelously concrete illustration of an abstract notion of group theory that makes the Magic Cube one of the most amazing things ever invented for teaching mathematical ideas."

Augustin Cauchy

You see that little young man? Well! He will supplant
all of us in so far as we are mathematicians.

Spoken by Lagrange to Laplace about the eleven-year-old Cauchy

AUGUSTIN LOUIS CAUCHY was born on August 21, 1789, in Paris, the eldest
of six children. By the time he was eleven both Laplace and Lagrange had recognized
Cauchy's extraordinary talent for mathematics. In school he won prizes for Greek,
Latin, and the humanities. At the age of twenty-one he was given a commission in
Napoleon's army as a civil engineer. For the next few years, Cauchy attended to his
engineering duties while carrying out brilliant mathematical research on the side.

In 1815, at the age of twenty-six, Cauchy was made Professor of Mathematics at
the École Polytechnique and was recognized as the leading mathematician in France.
Cauchy and his contemporary Gauss were the last men to know the whole of mathematics as known at their time, and both made important contributions to nearly every
branch, both pure and applied, as well as to physics and astronomy.

Cauchy introduced a new level of rigor into mathematical analysis. We owe our
contemporary notions of limit and continuity to him. He gave the first proof of the
fundamental theorem of calculus. Cauchy was the founder of complex function theory
and a pioneer in the theory of permutation groups and determinants. His total output
of mathematics fills twenty-four large quarto volumes and is second only to that of
Euler. He wrote over 500 research papers after the age of fifty. Cauchy died at the
age of sixty-eight on May 23, 1857.

Isomorphisms

6

Mathematicians do not study objects, but relations among objects; they are indifferent to the replacement of objects by others as long as relations do not change. Matter is not important, only form interests them.

Henri Poincaré

Motivation

Suppose two students, one American and another German, are asked to count a handful of objects. The American student says, "one, two, three, four, five . . ." while the German student says: "ein, zwei, drei, vier, fünf. . . ." Are the two students doing different things? No. They are both counting the objects but they are using different terminology to do it. Similarly, when one person says: "two plus three is five" and another says: "zwei und drei ist fünf," the two are in agreement on the *concept* they are describing, but they are using different terminology to describe the concept. An analogous situation often occurs with groups; the same group is described with different terminology. We have seen two examples of this so far. In Chapter 1 we described the symmetries of a square in geometrical terms (e.g., H, V, $R_{90°}$) while in Chapter 5 we described the *same* group by way of permutations of the corners. In both cases the underlying group was the symmetries of a square. In Chapter 4, we observed that when we have a cyclic group of order n generated by a the operation turns out to be essentially that of addition modulo n, since $a^r \cdot a^s = a^k$ where $k = (r + s)$ mod n.

Definition and Examples

In this chapter we give a formal method for determining whether two groups defined in different terms are really the *same*. When this is the case, we say there is an isomorphism between the two groups. This notion was first introduced by Galois about a century and a half ago. The term *isomorphism* is derived from the Greek words *isos*, "same" or "equal," and *morphe*, "form." R. Allenby has colorfully defined an algebraist as "a person who can't tell the difference between isomorphic systems."

> *DEFINITION*   Group Isomorphism
>
> An *isomorphism* ϕ from a group G to a group $\overline{G}$ is a one-to-one mapping (or function) from G onto $\overline{G}$ that preserves the group operation. That is,
>
> $$(ab)\phi = (a\phi)(b\phi) \qquad \text{for all } a, b \text{ in } G.$$
>
> If there is an isomorphism from G onto $\overline{G}$, we say that G and $\overline{G}$ are *isomorphic* and write $G \approx \overline{G}$.

The definition can be visualized as shown in Figure 6.1. The pairs of dashed arrows represent the group operations.

It is implicit in the definition of isomorphism that the operation of the left side of the equal sign is that of G while the operation on the right side is that of $\overline{G}$. The four cases involving $\cdot$ and $+$ are shown in Table 6.1.

There are four separate steps in proving that a group G is isomorphic to a group $\overline{G}$.

STEP 1   "Mapping." Define a candidate for the isomorphism; that is, define a function ϕ from G to $\overline{G}$.

STEP 2   "1–1." Prove that ϕ is one-to-one; that is, assume $a\phi = b\phi$ and prove $a = b$.

STEP 3   "Onto." Prove that ϕ is onto; that is, for any element $\overline{g}$ in $\overline{G}$, find an element g in G such that $g\phi = \overline{g}$.

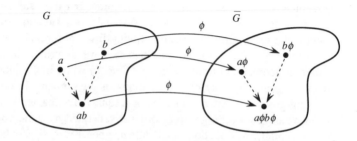

Figure 6.1

Table 6.1

G Operation	$\overline{G}$ Operation	Operation preservation
$\cdot$	$\cdot$	$(a \cdot b)\phi = a\phi \cdot b\phi$
$\cdot$	$+$	$(a \cdot b)\phi = a\phi + b\phi$
$+$	$\cdot$	$(a + b)\phi = a\phi \cdot b\phi$
$+$	$+$	$(a + b)\phi = a\phi + b\phi$

STEP 4   "O.P." Prove that ϕ is operation-preserving; that is, show $(ab)\phi = a\phi b\phi$ for all a and b in G.

None of these steps is unfamiliar to you. The only one that may appear novel is the fourth one. It requires that one be able to obtain the same result by multiplying two elements and then mapping, or by mapping two elements and then multiplying. Roughly speaking, this says that the two processes—operating and mapping—can be done in either order without affecting the result. This same concept arises in calculus when we say

$$\lim_{x \to a} (f(x) \cdot g(x)) = \lim_{x \to a} f(x) \lim_{x \to a} g(x)$$

or

$$\int_a^b (f + g)\,dx = \int_a^b f\,dx + \int_a^b g\,dx.$$

Before going any further, let's consider some examples.

Example 1   Let G be the real numbers under addition and $\overline{G}$ the positive real numbers under multiplication. Then G and $\overline{G}$ are isomorphic under the mapping $x\phi = 2^x$. Certainly, ϕ is a function from G to $\overline{G}$. To prove it is 1–1, suppose $2^x = 2^y$. Then $\log_2 2^x = \log_2 2^y$ and therefore $x = y$. For "onto," we must find for any positive real number y some real number x so that $x\phi = y$; that is, $2^x = y$. Well, solving for x gives $\log_2 y$. Finally,

$$(x + y)\phi = 2^{x+y} = 2^x \cdot 2^y = (x\phi)(y\phi)$$

for all x and y in G so that ϕ is operation-preserving as well. $\square$

Example 2   Any finite cyclic group of order n is isomorphic to Z_n. Any infinite cyclic group is isomorphic to Z. Indeed, in either case, if a is a generator of the cyclic group, the mapping $a^k \to k$ is an isomorphism. That this correspondence is a function and is one-to-one is the essence of Theorem 4.1. Obviously, the mapping is onto. That the mapping is operation-preserving follows from exercise 12 of Chapter 0 in the finite case and from the definitions in the infinite case. $\square$

Example 3 The mapping from **R** under addition to itself given by $x\phi = x^3$ is *not* an isomorphism. Although ϕ is one-to-one and onto, it is not operation-preserving since $(x + y)^3 \neq x^3 + y^3$. □

Example 4 $U(10) \approx Z_4 \approx U(5)$. To verify this, one need only observe that both $U(10)$ and $U(5)$ are cyclic of order 4. Then appeal to Example 2. □

Example 5 $U(10) \not\approx U(12)$. This is a bit more tricky to prove. First, note that $x^2 = 1$ for all x in $U(12)$. Now, suppose ϕ were an isomorphism from $U(10)$ onto $U(12)$. Then,

$$9\phi = (3 \cdot 3)\phi = (3\phi)(3\phi) = 1$$

and

$$1\phi = (1 \cdot 1)\phi = (1\phi)(1\phi) = 1.$$

Thus, $9\phi = 1\phi$, but $9 \neq 1$, a contradiction to the supposed one-to-one character of ϕ. □

Example 6 There is no isomorphism from Q, the group of rational numbers under addition, to $Q^\#$, the group of nonzero rational numbers under multiplication. If ϕ were such a mapping, there would be a rational number a such that $a\phi = -1$. But then

$$-1 = a\phi = \left(\frac{1}{2}a + \frac{1}{2}a\right)\phi = \left(\frac{1}{2}a\right)\phi \left(\frac{1}{2}a\right)\phi = \left[\left(\frac{1}{2}a\right)\phi\right]^2.$$

However, no rational number squared is -1. □

Example 7 Let $G = SL(2, \mathbf{R})$, the group of 2×2 real matrices with determinant 1. Let M be any 2×2 real matrix with nonzero determinant. Then we can define a mapping from G to G itself by, $A\phi_M = M^{-1}AM$, for all A in G. To verify that ϕ_M is an isomorphism, we carry out the four steps.

STEP 1 ϕ_M is a function from G to G. Here, we must show that $A\phi_M$ is indeed an element of G whenever A is. This follows from properties of determinants:

$$\det(M^{-1}AM) = (\det M)^{-1}(\det A)(\det M) = \det A = 1.$$

Thus, $M^{-1}AM$ is in G.

STEP 2 ϕ_M is one-to-one. Suppose $A\phi_M = B\phi_M$. Then $M^{-1}AM = M^{-1}BM$ and, by left and right cancellation, $A = B$.

STEP 3 ϕ_M is onto. Let B belong to G. We must find a matrix A in G such that $A\phi_M = B$. How shall we do this? If such a matrix A is to exist, it must have the property that $M^{-1}AM = B$. But this tells us exactly what A must be! For we may solve for A to obtain $A = MBM^{-1}$.

STEP 4 ϕ_M is operation-preserving. Let A and B belong to G. Then,

$$(AB)\phi_M = M^{-1}(AB)M = M^{-1}A(MM^{-1})BM$$
$$= (M^{-1}AM)(M^{-1}BM) = (A\phi_M)(B\phi_M).$$

The mapping ϕ_M is called *conjugation* by M.   □

Cayley's Theorem

Our next example is a classic theorem of Cayley. An important generalization of it will be given in Chapter 27.

Example 8   Cayley's Theorem (1854)
Every group is isomorphic to a group of permutations.
To prove this, let G be any group. We must find a group $\overline{G}$ of permutations that we believe is isomorphic to G. Since G is all we have to work with, we will have to use it to construct $\overline{G}$. For any g in G, define a function T_g from G to G by

$$xT_g = xg \qquad \text{for all } x \text{ in } G.$$

(In words, T_g is just multiplication by g on the right.) We leave it as an exercise to prove that T_g is a permutation on the set of elements of G. Now, let $\overline{G} = \{T_g \mid g \in G\}$. Then, $\overline{G}$ is a group under the operation of function composition. To verify this, we first observe that for any g and h in G we have $xT_gT_h = (xT_g)T_h = (xg)T_h = (xg)h = x(gh) = xT_{gh}$, so that $T_gT_h = T_{gh}$. From this it follows that T_e is the identity and $(T_g)^{-1} = T_{g^{-1}}$ (see exercise 8). Since function composition is associative, we have verified all the conditions for $\overline{G}$ to be a group.

The isomorphism ϕ between G and $\overline{G}$ is now ready-made. For every g in G, define $g\phi = T_g$. Clearly, $g = h$ implies $T_g = T_h$ so that ϕ is a function from G to $\overline{G}$. On the other hand, if $T_g = T_h$, then $eT_g = eT_h$ or $eg = eh$. Thus, ϕ is one-to-one. By the way $\overline{G}$ was constructed, we see that ϕ is onto. The only condition that remains to be checked is that ϕ is operation-preserving. To this end, let x and y belong to G. Then

$$(xy)\phi = T_{xy} = T_xT_y = (x\phi)(y\phi).$$ □

The group $\overline{G}$ constructed above is called the *right regular representation of* G. For concreteness, let us calculate the right regular representation $\overline{U(12)}$ for $U(12) = \{1, 5, 7, 11\}$. Writing the permutations of $U(12)$ in array form, we have (remember, T_x is just multiplication by x)

$$T_1 = \begin{bmatrix} 1 & 5 & 7 & 11 \\ 1 & 5 & 7 & 11 \end{bmatrix}, \quad T_5 = \begin{bmatrix} 1 & 5 & 7 & 11 \\ 5 & 1 & 11 & 7 \end{bmatrix},$$

$$T_7 = \begin{bmatrix} 1 & 5 & 7 & 11 \\ 7 & 11 & 1 & 5 \end{bmatrix}, \quad T_{11} = \begin{bmatrix} 1 & 5 & 7 & 11 \\ 11 & 7 & 5 & 1 \end{bmatrix}.$$

It is instructive to compare the Cayley table for $U(12)$ and its right regular representation $\overline{U(12)}$.

$U(12)$	1	5	7	11	$\overline{U(12)}$	T_1	T_5	T_7	T_{11}
1	1	5	7	11	T_1	T_1	T_5	T_7	T_{11}
5	5	1	11	7	T_5	T_5	T_1	T_{11}	T_7
7	7	11	1	5	T_7	T_7	T_{11}	T_1	T_5
11	11	7	5	1	T_{11}	T_{11}	T_7	T_5	T_1

It should be abundantly clear from these tables that $U(12)$ and $\overline{U(12)}$ are only notationally different.

Perhaps the most important aspect of Cayley's Theorem is that it shows the present-day set of axioms we have adopted for a group in the correct abstraction of its much earlier predecessor—a group of permutations. Indeed, Cayley's Theorem tells us that abstract groups are not different from permutation groups. Rather, it is the viewpoint that is different. It is this difference of viewpoint that has stimulated the tremendous progress in group theory and many other branches of mathematics in the twentieth century.

It is sometimes very difficult to prove or disprove, whichever the case may be, that two particular groups are isomorphic. For example, it requires somewhat sophisticated techniques to prove the surprising fact that the group of real numbers under addition is isomorphic to the group of complex numbers under addition. Likewise, it is not easy to prove that the group of nonzero complex numbers under multiplication is isomorphic to the group of complex numbers with absolute value of 1 under multiplication [1]. In geometric terms, this says that, as groups, the punctured plane is isomorphic to the unit circle.

Properties of Isomorphisms

Our first theorem in this chapter gives a catalog of properties of isomorphisms and isomorphic groups.

Theorem 6.1 *Properties of Isomorphisms*
Suppose ϕ is an isomorphism from a group G onto a group $\overline{G}$. Then

1. *ϕ carries the identity of G to the identity of $\overline{G}$.*
2. *For every integer n and for every group element a in G, $a^n\phi = (a\phi)^n$.*
3. *For any elements a and b in G, a and b commute if and only if $a\phi$ and $b\phi$ commute.*
4. *G is Abelian if and only if $\overline{G}$ is Abelian.*
5. *$|a| = |a\phi|$ for all a in G (isomorphisms preserve orders).*

6. *G is cyclic if and only if $\overline{G}$ is cyclic.*
7. *For a fixed integer k and a fixed group element b in G, the equation $x^k = b$ has the same number of solutions in G as does the equation $x^k = b\phi$ in $\overline{G}$.*
8. *ϕ^{-1} is an isomorphism from $\overline{G}$ onto G.*
9. *If K is a subgroup of G, then $K\phi = \{k\phi \mid k \in K\}$ is a subgroup of $\overline{G}$.*

Proof. We will restrict ourselves to proving only (1) and (5). Note, however, that (4) follows directly from (3), and (6) directly from (5). For convenience, let us denote the identity in G by e_G, and the identity in $\overline{G}$ by $e_{\overline{G}}$. Then $e_G = e_G e_G$ so that

$$e_G\phi = (e_G e_G)\phi = (e_G\phi)(e_G\phi).$$

But, $e_G\phi \in \overline{G}$ so that $e_G\phi = e_{\overline{G}}(e_G\phi)$, as well. Thus, by cancellation, we have $e_{\overline{G}} = e_G\phi$. This proves 1.

To prove (5), we note that $a^n = e$ if and only if $a^n\phi = e\phi$. So, by parts (1) and (2), $a^n = e$ if and only if $(a\phi)^n = e$. Thus, a has infinite order if and only if $a\phi$ has infinite order, and a has finite order n if and only if $a\phi$ has order n. ∎

Property (7) is quite useful for showing that two groups are *not* isomorphic. Often b is picked to be the identity. For example, consider $\mathbf{C}^{\#}$ and $\mathbf{R}^{\#}$. Because the equation $x^4 = 1$ has four solutions in $\mathbf{C}^{\#}$ but only two in $\mathbf{R}^{\#}$, no matter how one attempts to define an isomorphism from $\mathbf{C}^{\#}$ to $\mathbf{R}^{\#}$, property (7) cannot hold.

Theorem 6.1 shows that isomorphic groups have many properties in common. Actually, the definition is precisely formulated so that isomorphic groups have *all* group theoretic properties in common. By this we mean that if two groups are isomorphic, then any property that can be expressed in the language of group theory is true for one if and only if it is true for the other. This is why algebraists speak of isomorphic groups as "equal" or the "same." Admittedly, calling such groups equivalent, rather than the same, might be more appropriate, but we bow to long-standing tradition.

Automorphisms

Certain kinds of isomorphisms are referred to so often that they have been given special names.

DEFINITION Automorphism
An isomorphism from a group G onto itself is called an *automorphism* of G.

The isomorphism of Example 7 is an automorphism of $SL(2, \mathbf{R})$. Two more examples follow.

Example 9   The function ϕ from **C** to **C** given by $(a + bi)\phi = a - bi$ is an automorphism of the group of complex numbers under addition. The restriction of ϕ to $\mathbf{C}^{\#}$ is also an automorphism of the group of the nonzero complex numbers under multiplication.   □

Example 10   Let $\mathbf{R}^2 = \{(a, b) \mid a, b \in \mathbf{R}\}$. Then $(a, b)\phi = (b, a)$ is an automorphism of the group $\mathbf{R}^2$ under componentwise addition. Geometrically, ϕ reflects each point in the plane across the line $y = x$. More generally, any reflection across a line passing through the origin or any rotation of the plane about the origin is an automorphism of $\mathbf{R}^2$.   □

The automorphism defined in Example 7 is a particular instance of an automorphism that arises often enough to warrant a name and notation of its own.

DEFINITION   Inner Automorphism induced by a
Let G be a group, and let $a \in G$. The function ϕ_a defined by $x\phi_a = a^{-1}xa$ for all x in G is called the *inner automorphism of G induced by a.*

We leave it as an exercise to show that ϕ_a is actually an automorphism of G. (Use Example 7 as a model.)

Example 11   The action of the inner automorphism of D_4 induced by R_{90} is given below.

$$\phi_{R_{90}}$$

x	$\rightarrow R_{90}^{-1}$	$x R_{90}$		
R_0	$\rightarrow R_{90}^{-1}$	$R_0 R_{90}$	$= R_0$
R_{90}	$\rightarrow R_{90}^{-1}$	$R_{90} R_{90}$	$= R_{90}$
R_{180}	$\rightarrow R_{90}^{-1}$	$R_{180} R_{90}$	$= R_{180}$
R_{270}	$\rightarrow R_{90}^{-1}$	$R_{270} R_{90}$	$= R_{270}$
H	$\rightarrow R_{90}^{-1}$	$H R_{90}$	$= V$
V	$\rightarrow R_{90}^{-1}$	$V R_{90}$	$= H$
D	$\rightarrow R_{90}^{-1}$	$D R_{90}$	$= D'$
D'	$\rightarrow R_{90}^{-1}$	$D' R_{90}$	$= D$

□

When G is a group, we use Aut(G) to denote the set of all automorphisms of G; and Inn(G), the set of all inner automorphisms of G. The reason these sets are noteworthy is demonstrated by the next theorem.

Theorem 6.2   Aut(G) and Inn(G) Are Groups*
The set of automorphisms of a group and the set of inner automorphisms of a group are both groups.

---

*The group Aut(G) was first studied by O. Hölder in 1893 and, independently, by E. H. Moore in 1894.

Proof. Exercise. ◼

The determination of Inn(G) is routine. If $G = \{e, a, b, c, \ldots\}$, then Inn($G$) $= \{\phi_e, \phi_a, \phi_b, \phi_c, \ldots\}$. This latter list may have duplications, however, since ϕ_a may be equal to ϕ_b even though $a \neq b$ (see exercise 32). Thus, the only work involved in determining Inn(G) is in deciding which distinct elements give the distinct automorphisms. On the other hand, the determination of Aut(G) is, in general, quite involved.

Example 12 Inn(D_4)

To determine Inn(D_4), we first observe that the complete list of inner automorphisms is $\phi_{R_0}, \phi_{R_{90}}, \phi_{R_{180}}, \phi_{R_{270}}, \phi_H, \phi_V, \phi_D, \phi_{D'}$. Our job is to determine the repetitions in this list. Since $R_{180} \in Z(D_4)$, we have $x\phi_{R_{180}} = R_{180}^{-1}xR_{180} = x$ so that $\phi_{R_{180}} = \phi_{R_0}$. Also, $x\phi_{R_{270}} = R_{270}^{-1}xR_{270} = R_{90}^{-1}R_{180}^{-1}xR_{180}R_{90} = R_{90}^{-1}xR_{90} = x\phi_{R_{90}}$. Similarly, since $H = R_{180}V$ and $D' = R_{180}D$, we have $\phi_H = \phi_V$ and $\phi_D = \phi_{D'}$. This proves that the previous list can be pared down to $\phi_{R_0}, \phi_{R_{90}}, \phi_D, \phi_{D'}$. We leave it to the reader to show that these are distinct. ☐

Example 13 Aut(Z_{10})

To compute Aut(Z_{10}) we assume α is an element of Aut(Z_{10}) and try to discover enough information about α to determine how α must be defined. Because Z_{10} is so simple, this is not difficult to do. To begin with, observe that once we know 1α we know $k\alpha$ for any k. For

$$k\alpha = \underbrace{(1 + 1 + \cdots + 1)}_{k \text{ terms}}\alpha = \underbrace{1\alpha + 1\alpha + \cdots + 1\alpha}_{k \text{ terms}} = k(1\alpha).$$

So, we need only determine the choices for 1α that make α an automorphism of Z_{10}. Since part 5 of Theorem 6.1 tells us that $|1\alpha| = 10$, there are four candidates for 1α.

$$1\alpha = 1; \quad 1\alpha = 3; \quad 1\alpha = 7; \quad 1\alpha = 9. \qquad ☐$$

To distinguish among the four possibilities we refine our notation by denoting the mapping that sends 1 to 1 by α_1, 1 to 3 by α_3, 1 to 7 by α_7, and 1 to 9 by α_9. So the only possibilities for Aut(Z_{10}) are $\alpha_1, \alpha_3, \alpha_7, \alpha_9$. But are all these automorphisms? Clearly, α_1 is the identity. Let us check α_3. Since $1\alpha_3 = 3$ is a generator of Z_{10}, it follows that α_3 is onto (and, by exercise 10 of Chapter 5, it is also one-to-one). Finally, since $(a + b)\alpha_3 = 3(a + b) = 3a + 3b = a\alpha_3 + b\alpha_3$, we see that α_3 is operation-preserving as well. Thus, $\alpha_3 \in$ Aut(Z_{10}). The same argument shows that α_7 and α_9 are also automorphisms.

This gives us the elements of Aut(Z_{10}) but not the structure. For instance, what is $\alpha_3 \alpha_3$? Well, $1(\alpha_3 \alpha_3) = 3\alpha_3 = 3 \cdot 3 = 9 = 1\alpha_9$ so $\alpha_3 \alpha_3 = \alpha_9$. A similar calculation shows $\alpha_3^4 = \alpha_1$, so that $|\alpha_3| = 4$. Thus Aut(Z_{10}) is cyclic. Actually, the following Cayley tables reveal that Aut(Z_{10}) is isomorphic to $U(10)$.

$U(10)$	1	3	7	9
1	1	3	7	9
3	3	9	1	7
7	7	1	9	3
9	9	7	3	1

$\text{Aut}(Z_{10})$	α_1	α_3	α_7	α_9
α_1	α_1	α_3	α_7	α_9
α_3	α_3	α_9	α_1	α_7
α_7	α_7	α_1	α_9	α_3
α_9	α_9	α_7	α_3	α_1

With Example 13 as a guide, we are now ready to tackle the group $\text{Aut}(Z_n)$. The result is particularly nice since it relates the two kinds of groups we have most frequently encountered thus far—the cyclic groups Z_n and the U-groups $U(n)$.

Theorem 6.3 $\text{Aut}(Z_n) \approx U(n)$
For every positive integer n, $\text{Aut}(Z_n)$ is isomorphic to $U(n)$.

Proof.   As in Example 13, any automorphism α is determined by the value of 1α and $1\alpha \in U(n)$. Now consider the correspondence from $\text{Aut}(Z_n)$ to $U(n)$ given by $T\colon \alpha \rightarrow 1\alpha$. The fact that $k\alpha = k(1\alpha)$ (see Example 13) implies that T is a one-to-one mapping. For if α and β belong to $\text{Aut}(Z_n)$ and $1\alpha = 1\beta$, then $k\alpha = k\beta$ for all k in Z_n and therefore $\alpha = \beta$.

To prove that T is onto, let $r \in U(n)$ and consider the mapping α from Z_n to Z_n defined by $s\alpha = sr(\text{mod } n)$ for all s in Z_n. We leave it as an exercise to verify that α is an automorphism of Z_n. Then, since $\alpha T = 1\alpha = r$, T is onto $U(n)$.

Finally, we establish the fact that T is operation-preserving. Let α, $\beta \in \text{Aut}(Z_n)$. Then we have $(\alpha\beta)T = 1(\alpha\beta) = (1\alpha)\beta = \underbrace{(1 + 1 + \cdots + 1)}_{1\alpha}\beta =$

$\underbrace{1\beta + 1\beta + \cdots + 1\beta}_{} = (1\alpha)(1\beta) = \alpha T\beta T.$ This completes the proof. ∎

EXERCISES

No pain, no gain.
Arnold Schwarzennegger

1. Find an isomorphism from the group of integers under addition to the group of even integers under addition.
2. Find $\text{Aut}(Z)$.
3. Let $\mathbf{R}^+$ be the group of positive real numbers under multiplication. Show the mapping $x\phi = \sqrt{x}$ is an automorphism of $\mathbf{R}^+$.

4. Show $U(8)$ is not isomorphic to $U(10)$.

5. Show $U(8)$ is isomorphic to $U(12)$.

6. Prove that the relation isomorphism is an equivalence relation. (See Chapter 0 for definition of equivalence relation.)

7. Prove that S_4 is not isomorphic to D_{12}.

8. In the notation of Example 8, prove that $(T_g)^{-1} = T_{g^{-1}}$.

9. Explain why the three parts of exercise 1 of the Supplementary Exercises, Chapters 1–4, follow immediately from Theorem 6.1.

10. Let G be a group. Prove that the mapping $g\alpha = g^{-1}$ for all g in G is an automorphism if and only if G is Abelian.

11. Find two groups G and H such that $G \neq H$, but $\text{Aut}(G) \approx \text{Aut}(H)$.

12. Let G be the group given in exercise 4 of the Supplementary Exercises for Chapters 1–4. Find $\text{Inn}(G)$.

13. Let G be a group and let a belong to G. Prove the mapping of ϕ_a defined by $x\phi_a = a^{-1}xa$ is an automorphism of G.

14. Find $\text{Aut}(Z_6)$.

15. If G is a group, prove $\text{Aut}(G)$ and $\text{Inn}(G)$ are groups.

16. Let $r \in U(n)$. Prove that the mapping $\alpha : Z_n \to Z_n$ defined by $s\alpha = sr \bmod n$ for all s in Z_n is an automorphism of Z_n.

17. Prove that the mapping from $U(16)$ to itself given by $x \to x^3$ is an automorphism. What about $x \to x^5$ and $x \to x^7$? Generalize.

18. Suppose ϕ is an automorphism of D_4 that takes R_{90} to itself and H to V. Prove that ϕ is the inner automorphism induced by R_{90}.

19. Show that S_7 (the symmetric group on the integers $\{1, 2, 3, 4, 5, 6, 7\}$) is isomorphic to the subgroup of all those elements of S_8 that send 8 to 8.

20. Prove that the quaternion group (see exercise 4, Supplementary Exercises for Chapters 1–4) is not isomorphic to the dihedral group D_4.

21. Find $\text{Inn}(G)$, where G is the group of quaternions (see exercise 4, Supplementary Exercises for Chapters 1–4).

22. Referring to Example 8, prove that T_g is indeed a permutation on the set G.

23. Show that the mapping $(a + bi)\phi = a - bi$ is an automorphism of the group of complex numbers under addition. Show that ϕ preserves complex multiplication as well—that is, $(xy)\phi = (x\phi)(y\phi)$ for all x and y in $\mathbf{C}$.

24. Let

$$G = \{a + b\sqrt{2} \mid a, b \text{ rational}\}$$

and

$$H = \left\{ \begin{bmatrix} a & 2b \\ b & a \end{bmatrix} \,\middle|\, a, b \text{ rational} \right\}.$$

Show that G and H are isomorphic under addition. Prove that G and H are closed under multiplication. Does your isomorphism preserve multiplication as well as addition? (G and H are examples of rings—a topic we will take up later.)

25. Prove that Z under addition is not isomorphic to Q under addition.

26. Look up the words *isobar, isomer,* and *isotope* in a dictionary. Relate their meanings to the meaning of isomorphism.

27. Let **C** be the complex numbers and

$$M = \left\{ \begin{bmatrix} a & -b \\ b & a \end{bmatrix} \,\middle|\, a, b \in \mathbf{R} \right\}.$$

Prove **C** and M are isomorphic under addition and $\mathbf{C}^{\#}$ and $M^{\#}$ are isomorphic under multiplication.

28. Let $\mathbf{R}^n = \{(a_1, a_2, \ldots, a_n) \mid a_i \in \mathbf{R}\}$. Show the mapping $(a_1, a_2, \ldots, a_n)\phi = (-a_1, -a_2, \ldots, -a_n)$ is an automorphism of the group $\mathbf{R}^n$ under componentwise addition. This automorphism is called *inversion*. Describe the action of ϕ geometrically.

29. Consider the following statement: The order of a subgroup divides the order of the group. Suppose you could prove this for finite permutation groups. Would the statement then be true for all finite groups? Explain.

30. Prove that for any integer n, there are only a finite number of nonisomorphic groups of order n.

31. Show that the mapping $a \rightarrow \log_{10} a$ is an isomorphism from $\mathbf{R}^+$ under multiplication to $\mathbf{R}$ under addition. Explain how this isomorphism is implicitly employed when one uses a slide rule to multiply or divide real numbers.

32. Let G be a group and let $g \in G$. If $z \in Z(G)$, show that the inner automorphism induced by g is the same as the inner automorphism induced by zg (that is, the mappings ϕ_g and ϕ_{zg} are equal).

33. Suppose g and h induce the same inner automorphism of a group G. Prove that $gh^{-1} \in Z(G)$.

34. Combine the results of exercises 32 and 33 into a single "if and only if" theorem.

35. Explain why S_n ($n \geq 3$) contains a subgroup isomorphic to D_n.

36. Let $G = \{0, \pm 2, \pm 4, \pm 6, \ldots\}$ and $H = \{0, \pm 3, \pm 6, \pm 9, \ldots\}$. Show that G and H are isomorphic groups under addition.

REFERENCE

1. J. R. Clay, "The Punctured Plane Is Isomorphic to the Unit Circle," *Journal of Number Theory* 1 (1964): 500–501.

SUGGESTED FILM

Nim and Other Oriented Graph Games with Andrew Gleason, International Film Bureau, Inc., 63 minutes.

This filmed lecture describes Nim and related games and constructs an algebraic theory that gives information about games more complicated than Nim. The film gives a good example of a nontrivial, nonobvious application of group theory.

Arthur Cayley

Cayley is forging the weapons for future generations of physicists.

Peter Tait

ARTHUR CAYLEY was born on August 16, 1821, in England. His genius showed itself at an early age. At Trinity College, Cambridge, he was senior wrangler in the mathematical tripos and placed first in the competition for the Smith Prize. He published his first research paper while an undergraduate of twenty and the next year he published eight papers. While still in his early twenties, he originated the concept of n-dimensional geometry.

After graduating, Cayley stayed on at Trinity College for three years as a tutor. At the age of twenty-five he began a fourteen-year career as a lawyer. During this period he published approximately 200 mathematical papers, many of which are now classics.

In 1863 Cayley accepted the newly established Sadlerian professorship of mathematics at Cambridge University. He spent the rest of his life in that position. One of his notable nonmathematical accomplishments was his role in the successful effort to have women admitted to Cambridge.

Among Cayley's many innovations in mathematics were the notions of an abstract group and a group algebra, the matrix concept, and the theory of determinants. He made major contributions to geometry and linear algebra. Cayley and his life-long friend and collaborator J. J. Sylvester were the founders of the theory of invariants, which was later to play an important role in the theory of relativity.

Cayley's collected works comprise thirteen volumes, each about 600 pages in length. He died on January 26, 1895. One of his students wrote of him:

> But he was more than a mathematician with a singleness of aim, which Wordsworth would have chosen for his "Happy Warrior," he persevered to the last in his nobly lived ideal. His life had a significant influence on those who knew him; they admired his character as much as they respected his genius; and they felt that, at his death, a great man had passed from the world.

External Direct Products

7

By an irony of fate group theory later grew into one of the central themes of physics, and it now dominates the thinking of all of us who are struggling to understand the fundamental particles of nature.

Freeman Dyson, *Scientific American*

Definition and Examples

In this chapter, we show how one may piece together groups to make larger groups. In the next chapter, we will show that one can often start with one large group and decompose it into a product of smaller groups in much the same way a composite positive integer can be broken down into a product of primes. These methods will later be used to give us a simple way to construct all finite Abelian groups.

DEFINITION   External Direct Product

Let $G_1, G_2, \ldots, G_n$ be a collection of groups. The *external direct product* of $G_1, G_2, \ldots, G_n$, written as $G_1 \oplus G_2 \oplus \cdots \oplus G_n$, is the set of all n-tuples for which the ith component is an element of G_i, and the operation is componentwise. In symbols,

$$G_1 \oplus G_2 \oplus \cdots \oplus G_n = \{(g_1, g_2, \ldots, g_n) \mid g_i \in G_i\},$$

where $(g_1, g_2, \ldots, g_n)(g_1', g_2', \ldots, g_n')$ is defined to be $(g_1 g_1', g_2 g_2', \ldots, g_n g_n')$. It is understood that each product $g_i g_i'$ is performed with the operation of G_i. We leave it to the reader to show that the external direct product of groups is itself a group.*

---

*This group was introduced by O. Hölder.

This construction is not new to students who have had linear algebra or physics. Indeed, $\mathbf{R}^2 = \mathbf{R} \oplus \mathbf{R}$ and $\mathbf{R}^3 = \mathbf{R} \oplus \mathbf{R} \oplus \mathbf{R}$—the operation being componentwise addition. Of course, there is also scalar multiplication, but we ignore this for the time being, since we are interested only in the group structure at this point.

Example 1

$$U(8) \oplus U(10) = \{(1, 1), (1, 3), (1, 7), (1, 9), (3, 1), (3, 3), (3, 7),$$
$$(3, 9), (5, 1), (5, 3), (5, 7), (5, 9), (7, 1), (7, 3),$$
$$(7, 7), (7, 9)\}.$$

The product $(3, 7)(7, 9) = (5, 3)$, since the first two components are combined by multiplication modulo 8, while the second two components are combined by multiplication modulo 10. □

Example 2

$$Z_2 \oplus Z_3 = \{(0, 0), (0, 1), (0, 2), (1, 0), (1, 1), (1, 2)\}.$$

Clearly, this is an Abelian group of order 6. Is this group related to another Abelian group of order 6 that we know, namely, Z_6? Consider the subgroup of $Z_2 \oplus Z_3$ generated by $(1, 1)$. Since the operation in each component is addition, we have $(1, 1) = (1, 1)$, $2(1, 1) = (0, 2)$, $3(1, 1) = (1, 0)$, $4(1, 1) = (0, 1)$, $5(1, 1) = (1, 2)$, and $6(1, 1) = (0, 0)$. Hence $Z_2 \oplus Z_3$ is cyclic. It follows then that $Z_2 \oplus Z_3$ is isomorphic to Z_6. □

On the basis of Example 2, one might be tempted to conjecture that $Z_m \oplus Z_n \approx Z_{mn}$ for all positive integers m and n. This is easily seen to be false by looking at $Z_2 \oplus Z_2$. Theorem 7.2 shows, however, that these groups are isomorphic in certain cases.

Properties of External Direct Products

Our first theorem gives a simple method for computing the order of an element in a direct product in terms of the orders of the component pieces.

> **Theorem 7.1** *Order of an Element in a Direct Product*
> *The order of an element of a direct product of finite groups is the least common multiple of the orders of the components of the element. In symbols,*
>
> $$|(g_1, g_2, \ldots, g_n)| = \mathrm{lcm}\{|g_1|, |g_2|, \ldots, |g_n|\}.$$

Proof.   To simplify matters, we first consider the special case for which the direct product has only two factors. Let (g_1, g_2) be an arbitrary element of $G_1 \oplus G_2$. Let $s = \mathrm{lcm}\{|g_1|, |g_2|\}$ and $t = |(g_1, g_2)|$. Clearly then,

$$(g_1, g_2)^s = (g_1^s, g_2^s) = (e, e),$$

so, by the corollary to Theorem 4.1, t must divide s. In particular, $t \leq s$. But, $(g_1^t, g_2^t) = (g_1, g_2)^t = (e, e)$, and it follows, for the same reason, that both $|g_1|$ and $|g_2|$ must divide t. Thus, t is a common multiple of $|g_1|$ and $|g_2|$; and, therefore, $s \leq t$, since s is the *least* common multiple of $|g_1|$ and $|g_2|$. Putting this information about s and t together, we find $s = t$.

We leave it to the reader to verify that the argument extends to the general case. ∎

The next two examples are applications of Theorem 7.1.

Example 3 We determine the number of elements of order 5 in $Z_{25} \oplus Z_5$. By Theorem 7.1, we may count the number of elements (a, b) in $Z_{25} \oplus Z_5$ with the property that $5 = |(a, b)| = \text{lcm}\{|a|, |b|\}$. Clearly this requires that either $|a| = 5$ and $|b| = 1$ or 5, or $|b| = 5$ and $|a| = 1$ or 5. We consider three mutually exclusive cases.

CASE 1 $|a| = 5$ and $|b| = 5$.

Here there are four choices for a and four choices for b. This gives sixteen elements of order 5.

CASE 2 $|a| = 5$ and $|b| = 1$.

In this case there are four choices for a and only one for b. This gives four more elements of order 5.

CASE 3 $|a| = 1$ and $|b| = 5$.

This time there is one choice for a and four choices for b, so we obtain four more elements of order 5.

Thus, $Z_{25} \oplus Z_5$ has 24 elements of order 5. ☐

Example 4 We determine the number of cyclic subgroups of order 10 in $Z_{100} \oplus Z_{25}$. We begin by counting the number of elements (a, b) of order 10.

CASE 1 $|a| = 10$ and $|b| = 1$ or 5

Since Z_{100} has a unique cyclic subgroup of order 10 and any cyclic group of order 10 has 4 generators (Theorem 4.4), there are 4 choices for a. Similarly, there are 5 choices for b. This gives 20 possibilities for (a, b).

CASE 2 $|a| = 2$ and $|b| = 5$.

Since any finite cyclic group has a unique subgroup of order 2, there is only one choice for a. Obviously, there are 4 choices for b. So, this case yields 4 more possibilities for (a, b).

Thus, $Z_{100} \oplus Z_{25}$ has 24 elements of order 10. Because each cyclic subgroup of order 10 has 4 elements of order 10 and no two of them can have an element of order 10 in common, there must be $24/4 = 6$ cyclic subgroups of order 10. (This method is analogous to determining the number of sheep in a flock by counting legs and dividing by 4.) ☐

The next theorem and its corollary characterize those direct products of cyclic groups that are themselves cyclic.

Theorem 7.2 *Criterion for $G \oplus H$ to Be Cyclic*
Let G and H be finite cyclic groups. Then $G \oplus H$ is cyclic if and only if $|G|$ and $|H|$ are relatively prime.

Proof. Let $|G| = m$ and $|H| = n$ so that $|G \oplus H| = mn$. To prove the first half of the theorem, we assume $G \oplus H$ is cyclic and show that m and n are relatively prime. Since $G \oplus H$ is cyclic, there is an element (g, h) in $G \oplus H$ of order mn. From Theorem 7.1, we obtain $mn = |(g, h)| = \text{lcm}(|g|, |h|)$. On the other hand, since $|g|$ divides m and $|h|$ divides n (see Theorem 4.3), we know $\text{lcm}(|g|, |h|)$ divides $\text{lcm}(m, n)$. Since it is always true that $\text{lcm}(m, n) \le mn$, we conclude $\text{lcm}(m, n) = mn$ and, therefore, $\gcd(m, n) = 1$ (see exercise 7 of Chapter 0).

To prove the other half of the theorem, let $G = \langle g \rangle$ and $H = \langle h \rangle$ and suppose $\gcd(m, n) = 1$. Then, $|(g, h)| = \text{lcm}(m, n) = mn = |G \oplus H|$ so that (g, h) is a generator of $G \oplus H$. ∎

As a consequence of Theorem 7.2 and an induction argument, we obtain the following extension of Theorem 7.2.

Corollary 1 *Criterion for $G_1 \oplus G_2 \oplus \cdots \oplus G_n$ to Be Cyclic*
An external direct product $G_1 \oplus G_2 \oplus \cdots \oplus G_n$ of finite cyclic groups is cyclic if and only if $|G_i|$ and $|G_j|$ are relatively prime when $i \ne j$.

Corollary 2 *Criterion for $Z_{n_1 n_2 \cdots n_k} \approx Z_{n_1} \oplus Z_{n_2} \oplus \cdots \oplus Z_{n_k}$*
Let $m = n_1 n_2 \cdots n_k$. Then Z_m is isomorphic to $Z_{n_1} \oplus Z_{n_2} \oplus \cdots \oplus Z_{n_k}$ if and only if n_i and n_j are relatively prime when $i \ne j$.

By using the above results in an iterative fashion, one can express the same group (up to isomorphism) in many different forms. For example, we have

$$Z_2 \oplus Z_2 \oplus Z_3 \oplus Z_5 \approx Z_2 \oplus Z_6 \oplus Z_5 \approx Z_2 \oplus Z_{30}.$$

Similarly,

$$Z_2 \oplus Z_2 \oplus Z_3 \oplus Z_5 \approx Z_2 \oplus Z_6 \oplus Z_5 \approx Z_2 \oplus Z_3 \oplus Z_2 \oplus Z_5 \approx Z_6 \oplus Z_{10}.$$

Thus, $Z_2 \oplus Z_{30} \approx Z_6 \oplus Z_{10}$. Note, however, that $Z_2 \oplus Z_{30} \not\approx Z_{60}$.

EXERCISES

It is true that Fourier has the opinion that the principal object of mathematics is the public utility and the explanation of natural phenomena; but a scientist like him ought to know that the unique object of science is the honor of the human spirit and on this basis a question of the theory of numbers is worth as much as a question about the planetary system.

C. J. Jacobi

1. Find the order of each element in $Z_2 \oplus Z_4$.
2. Show that $G \oplus H$ is Abelian if and only if G and H are Abelian.
3. Show that $Z_2 \oplus Z_2 \oplus Z_2$ has seven subgroups of order 2.
4. Determine the subgroup lattice of $Z_2 \oplus Z_2$.
5. Prove or disprove that $Z \oplus Z$ is a cyclic group.
6. Prove, by comparing orders of elements, that $Z_8 \oplus Z_2$ is not isomorphic to $Z_4 \oplus Z_4$.
7. Prove that $G_1 \oplus G_2$ is isomorphic to $G_2 \oplus G_1$.
8. Is $Z_3 \oplus Z_9$ isomorphic to Z_{27}? Why?
9. Is $Z_3 \oplus Z_5$ isomorphic to Z_{15}? Why?
10. How many elements of order 9 does $Z_3 \oplus Z_9$ have? (Do not do this exercise by brute force.)
11. Explain why $Z_8 \oplus Z_4$ and $Z_{8000000} \oplus Z_{4000000}$ must have the same numbers of elements of order 4.
12. The dihedral group D_n of order $2n$ has a subgroup of n rotations and a subgroup of order 2. Explain why D_n cannot be isomorphic to the external direct product of two such groups.
13. Prove that the group of complex numbers under addition is isomorphic to **R** $\oplus$ **R**.
14. Suppose $G_1 \approx G_2$ and $H_1 \approx H_2$. Prove that $G_1 \oplus H_1 \approx G_2 \oplus H_2$.
15. Construct a Cayley table for $Z_2 \oplus Z_3$.
16. How many elements of order 4 does $Z_4 \oplus Z_4$ have? (Do not do this exercise by checking each member of $Z_4 \oplus Z_4$.)
17. What is the order of any nonidentity element of $Z_3 \oplus Z_3 \oplus Z_3$?
18. How many subgroups of order 4 does $Z_4 \oplus Z_2$ have?
19. Let M be the group of all real 2×2 matrices under addition. Let $N = \mathbf{R} \oplus \mathbf{R} \oplus \mathbf{R} \oplus \mathbf{R}$ under componentwise addition. Prove that M and N are isomorphic. What is the corresponding theorem for the group of $n \times n$ matrices under addition?
20. The group $S_3 \oplus Z_2$ is isomorphic to which of the following groups: Z_{12}, $Z_6 \oplus Z_2$, A_4, D_6?
21. Let G be a group. Let $H = \{(g, g) \mid g \in G\}$. Show that H is a subgroup of $G \oplus G$. (This subgroup is called the *diagonal* of $G \oplus G$.) When G is the set of real numbers under addition, describe $G \oplus G$ and H geometrically.
22. Construct a Cayley table for $U(6) \oplus Z_3$.

23. Find all subgroups of order 3 in $Z_9 \oplus Z_3$.

24. Find all subgroups of order 4 in $Z_4 \oplus Z_4$.

25. Show that G is isomorphic to a subgroup of $G \oplus H$.

26. Let

$$H = \left\{ \begin{bmatrix} 1 & a & b \\ 0 & 1 & 0 \\ 0 & 0 & 1 \end{bmatrix} \middle| a, b \in Z_3 \right\}.$$

(See exercise 40 of Chapter 2 for the definition of multiplication.) Show that H is an Abelian group of order 9. Is H isomorphic to Z_9 or $Z_3 \oplus Z_3$?

27. Let $G = \{3^m 6^n \mid m, n \in Z\}$ under multiplication. Prove that G is isomorphic to $Z \oplus Z$.

28. Let $(a_1, a_2, \ldots, a_n) \in G_1 \oplus G_2 \oplus \cdots \oplus G_n$. Give a necessary and sufficient condition for $|(a_1, a_2, \ldots, a_n)| = \infty$.

29. Prove $D_3 \oplus D_4 \neq D_{24}$.

30. Determine the number of cyclic subgroups of order 15 in $Z_{90} \oplus Z_{36}$.

31. For any Abelian group G and any positive integer n, let $G^n = \{g^n \mid g \in G\}$ (see exercise 13, Supplementary Exercises for Chapters 1–4). If H and K are Abelian, show that $(H \oplus K)^n = H^n \oplus K^n$.

32. Suppose G is a group of order 4 and $x^2 = e$ for all x in G. Prove that G is isomorphic to $Z_2 \oplus Z_2$. (This exercise is referred to in Chapter 9.)

33. If a finite Abelian group has exactly 24 elements of order 6, how many subgroups of order 6 does it have?

PROGRAMMING EXERCISES

In a few minutes, a computer can make a mistake so great that it would take many men many months to equal it.

Merle L. Meacham

1. Write a program that will print out the Cayley table for $Z_m \oplus Z_n$. Assume $m \leq 8, n \leq 8$. Run your program for $m = 5, n = 6; m = 3, n = 5; m = 4, n = 8$.

2. Write a program to compute the order of any element of $Z_m \oplus Z_n$. Run your program for $(2, 3)$, $(4, 6)$, and $(3, 2)$ from $Z_{12} \oplus Z_{15}$.

Internal Direct Products

8

The universe is an enormous direct product of representations of symmetry groups.

Steven Weinberg*

Definition and Examples

As we have seen, the external direct product provides a way to put groups together into a larger group. It would be quite useful to be able to reverse this process, that is, to be able to start with a large group and break it down into a product of smaller groups. It is occasionally possible to do this. To this end, suppose H and K are subgroups of some group G. We define the set $HK = \{hk \mid h \in H, k \in K\}$.

Example 1 In $U(24) = \{1, 5, 7, 11, 13, 17, 19, 23\}$, let $H = \{1, 17\}$ and $K = \{1, 13\}$. Then, $HK = \{1, 13, 17, 5\}$; since $5 = 17 \cdot 13 \bmod 24$. $\qquad\square$

Example 2 In S_3, let $H = \{(1), (12)\}$ and $K = \{(1), (13)\}$. Then, $HK = \{(1), (13), (12), (12)(13)\} = \{(1), (13), (12), (123)\}$. $\qquad\square$

The student should be careful not to assume that the set HK is a subgroup of G; in Example 1 it is, but in Example 2 it is not.

DEFINITION Internal Direct Product of H and K
Let H and K be subgroups of a group G. We say G is the *internal direct product of H and K* and write $G = H \times K$ if

---

*Weinberg received the 1979 Nobel Price in physics with Sheldon Glashow and Abdus Salam for their construction of a single theory incorporating weak and electromagnetic interactions.

1. $G = HK$
2. $hk = kh$ for all $h \in H$, $k \in K$
3. $H \cap K = \{e\}$.

The reason for the phrase "internal direct product" is easy to justify. We want to call G the internal direct product of H and K if H and K are subgroups of G, and G is naturally isomorphic to the external direct product of H and K. One forms the internal direct product by *starting* with a group G and then proceeding to produce two subgroups H and K within G, such that G is *isomorphic* to the external direct product of H and K. (The three conditions ensure that this is the case—see Theorem 8.1.) On the other hand, one forms an external direct product by *starting* with any two groups H and K, related or not, and proceeding to produce the larger group $H \oplus K$. The difference between the two products is that the internal direct product can be formed within G itself, using subgroups of G and the operation of G, while the external direct product can be formed with totally unrelated groups by creating a new set and a new operation. (See Figures 8.1 and 8.2.)

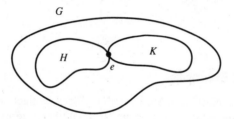

Figure 8.1 For the internal direct product, H and K must be subgroups of the same group.

Perhaps the following analogy with integers is useful in clarifying the distinction between the two products of groups discussed in the preceding paragraph. Just as one may take any (finite) collection of integers and form their product, one may also take any collection of groups and form their external direct product. Conversely, just as one may start with a particular integer and express it as a product of certain of its divisors, one may be able to start with a particular group and factor it as an internal direct product of certain of its subgroups.

Example 3 In D_6, the dihedral group of order 12, let F denote some reflection

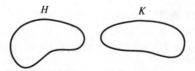

Figure 8.2 For the external direct product, H and K can be any groups.

and let R_k denote a rotation of k degrees. Then,

$$D_6 = \{R_0, R_{120}, R_{240}, F, R_{120}F, R_{240}F\} \times \{R_0, R_{180}\}. \qquad \square$$

Students should be cautioned about the necessity of having all three conditions of the definition of internal direct product satisfied to ensure that $HK \approx H \oplus K$. For example, if we take

$$G = S_3, \qquad H = \langle (123) \rangle \qquad \text{and} \qquad K = \langle (12) \rangle,$$

then $G = HK$, and $H \cap K = \{e\}$. But, G is *not* isomorphic to $H \oplus K$, since, by Theorem 7.2, $H \oplus K$ is cyclic, while S_3 is not. Note that H and K do not satisfy the second condition of the definition.

A group G can also be the internal direct product of a collection of subgroups, as well.

DEFINITION Internal Direct Product of $H_1 \times H_2 \times \cdots \times H_n$
Let $H_1, H_2, \ldots, H_n$ be subgroups of G. We say G is the *internal direct product of $H_1, H_2, \ldots, H_n$* and write $G = H_1 \times H_2 \times \cdots \times H_n$, if

1. $G = H_1H_2 \cdots H_n = \{h_1h_2 \cdots h_n \mid h_i \in H_i\}$
2. $h_ih_j = h_jh_i$ for all $h_i \in H_i$ and $h_j \in H_j$ with $i \neq j$
3. $(H_1H_2 \cdots H_i) \cap H_{i+1} = \{e\}$ for $i = 1, 2, \ldots, n - 1$

This definition is somewhat more complicated than the one given for two subgroups. The student may wonder about the motivation for it; that is, why should we want elements from different components to commute, and why is it desirable for each subgroup to be disjoint from the product of all the others? The reason is quite simple. We want the internal direct product to be isomorphic to the external direct product. As the next theorem shows, the three conditions in the definition of internal direct product were chosen to ensure that this same property holds.

Theorem 8.1 $H_1 \times H_2 \times \cdots \times H_n \approx H_1 \oplus H_2 \oplus \cdots \oplus H_n$
If a group G is the internal direct product of subgroups, $H_1, H_2, \ldots, H_n$, then G is isomorphic to the external direct product of $H_1, H_2, \ldots, H_n$.

Proof. It follows from the definition of internal direct product that each member of G can be expressed in the form $h_1h_2 \cdots h_n$, where $h_i \in H_i$. We claim that each member of G has a unique such representation. To prove this, suppose $g = h_1h_2 \cdots h_n$ and $g = h_1'h_2' \cdots h_n'$, where h_i and h_i' belong to H_i for $i = 1, \ldots, n$. Then, using the fact that the h's from different H_i's commute, we can solve the equation

(*) $$h_1h_2 \cdots h_n = h_1'h_2' \cdots h_n'$$

for $h_n'h_n^{-1}$ to obtain

$$h_n'h_n^{-1} = (h_1')^{-1}h_1(h_2')^{-1}h_2 \cdots (h_{n-1}')^{-1}h_{n-1}.$$

But then,

$$h'_n h_n^{-1} \in H_1 H_2 \cdots H_{n-1} \cap H_n = \{e\}$$

so that $h'_n h_n^{-1} = e$ and, therefore, $h'_n = h_n$. At this point, we can cancel h_n and h'_n from opposite sides of the equal sign in (*) and repeat the preceding argument to obtain $h_{n-1} = h'_{n-1}$. Continuing in this fashion, we eventually have $h_i = h'_i$ for $i = 1, \ldots, n$. With our claim established, we may now define a function ϕ from G to $H_1 \oplus H_2 \oplus \cdots \oplus H_n$ by $(h_1 h_2 \cdots h_n)\phi = (h_1, h_2, \ldots, h_n)$. It is an easy exercise to verify that ϕ is an isomorphism. ∎

The topic of direct products, like that of mappings, is one where notation and terminology vary widely. Many authors use $H \times K$ to denote the internal direct product and the external direct product of H and K, making no notational distinction between the two products. A few authors define only the external direct product. Many people reserve the notation $H \oplus K$ for the situation where H and K are Abelian groups under addition and call it the *direct sum* of H and K. In fact, we will adopt this terminology in the section on ring theory, since rings are always Abelian groups under addition.

The Group of Units Modulo *n* As an Internal and External Direct Product

The U-groups provide a convenient way to illustrate the preceding ideas and to clarify the distinction between internal and external direct products. We first introduce some notation. If k is a divisor of n, let

$$U_k(n) = \{x \in U(n) \mid x = 1 \bmod k\}.$$

For example, $U_7(105) = \{1, 8, 22, 29, 43, 64, 71, 92\}$. It can be readily shown that $U_k(n)$ is indeed a subgroup of $U(n)$. (See exercise 12 of Chapter 3.)

Theorem 8.2   *U(n) as an Internal and External Direct Product*
Suppose s and t are relatively prime. Then $U(st)$ is the internal direct product of $U_s(st)$ and $U_t(st)$, and $U(st)$ is isomorphic to the external direct product of $U(s)$ and $U(t)$. Moreover, $U_s(st)$ is isomorphic to $U(t)$, and $U_t(st)$ is isomorphic to $U(s)$. In short,

$$U(st) = U_s(st) \times U_t(st) \approx U(t) \oplus U(s).$$

The interested student can find a proof of this theorem in [1].

As a consequence of Theorem 8.2, we have the following result.

Corollary   Let $m = n_1 n_2 \cdots n_k$ where $\gcd(n_i, n_j) = 1$ for $i \neq j$. Then,

$$U(m) = U_{m/n_1}(m) \times U_{m/n_2}(m) \times \cdots \times U_{m/n_k}(m)$$
$$\approx U(n_1) \oplus U(n_2) \oplus \cdots \oplus U(n_k).$$

To see how these results work, let's apply them to $U(105)$. We obtain

$$U(105) = U(15 \cdot 7) = U_{15}(105) \times U_7(105)$$
$$= \{1, 16, 31, 46, 61, 76\} \times \{1, 8, 22, 29, 43, 64, 71, 92\}$$
$$\approx U(7) \oplus U(15),$$
$$U(105) = U(5 \cdot 21) = U_5(105) \times U_{21}(105)$$
$$= \{1, 11, 16, 26, 31, 41, 46, 61, 71, 76, 86, 101\} \times \{1, 22, 43, 64\}$$
$$\approx U(21) \oplus U(5),$$
$$U(105) = U(3 \cdot 5 \cdot 7) = U_{35}(105) \times U_{21}(105) \times U_{15}(105)$$
$$= \{1, 71\} \times \{1, 22, 43, 64\} \times \{1, 16, 31, 46, 61, 76\}$$
$$\approx U(3) \oplus U(5) \oplus U(7).$$

Let's use this last example to illustrate Theorem 8.1. Notice that 38 belongs to $U(105)$. Thus, Theorem 8.1 tells us that there must be exactly one element from each of the three subgroups on the right such that their product is 38. Indeed, exhaustion reveals that the only such product is $71 \cdot 43 \cdot 31$.

Among all groups, surely the cyclic groups Z_n have the simplest structure and, at the same time, are the easiest to compute with. Direct products of groups of the form Z_n are only slightly more complicated in structure and computability. Because of this, algebraists endeavor to describe a finite Abelian group as such a direct product. Indeed, we shall soon see that every finite Abelian group can be so represented. With this goal in mind, let us reexamine the U-groups. Using the corollary to Theorem 8.2 and the fact (see [2, p. 93]), first proved by Carl Gauss in 1801, that

$$U(2) \approx \{1\}, \qquad U(4) \approx Z_2, \qquad U(2^n) \approx Z_2 \oplus Z_{2^{n-2}} \qquad \text{for} \qquad n \geq 3,$$

and

$$U(p^n) \approx Z_{p^n - p^{n-1}} \qquad \text{for } p \text{ an odd prime,}$$

we now can write U-groups as an external direct product of cyclic groups. For example,

$$U(105) = U(3 \cdot 5 \cdot 7) \approx U(3) \oplus U(5) \oplus U(7)$$
$$\approx Z_2 \oplus Z_4 \oplus Z_6$$

and

$$U(720) = U(16 \cdot 9 \cdot 5) \approx U(16) \oplus U(9)$$
$$\approx Z_2 \oplus Z_4 \oplus Z_6 \oplus Z_4.$$

What is the advantage of expressing a group in this form? Well, for one thing, we immediately see that the orders of the elements $U(720)$ can only be 1, 2, 3, 4, 6, or 12. For another thing, we can readily determine how many elements of order 12, say, that $U(720)$ has. Because $U(720)$ is isomorphic to $Z_2 \oplus Z_4 \oplus Z_6 \oplus Z_4$, it suffices to calculate the number of elements of order 12 in $Z_2 \oplus Z_4 \oplus Z_6 \oplus Z_4$, and this is easy. By Theorem 7.1 an element (a, b, c, d) has order 12 if and only if $|c| = 3$ or 6, and $|b| = 4$ or $|c| = 3$ or 6, and $|d| = 4$. So, in the first case we may take c to be 1, 2, 4, or 5 and b to be 1 or 3. Thus, there are eight choices in all for b and c. Once these are made, we may choose a and d arbitrarily. This is eight more choices. Thus, there are 64 elements of the form (a, b, c, d) where $|c| = 3$ or 6 and $|b| = 4$.

In addition to these 64, we must include elements of the form (a, b, c, d) where $|c| = 3$ or 6, $|d| = 4$, $|b| = 1$ or 2, and a is arbitrary. Here there are $4 \cdot 2 \cdot 2 \cdot 2 = 32$ choices. So, altogether, $U(720)$ has 96 elements of order 12. These calculations tell us more. Since $\text{Aut}(Z_{720})$ is isomorphic to $U(720)$, we also know that there are 96 automorphisms of Z_{720} of order 12. Imagine trying to deduce this information directly from $U(720)$ or, worse yet, from $\text{Aut}(Z_{720})$! These results beautifully illustrate the advantage of being able to represent a finite Abelian group as a direct product of cyclic groups. They also illustrate the value of our theorems about $\text{Aut}(Z_n)$ and $U(n)$. After all, theorems are just labor-saving devices. If you want to convince yourself of this, try to prove directly from the definitions that $\text{Aut}(Z_{720})$ has exactly 96 elements of order 12.

We conclude this chapter with an application of the preceeding results to number theory.

Example 4 We determine the last two digits of 49^{111}. First, observe that the value we seek is simply 49^{111} mod 100 and that 49 belongs to $U(100)$. Since $U(100) \approx U(4) \oplus U(25) \approx Z_2 \oplus Z_{20}$, we see that $x^{20} = 1$ for all x in $U(100)$. Thus, modulo 100, $49^{111} = (49^{20})^5 49^{11} = 49^{11} = (7^2)^{11} = 7^{22} = 7^{20} \cdot 7^2 = 49$.   □

EXERCISES

The pursuit of an idea is as exciting as the pursuit of a whale.

Henry Norris Russell

1. Let $\mathbf{R}^{\#}$ denote the group of all nonzero real numbers under multiplication. Let $\mathbf{R}^+$ denote the group of positive real numbers under multiplication. Prove that $\mathbf{R}^{\#}$ is the internal direct product of $\mathbf{R}^+$ and the subgroup $\{1, -1\}$.

2. Recall that if H and K are subgroups of a group G, then $HK = \{hk \mid h \in H, k \in K\}$. In $U(24)$, determine the set $U_4(24)U_6(24)$. Is it the same as $U(24)$? Is the product direct? Why?

3. Find the set $U_8(40)U_5(40)$ by multiplying everything out. Is it the same as $U(40)$? Should it be? Why?

4. Suppose s and t are relatively prime. What mapping would you use to prove that $U_s(st)$ is isomorphic to $U(t)$?

5. Is $U(20)$ the internal direct product of $U_4(20)$ and $U_{10}(20)$? Does this contradict Theorem 8.2?

6. Express $U(165)$ as an internal direct product of proper subgroups in three different ways.

7. Express $U(165)$ as an external direct product of U-groups in three different ways.

8. Express $U(165)$ as an external direct product of cyclic additive groups of the form Z_n.

9. Prove that D_4 cannot be expressed as an internal direct product of two proper subgroups.

10. According to Theorem 8.2, $U(105)$ is the direct product of $U_{15}(105)$ and $U_7(105)$. Find elements a in $U_{15}(105)$ and b in $U_7(105)$ so that $38 = ab$. Without trying all possibilities, explain why a and b are uniquely determined by 38.

11. Without doing any calculating in $U(27)$, decide how many subgroups $U(27)$ has.

12. Without doing any calculations in $\text{Aut}(Z_{20})$, determine how many elements of $\text{Aut}(Z_{20})$ have order 4. How many have order 2?

13. Without doing any calculations in $\text{Aut}(Z_{720})$, determine how many elements of $\text{Aut}(Z_{720})$ have order 6.

14. Without doing any calculations in $\text{Aut}(Z_{50})$, prove that $\text{Aut}(Z_{50})$ is cyclic.

15. Let H and K be subgroups of a group G. If $G = HK$ and $g = hk$ where $h \in H$ and $k \in K$, is there any relationship between $|g|$ and $|h|$ and $|k|$? What if $G = H \times K$?

16. Verify that the mapping ϕ, defined in the proof of Theorem 8.1, is an isomorphism.

17. What is the largest order of any element in $U(900)$?

18. Let p and q be odd primes and m and n be positive integers. Explain why $U(p^m) \oplus U(q^n)$ is not cyclic.

19. Use the results presented in the last section of this chapter to prove that $U(55)$ is isomorphic to $U(75)$.

20. Use the results presented in the last section of this chapter to prove that $U(144)$ is isomorphic to $U(140)$.

21. For every $n > 2$, prove that $U(n)^2 = \{x^2 \mid x \in U(n)\}$ is a proper subgroup of $U(n)$.

22. Show that $U(55)^3 = \{x^3 \mid x \in U(55)\}$ is $U(55)$.

23. Let $G = \{3^a 6^b 10^c \mid a, b, c, \in Z\}$ under multiplication and $H = \{3^a 6^b 12^c \mid a, b, c, \in Z\}$ under multiplication. Prove that $G = \langle 3 \rangle \times \langle 6 \rangle \times \langle 10 \rangle$, while $H \neq \langle 3 \rangle \times \langle 6 \rangle \times \langle 12 \rangle$.

24. Show that no U-group has order 14.

25. Show that there is a U-group containing a subgroup isomorphic to Z_{14}.

26. Show that no U-group is isomorphic to $Z_4 \oplus Z_4$.

27. Show that there is a U-group containing a subgroup isomorphic to $Z_4 \oplus Z_4$.

28. Show that there is a U-group containing a subgroup isomorphic to $Z_3 \oplus Z_3$.

29. In Z, let $H = \langle 4 \rangle$ and $K = \langle 10 \rangle$. Express HK in the form $\langle g \rangle$. Generalize to the case where $H = \langle a \rangle$ and $K = \langle b \rangle$.

30. In Z, let $H = \langle 5 \rangle$ and $K = \langle 7 \rangle$. Prove that $Z = HK$. Is $Z = H \times K$?

31. Express $\text{Aut}(Z_2 \oplus Z_3 \oplus Z_5)$ as an external direct product of groups of the form Z_n.

32. Determine the last two digits of 97^{84}.

33. Let $G = \left\{ \begin{bmatrix} 1 & a & b \\ 0 & 1 & c \\ 0 & 0 & 1 \end{bmatrix} \middle| a, b, c \in Z_3 \right\}, H = \left\{ \begin{bmatrix} 1 & a & 0 \\ 0 & 1 & 0 \\ 0 & 0 & 1 \end{bmatrix} \middle| a \in Z_3 \right\},$

$K = \left\{ \begin{bmatrix} 1 & 0 & b \\ 0 & 1 & 0 \\ 0 & 0 & 1 \end{bmatrix} \middle| b \in Z_3 \right\}, L = \left\{ \begin{bmatrix} 1 & 0 & 0 \\ 0 & 1 & c \\ 0 & 0 & 1 \end{bmatrix} \middle| c \in Z_3 \right\}.$ Prove that

$G = HKL$. Is $G = H \times K \times L$?

PROGRAMMING EXERCISES

The useful and beautiful are never separated.

Periander

1. Write a program that will print out $U(n)$ as an internal direct product, as indicated in Theorem 8.2. Assume the relatively prime factors of n are given by the user. Test your program for $105 = 15 \cdot 7$, $105 = 3 \cdot 5 \cdot 7$, $105 = 35 \cdot 3$, $72 = 9 \cdot 8$, $360 = 9 \cdot 40$, $360 = 45 \cdot 8$.

2. Let $U(n)^k = \{x^k \mid x \in U(n)\}$. Write a program that computes $U(n)^k$. Make reasonable assumptions about n and k. Run your program for $(n, k) = (45, 3)$, $(65, 3)$, $(144, 3)$, $(200, 2)$, and $(200, 5)$. Do you see a relationship between $|U(n)|$ and k? Use the theory developed in this chapter to analyze these groups as external direct products of cyclic groups of the form Z_n.

3. Write a program that will print the elements of $U(st)^k$ and $U(s)^k \oplus U(t)^k$, where s and t are relatively prime positive integers greater than 1. (From Theorem 8.2 and exercise 31 of Chapter 7, these groups are isomorphic). Run your program for $(s, t, k) = (9, 8, 2)$, $(9, 8, 3)$, $(9, 8, 4)$, $(5, 11, 2)$, $(5, 11, 3)$, $(5, 11, 4)$, $(5, 11, 5)$, $(20, 7, 2)$, $(20, 7, 4)$, and $(20, 7, 5)$.

4. Implement the algorithm given on page 123 to express $U(n)$ as an external direct product of groups of the form Z_k. Assume n is given in prime-power factorization form. Run your program for $3 \cdot 5 \cdot 7$, $16 \cdot 9 \cdot 5$, $8 \cdot 3 \cdot 25$, $9 \cdot 5 \cdot 11$, $2 \cdot 27 \cdot 125$.

REFERENCES

1. J. A. Gallian and D. Rusin, "Factoring Groups of Integers Modulo *n*," *Mathematics Magazine* 53 (1980): 33–36.

2. D. Shanks, *Solved and Unsolved Problems in Number Theory,* 2d ed., New York: Chelsea, 1978.

SUGGESTED READINGS

Richard W. Ball, "On the Order of an Element in a Group," *American Mathematical Monthly* 71 (1964): 784–785.

A number of results about the orders of elements of a group are given.

A. J. Pettofrezzo and D. R. Byrkit, *Elements of Number Theory,* Englewood Cliffs, N.J.: Prentice-Hall, 1970.

Section 2.14 (pages 81–82) of this book gives an algorithm for solving the equation $|U(x)| = m$.

SUPPLEMENTARY EXERCISES for Chapters 5–8

> My mind rebels at stagnation. Give me problems, give me work, give me the
> most obstruse cryptogram, or the most intricate analysis, and I am in my own
> proper atmosphere.
>
> Sherlock Holmes, *The Sign of the Four*

1. A subgroup N of a group G is called a *characteristic subgroup* if $N\phi = N$ for all automorphisms ϕ of G. (The term *characteristic* was coined by G. Frobenius in 1895.) Prove that every subgroup of a cyclic group is characteristic.

2. Prove that the center of a group is characteristic.

3. The commutator subgroup of G' of a group G is the subgroup generated by the set $\{x^{-1}y^{-1}xy \mid x, y \in G\}$. (That is, every element of G' has the form $a_1^{i_1}a_2^{i_2} \cdots a_k^{i_k}$ where each a_j has the form $x^{-1}y^{-1}xy$, each $i_j = \pm 1$, and k is any positive integer.) Prove that G' is a characteristic subgroup of G. This subgroup was first introduced by G. A. Miller in 1898.

4. Prove that the characteristic property is transitive. That is, if N is a characteristic subgroup of K and K is a characteristic subgroup of G, then N is a characteristic subgroup of G.

5. Let $G = Z_3 \oplus Z_3 \oplus Z_3$ and H be the subgroup of $SL(2, Z_3)$, consisting of

$$\left\{ \begin{bmatrix} 1 & a & b \\ 0 & 1 & c \\ 0 & 0 & 1 \end{bmatrix} \;\middle|\; a, b, c \in Z_3 \right\}.$$

(See exercise 40 of Chapter 2 for the definition of multiplication.) Determine the number of elements of each order in G and H. Are G and H isomorphic?

6. Let H and K be subgroups of G. Prove that HK is a group if and only if $HK = KH$.

7. Let H and K be subgroups of a finite group G. Prove that

$$|HK| = \frac{|H|\,|K|}{|H \cap K|}.$$

(This exercise is referred to in Chapter 26.)

8. The *exponent* of a group is the smallest positive integer n such that $x^n = e$ for all x in the group. Prove that every finite Abelian group has an exponent that divides the order of the group.

9. Determine all U-groups of exponent 2.

10. Can a group have more subgroups than it has elements?

11. Let $\mathbf{R}^+$ denote the multiplicative group of positive reals and $T = \{z \in \mathbf{C} \mid |z|$

$= 1\}$ be the multiplicative group of complex numbers of norm 1. (Recall $|a + bi| = \sqrt{a^2 + b^2}$.) Prove $\mathbf{C}^{\#} = \mathbf{R}^{+} \times T$. T is called the *circle group*.

12. Prove that $Q^{\#}$ under multiplication is not isomorphic to $\mathbf{R}^{\#}$ under multiplication.

13. Prove that Q under addition is not isomorphic to $\mathbf{R}$ under addition.

14. Prove that $\mathbf{R}$ under addition is not isomorphic to $\mathbf{R}^{\#}$ under multiplication.

15. Show that Q^{+} (the set of positive rational numbers) under multiplication is not isomorphic to Q under addition.

16. Suppose $G = \{e, x, x^2, y, yx, yx^2\}$ is a non-Abelian group with $|x| = 3$ and $|y| = 2$. Show $xy = yx^2$.

17. If $G = H \times K$ and H and K are Abelian, prove that G is Abelian.

18. Let G be an Abelian group under addition. Let n be a fixed positive integer and let $H = \{(g, ng) \mid g \in G\}$. Show H is a subgroup of $G \oplus G$. When G is the set of real numbers under addition, describe H geometrically.

19. Prove that D_{2m}, where m is an odd integer, can be expressed as in internal direct product of two proper subgroups.

20. Suppose $G = \oplus_{i=1}^{n} G_i$. Prove $Z(G) = \oplus_{i=1}^{n} Z(G_i)$.

21. Give an example of a group G with a proper subgroup H such that G and H are isomorphic.

22. What is the order of the largest cyclic subgroup in $\text{Aut}(Z_{720})$? (Hint: It is not necessary to consider automorphisms of Z_{720}.)

23. Let G be a nontrivial finite group. Show that G has an element of prime order.

24. Let G be a group and $g \in G$. Show $Z(G)\langle g \rangle$ is a subgroup of G.

25. Show that $D_{11} \oplus Z_3 \neq D_3 \oplus Z_{11}$. (This exercise is referred to in Chapter 26.)

26. Show that $D_{33} \neq D_{11} \oplus Z_3$. (This exercise is referred to in Chapter 26.)

27. Show that $D_{33} \neq D_3 \oplus Z_{11}$. (This exercise is referred to in Chapter 26.)

28. Exhibit four nonisomorphic groups of order 66. (This exercise is referred to in Chapter 26.)

29. Prove that $|\text{Inn}(G)| = 1$ if and only if G is Abelian.

Cosets and Lagrange's Theorem

9

The next theorem, attributed to J. L. Lagrange, is of fundamental importance for it introduces arithmetic relationships into group theory.

Richard A. Dean, *Elements of Abstract Algebra*

Properties of Cosets

In this chapter we will prove the single most important theorem in finite group theory—Lagrange's theorem. But first, we introduce a new and powerful tool for analyzing a group—the notion of a coset. This notion was invented by Galois in 1830, although the term was coined by G. A. Miller in 1910.

> *DEFINITION*   Coset of H in G
>
> Let G be a group and H be a subgroup of G. For any $a \in G$, the set $aH = \{ah \mid h \in H\}$ is called the *left coset of H in G containing a*. Analogously, $Ha = \{ha \mid h \in H\}$ is called the *right coset of H in G containing a*.

Example 1   Let $G = S_3$ and $H = \{(1), (13)\}$. Then,

$$(1)H = H,$$
$$(12)H = \{(12), (12)(13)\} = \{(12), (123)\} = (123)H,$$
$$(13)H = \{(13), (1)\} = H,$$
$$(23)H = \{(23), (23)(13)\} = \{(23), (132)\} = (132)H. \qquad \square$$

Example 2   Let $\mathcal{H} = \{R_0, R_{180}\}$ in D_4, the dihedral group of order 8. Then,

$$R_0 \mathcal{H} = \mathcal{H},$$
$$R_{90} \mathcal{H} = \{R_{90}, R_{270}\} = R_{270} \mathcal{H},$$
$$R_{180} \mathcal{H} = \{R_{180}, R_0\} = \mathcal{H},$$
$$V \mathcal{H} = \{V, VR_{180}\} = \{V, H\} = H \mathcal{H},$$
$$D \mathcal{H} = \{D, D'\} = D' \mathcal{H}.$$

☐

Example 3   Let $H = \{0, 3, 6\}$ in Z_9 under addition. In the case that the group operation is addition, we use the notation $a + H$ instead of aH. Then the cosets of H in Z_9 are

$$0 + H = \{0, 3, 6\} = 3 + H = 6 + H,$$
$$1 + H = \{1, 4, 7\} = 4 + H = 7 + H,$$
$$2 + H = \{2, 5, 8\} = 5 + H = 8 + H.$$

☐

These three examples illustrate a few facts about cosets that are worthy of our attention. First, cosets are usually not subgroups. Second, aH may be the same as bH, even though a is not the same as b. Third, since $(12)H = \{(12), (123)\}$ while $H(12) = \{(12), (132)\}$, aH need not be the same as Ha.

These examples and observations raise many questions. When does $aH = bH$? Do aH and bH have any elements in common? When does $aH = Ha$? Which cosets are subgroups? Why are cosets important? The next lemma and theorem answer these questions.

Lemma   *Properties of Cosets*
Let H be a subgroup of G, and let a and b belong to G. Then,

1. $a \in aH$,
2. $aH = H$ *if and only if* $a \in H$,
3. $aH = bH$ *or* $aH \cap bH = \varnothing$,
4. $aH = bH$ *if and only if* $a^{-1}b \in H$,
5. $|aH| = |bH|$,
6. $aH = Ha$ *if and only if* $H = a^{-1}Ha$,
7. aH *is a subgroup of G if and only if* $a \in H$.

Proof.
1. $a = ae \in aH$.
2. Part 2 is left as an exercise.
3. To prove part 3, we suppose that $aH \cap bH \neq \varnothing$ and prove $aH = bH$. Let $x \in aH \cap bH$. Then there exist h_1, h_2, in H such that $x = ah_1$ and $x = bh_2$. Thus, $a = bh_2h_1^{-1}$ and $aH = bh_2h_1^{-1} H = bH$, by part 2.
4. Observe that $aH = bH$ if and only if $H = a^{-1}bH$. The result now follows from part 2.
5. We leave it as an exercise for the student to prove the correspondence $ah \rightarrow bh$ for all h in H is a one-to-one, onto function from aH to bH.

6. Note that $aH = Ha$ if and only if $a^{-1}(aH) = a^{-1}(Ha)$, that is, if and only if $H = a^{-1}Ha$.

7. If aH is a subgroup, then it contains the identity e. Thus, $aH \cap eH \neq \emptyset$; and, by part 3, we have $aH = eH = H$. Thus, from part 2, we have $a \in H$. Conversely, if $a \in H$, then again by part 2, $aH = H$. ∎

Although most mathematical theorems are written in symbolic form, one should also know *in words,* what they say. In the preceding lemma, part 1 says simply that the left coset of H containing a *does* contain a. Part 2 says that the H "absorbs" an element if and only if the element belongs to H. Part 3 says—and this is very important—that two left cosets are either identical or disjoint. Thus, a left coset is uniquely determined by any one of its elements. Part 4 shows how we may transfer a question about equality of left cosets of H to a question about H itself and vice versa. Part 5 says that all cosets have the same size. Part 6 is analogous to part 4 in that it shows how a question about the equality of the left and right cosets of H containing a is equivalent to a question about the equality of two subgroups of H. The last part of the lemma says that H itself is the only coset of H that is a subgroup of G. We also remark that the properties in the lemma hold just as well for right cosets.

Note that parts 1, 3, and 5 of the lemma guarantee that the left cosets of a subgroup H of G partition G into equal size blocks. In practice, the subgroup H is often chosen so that the cosets partition the group in some highly desirable fashion. For example, if G is $\mathbf{R}^3$ (i.e., $\mathbf{R} \oplus \mathbf{R} \oplus \mathbf{R}$) and H is a plane through the origin, then the coset $(a, b, c) + H$ is the plane passing through the point (a, b, c) and parallel to H. Thus, the cosets of H constitute a partition of 3-space into planes parallel to H. If $G = GL(2, \mathbf{R})$ and $H = SL(2, \mathbf{R})$, then for any matrix A in G, the coset AH is the set of *all* 2×2 matrices with the same determinant as A. Thus,

$$\begin{bmatrix} 2 & 0 \\ 0 & 1 \end{bmatrix} H \qquad \text{is the set of all } 2 \times 2 \text{ matrices of determinant 2}$$

and

$$\begin{bmatrix} 1 & 2 \\ 2 & 1 \end{bmatrix} H \qquad \text{is the set of all } 2 \times 2 \text{ matrices of determinant } -3.$$

Lagrange's Theorem and Consequences

We are now ready to prove a theorem that has been around for over 200 years—longer than group theory itself! (This theorem was not originally stated in group theoretic terms.) At this stage, it should come as no surprise.

Theorem 9.1 *Lagrange's Theorem** $|H|$ *Divides* $|G|$
If G is a finite group and H is a subgroup of G, then $|H|$ *divides* $|G|$. *Moreover, the number of distinct left (right) cosets of H in G is* $|G|/|H|$.

Proof. Let $a_1H, a_2H, \ldots, a_rH$ denote the distinct left cosets of H in G. Then, for each a in G, we have $aH = a_iH$ for some i. Also, by part 1 of the lemma, $a \in aH$. Thus, each member of G belongs to one of the cosets a_iH. In symbols,

$$G = a_1H \cup a_2H \cup \cdots \cup a_rH.$$

Now, part 3 of the lemma shows that this union is disjoint so that

$$|G| = |a_1H| + |a_2H| + \cdots + |a_rH|.$$

Finally, since $|a_iH| = |H|$ for each i, we have $|G| = r|H|$. ∎

We pause to emphasize that Lagrange's theorem is a candidate criterion; that is, it provides a list of candidates for the orders of the subgroups of a group. Thus, a group of order 12 may have subgroups of orders 12, 6, 4, 3, 2, 1 but no others. *Warning!* The converse of Lagrange's theorem is false. For example, a group of order 12 need not have a subgroup of order 6. We prove this in the next chapter.

A special name and notation have been adopted for the number of left (or right) cosets of a subgroup in a group. The *index* of a subgroup H in G is the number of left cosets of H in G. This number is denoted by $|G{:}H|$. When G is finite, Lagrange's Theorem tells us that $|G{:}H| = |G|/|H|$.

Corollary 1 $|a|$ *Divides* $|G|$
In a finite group, the order of each element of the group divides the order of the group.

Proof. Recall that the order of an element is the order of the subgroup generated by that element. ∎

Corollary 2 *Groups of Prime Order Are Cyclic*
A group of prime order is cyclic.

Proof. Suppose G has prime order. Let $a \in G$ and $a \neq e$. Then, $|\langle a \rangle|$ divides $|G|$ and $|\langle a \rangle| \neq 1$. Thus, $|\langle a \rangle| = |G|$ and the corollary follows. ∎

Corollary 3 $a^{|G|} = e$
Let G be a finite group, and let $a \in G$. *Then,* $a^{|G|} = e$.

Proof. By Corollary 1, $|G| = |a|k$. Thus, $a^{|G|} = a^{|a|k} = e^k = e$. ∎

---

*Lagrange stated his version of this theorem in 1770, but the first complete proof is due to Pietro Abbati some 30 years later.

Corollary 4   *Fermat's Little Theorem*
For every integer a and every prime p, a^p modulo $p = a$ modulo p.

Proof. By the division algorithm, $a = pm + r$ where $0 \leqslant r < p$. Thus, $a = r$ modulo p, and it suffices to prove $r^p = r$ modulo p. If $r = 0$, the result is trivial, so we may assume $r \in U(p)$. (Recall, $U(p) = \{1, 2, \ldots, p - 1\}$ under multiplication modulo p.) Then, by the previous corollary, $r^{p-1} = 1$ and, therefore, $r^p = r$.   ∎

Fermat's Little Theorem has been used in conjunction with computers to test for primality of certain numbers. One case concerned the number $p = 2^{257} - 1$. If p is a prime, then we know from Fermat's Little Theorem that $10^p = 10$ mod p and, therefore, $10^{p+1} = 100$ mod p. Using multiple precision and a simple loop, a computer was able to calculate $10^{p+1} = 10^{2^{257}}$ mod p in a few seconds. The result was not 100, and so p is not prime.

As an immediate consequence of the above corollaries, we may classify all groups of order at most 7.

Example 4   Classification of Groups of Order at Most 7
The only groups (up to isomorphism) of order at most 7 are Z_1, Z_2, Z_3, Z_4, $Z_2 \oplus Z_2$, Z_5, Z_6, D_3, and Z_7. Clearly, any group order 1 is isomorphic to Z_1. Now, Corollary 2 shows that we need discuss only groups of order 4 and 6. If G is a noncyclic group of order 4, then, by Corollary 1, for any x in G, we have $x^2 = e$. Thus, by exercise 38 of Chapter 2, G is Abelian. Proving that G is isomorphic to $Z_2 \oplus Z_2$ is an easy exercise (exercise 32 of Chapter 7).

Now let G be a group of order 6. We claim that G has an element of order 3. By Corollary 1 of Lagrange's Theorem, any nonidentity element of G has order 2, 3, or 6. If there is an element a of order 6, then $|a^2| = 3$. Thus, to verify our claim, we may assume that every nonidentity element of G has order 2. In this case, exercise 38 of Chapter 2 says that G is Abelian. Let a, b belong to G. Then $\{e, a, b, ab\}$ is a subgroup of order 4 of a group of order 6. Since this is impossible, G must have an element of order 3; call it x. Now, let $y \in G$, but $y \notin \langle x \rangle$; then, $|y| = 2$, for if $|y| = 3$, then $\langle x \rangle$, $y\langle x \rangle$, and $y^2\langle x \rangle$ are distinct cosets containing nine elements. It follows that $G = \{e, x, x^2, y, yx, yx^2\}$. If G is Abelian, then $|xy| = 6$ and G is isomorphic to Z_6. If G is non-Abelian, we will show that there is only one possible way to define multiplication for G. This will prove that there is only one non-Abelian group of order 6 (up to isomorphism).

To this end, consider the product xy. Which of the six elements of G is it? Well, since $y \notin \langle x \rangle$, we see that xy cannot be equal to e, x, or x^2; cancellation shows that $xy \neq y$; and the fact that G is non-Abelian shows that $xy \neq yx$. Thus, $xy = yx^2$. But, this relation completely determines the multiplication table for G. For example, consider the row headed by x. We know $x \cdot e = x$, $x \cdot x = x^2$, $x \cdot x^2 = e$, $xy = yx^2$, $x(yx) = (xy)x = (yx^2)x = y$, and $x(yx^2) = (xy)x^2 =$

$(yx^2)x^2 = yx$. The rest of the multiplication table can be completed in a similar fashion (see exercise 44). □

An Application of Cosets to Permutation Groups

Lagrange's Theorem and its corollaries dramatically demonstrate the fruitfulness of the coset concept. We next consider an application of cosets to permutation groups.

DEFINITION Stabilizer of a Point

Let G be a group of permutations on the set $\{1, 2, \ldots, n\}$. For any i from 1 to n, let $\text{stab}_G(i) = \{\phi \in G \mid i\phi = i\}$. We call $\text{stab}_G(i)$ the *stabilizer of i in G*. The student should verify that $\text{stab}_G(i)$ is a subgroup of G. (See exercise 27 of Chapter 5.)

DEFINITION Orbit of a Point

Let G be a group of permutations of a set S. For each s in S, let $\text{orb}_G(s) = \{s\phi \mid \phi \in G\}$. The set $\text{orb}_G(s)$ is a subset of S called the *orbit of s under G*.

Example 5 should clarify these two definitions.

Example 5 Let

$G = \{(1), (132)(465)(78), (132)(465), (123)(456), (123)(456)(78), (78)\}$

Then,

$\text{orb}_G(1) = \{1, 3, 2\},$ $\quad$ $\text{stab}_G(1) = \{(1), (78)\},$
$\text{orb}_G(2) = \{2, 1, 3\},$ $\quad$ $\text{stab}_G(2) = \{(1), (78)\},$
$\text{orb}_G(4) = \{4, 6, 5\},$ $\quad$ $\text{stab}_G(4) = \{(1), (78)\},$
$\text{orb}_G(7) = \{7, 8\},$ $\quad$ $\text{stab}_G(7) = \{(1), (132)(465), (123)(456)\}.$ □

Example 6 We may view D_4 as a group of permutations of the points making up a square. Figure 9.1a illustrates the orbit of the point p, and Figure 9.1b illustrates the orbit of the point q under D_4. Observe that $\text{stab}_{D_4}(p) = \{R_0, D\}$, while $\text{stab}_{D_4}(q) = \{R_0\}$. □

The preceding two examples illustrate the following theorem also.

Theorem 9.2 *Orbit-Stabilizer Theorem*
Let G be a finite group of permutations on the set $\{1, 2, \ldots, n\}$. Then, for any i from 1 to n, $|G| = |\text{orb}_G(i)| \, |\text{stab}_G(i)|$.

Proof. By Lagrange's Theorem, $|G|/|\text{stab}_G(i)|$ is the number of distinct right cosets of $\text{stab}_G(i)$ in G. (We use right cosets here because our per-

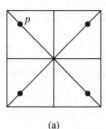
(a)

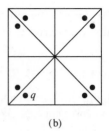
(b)

Figure 9.1

mutations are written on the right.) Thus, it suffices to establish a one-to-one correspondence between the right cosets of $\text{stab}_G(i)$ and the integers in the orbit of i. To do this, we define a correspondence by $\text{stab}_G(i)\phi \xrightarrow{T} i\phi$. To show T is a well-defined function, we must show $\text{stab}_G(i)\alpha = \text{stab}_G(i)\beta$ implies $i\alpha = i\beta$. But, $\text{stab}_G(i)\alpha = \text{stab}_G(i)\beta$ implies $\alpha\beta^{-1} \in \text{stab}_G(i)$ so that $i\alpha\beta^{-1} = i$ and, therefore, $i\alpha = i\beta$. Reading this argument backwards shows that T is also one-to-one. We complete the proof by showing that T is onto $\text{orb}_G(i)$. Let $j \in \text{orb}_G(i)$. Then $j = i\alpha$ for some $\alpha \in G$ and clearly $(\text{stab}_G(i)\alpha)T = i\alpha = j$ so that T is onto. ∎

We leave as an exercise the important fact that the orbits of the elements of a set S under a group partition S (exercise 31).

The Rotation Group of a Cube and a Soccer Ball

It cannot be overemphasized that Theorem 9.2 and Lagrange's Theorem are *counting* theorems.* They enable one to determine the number of elements in various sets. To see how Theorem 9.2 works, we will determine the order of the rotation group of a cube and a soccer ball. That is, we wish to find the number of essentially different ways that we can take a cube or a soccer ball in a certain location in space, physically rotate it, and then return it to its original position in space.

Example 7 Let G be the rotation group of a cube. Label the six faces of the cube 1 through 6. Since any rotation of the cube must carry each face of the cube to exactly one other face of the cube and different rotations induce different permutations of the faces, G can be viewed as a group of permutations on the set $\{1, 2, 3, 4, 5, 6\}$. Clearly, there is some rotation that carries face number 1 to any other face so that $|\text{orb}_G(1)| = 6$. Next, we consider $\text{stab}_G(1)$. Here, we

---

People who don't count, won't count (Anatol France).

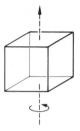

Figure 9.2 Axis of rotation of a cube.

are asking for all rotations of a cube that leave face number 1 where it is. Surely, there are only four such motions—rotations of 0°, 90°, 180°, 270°—about the line perpendicular to the face and passing through its center (see Figure 9.2). Thus, by Theorem 9.2, $|G| = |\text{orb}_G(1)|\,|\text{stab}_G(1)| = 6 \cdot 4 = 24$. $\qquad\square$

Now that we know how many rotations a cube has, it is simple to determine the actual structure of the rotation group of a cube. Recall, S_4 is the symmetric group of degree 4.

Theorem 9.3 *The Rotation of a Cube*
The group of rotations of a cube is isomorphic to S_4.

Proof. Since the group of rotations of a cube has the same order as S_4, we need only prove that the group of rotations is isomorphic to a subgroup of S_4. To this end, observe that a cube has four diagonals and any rotation of the cube must carry a diagonal to a diagonal. Thus, the rotation group induces a group of permutations on the four diagonals. But we must be careful not to assume that different rotations correspond to different permutations. To see that this is so, all we need do is show that all 24 permutations of the diagonals arise from rotations. Labeling the consecutive diagonals 1, 2, 3, and 4, it is obvious that there is a 90° rotation that gives the permutation $\alpha = (1234)$; another 90° rotation perpendicular to our first one gives the permutation $\beta = (1423)$. See Figure 9.3. So, the group of permutations induced by the rotations contains the eight element subgroup $\{\varepsilon, \alpha, \alpha^2, \alpha^3, \beta^2, \beta^2\alpha, \beta^2\alpha^2, \beta^2\alpha^3\}$ and $\alpha\beta$, which has order 3. Clearly, then, the rotations yield all 24 permutations. $\qquad\blacksquare$

Taking a bit of geometry for granted, we can also use Theorem 9.2 to determine the number of rotations of a four-dimensional cube (sometimes called a hypercube). A four-dimensional cube has eight three-dimensional cubes as its faces.* Let G denote the rotation group of a four-dimensional cube, and label one of the eight three-dimensional bounding cubes with the number 1. Analogous to the three-dimensional case, there is a rotation that takes the three-dimensional

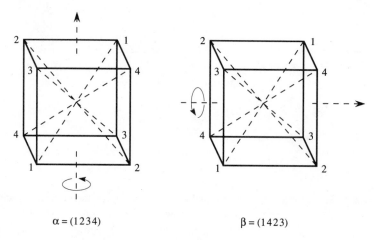

$$\alpha = (1234) \qquad\qquad \beta = (1423)$$

Figure 9.3

cube number 1 to any of the eight three-dimensional cubes. So we have $|\mathrm{orb}_G(1)| = 8$. But $\mathrm{stab}_G(1)$ is the subgroup of G that does not change the location of the three-dimensional cube number 1, and our previous argument shows that there are 24 such rotations. Thus, $|G| = |\mathrm{orb}_G(1)|\,|\mathrm{stab}_G(1)| = 8 \cdot 24 = 192$.

For completeness, we will include the following information about the n-dimensional cube.

dimension	number of vertices	number of $(n-1)$-dimensional faces	order of rotation group
n	2^n	$2 \cdot n$	$n\,!\,2^{n-1}$

Example 8 A soccer ball has 20 faces that are regular hexagons and 12 faces that are regular pentagons. (The technical term for this solid is *truncated icosahedron*.) To determine the number of rotational symmetries of a soccer ball, we simply observe that any pentagon can be carried to any other pentagon by a

---

*You might persuade yourself of this by observing that a two-dimensional "cube"—that is, a square—has $2 \cdot 2$ line segments as its faces; a three-dimensional cube has $2 \cdot 3$ squares as its faces; and, in general, an n-dimensional cube has $2 \cdot n$ cubes of dimension $n - 1$ as its faces. Better yet, read the delightful book *Flatland* [1].

rotation, and there are five rotations that fix (stabilize) the location of any particular pentagon. Thus, by the orbit-stabilizer theorem, there are $12 \cdot 5 = 60$ rotational symmetries. (In case you are interested, the rotation group of a soccer ball is isomorphic to A_5.) $\square$

In nature, the helix is the structure that occurs most often. Running a close second are polyhedrons made from pentagons and hexagons, the dodecahedron and truncated icosahedron being two instances. Although it is impossible to enclose space with hexagons alone, adding 12 pentagons will be sufficient to enclose the space. Many viruses have this kind of structure (see [2]).

EXERCISES

If but the will be firmly bent, no stuff resists the mind's intent.

Oliver St. John Gogarty, *The Image-Maker*

1. Let $H = \{(1), (12)(34), (13)(24), (14)(23)\}$. Find the left cosets of H in A_4 (see Table 5.1 on page 92).

2. Let H be as in exercise 1. How many left cosets of H in S_4 are there? (Do this without listing them.)

3. Let $H = \{0, \pm 3, \pm 6, \pm 9, \ldots\}$. Find all the left cosets of H in Z.

4. Rewrite the condition $a^{-1}b \in H$ given in part 4 of the lemma in additive notation. Assume the group is Abelian.

5. Let H be as in exercise 3. Use the previous exercise to decide whether or not the following cosets of H are the same.
   a. $11 + H$ and $17 + H$
   b. $-1 + H$ and $5 + H$
   c. $7 + H$ and $23 + H$

6. Let n be an integer greater than 1. Let $H = \{0, \pm n, \pm 2n, \pm 3n, \ldots\}$. Find all left cosets of H in Z. How many are there?

7. Find all of the left cosets of $\{1, 11\}$ in $U(30)$.

8. Suppose a has order 15. Find all of the left cosets of $\langle a^5 \rangle$ in $\langle a \rangle$.

9. Let $|a| = 30$. How many left cosets of $\langle a^4 \rangle$ in $\langle a \rangle$ are there? List them.

10. Let G be a group and H be a subgroup of G. Let $a \in G$. Prove that $aH = H$ if and only if $a \in H$.

11. Let G be a group and H a subgroup of G. Let $a, b \in G$. Prove that the number of elements in aH is the same as the number of elements in bH.

12. In $\mathbf{R} \oplus \mathbf{R}$ under componentwise addition, let $H = \{(x, 3x) \mid x \in \mathbf{R}\}$. (Note that H is the subgroup of all points on the line $y = 3x$.) Show that $(2, 5) + H$ is a straight line passing through the point $(2, 5)$ and parallel to the line $y = 3x$.

13. In $\mathbf{R} \oplus \mathbf{R}$, suppose H is the subgroup of all points lying on a line through the origin. Show that any left coset of H is just a line parallel to H.

14. In $\mathbf{R} \oplus \mathbf{R} \oplus \mathbf{R}$, let H be a subgroup of points lying in a plane through the origin. Show that every left coset of H is a plane parallel to H.

15. The set of all solutions of the linear system

$$3x + 2y - 3z = 1$$
$$5x + y + 4z = -3$$

is a coset of some subgroup of $\mathbf{R} \oplus \mathbf{R} \oplus \mathbf{R}$. Describe this subgroup.

16. Let G be the group of nonzero complex numbers under multiplication, and let $H = \{x \in G \mid |x| = 1\}$. (Recall $|a + bi| = \sqrt{a^2 + b^2}$.) Give a geometrical description of the cosets of H.

17. Let G be a group of order 60. What are the possible orders for the subgroups of G?

18. Suppose K is a proper subgroup of H, and H is a proper subgroup of G. If $|K| = 42$ and $|G| = 420$, what are the possible orders of H?

19. Suppose $|G| = pq$ where p and q are prime. Prove that every proper subgroup of G is cyclic.

20. Recall that, for any integer greater than 1, $\phi(n)$ denotes the number of integers less than n and relatively prime to n. Prove that if a is any integer relatively prime to n, then $a^{\phi(n)} = 1$ modulo n.

21. Compute 5^{15} modulo 7 and 7^{13} modulo 11.

22. Use Corollary 1 of Lagrange's Theorem to prove that the order of $U(n)$ is even when $n > 2$.

23. Prove that a non-Abelian group of order 10 must have 5 elements of order 2. Generalize this to the case of a non-Abelian group of order $2p$ where p is prime and $p \neq 2$. How many elements of order 2 does an Abelian group of order $2p$ have when p is prime and $p \neq 2$?

24. Without checking the group axioms, explain why the following table cannot be a group table.

	a	b	c	d	f
a	a	b	c	d	f
b	b	a	d	f	c
c	c	f	a	b	d
d	d	c	f	a	b
f	f	d	b	c	a

25. Find all the left cosets of $\{(0, 1), (1, 2), (2, 4), (3, 3)\}$ in $Z_4 \oplus U(5)$.

26. Suppose G is a group with more than one element and G has no proper, nontrivial subgroups. Prove that $|G|$ is prime. (Do not assume at the outset that G is finite.)

27. Let $|G| = 15$. If G has only one subgroup of order 3 and only one of order 5, prove that G is cyclic. Generalize to $|G| = pq$, where p and q are prime.

28. Let G be a group of order 25. Prove that G is cyclic or $g^5 = e$ for all g in G.

29. Let $|G| = 33$. What are the possible orders for the elements of G? Show that G must have an element of order 3.

30. Let $|G| = 8$. Show that G must have an element of order 2. Show by example that G need not have an element of order 4.

31. Let G be a group of permutations of a set S. Prove that the orbits of the members of S constitute a partition of S. (This exercise is referred to in Chapter 26.)

32. Explain why S_3 is not listed in Example 4 as one of the groups of order at most 7.

33. Let $G = \{(1), (12)(34), (1234)(56), (13)(24), (1432)(56), (56)(13), (14)(23), (24)(56)\}$.
    a. Find the stabilizer of 1 and orbit of 1.
    b. Find the stabilizer of 3 and the orbit of 3.
    c. Find the stabilizer of 5 and the orbit of 5.

34. Suppose G is a finite group of order n and G has exactly 2^{n-1} subgroups. Show that $n = 1$ or 2.

35. Determine the complete subgroup lattice for D_5, the dihedral group of order 10.

36. Prove that a group of order 10 must have an element of order 2.

37. Suppose a group contains elements of orders 1 through 10. What is the minimum possible order of the group?

38. Let G be a finite Abelian group and n a positive integer that is relatively prime to $|G|$. Show that the mapping $a \rightarrow a^n$ is an automorphism of G.

39. Suppose $G = H \times K$, where neither H nor K is trivial. If G is non-Abelian, show that $|G| \geqslant 12$.

40. Let G be the group of plane rotations about a point P in the plane. Thinking of G as a group of permutations of the plane, describe the orbit of a point Q in the plane. (This is the motivation for the name *orbit*.)

41. Let G be the rotation group of a cube. Label the faces of the cube 1 through 6, and let H be the subgroup of elements of G that carry face 1 to itself. If σ is a rotation that carries face 2 to face 1, give a physical description of the coset σH.

42. The group D_4 acts as a group of permutations of the points making up the squares shown on page 142. (The axes of symmetry are drawn for reference purposes.) For each square, locate the points in the orbit of the indicated point under D_4. In each case, determine the stabilizer of the indicated point.

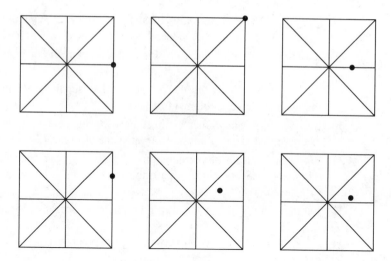

43. Let $G = GL(2, \mathbf{R})$, the group of 2×2 matrices over $\mathbf{R}$ with nonzero determinant. Let H be the subgroup of matrices of determinant ± 1. If a, $b \in G$ and $aH = bH$, what can be said about $\det(a)$ and $\det(b)$? Is the converse true? (Determinants have the property that $\det(xy) = \det(x)\det(y)$.)

44. Complete the multiplication table for the group of order 6 in Example 4.

45. If G is a finite group with fewer than 100 elements and G has subgroups of orders 10 and 25, what is the order of G?

46. Calculate the orders of the following (refer to Figure 29.4(a) for illustrations):
    a. The group of rotations of a regular tetrahedron (a solid with 4 congruent triangles as faces)
    b. The group of rotations of a regular octahedron (a solid with 8 congruent triangles as faces)
    c. The group of rotations of a regular dodecahedron (a solid with 12 congruent pentagons as faces)
    d. The group of rotations of a regular icosahedron (a solid with 20 congruent triangles as faces)*

47. A soccer ball has 20 faces that are regular hexagons and 12 faces that are regular pentagons. Use Theorem 9.2 to explain why a soccer ball cannot have a 60° rotational symmetry about a line through the centers of two opposite hexagonal faces.

48. Determine the number of rotation symmetries each of the following solids have. (The figure below each solid is an "unfolded" version of the solid.)

---

*The icosahedron is the logo of the Mathematical Association of America.

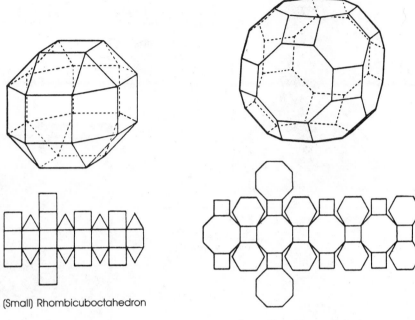

(Small) Rhombicuboctahedron

Great Rhombicuboctahedron or
Truncated Cuboctahedron

REFERENCES

1. E. A. Abbott, *Flatland*, 6th ed., New York: Dover, 1952.
2. John Galloway, "Nature's Second-Favourite Structure," *New Scientist* 114 (April 1987): 36–39.

SUGGESTED FILMS

The Hypercube: Projections and Slicing, Banchoff/Strauss Productions, 12 min., color.

In this award-winning computer-generated film, the four-dimensional cube appears first as a square that rotates about axes to appear as a 3-cube, then about different planes in 4-space to show the full structure of the 4-cube. The same sequence then appears in perspective. Finally the computer calculates and displays the slices of a square by parallel lines, the slices of a 3-cube by three sets of parallel planes, and the slices of a 4-cube by four different sets of parallel hyperplanes.

Symmetries of the Cube with H. S. M. Coxeter and W. O. J. Moser, International Film Bureau, $13\frac{1}{2}$ min., color.

The aim of this film is to exhibit the interesting symmetry properties of the square, cube, and octahedron. The viewer is shown how the cube can be generated by means of three reflections, and that the group of symmetries of a cube has order 48. This is done by showing that an orthoscheme (which is 1/48 of the cube) placed inside of a kaleidoscope made of three mirrors produces the entire cube. The film also demonstrates that the cube and the octahedron have the same symmetry group.

Joseph Lagrange

Lagrange is the Lofty Pyramid of the Mathematical Sciences.

Napoleon Bonaparte

This stamp was issued by France in Lagrange's honor in 1958.

JOSEPH LOUIS LAGRANGE was born in Italy of French ancestry on January 25, 1736. He became captivated by mathematics at an early age when he read an essay by Halley on Newton's calculus. At the age of nineteen he became a professor of mathematics at the Royal Artillery School in Turin. Lagrange made significant contributions to many branches of mathematics and physics, among these the theory of numbers, the theory of equations, ordinary and partial differential equations, the calculus of variations, analytic geometry, fluid dynamics, and celestial mechanics. His methods for solving third- and fourth-degree polynomial equations by radicals laid the groundwork for the group-theoretic approach to solving polynomials taken by Galois. Lagrange was a very careful writer with a clear and elegant style.

At the age of forty Lagrange was appointed Head of the Berlin Academy, succeeding Euler. In offering this appointment, Frederick The Great proclaimed that the "greatest king in Europe" ought to have the "greatest mathematician in Europe" at his court. In 1787, Lagrange was invited to Paris by Louis XVI and became a good friend of the king and his wife, Marie Antoinette. In 1793, Lagrange headed a commission, which included Laplace and Lavoisier, to devise a new system of weights and measures. Out of this came the metric system. Late in his life he was made a count by Napoleon. Lagrange died on April 10, 1813.

Normal Subgroups and Factor Groups

10

It is tribute to the genius of Galois that he recognized that those subgroups for which the left and right cosets coincide are distinguished ones. Very often in mathematics the crucial problem is to recognize and to discover what are the relevant concepts; once this is accomplished the job may be more than half done.

I. N. Herstein, *Topics in Algebra*

Normal Subgroups

As we have seen in Chapter 9, if G is a group and H is a subgroup of G, it is not always true that $aH = Ha$ for all a in G. There are certain situations where this does hold, however, and these cases turn out to be of critical importance in the theory of groups. It was Galois, about 150 years ago, who first recognized that such subgroups were worthy of special attention.

> *DEFINITION*   Normal Subgroup
> A subgroup H of a group G is called a *normal* subgroup of G if $aH = Ha$ for all a in G. We denote this by $H \lhd G$.

There are several equivalent formulations of the definition of normality. We have chosen the one that is the easiest to use in applications. However, to *verify* that a subgroup is normal, it is usually better to use the next theorem. It allows us to substitute a condition about two subgroups of G for a condition about two cosets of G. The proof is left as an exercise.

> *Theorem 10.1*   *Normal Subgroup Test*
> *A subgroup H of G is normal in G if and only if $x^{-1} Hx \subseteq H$ for all x in G.*

Many students make the mistake of thinking that H is normal in G means $ah = ha$ for $a \in G$ and $h \in H$. This is not what normality of H means; rather, it means that if $a \in G$ and $h \in H$, then there exists some $h' \in H$ such that $ah = h'a$.

Example 1   The center $Z(G)$ of a group is always normal.   □

Example 2   The alternating group A_n of even permutations is a normal subgroup of S_n.   □

Example 3   The subgroup of rotations in D_n is normal in D_n.   □

Example 4   The group $SL(2, \mathbf{R})$ of 2×2 matrices with determinant 1 is a normal subgroup of $GL(2, \mathbf{R})$, the group 2×2 matrices with nonzero determinant.   □

Example 5   Referring to the group table for A_4 given in Table 5.1 on page 92 we may observe that $H = \{\alpha_1, \alpha_2, \alpha_3, \alpha_4\}$ is a normal subgroup of A_4 while $K = \{\alpha_1, \alpha_5, \alpha_9\}$ is *not* a normal subgroup of A_4. To see that H is normal simply note that for any β in A_4, $\beta^{-1}H\beta$ is a subgroup of order 4 and H is the only subgroup of A_4 of order 4. Thus, $\beta^{-1}H\beta = H$. In contrast, $\alpha_2^{-1}\alpha_5\alpha_2 = \alpha_7$ so that $\alpha_2^{-1}K\alpha_2 \not\subseteq K$.   □

Examples 2 and 3 are special cases of our next example.

Example 6   If H has only two left cosets in G, then H is normal in G.   □

Factor Groups

We have yet to explain why normal subgroups are of special significance. The reason is simple. When the subgroup H of G is normal, then the set of left (or right) cosets of H in G is itself a group—called the *factor group of G by H* (or the *quotient group of G by H*). Quite often, one can obtain information about a group by studying one of its factor groups. This method will be illustrated in the examples.

> **Theorem 10.2**   (*O. Hölder, 1889) Factor Groups*
> *Let G be a group and H a normal subgroup of G. The set $G/H = \{aH \mid a \in G\}$ is a group under the operation $(aH)(bH) = abH$.**

---

*The notation G/H was first used by C. Jordan.

Proof. Our first task is to show that the operation is well defined; that is, we must show the correspondence defined above from $G/H \times G/H$ into G/H is actually a function. To do this, assume $aH = a'H$ and $bH = b'H$. Then $a' = ah_1$ and $b' = bh_2$ for some h_1, h_2 in H, and therefore, $a'b'H = ah_1bh_2 H = ah_1bH = ah_1Hb = aHb = abH$. Here we have used part 2 of the lemma from Chapter 9 and the fact that $H \lhd G$. The rest is easy: $eH = H$ is the identity; $a^{-1}H$ is the inverse of aH; and $(aHbH)cH = (ab)HcH = (ab)cH = a(bc)H = aH(bc)H = aH(bHcH)$. This proves that G/H is a group. ∎

Although it is merely a curiosity, we point out that the converse of Theorem 10.2 is also true; that is, if the correspondence $aHbH = abH$ defines a group operation on the set of left cosets of H in G, then H is normal in G.

The next few examples illustrate the factor group concept.

Example 7 Let $4Z = \{0, \pm 4, \pm 8, \ldots\}$. To construct $Z/4Z$, we first must determine the left cosets of $4Z$ in Z. Consider the following four cosets:

$$0 + 4Z = 4Z = \{0, \pm4, \pm8, \ldots\},$$
$$1 + 4Z = \{1, 5, 9, \ldots; -3, -7, -11, \ldots\},$$
$$2 + 4Z = \{2, 6, 10, \ldots; -2, -6, -10, \ldots\},$$
$$3 + 4Z = \{3, 7, 11, \ldots; -1, -5, -9, \ldots\}.$$

We claim that there are no others. For if $k \in Z$, then $k = 4q + r$ where $0 \leq r < 4$; and, therefore, $k + 4Z = r + 4q + 4Z = r + 4Z$. Now that we know the elements of the factor group, our next job is to determine the structure of $Z/4Z$. Its Cayley table is

	$0 + 4Z$	$1 + 4Z$	$2 + 4Z$	$3 + 4Z$
$0 + 4Z$	$0 + 4Z$	$1 + 4Z$	$2 + 4Z$	$3 + 4Z$
$1 + 4Z$	$1 + 4Z$	$2 + 4Z$	$3 + 4Z$	$0 + 4Z$
$2 + 4Z$	$2 + 4Z$	$3 + 4Z$	$0 + 4Z$	$1 + 4Z$
$3 + 4Z$	$3 + 4Z$	$0 + 4Z$	$1 + 4Z$	$2 + 4Z$

Clearly, then, $Z/4Z \approx Z_4$. More generally, if for any $n > 0$ we let $nZ = \{0, \pm n, \pm 2n, \pm 3n, \ldots\}$, then Z/nZ is isomorphic to Z_n. ☐

Example 8 Let $G = Z_{18}$ and $H = \langle 6 \rangle = \{0, 6, 12\}$. Then $G/H = \{0 + H, 1 + H, 2 + H, 3 + H, 4 + H, 5 + H\}$. To illustrate how the group elements are combined, consider $(5 + H) + (4 + H)$. This should be one of the six elements listed in the set G/H. Well, $(5 + H) + (4 + H) = 5 + 4 + H = 9 + H = 3 + 6 + H = 3 + H$ since H absorbs all multiples of 6. ☐

A few words of caution about notation are warranted here. When H is a normal subgroup of G, the expression $|aH|$ has two possible interpretations. One could be thinking of aH as a *set* of elements and $|aH|$ as the size of the set; or as is more often the case, one could be thinking of aH as a group element of the factor group G/H and $|aH|$ as the order of the *element aH* in G/H. In Example 8, for instance, the *set* $3 + H$ has order 3 since $3 + H = \{3, 9, 15\}$. But the *group element* $3 + H$ has order 2 since $(3 + H) + (3 + H) = 6 + H = 0 + H$. As is usually the case when one notation has more than one meaning, the appropriate interpretation will be clear from the context.

Example 9 Let $\mathcal{H} = \{R_0, R_{180}\}$, and consider the factor group of the dihedral group D_4 by $\mathcal{H}$,

$$D_4/\mathcal{H} = \{\mathcal{H}, R_{90}\,\mathcal{H}, H\mathcal{H}, D\mathcal{H}\}.$$

The multiplication table for $D_4/\mathcal{H}$ is given in Table 10.1. (Notice that even though $H\mathcal{H}R_{90}\mathcal{H} = D'\mathcal{H}$, we have used $D\mathcal{H}$ in Table 10.1 because $D'\mathcal{H} = D\mathcal{H}$, and we use the same symbols in the table as in the heading for the table.)

Table 10.1

	$\mathcal{H}$	$R_{90}\mathcal{H}$	$H\mathcal{H}$	$D\mathcal{H}$
$\mathcal{H}$	$\mathcal{H}$	$R_{90}\mathcal{H}$	$H\mathcal{H}$	$D\mathcal{H}$
$R_{90}\mathcal{H}$	$R_{90}\mathcal{H}$	$\mathcal{H}$	$D\mathcal{H}$	$H\mathcal{H}$
$H\mathcal{H}$	$H\mathcal{H}$	$D\mathcal{H}$	$\mathcal{H}$	$R_{90}\mathcal{H}$
$D\mathcal{H}$	$D\mathcal{H}$	$H\mathcal{H}$	$R_{90}\mathcal{H}$	$\mathcal{H}$

$D_4/\mathcal{H}$ provides a good opportunity to demonstrate how a factor group of G is related to G itself. Suppose we arrange the heading of the Cayley table for D_4 in such a way that elements from the same coset of $\mathcal{H}$ are in adjacent columns. Then, the multiplication table for D_4 can be blocked off into boxes that are cosets of $\mathcal{H}$, and the substitution that replaces a box containing the element x with the coset $x\mathcal{H}$ yields the Cayley table for $D_4/\mathcal{H}$ (Table 10.2).

Table 10.2

	R_0	R_{180}	R_{90}	R_{270}	H	V	D	D'
R_0	R_0	R_{180}	R_{90}	R_{270}	H	V	D	D'
R_{180}	R_{180}	R_0	R_{270}	R_{90}	V	H	D'	D
R_{90}	R_{90}	R_{270}	R_{180}	R_0	D	D'	V	H
R_{270}	R_{270}	R_{90}	R_0	R_{180}	D'	D	H	V
H	H	V	D'	D	R_0	R_{180}	R_{270}	R_{90}
V	V	H	D	D'	R_{180}	R_0	R_{90}	R_{270}
D	D	D'	H	V	R_{90}	R_{270}	R_0	R_{180}
D'	D'	D	V	H	R_{270}	R_{90}	R_{180}	R_0

Thus, when we pass from D_4 to $D_4/\mathcal{H}$, the box

$$\begin{array}{|cc|} \hline H & V \\ V & H \\ \hline \end{array}$$

in Table 10.2 becomes the element $H\mathcal{H}$ in Table 10.1. Similarly, the box

$$\begin{array}{|cc|} \hline D & D' \\ D' & D \\ \hline \end{array}$$

becomes the element $D\mathcal{H}$, and so on. $\qquad\square$

In this way, one can see that the formation of a factor group G/H causes a systematic collapsing of the elements of G. In particular, all the elements in the coset of H containing a collapse to the single group element aH in G/H.

Example 10 Consider the group A_4 as represented by Table 5.1 on page 92. (Here i denotes the permutation α_i). Let $H = \{1, 2, 3, 4\}$. Then the three cosets of H are H, $5H = \{5, 6, 7, 8\}$, and $9H = \{9, 10, 11, 12\}$. (In this case, rearrangement of the headings is unnecessary.) Blocking off the table for A_4 into boxes that are cosets of H and replacing the boxes with 1, 5, and 9 with the cosets $1H$, $5H$ and $9H$ (see Table 10.3), we obtain the Cayley table for G/H given in Table 10.4.

Table 10.3

1	2	3	4	5	6	7	8	9	10	11	12
2	1	4	3	8	7	6	5	11	12	9	10
3	4	1	2	6	5	8	7	12	11	10	9
4	3	2	1	7	8	5	6	10	9	12	11
5	6	7	8	9	10	11	12	1	2	3	4
6	5	8	7	12	11	10	9	3	4	1	2
7	8	5	6	10	9	12	11	4	3	2	1
8	7	6	5	11	12	9	10	2	1	4	3
9	10	11	12	1	2	3	4	5	6	7	8
10	9	12	11	4	3	2	1	7	8	5	6
11	12	9	10	2	1	4	3	8	7	6	5
12	11	10	9	3	4	1	2	6	5	8	7

Table 10.4

	$1H$	$5H$	$9H$
$1H$	$1H$	$5H$	$9H$
$5H$	$5H$	$9H$	$1H$
$9H$	$9H$	$1H$	$5H$

This procedure can be illustrated more vividly with colors. Let's say we had printed the elements of H in green, the elements of $5H$ in red, and the elements of $9H$ in blue. Then in Table 10.3, each box would consist of elements of a uniform color. We could then think of the factor group as consisting of the three colors that define a group table isomorphic to G/H.

	Green	Red	Blue
Green	Green	Red	Blue
Red	Red	Blue	Green
Blue	Blue	Green	Red

It is instructive to see what happens if we attempt the same procedure with a group G and a subgroup H that is not normal in G—that is, arrange the heading of the Cayley table so that the elements from the same coset of H are in adjacent columns and attempt to block off the table into boxes that are also cosets of H to produce a Cayley table for the set of cosets. Say, for instance, we take G to be A_4 and $H = \{1, 5, 9\}$. The cosets of H are H, $2H = \{2, 8, 11\}$, $3H = \{3, 6, 12\}$, $4H = \{4, 7, 10\}$. Then the first three rows of the rearranged Cayley table for A_4 are

	1	5	9	2	8	11	3	6	12	4	7	10
1	1	5	9	2	8	11	3	6	12	4	7	10
5	5	9	1	6	12	3	7	10	4	8	11	2
9	9	1	5	10	4	7	11	2	8	12	3	6

But already we are in trouble, for blocking these off into 3×3 boxes yields boxes that contain elements of different cosets. Hence, it is impossible to represent an entire box by a single element of the box the way we could for boxes made from the cosets of a normal subgroup. Had we printed the rearranged table in four colors with members of the same coset having the same color, we would see multicolored boxes rather than the uniformly colored boxes produced by a normal subgroup. $\qquad\square$

In Chapter 12 we will prove that every finite Abelian group is isomorphic to a direct product of cyclic groups. In particular, an Abelian group of order 8 is isomorphic to one of Z_8, $Z_4 \oplus Z_2$, or $Z_2 \oplus Z_2 \oplus Z_2$. In the next two examples, we examine Abelian factor groups of order 8 and determine the isomorphism type of each.

Example 11 Let

$$G = U(32) = \{1, 3, 5, 7, 9, 11, 13, 15, 17, 19, 21, 23, 25, 27, 29, 31\}$$

and $H = U_{16}(32) = \{1, 17\}$. Then G/H is an Abelian group of order $16/2 =$

8. Which of the three Abelian groups of order 8 is it, Z_8, $Z_4 \oplus Z_2$, or $Z_2 \oplus Z_2 \oplus Z_2$? To answer this question, we need only determine the elements of G/H and their orders. Observe that the eight cosets

$$1H = \{1, 17\}, \quad 3H = \{3, 19\}, \quad 5H = \{5, 21\}, \quad 7H = \{7, 23\},$$
$$9H = \{9, 25\}, \quad 11H = \{11, 27\}, \quad 13H = \{13, 29\}, \quad 15H = \{15, 31\}$$

are all distinct so they form the factor group G/H. Clearly, $(3H)^2 = 9H \neq H$ so $3H$ has order at least 4. Thus, G/H is not $Z_2 \oplus Z_2 \oplus Z_2$. On the other hand, direct computations show that both $7H$ and $9H$ have order 2 so that G/H cannot be Z_8 either. This proves that $U(32)/U_{16}(32) \approx Z_4 \oplus Z_2$, which (not so incidentally!) is isomorphic to $U(16)$. □

Example 12 Let $G = U(32)$ and $K = \{1, 15\}$. Then $|G/K| = 8$, and we ask which of the three Abelian groups of order 8 is G/K? Since $(3K)^4 = 81K = 17K \neq K$, $|3K| = 8$. Thus, $G/K \approx Z_8$. □

It is crucial to understand that when one factors out by a normal subgroup H, what one is really doing is defining every element in H to be the *identity*. Thus, in Example 9, we are making $R_{180} \mathcal{H} = \mathcal{H}$ the identity. Likewise $R_{270} \mathcal{H} = R_{90} R_{180} \mathcal{H} = R_{90} \mathcal{H}$. Similarly, in Example 7, we are declaring any multiple of 4 to be 0 in the factor group $Z/4Z$. This is why $5 + 4Z = 1 + 4 + 4Z = 1 + 4Z$, and so on. In Example 11, we have $3H = 19H$, since $19 = 3 \cdot 17$ in $U(32)$ and going to the factor group makes 17 the identity. Algebraists often refer to the process of creating the factor group G/H as "killing" H.

Applications of Factor Groups

Why are factor groups important? Well, when G is finite and $H \neq \{e\}$, G/H is smaller than G, and its structure is usually less complicated than that of G. At the same time, G/H simulates G in many ways. In fact, we may think of a factor group of G as a less complicated approximation of G (similar to using the rational number 3.14 for the irrational number π). What makes factor groups important is that one can often deduce properties of G by examining the less complicated group G/H instead. An excellent illustration of this is given in the next example.

Example 13 Converse of LaGrange's Theorem Is False
The group A_4 of even permutations on the set $\{1, 2, 3, 4\}$ has no subgroup H of order 6. (Thus, A_4 is a counterexample to the converse of Lagrange's Theorem.*)

---

*The first counterexample to the converse of Lagrange's Theorem was given by Paolo Ruffini in 1799.

To see this, suppose that A_4 does have a subgroup H of order 6. By Example 6, we know $H \lhd A_4$. Thus, the factor group A_4/H exists and has order 2. Since the order of an element divides the order of the group, we have $(\alpha H)^2 = \alpha^2 H = H$ for all $\alpha \in A_4$. Thus, $\alpha^2 \in H$ for all α in A_4. Referring to the main diagonal of the group table for A_4 given in Table 5.1 on page 92, however, we observe that A_4 has nine different elements of the form α^2, all of which must belong to H, a subgroup of order 6. This is clearly impossible, so a subgroup of order 6 cannot exist in A_4.* ☐

We conclude this chapter with two additional examples to illustrate how knowledge of a factor group of G reveals information about G itself.

Theorem 10.3   The G/Z Theorem
Let G be a group and $Z(G)$ the center of G. If $G/Z(G)$ is cyclic, then G is Abelian.

Proof. Let $gZ(G)$ be a generator of the factor group $G/Z(G)$, and let a, $b \in G$. Then there exist integers i and j such that

$$aZ(G) = (gZ(G))^i = g^i Z(G)$$

and

$$bZ(G) = (gZ(G))^j = g^j Z(G).$$

Thus, $a = g^i x$ for some x in $Z(G)$ and $b = g^j y$ for some y in $Z(G)$. It follows then that

$$ab = (g^i x)(g^j y) = g^i(xg^j)y = g^i(g^j x)y$$
$$= (g^i g^j)(xy) = (g^j g^i)(yx) = (g^j y)(g^i x) = ba. ∎$$

A few remarks about Theorem 10.3 are in order. First, our proof shows that a better result is possible: if G/H is cyclic where H is a subgroup of $Z(G)$, then G is Abelian. Second, in practice, it is the contrapositive of the theorem that is most often used: that is, if G is non-Abelian, then $G/Z(G)$ is not cyclic. For example, it follows immediately from this statement and Lagrange's Theorem that a non-Abelian group of order pq where p and q are primes must have a trivial center.

Theorem 10.4 $G/Z(G) \approx \text{Inn}(G)$
For any group G, $G/Z(G)$ is isomorphic to $\text{Inn}(G)$.

Proof. Consider the correspondence from $G/Z(G)$ to $\text{Inn}(G)$ given by T: $Z(G)g \to \phi_g$ (where, recall, $x\phi_g = g^{-1}xg$ for all x in G). First, we show

---

*How often have I said to you that when you have eliminated the impossible, whatever remains, however improbable, must be the truth. Sir Arthur Conan Doyle.

that T is a function. Suppose $Z(G)g = Z(G)h$ so that $hg^{-1} \in Z(G)$. Then, for all x in G, $x = (hg^{-1})^{-1}xhg^{-1} = gh^{-1}xhg^{-1}$. Thus, $g^{-1}xg = h^{-1}xh$ for all x in G, and, therefore, $\phi_g = \phi_h$. Reversing this argument shows that T is one-to-one, as well. Clearly, T is onto.

That T is operation-preserving follows directly from the fact that $\phi_g\phi_h = \phi_{gh}$ for all g and h in G.   ∎

As an application of Theorems 10.3 and 10.4, we may easily determine Inn(D_6) without looking at Inn(D_6)!

Example 14   We know from Example 11 of Chapter 3 that $|Z(D_6)| = 2$. Thus, $|D_6/Z(D_6)| = 6$. So, by our classification of groups of order 6 (Example 4 of Chapter 9), we know that Inn(D_6) is isomorphic to D_3 or Z_6. Now, if Inn(D_6) were cyclic, then by Theorem 10.4, $D_6/Z(D_6)$ would be also. But then, Theorem 10.3 would tell us that D_6 is Abelian. So, Inn(D_6) is isomorphic to D_3.   □

EXERCISES

Error is a hardy plant: it flourisheth in every soil.

Martin Farquhar Tupper, *Proverbial Philosophy* [1838–1842].
Of Truth in Things False

1. Explain why the center of a group is always a normal subgroup of the group.
2. Prove Theorem 10.1.
3. Prove that A_n is normal in S_n.
4. Show that the n rotations in D_n form a normal subgroup of D_n.
5. Prove that $SL(2, \mathbf{R})$ is a normal subgroup of $GL(2, \mathbf{R})$.
6. Viewing $\langle 3 \rangle$ and $\langle 12 \rangle$ as subgroups of Z, prove that $\langle 3 \rangle/\langle 12 \rangle$ is isomorphic to Z_4. Similarly, prove that $\langle 8 \rangle/\langle 48 \rangle$ is isomorphic to Z_6. Generalize to arbitrary integers k and n.
7. Prove that if H has only two distinct cosets in G, then H is normal in G. (This exercise is referenced to in Chapters 26 and 27.)
8. Let $H = \{(1), (23)(34)\}$ in A_4.
   a. Show that H is not normal in A_4.
   b. Referring to the multiplication table for A_4 in Table 5.1 on page 92, show that, although $\alpha_7 H = \alpha_8 H$ and $\alpha_9 H = \alpha_{10} H$, it is not true that $\alpha_7\alpha_9 H = \alpha_8\alpha_{10} H$.
   Explain why this proves that the left cosets of H do not form a group under coset multiplication.
9. Let $G = Z_4 \oplus U(4)$. Let $H = \langle (2, 3) \rangle$, and $K = \langle (2, 1) \rangle$. Determine the isomorphism class of G/H and G/K. (This shows that $H \approx K$ does not imply $G/H \approx G/K$.)

10. Prove that a factor group of a cyclic group is cyclic.

11. What is the order of element $5 + \langle 6 \rangle$ in the factor group $Z_{18}/\langle 6 \rangle$?

12. What is the order of the element $14 + \langle 8 \rangle$ in the factor group $Z_{24}/\langle 8 \rangle$?

13. What is the order of the element $4U_5(105)$ in the factor group $U(105)/U_5(105)$?

14. Recall that $Z(D_6) = \{R_0, R_{180}\}$. What is the order of the element $R_{60} Z(D_6)$ in the factor group $D_6/Z(D_6)$?

15. Let $G = Z/\langle 20 \rangle$ and $H = \langle 4 \rangle/\langle 20 \rangle$. List the elements of H and G/H.

16. What is the order of the factor group $Z_{60}/\langle 15 \rangle$?

17. What is the order of the factor group $Z_{10} \oplus U(10)/\langle (2, 9) \rangle$?

18. Construct the Cayley table for $U(20)/U_5(20)$.

19. Is $U(30)/U_5(30)$ isomorphic to $Z_2 \oplus Z_2$ or Z_4?

20. Determine the order of $Z \oplus Z/\langle (2, 2) \rangle$. Is the group cyclic?

21. Determine the order of $Z \oplus Z/\langle (4, 2) \rangle$. Is the group cyclic?

22. The group $Z_4 \oplus Z_{12}/\langle (2, 2) \rangle$ is isomorphic to one of Z_8, Z_4, $\oplus Z_2$, or $Z_2 \oplus Z_2 \oplus Z_2$. Which one?

23. Let $G = U(32)$ and $H = \{1, 31\}$. The group G/H is isomorphic to one of Z_8, $Z_4 \oplus Z_2$, or $Z_2 \oplus Z_2 \oplus Z_2$. Which one?

24. Let G be the group given by the table in exercise 4 of the Supplementary Exercises for Chapters 1–4 on page 80, and H the subgroup $\{e, a^2\}$. Is G/H isomorphic to Z_4 or $Z_2 \oplus Z_2$?

25. Let $G = U(16)$, $H = \{1, 15\}$, and $K = \{1, 9\}$. Are H and K isomorphic? Are G/H and G/K isomorphic?

26. Let $G = Z_4 \oplus Z_4$, $H = \{(0, 0), (2, 0), (0, 2), (2, 2)\}$, and $K = \langle (1, 2) \rangle$. Is G/H isomorphic to Z_4 or $Z_2 \oplus Z_2$? Is G/K isomorphic to Z_4 or $Z_2 \oplus Z_2$?

27. Let $G = GL(2, R)$ and $H = \{A \in G \mid \det A = 3^k, k \in Z\}$. Prove that H is a normal subgroup of G.

27. Determine all subgroups of $R^\#$ (nonzero reals under multiplication) of index 2.

29. Show, by example, that in a factor group G/H it can happen that $aH = bH$ but $|a| \neq |b|$.

30. Let H be a normal subgroup of G and let a belong to G. If the element aH has order 3 in the group G/H and $|H| = 10$, what are the possibilities for the order of a?

31. Prove that a factor group of an Abelian group is Abelian.

32. An element is called a *square* if it can be expressed in the form b^2 for some b. Suppose G is an Abelian group and H is a subgroup of G. If every element of H is a square and every element of G/H is a square, prove that every element of G is a square.

33. Observe from the table for A_4 given in Table 5.1 on page 92 that the subgroup given in Example 5 is the only subgroup of A_4 of order 4. Why does this

imply that this subgroup must be normal in A_4? Generalize this to arbitrary finite groups.

34. Let G be a finite group and H a normal subgroup of G. Prove that the order of the element gH in G/H must divide the order of g in G.

35. Suppose G is a non-Abelian group of order p^3 (where p is a prime) and $Z(G) \neq \{e\}$. Prove $|Z(G)| = p$.

36. If $|G| = pq$, where p and q are not necessarily distinct primes, prove $|Z(G)| = 1$ or pq.

37. Let N be a normal subgroup of G and H a subgroup of G. If N is a subgroup of H, prove H/N is a normal subgroup of G/N if and only if H is a normal subgroup of G.

38. Let $G = \{\pm 1, \pm i, \pm j, \pm k\}$, where $i^2 = j^2 = k^2 = -1$, $-i = (-1)i$, $1^2 = (-1)^2 = 1$, $ij = -ji = k$, $jk = -kj = i$, $ki = -ik = j$.
a. Construct the Cayley table for G.
b. Show that $H = \{1, -1\} \triangleleft G$.
c. Construct the Cayley table for G/H.
(The rules involving i, j, and k can be remembered by using the circle

Going clockwise, the product of two consecutive elements is the third one. The same is true for going counterclockwise, except that we obtain the negative of the third element.) This group is called the *quaternions* and was invented by William Hamilton in 1843. The quaternions are used to describe rotations in three-dimensional space, and they are used in physics. The quaternions can be used to extend the complex numbers in a natural way.

39. In D_4, let $K = \{R_0, D\}$ and let $L = \{R_0, D, D', R_{180}\}$. Show that $K \triangleleft L \triangleleft D_4$, but K is not normal in D_4. (Normality is not transitive.)

40. If N is a normal subgroup of G and H is any subgroup of G, prove that NH is a subgroup of G. (This exercise is referred to in Chapter 26.)

41. Show that the intersection of two normal subgroups of G is a normal subgroup of G.

42. If N and M are normal subgroups of G, prove that NM is also a normal subgroup of G.

43. If N and M are normal subgroups of G and $N \cap M = \{e\}$, prove that $nm = mn$ for all $n \in N$, $m \in M$. (This exercise is referred to in Chapter 26.)

44. Let N be a normal subgroup of a finite group G. If N is cyclic, prove that every subgroup of N is also normal in G. (This exercise is referred to in Chapter 26.)

45. Let H be a normal subgroup of a finite group G. If $\gcd(|x|, |G/H|) = 1$, show that $x \in H$. (This exercise is referred to in Chapter 27.)

46. Let G be a group and let G' be the subgroup of G generated by the set $S = \{x^{-1}y^{-1}xy \mid x, y \in G\}$. (See exercise 3, Supplementary Exercises for Chapters 5–8, for a more complete description of G'.)
  a. Prove that G' is normal in G.
  b. Prove that G/G' is Abelian.
  c. If G/N is Abelian, prove that $N \geqslant G'$.
  d. Prove that if H is a subgroup of G and $H \geqslant G'$, then H is normal in G.

47. If N is a normal subgroup of G and $|G/N| = m$, show that $x^m \in N$ for all x in G.

48. Suppose G has a subgroup of order n. Prove that the intersection of all subgroups of G of order n is a normal subgroup of G.

49. Prove that if $G = HK$ where $H \triangleleft G$, $K \triangleleft G$, and $H \cap K = \{e\}$, then G is the internal direct product of H and K.

50. If $G = H \times K$, prove that $H \triangleleft G$ and $K \triangleleft G$.

51. If G is non-Abelian, show that $\text{Aut}(G)$ is not cyclic.

52. Let $|G| = p^n m$, where p is prime and $\gcd(p, m) = 1$. Suppose H is a normal subgroup of G of order p^n. If K is a subgroup of G of order p^k, show that $K \subseteq H$.

53. Suppose H is a normal subgroup of a finite group G. If G/H has an element of order n, show that G has an element of order n. Show, by example, that the assumption that G is finite is necessary.

54. Recall that a subgroup N of a group G is called characteristic if $N\phi = N$ for all automorphisms ϕ of G. (See exercises 1–4, Supplementary Exercises for Chapters 5–8.) If N is a characteristic subgroup of G, show that N is a normal subgroup of G.

55. In D_4, let $\mathcal{H} = \{R_0, H\}$. Form an operation table for the cosets $\mathcal{H}, D\mathcal{H}, V\mathcal{H}, D'\mathcal{H}$. Is the result a group table? Does your answer contradict Theorem 10.2?

56. Show that S_4 has a unique subgroup of order 12.

57. Let G be the group of movements of the Rubik's cube. Let H be the subgroup of movements that leave one face in the clean state (that is, the face shows a single color). Is H a normal subgroup of G?

58. If H is a normal subgroup of G and $|H| = 2$, prove that H is contained in the center of G.

59. Prove that A_5 cannot have a normal subgroup of order 2.

60. Let G be a finite group and H an odd order subgroup of G of index 2. Show that the product of all the elements of G (taken in any order) cannot belong to H.

61. Without looking at inner automorphisms of D_n, determine the number of inner automorphisms of D_n.

REFERENCE

1. Tony Rothman, "Genius and Biographers: The Fictionalization of Évariste Galois," *The American Mathematical Monthly* 89 (1982): 84–106.

SUGGESTED READING

K. R. McLean, "When Isomorphic Groups Are Not the Same," *Mathematical Gazette* 57 (1973): 207–208.

This article gives a simple example to show that two groups may be isomorphic, but behave differently, when they are subgroups of a larger group.

Évariste Galois

Galois at seventeen was making discoveries of epochal significance in the theory of equations, discoveries whose consequences are not yet exhausted after more than a century.

E. T. Bell, *Men of Mathematics*

This French stamp was issued in the 1984 "Celebrity Series" in support of the Red Cross Fund.

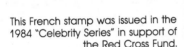

ÉVARISTE GALOIS (pronounced GAL-wah) was born on October 25, 1811, near Paris. He took his first mathematics course when he was fifteen and quickly mastered the works of Legendre and Lagrange. At eighteen, Galois wrote his important research on the theory of equations and submitted it to the French Academy of Sciences for publication. The paper was given to Cauchy for refereeing. Cauchy, impressed by the paper, agreed to present it to the academy, but never did. At the age of nineteen, Galois entered a paper of the highest quality in the competition for the Grand Prize in Mathematics, given by the French Academy of Sciences. The paper was given to Fourier, who died shortly thereafter. Galois's paper was never seen again.

Galois twice failed his entrance examination to l'École Polytechnique. He did not know some basic mathematics, and he did mathematics almost entirely in his head, to the annoyance of the examiner. Legend has it that Galois became so enraged at the stupidity of the examiner that he threw an eraser at him.

Galois spent most the last year and half of his life in prison for revolutionary political offenses. While in prison, he attempted suicide, and prophesied his death in a duel. On May 30, 1832, Galois was shot in a duel and died the next day at the age of twenty. The life and death of Galois have long been a source of fascination and speculation for mathematics historians. One article [1] argues convincingly that three of the most widely read accounts of Galois's life are highly fictitious.

Among the many concepts introduced by Galois are normal subgroups, isomorphisms, simple groups, finite fields, and Galois theory. His work provided a method for disposing of several famous constructability problems, such as trisecting an arbitrary angle and doubling a cube. Galois's entire collected works fill only sixty pages.

Group Homomorphisms

$$11$$

To see what is general in what is particular and what is permanent in what is transitory is the aim of scientific thought.

Alfred North Whitehead

Definition and Examples

In this chapter, we consider one of the most fundamental ideas of algebra—homomorphisms. The term *homomorphism* comes from the Greek words *homo*, "like," and *morphe*, "form." We will see that a homomorphism is a natural generalization of an isomorphism and that there is an intimate connection between factor groups of a group and homomorphisms of a group. The concept was introduced by Camille Jordan in 1870, in his influential book *Traité des Substitutions*.

DEFINITION Group Homomorphism
A homomorphism ϕ from a group G to a group $\overline{G}$ is a mapping from G into $\overline{G}$ that preserves the group operation; that is, $(ab)\phi = (a\phi)(b\phi)$ for all a, b in G.

Before giving examples and stating numerous properties of homomorphisms, it is convenient to introduce an important subgroup that is intimately related to the image of a homomorphism.

DEFINITION Kernel of a Homomorphism
The *kernel* of a homomorphism ϕ from a group G to a group with identity e is the set $\{x \in G \mid x\phi = e\}$. The kernel of ϕ is denoted by Ker ϕ.

Example 1   Any isomorphism is a homomorphism that is also onto and one-to-one. The kernel of an isomorphism is the identity.   □

Example 2   Let $G = GL(2, \mathbf{R})$, and let $\mathbf{R}^{\#}$ be the group of nonzero real numbers under multiplication. Then the determinant mapping $A \to \det A$ is a homomorphism from G to $\mathbf{R}^{\#}$. The kernel of the determinant mapping is $SL(2, \mathbf{R})$.   □

Example 3   The mapping from $\mathbf{R}^{\#}$ to $\mathbf{R}^{\#}$ that sends x to the absolute value of x is a homomorphism. The kernel of the absolute value mapping is $\{+1, -1\}$.   □

Example 4   Let $\mathbf{R}[x]$ denote the group of all polynomials with real coefficients under addition. For any f in $\mathbf{R}[x]$, let f' denote the derivative of f. Then the mapping $f \to f'$ is a homomorphism from $\mathbf{R}[x]$ to itself. The kernel of the derivative mapping is the set of all constant functions.   □

Example 5   The mapping of ϕ from Z to Z_n, defined by $m\phi = r$ where r is the remainder of m when divided by n, is a homomorphism. The kernel of this mapping is $\langle n \rangle$.   □

The natural homomorphism from Z to Z_n given in Example 5 has many applications in number theory. In 1770, Lagrange proved that every positive integer can be written as the sum of four squares (that is, in the form $a^2 + b^2 + c^2 + d^2$). Our next example shows that there are infinitely many integers that are not the sum of three squares.

Example 6   No integer equal to 7 modulo 8 can be written in the form $a^2 + b^2 + c^2$. If this were so, then $7 = a^2 \bmod 8 + b^2 \bmod 8 + c^2 \bmod 8$. Now observe that the square of any even integer is 0 or 4 mod 8, while the square of any odd integer is 1 mod 8 (see exercise 37 of Chapter 0). But no three numbers chosen from 0, 1, and 4 add up to 7 mod 8.   □

Example 7   The mapping from the group of real numbers under addition to itself given by $x \to [x]$, the greatest integer less than or equal to x, is *not* a homomorphism.   □

When defining a homomorphism from a group in which there are several ways to represent the elements, caution must be exercised to ensure that the correspondence is a function. (The term *well defined* is often used in this context.) For example, since $3(x + y) = 3x + 3y$ in Z_6, one might believe the correspondence $x \to 3x$ from Z_3 to Z_6 is a homomorphism. But it is not a function since $0 = 3$ in Z_3, but $3 \cdot 0 \neq 3 \cdot 3$ in Z_6.

Properties of Homomorphisms

Theorem 11.1   *Properties of Homomorphisms*
Let ϕ be a homomorphism from a group G to a group $\overline{G}$. Let g be an element of G and H a subgroup of G. Then

1. *ϕ carries the identity of G to the identity of $\overline{G}$.*
2. *$g^n\phi = (g\phi)^n$.*
3. *$H\phi = \{h\phi \mid h \in H\}$ is a subgroup of $\overline{G}$.*
4. *If H is cyclic, then $H\phi$ is cyclic.*
5. *If H is Abelian, then $H\phi$ is Abelian.*
6. *If H is normal in G, then $H\phi$ is normal in $G\phi$.*
7. *If $|g| = n$, then $|g\phi|$ divides n.*
8. *If $g\phi = g'$, then $g'\phi^{-1} = \{x \in G \mid x\phi = g'\} = g\text{Ker } \phi$.*
9. *If $|H| = n$, then $|H\phi|$ divides n.*
10. *If $|\text{Ker } \phi| = n$, then ϕ is an n-to-1 mapping from G onto $G\phi$.*
11. *If $\overline{K}$ is a subgroup of $\overline{G}$, then $\overline{K}\phi^{-1} = \{k \in G \mid k\phi \in \overline{K}\}$ is a subgroup of G.*
12. *If $\overline{K}$ is a normal subgroup of $\overline{G}$, then $\overline{K}\phi^{-1}$ is a normal subgroup of G.*
13. *If ϕ is onto and Ker $\phi = \{e\}$, then ϕ is an isomorphism from G to $\overline{G}$.*

Proof.   The proofs of these properties of homomorphisms are straightforward and are left as exercises.   ∎

A few remarks about Theorem 11.1 are in order here. Students should remember the various parts of the theorem in words. For example, part 4 says the homomorphic image of a cyclic group is cyclic. Part 6 says the homomorphic image of a normal subgroup of G is normal in the image of G. Part 10 says that if ϕ is a homomorphism from G to $\overline{G}$, then every element of $\overline{G}$ that gets "hit" by ϕ gets hit the same number of times as does the identity. The set $g'\phi^{-1}$ defined in part 8 is called the *inverse image* of g' (or *pullback*) of g'. Note that the inverse image of an element is a coset of the kernel. Similarly, the set $\overline{K}\phi^{-1}$ defined in part 11 is called the *inverse image* of $\overline{K}$ (or *pullback* of $\overline{K}$).

Part 8 of Theorem 11.1 is reminiscent of something from linear algebra and differential equations. Recall that if x is a particular solution to a system of linear equations and S is the entire solution set of the corresponding homogeneous system of linear equations, then $x + S$ is the entire solution set of the nonhomogeneous system. In reality, this statement is just a special case of part 8. Parts 1, 8, and 10 of Theorem 11.1 are pictorially represented in Figure 11.1.

The special case of part 12 where $K = \{e\}$ is of such importance that we single it out.

Corollary   *Kernels Are Normal*
Let ϕ be a group homomorphism from G to $\overline{G}$. Then Ker ϕ is a normal subgroup of G.

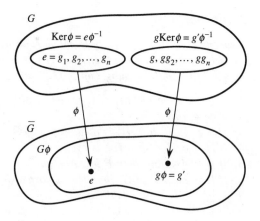

Figure 11.1

The next example illustrates how one can easily determine all homomorphisms from a cyclic group to a cyclic group.

Example 8 We determine all homomorphisms from Z_{12} to Z_{30}. By part 2 of Theorem 11.1, a homomorphism is completely specified by the image of 1. That is, if 1 maps to a then x maps to xa. Lagrange's Theorem and part 7 of Theorem 11.1 require that $|a|$ divide both 12 and 30. So, $|a| = 1, 2, 3, 6$. Thus, $a = 0$, 15, 10, 20, 5, or 25. This gives us a list of candidates for the homomorphisms. That each of these six possibilities yields an operation preserving function can now be verified by direct calculations. (Note that gcd(12, 30) = 6. This is not a coincidence!) ☐

Example 9 The mapping from S_n to Z_2 that takes an even permutation to 0 and an odd permutation to 1 is a homomorphism. Figure 11.2 illustrates the telescoping nature of the mapping. ▣

The First Isomorphism Theorem

In Chapter 10 we showed that for a group G and a normal subgroup H, we could arrange the Cayley table of G into boxes that represented the cosets of H in G, and these boxes then were a Cayley table for G/H. The next theorem shows that for any homomorphism ϕ of G and the normal subgroup Ker ϕ, the same process produces a Cayley table isomorphic to the homomorphic image of G. Thus, homomorphisms, like factor groups, cause a *systematic* collapsing of a group to be simpler but closely related group. This can be likened to viewing a group through the reverse end of a telescope—the general feature of the groups are present, but the apparent size is diminished. The important relationship between

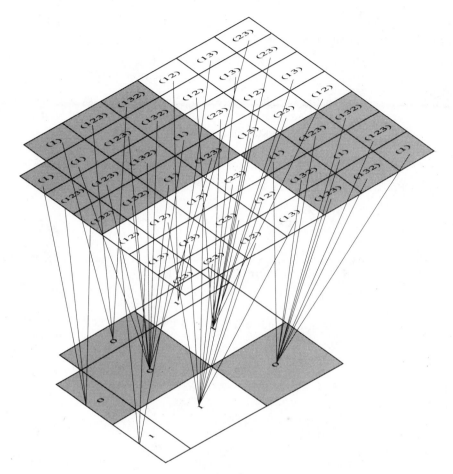

Figure 11.2 Homomorphism from S_3 to Z_2.

homomorphisms and factor groups given below is often called the Fundamental Theorem of Group Homomorphisms.

Theorem 11.2 *(Jordan, 1870) First Isomorphism Theorem*
Let ϕ be a group homomorphism from G to $\overline{G}$. Then the mapping from $G/\text{Ker } \phi$ to $G\phi$, given by $g\text{Ker } \phi \to g\phi$ is an isomorphism. In symbols, $G/\text{Ker } \phi \approx G\phi$.

Proof.   Let us use ψ to denote the mapping $g\text{Ker } \phi \to g\phi$. First, we show ψ is well defined (that is, the mapping is independent of the particular coset representative chosen). Suppose $x\text{Ker } \phi = y\text{Ker } \phi$. Then $y^{-1}x \in \text{Ker } \phi$ and

$e = (y^{-1}x)\phi = (y\phi)^{-1}x\phi$. Thus, $x\phi = y\phi$ and ψ is indeed a function. To show that ψ is operation-preserving, observe that

$$(x\text{Ker } \phi \; y\text{Ker } \phi)\psi = (xy\text{Ker } \phi)\psi = (xy)\phi = x\phi y\phi$$
$$= (x\text{Ker } \phi)\psi(y\text{Ker } \phi)\psi.$$

Finally, we note that $g\text{Ker } \phi \in \text{Ker } \psi$ only if $g\phi = e$; that is, $g \in \text{Ker } \phi$. This shows that $\text{Ker } \psi$ is the identity in $G/\text{Ker } \phi$ and by part 10 of Theorem 11.1, ψ is one-to-one. ■

Example 10 To illustrate Theorem 11.2 and its proof, consider the homomorphism ϕ from D_4 to itself given by

$$
\begin{array}{cccccccc}
R_0 & R_{180} & R_{90} & R_{270} & H & V & D & D' \\
\searrow & \swarrow & \searrow & \swarrow & \searrow & \swarrow & \searrow & \swarrow \\
& R_0 & & R_{90} & & H & & D
\end{array}
$$

Then $\text{Ker } \phi = \{R_0, R_{180}\}$ and the mapping ψ in Theorem 11.2 is $R_0\text{Ker } \phi \to R_0$, $R_{90}\text{Ker } \phi \to R_{90}$, $H\text{Ker } \phi \to H$, $D\text{Ker } \phi \to D$. Since we multiply cosets by multiplying their representatives, $G/\text{Ker } \phi$ is obviously isomorphic to $G\phi$. □

Mathematicians often given a pictorial representation of Theorem 11.2 as follows:

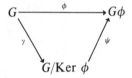

where $\gamma: G \to G/\text{Ker } \phi$ is defined as $g\gamma = g\text{Ker } \phi$. The mapping γ is called the *natural mapping* from G to $G/\text{Ker } \phi$. Our proof of Theorem 11.2 shows that $\gamma\psi = \phi$. In this case, one says that the above diagram is commutative.

As a consequence of Theorem 11.2, we see that all homomorphic images of G can be determined without leaving G. For we may simply consider the various factor groups of G. For example, we know the homomorphic image of an Abelian group is Abelian because the factor group of an Abelian group is Abelian. We know the number of homomorphic images of a cyclic group G of order n is the number of divisors of n, since there is exactly one subgroup of G (and therefore one factor group of G) for each divisor of n. (Be careful, the number of homomorphisms of a cyclic group of order n need not be the same as the number of divisors of n since different homomorphisms can have the same image.)

An appreciation for Theorem 11.2 can be gained by looking at a few examples.

Example 11 $Z/\langle n \rangle \approx Z_n$

Consider the mapping from Z to Z_n defined in Example 5. Clearly, its kernel is $\langle n \rangle$. So, by Theorem 11.2, $Z/\langle n \rangle \approx Z_n$. $\qquad\square$

Example 12 The Wrapping Function

Recall the wrapping function W from trigonometry. The real line is wrapped around a unit circle in the plane centered at $(0, 0)$ in the counterclockwise direction with the number 0 at the point $(1, 0)$ as in Figure 11.3. The function W assigns to each real number a, the point a radians from $(1, 0)$ on the circle. This mapping is a homomorphism from **R** onto the circle group (the group of complex numbers of magnitude 1 under multiplication—see exercise 11, Supplementary Exercises for Chapters 5–8). Indeed, it follows from elementary facts of trigonometry that $xW = \cos x + i \sin x$ and $(x + y)W = xWyW$. Since W is periodic of period 2π, Ker $W = \langle 2\pi \rangle$. So, from the First Isomorphism Theorem, we see that $\mathbf{R}/\langle 2\pi \rangle$ is isomorphic to the circle group. $\qquad\square$

Our next example is a theorem that is used repeatedly in Chapters 26 and 27.

Example 13 The N/C Theorem

Let H be a subgroup of a group G. Recall, the normalizer of H in G is $N(H) = \{x \in G \mid x^{-1}Hx = H\}$ and the centralizer of H in G is $C(H) = \{x \in G \mid x^{-1}hx = h \text{ for all } h \text{ in } H\}$. Consider the mapping from $N(H)$ to Aut(H) given by $g \rightarrow \phi_g$ where ϕ_g is the inner automorphism of H induced by g (that is, $h\phi_g = g^{-1}hg$ for all h in H.) This mapping is a homomorphism with kernel $C(H)$. So, by Theorem 11.2, $N(H)/C(H)$ is isomorphic to a subgroup of Aut(H). $\qquad\square$

The corollary of Theorem 11.1 says that the kernel of every homomorphism of a group is a normal subgroup of the group. We conclude this chapter by verifying that the converse of this statement is also true.

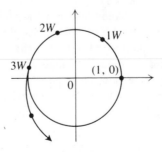

Figure 11.3

Theorem 11.3   *Normal Subgroups Are Kernels*
*Every normal subgroup of a group G is the kernel of a homomorphism of
G. In particular, a normal subgroup N is the kernel of the mapping g → gN
from G to G/N.*

Proof.   Define $\phi:G \to G/N$ by $g\phi = gN$. (This mapping is called the *natural homomorphism* from G to G/N.) Then, $(xy)\phi = (xy)N = xNyN = x\phi y\phi$. ∎

Examples 11, 12, and 13 illustrate the utility of the First Isomorphism
Theorem. But what about homomorphisms in general? Why would one care to
study a homomorphism of a group? The answer is that, just as was the case with
factor groups of a group, homomorphic images of a group tell us *some* of the
properties of the original group. One measure of the likeness of a group and its
homomorphic image is the size of the kernel. If the kernel of the homomorphism
of group G is the identity, then the image of G tells us everything (group
theoretically) about G (the two being isomorphic). On the other hand, if the
kernel of the homomorphism is G itself, then the image tells us nothing about
G. Between these two extremes, some information about G is preserved and
some is lost. The utility of a particular homomorphism lies in its ability to
preserve the properties of the group we want, while losing some inessential ones.
In this way, we have replaced G by a group less complicated (and therefore
easier to study) than G; but in the process, we have saved enough information
to answer questions that we have about G itself. For example, if G is a group
of order 60 and G has a homomorphic image of order 12 that is Abelian, then
we know from properties 10, 11, and 12 of Theorem 11.1 that G has normal
subgroups of orders 5, 10, 15, 20, 30, and 60.

Perhaps the following analogy between homomorphisms and photography
is instructive.* A photograph of a person cannot tell us exactly the person's
height, weight, or age. Nevertheless, it *may* be possible to decide from a photograph whether the person is tall or short, heavy or thin, old or young. In the
same way, a homomorphic image of a group gives us *some* information about
the group.

In certain branches of group theory, and especially in physics and chemistry,
one often wants to know all homomorphic images of a group that are matrix
groups over the complex numbers (these are called *group representations*). Here,
we may carry our analogy with photography one step further by saying that this
is like wanting photographs of a person from many different angles (front view,
profile, head-to-toe view, close-up, etc.), as well as x-rays! Just as this composite
information from the photographs reveals much about the person, a number of
homomorphic images of a group reveals much about the group.

All perception of truth is the detection of an analogy. Henry David Thoreau, *Journal.*

EXERCISES

> The greater the difficulty, the more glory in surmounting it. Skillful pilots gain their reputation from storms and tempests.
>
> Epicurus

1. Let $\mathbf{R}^\#$ be the group of nonzero real numbers under multiplication, and let r be a positive integer. Show that the mapping that takes x to x^r is a homomorphism from $\mathbf{R}^\#$ to $\mathbf{R}^\#$.

2. Let G be the group of all polynomials with real coefficients under addition. Let $\overline{G}$ be the group of all functions form $\mathbf{R}$ to $\mathbf{R}$ under addition. Show that the mapping $f \rightarrow \int f$ from G to $\overline{G}$ is a homomorphism. (Assume the constant of integration is 0.) What is the kernel of this mapping? Is this mapping a homomorphism if we assume the constant of integration is 1?

3. Prove that the mapping given in Example 2 is a homomorphism.

4. Prove that the mapping given in Example 3 is a homomorphism.

5. Prove that the mapping given in Example 4 is a homomorphism.

6. Let G be a group of permutations. For each σ in G, define

$$\text{sgn}(\sigma) = \begin{cases} +1 & \text{if } \sigma \text{ is an even permutation,} \\ -1 & \text{if } \sigma \text{ is an odd permutation.} \end{cases}$$

   Prove that sgn is a homomorphism from G to $\{+1, -1\}$. What is the kernel?

7. Prove that the mapping from $G \oplus H$ into G given by $(g, h) \rightarrow g$ is a homomorphism. What is the kernel? This mapping is called the *projection* of $G \oplus H$ onto G.

8. Let G be a subgroup of some dihedral group. For each x in G, define

$$x\phi = \begin{cases} +1 & \text{if } x \text{ is a rotation,} \\ -1 & \text{if } x \text{ is a reflection.} \end{cases}$$

   Prove that ϕ is a homomorphism from G to $\{+1, -1\}$. What is the kernel of ϕ?

9. Verify that the mapping given in Example 7 is not a homomorphism.

10. Suppose k is a divisor of n. Prove that $Z_n/\langle k \rangle \approx Z_k$.

11. Prove that $A \oplus B/A \oplus \{e\} \approx B$.

12. Explain why the correspondence $x \rightarrow 3x$ from Z_{12} to Z_{10} is not a homomorphism.

13. Suppose ϕ is a homomorphism from Z_{30} to Z_{30} and Ker $\phi = \{0, 10, 20\}$. If $23\phi = 6$, determine all elements that map to 6.

14. Prove that there is no homomorphism from $Z_8 \oplus Z_2$ onto $Z_4 \oplus Z_4$.

15. Prove that there is no homomorphism from $Z_{16} \oplus Z_2$ onto $Z_4 \oplus Z_4$.

16. Can there be a homomorphism from $Z_4 \oplus Z_4$ onto Z_8? Can there be a homomorphism from Z_{16} onto $Z_2 \oplus Z_2$? Explain your answers.

17. Suppose that there is a homomorphism ϕ from Z_{17} to some group and ϕ is not one-to-one. Determine ϕ.

18. How many homomorphisms are there from Z_{20} onto Z_8? How many are there to Z_8?

19. How many homomorphisms are there from Z_{20} onto Z_{10}? How many are there to Z_{10}?

20. Determine all homomorphisms from Z_4 into $Z_2 \oplus Z_2$.

21. Determine all homomorphisms from Z_n to itself.

22. Suppose ϕ is a homomorphism from S_4 onto Z_2. Determine Ker ϕ. Determine all homomorphisms from S_4 to Z_2.

23. Suppose there is a homomorphism from G onto Z_{10}. Prove that G has normal subgroups of index 2 and 5.

24. Suppose ϕ is a homomorphism from a group G onto Z_{12} and the kernel of ϕ has order 5. Explain why G must have normal subgroups of orders 5, 10, 15, 20, 30, and 60.

25. Suppose ϕ is a homomorphism from $U(30)$ to $U(30)$ and Ker $\phi = \{1, 11\}$. If $7\phi = 7$, find all elements of $U(30)$ that map to 7.

26. Find a homomorphism ϕ from $U(30)$ to $U(30)$ with kernel $\{1, 11\}$ and $7\phi = 7$.

27. Suppose ϕ is a homomorphism from $U(40)$ to $U(40)$ and Ker $\phi = \{1, 9, 17, 33\}$. If $11\phi = 11$, find all elements of $U(40)$ that map to 11.

28. Find a homomorphism ϕ from $U(40)$ to $U(40)$ with kernel $\{1, 9, 17, 33\}$ and $11\phi = 11$.

29. Prove that the mapping $\phi: Z \oplus Z \to Z$ given by $(a, b) \to a - b$ is a homomorphism. What is the kernel of ϕ? Describe the set $3\phi^{-1}$ (that is, all elements of that map to 3).

30. Suppose that there is a homomorphism ϕ from $Z \oplus Z$ into a group G such that $(3, 2)\phi = a$ and $(2, 1)\phi = b$. Determine $(4, 4) \phi$ in terms of a and b. Assume the operation of G is addition.

31. Prove that the mapping $x \to x^4$ from $C^\#$ to $C^\#$ is a homomorphism. What is the kernel?

32. For each pair of positive integers m and n, we can define a homomorphism from Z to $Z_m \oplus Z_n$ by $x \to (x \bmod m, x \bmod n)$. What is the kernel when $(m, n) = (3, 4)$? What is the kernel when $(m, n) = (6, 4)$? Generalize.

33. (Second Isomorphism Theorem.) If K is a subgroup of G and N is a normal subgroup of G, prove that $K/K \cap N$ is isomorphic to KN/N.

34. (Third Isomorphism Theorem.) If M and N are normal subgroups of G and $N \leq M$, prove that $(G/N)/(M/N) \approx G/M$.

35. Let $\phi(d)$ denote the Euler phi function of d (see page 71). Show that the number of homomorphisms from Z_n to Z_k is $\Sigma\phi(d)$ where the sum runs over all common divisors d of n and k. (It follows from number theory that this sum is actually gcd(n, k).)

36. Let k be a divisor of n. Consider the homomorphism from $U(n)$ to $U(k)$ given by $x \to x \bmod k$. What is the relationship between this homomorphism and the subgroup $U_k(n)$ of $U(n)$?

37. Determine all homomorphic images of D_4 (up to isomorphism).

38. Let N be a normal subgroup of a finite group G. Use the theorems of this chapter to prove that the order of the group element gN in G/N divides the order of g.

39. Suppose G is a finite group and Z_{10} is a homomorphic image of G. What can we say about $|G|$?

40. Suppose Z_{10} and Z_{15} are both homomorphic images of a finite group G. What can be said about $|G|$?

41. Suppose that for each prime p, Z_p is the homomorphic image of a group G. What can we say about $|G|$?

42. (For students who have had linear algebra.) Suppose x is a particular solution to a system of linear equations and S is the entire solution set of the corresponding homogeneous system of linear equations. Explain why part 8 of Theorem 11.1 guarantees that $x + S$ is the entire solution set of the nonhomogeneous system. In particular, describe the relevant groups and the homomorphism between them.

43. Let N be a normal subgroup of a group G. Use part 12 of Theorem 11.1 to prove that every subgroup of G/N has the form H/N, where H is a subgroup of G. (This exercise is referred to in Chapter 26.)

44. Show that a homomorphism defined on a cyclic group is completely determined by its action on a generator of the group.

45. Use the First Isomorphism Theorem to prove Theorem 10.4.

46. Suppose $\overline{G}$ is the homomorphic image of a finite group G. If $\overline{G}$ has an element of order n, show that G has an element of order n.

47. Let $Z[x]$ be the group of polynomials in x with integer coefficients under addition. Prove that the mapping from $Z[x]$ into Z given by $f(x) \to f(3)$ is a homomorphism. Give a geometrical description of the kernel of this homomorphism.

48. If H and K are normal subgroups of G and $H \cap K = \{e\}$, prove that G is isomorphic to a subgroup of $G/H \oplus G/K$.

49. Suppose H and K are distinct subgroups of G of index 2. Prove that $H \cap K$ is a normal subgroup of G of index 4 and $G/H \cap K$ is not cyclic.

50. Suppose that the number of homomorphisms from G to H is n. How many

homomorphisms are there from $G \oplus G \oplus \cdots \oplus G$ (s terms) to H? How many homomorphisms are there from G to $H \oplus H \oplus \cdots \oplus H$ (s terms)?

SUGGESTED READINGS

Loren Larson, "A Theorem About Primes Proved on a Chessboard," *Mathematics Magazine* 50 (1977): 69–74.

This paper gives a chessboard interpretation of a number of algebraic concepts, such as cosets and group homomorphisms. These ideas are used to solve the "n-queens" problem and to prove Fermat's Two Square Theorem: every prime $p = 1 \bmod 4$ is the sum of two unique squares.

Camille Jordan

Although these contributions [to analysis and topology] would have been enough to rank Jordan very high among his mathematical contemporaries, it is chiefly as an algebraist that he reached celebrity when he was barely thirty; and during the next forty years he was universally regarded as the undisputed master of group theory.

J. Dieudonné, *Dictionary of Scientific Biography*

CAMILLE JORDAN was born into a well-to-do family on January 5, 1838, in Lyons, France. Like his father, he graduated from the École Polytechnique and became an engineer. Nearly all of his 120 research papers in mathematics were written before his retirement from engineering in 1885. From 1873 until 1912, Jordan taught simultaneously at the École Polytechnique and the College of France.

In the great French tradition, Jordan was a universal mathematician who published in nearly every branch of mathematics. Among the concepts named after him are the Jordan canonical form in matrix theory, the Jordan curve theorem from topology, the Jordan-Hölder theorem from group theory, and Jordan algebras. His classic book *Traité des Substitutions,* published in 1870, was the first to be devoted solely to group theory and its applications to other branches of mathematics. This book provided the first clear and complete account of the theory invented by Galois to determine which polynomials were solvable by radicals, and it was the first major investigation of infinite groups. In the book, Jordan coined the word *Abelian* to describe commutative groups and, although Galois had introduced the term *group,* it was through the influence of Jordan's book that the term became standard.

Another book that had great influence and set a new standard for rigor was his *Cours d'analyse.* This book gave the first clear definitions of the notions of *volume* and *multiple integral.* It also gave conditions under which a multiple integral can be evaluated by successive integrations. Nearly one hundred years after this book appeared, the distinguished mathematician and mathematical historian B. L. van der Waerden wrote "For me, every single chapter of the *Cours d'analyse* is a pleasure to read." Jordan died in Paris on January 22, 1922.

Fundamental Theorem of Finite Abelian Groups

12

By a small sample we may judge of the whole piece
Miguel de Cervantes, *Don Quixote*

The Fundamental Theorem

In this chapter, we present a theorem that describes to an algebraist's eyes (that is, up to isomorphism) all finite Abelian groups in a standardized way. Before giving the proof, which is long and difficult, we discuss some consequences of the theorem and its proof. The first proof of the theorem was given by Leopold Kronecker in 1858.

> **Theorem 12.1** *Fundamental Theorem of Finite Abelian Groups*
> *Every finite Abelian group is a direct product of cyclic groups of prime-power order. Moreover, the factorization is unique except for rearrangement of the factors.*

Since a cyclic group of order n is isomorphic to Z_n, Theorem 12.1 shows that every finite Abelian group G is isomorphic to a group of the form

$$Z_{p_1^{n_1}} \oplus Z_{p_2^{n_2}} \oplus \cdots \oplus Z_{p_k^{n_k}}$$

where the p_i are not necessarily distinct primes and the prime-powers $p_1^{n_1}$, $p_2^{n_2}, \ldots, p_k^{n_k}$ are uniquely determined by G. Writing a group in this form is called *determining the isomorphism class of G.*

The Isomorphism Classes of Abelian Groups

The Fundamental Theorem is extremely powerful. As an application, we can use it as an algorithm for constructing all Abelian groups of any order. Let's look at groups whose orders have the form p^k where p is prime and $k \leq 4$.

Order of G	Possible direct products for G
p	Z_p
p^2	Z_{p^2}
	$Z_p \oplus Z_p$
p^3	Z_{p^3}
	$Z_{p^2} \oplus Z_p$
	$Z_p \oplus Z_p \oplus Z_p$
p^4	Z_{p^4}
	$Z_{p^3} \oplus Z_p$
	$Z_{p^2} \oplus Z_{p^2}$
	$Z_{p^2} \oplus Z_p \oplus Z_p$
	$Z_p \oplus Z_p \oplus Z_p \oplus Z_p$

In general, there is one group of order p^k for each set of positive integers whose sum is k (such a set is called a *partition* of k); that is, if k can be written as

$$k = n_1 + n_2 + \cdots + n_t$$

where each n_i is a positive integer, then

$$Z_{p^{n_1}} \oplus Z_{p^{n_2}} \oplus \cdots \oplus Z_{p^{n_t}}$$

is an Abelian group of order p^k. Furthermore, the uniqueness portion of the Fundamental Theorem guarantees that distinct partitions of k yield distinct isomorphism classes. Thus, for example, $Z_9 \oplus Z_3$ is not isomorphic to $Z_3 \oplus Z_3 \oplus Z_3$. A reliable mnemonic for comparing external direct products is the cancellation property: If A is *finite,* then

$$A \oplus B \approx A \oplus C \quad \text{if and only if} \quad B \approx C \quad \text{(see [1])}.$$

Thus $Z_4 \oplus Z_4$ is not isomorphic to $Z_4 \oplus Z_2 \oplus Z_2$ because Z_4 is not isomorphic to $Z_2 \oplus Z_2$.

To fully appreciate the potency of the Fundamental Theorem, contrast the ease with which the Abelian groups of order p^k, $k \leq 4$, were determined with the corresponding problem for non-Abelian groups. Even a description of the two non-Abelian groups of order 8 is a challenge (see Chapter 28), and a description of the nine non-Abelian groups of order 16 is well beyond the level of this text.

Now that we know how to construct all the Abelian groups of prime-power order, we move to the problem of constructing all Abelian groups of a certain order n where n has two or more distinct prime divisors. We begin by writing n in prime-power decomposition form $n = p_1^{n_1} p_2^{n_2} \cdots p_k^{n_k}$. Next, individually form all Abelian groups of order $p_1^{n_1}$, then $p_2^{n_2}$, and so on, as described earlier. Finally, form all possible external direct products of these. For example, let $n = 1176 = 2^3 \cdot 3 \cdot 7^2$. Then, the complete list of the distinct isomorphism classes of Abelian groups of order 1176 is

$$Z_8 \oplus Z_3 \oplus Z_{49},$$
$$Z_4 \oplus Z_2 \oplus Z_3 \oplus Z_{49},$$
$$Z_2 \oplus Z_2 \oplus Z_2 \oplus Z_3 \oplus Z_{49},$$
$$Z_8 \oplus Z_3 \oplus Z_7 \oplus Z_7,$$
$$Z_4 \oplus Z_2 \oplus Z_3 \oplus Z_7 \oplus Z_7,$$
$$Z_2 \oplus Z_2 \oplus Z_2 \oplus Z_3 \oplus Z_7 \oplus Z_7.$$

If we are given any particular Abelian group G of order 1176, the question we want to answer about G is which of the preceding six isomorphism classes represents the structure of G? Again, we can answer this question by comparing the orders of the elements of G with the orders of the elements in the six direct products, since it can be shown that two Abelian groups are isomorphic if and only if they have the same number of elements of each order. For instance, we could determine whether G has any elements of order 8. If so, then G must be isomorphic to the first or fourth group above, since these are the only ones with elements of order 8. To narrow G down to a single choice, we now need only check whether or not G has an element of order 49, since the first product above has such an element, while the fourth one does not.

What if we have some specific Abelian group G of order $p_1^{n_1} p_2^{n_2} \cdots p_k^{n_k}$ where the p_i's are distinct primes? How can G be expressed as an *internal* direct product of cyclic groups of prime-power order? For simplicity, let us say the group has 2^n elements. First, we must compute the orders of the elements. After this is done, pick an element of maximum order 2^r, call it a_1. Then $\langle a_1 \rangle$ is one of the factors in the desired internal direct product. If $G \neq \langle a_1 \rangle$ choose an element a_2 of maximum order 2^s such that $s \leq n - r$ and none of $a_2, a_2^2, a_2^4, \ldots,$ $a_2^{2^{s-1}}$ is in $\langle a_1 \rangle$. Then $\langle a_2 \rangle$ is a second direct factor. If $n \neq r + s$, select an element a_3 of maximum order 2^t such that $t \leq n - r - s$ and none of $a_3, a_3^2,$ $a_3^4, \ldots, a_3^{2^{t-1}}$ is in $\langle a_1 \rangle \times \langle a_2 \rangle = \{a_1^i a_2^j \mid 0 \leq i < r, 0 \leq j < s\}$. Then $\langle a_3 \rangle$ is another direct factor. We continue in this fashion until our direct product has the same order as G.

A formal presentation of this algorithm for any Abelian group G of prime-power order p^n is as follows:

Greedy Algorithm for an Abelian Group of Order p^n

1. *Compute the orders of the elements of the group G.*
2. *Select an element a_1 of maximum order and define $G_1 = \langle a_1 \rangle$. Set $i = 1$.*

3. *If* $|G| = |G_i|$, *stop. Otherwise, replace* i *by* $i + 1$.
4. *Select an element* a_i *of maximum order* p^k *such that* $p^k \leq |G|/|G_{i-1}|$ *and none of* a_i, a_i^p, $a_i^{p^2}$, . . . , $a_i^{p^{k-1}}$ *is in* G_{i-1}, *and define* $G_i = G_{i-1} \times \langle a_i \rangle$.
5. *Return to step 3.*

In the general case where $|G| = p_1^{n_1} p_2^{n_2} \cdots p_k^{n_k}$ one simply uses the algorithm to build up a direct product of order $p_1^{n_1}$, then another of order $p_2^{n_2}$, and so on. The direct product of all of these pieces is the desired factorization of G. The following example is small enough that we can compute the appropriate internal and external direct products by hand.

Example 1 Let $G = \{1, 8, 12, 14, 18, 21, 27, 31, 34, 38, 44, 47, 51, 53, 57, 64\}$ under multiplication modulo 65. Since G has order 16, we know it is isomorphic to one of

$$Z_{16},$$
$$Z_8 \oplus Z_2,$$
$$Z_4 \oplus Z_4,$$
$$Z_4 \oplus Z_2 \oplus Z_2,$$
$$Z_2 \oplus Z_2 \oplus Z_2 \oplus Z_2.$$

To decide which one, we dirty our hands to calculate the orders of the elements of G.

Element	1	8	12	14	18	21	27	31	34	38	44	47	51	53	57	64
Order	1	4	4	2	4	4	4	4	4	4	4	4	2	4	4	2

From the table of orders we can instantly rule out all but $Z_4 \oplus Z_4$ and $Z_4 \oplus Z_2 \oplus Z_2$ as possibilities. Finally, we observe that this latter group has only 8 elements of order 4 (exercise 4) so that $G \approx Z_4 \oplus Z_4$.

Expressing G as an internal direct product is even easier. Pick an element of maximum order, say, the element 8. Then $\langle 8 \rangle$ is a factor in the product. Next, choose a second element, say, a, so that a has order 4 and a and a^2 are not in $\langle 8 \rangle = \{1, 8, 64, 57\}$. Since 12 has this property, we have $G = \langle 8 \rangle \times \langle 12 \rangle$. $\square$

Example 1 illustrates how quick and easy it is to write an Abelian group as a direct product given the orders of the elements of the group. But calculating all those orders is certainly not an appealing prospect! The good news is that, in practice, a combination of theory and calculating the orders of a few elements will usually suffice.

Example 2 Let $G = \{1, 8, 17, 19, 26, 28, 37, 44, 46, 53, 62, 64, 71, 73, 82, 89, 91, 98, 107, 109, 116, 118, 127, 134\}$ under multiplication modulo 135.

Since G has order 24, it is isomorphic to one of

$$Z_8 \oplus Z_3 \approx Z_{24}$$
$$Z_4 \oplus Z_2 \oplus Z_3 \approx Z_{12} \oplus Z_2$$
$$Z_2 \oplus Z_2 \oplus Z_2 \oplus Z_3 \approx Z_6 \oplus Z_2 \oplus Z_2$$

Consider the element 8. Direct calculations show that $8^6 = 109$ and $8^{12} = 1$. (Be sure to mod as you go. For example, $8^3 = 512 = 107 \bmod 135$, so compute 8^4 as $8 \cdot 107$ rather than $8 \cdot 512$.) But now we know G. Why? Clearly, $|8| = 12$ rules out the third group on the list. At the same time, $|109| = 2 = |134|$ (remember, $134 = -1 \bmod 135$) implies that G is not Z_{24} (see Theorem 4.4). Thus, $G \approx Z_{12} \oplus Z_2$, and $G = \langle 8 \rangle \times \langle 134 \rangle$. $\qquad\square$

Rather than express an Abelian group as a direct product of cyclic groups of prime-power orders, it is often more convenient to combine the cyclic factors of relatively prime order, as we did in Example 2, to obtain a direct product of the form $Z_{n_1} \oplus Z_{n_2} \oplus \cdots \oplus Z_{n_k}$ where n_i divides n_{i-1}. For example, $Z_4 \oplus Z_4 \oplus Z_2 \oplus Z_9 \oplus Z_3 \oplus Z_5$ would be written as $Z_{180} \oplus Z_{12} \oplus Z_2$ (see exercise 10). The algorithm above is easily adapted to accomplish this by replacing step 4 by 4': Select an element a_i of maximum order m such that $m \leq |G|/|G_{i-1}|$ and none of $a_i, a_i^2, \ldots, a_i^{m-1}$ is in G_{i-1}, and define $G_i = G_{i-1} \times \langle a_i \rangle$.

As a consequence of the Fundamental Theorem of Finite Abelian Groups, we have the following corollary.

Corollary *Existence of Subgroups of Abelian Groups*
If m divides the order of a finite Abelian group G, then G has a subgroup of order m.

It is instructive to verify this corollary for a specific case. Let us say G is an Abelian group of order 72 and we wish to produce a subgroup of order 12. According to the Fundamental Theorem, G is isomorphic to one of the following six groups:

$Z_8 \oplus Z_9,$	$Z_8 \oplus Z_3 \oplus Z_3,$
$Z_4 \oplus Z_2 \oplus Z_9,$	$Z_4 \oplus Z_2 \oplus Z_3 \oplus Z_3,$
$Z_2 \oplus Z_2 \oplus Z_2 \oplus Z_9,$	$Z_2 \oplus Z_2 \oplus Z_2 \oplus Z_3 \oplus Z_3.$

Obviously, $Z_8 \oplus Z_9 \approx Z_{72}$ and $Z_4 \oplus Z_2 \oplus Z_3 \oplus Z_3 \approx Z_{12} \oplus Z_6$ have a subgroup of order 12. To construct a subgroup of order 12 in $Z_4 \oplus Z_2 \oplus Z_9$, we simply piece together all of Z_4 and the subgroup of order 3 in Z_9; that is, $\{(a, 0, b) \mid a \in Z_4, b \in \{0, 3, 6\}\}$. A subgroup of order 12 in $Z_8 \oplus Z_3 \oplus Z_3$ is given by $\{(a, b, 0) \mid a \in \{0, 2, 4, 6\}, b \in Z_3\}$. An analogous procedure applies to the remaining cases and indeed to any finite Abelian group.

Proof of the Fundamental Theorem

Because of the length and complexity of the proof of the Fundamental Theorem of Finite Abelian groups, we will break it up into a series of lemmas.

Lemma 1 *Let G be a finite Abelian group of order mn where m and n are relatively prime. If $H = \{x \in G \mid x^m = e\}$ and $K = \{x \in G \mid x^n = e\}$, then $G = H \times K$.*

Proof. It is an easy exercise to prove that H and K are subgroups of G (see exercise 21 of Chapter 3). Because G is Abelian we need only prove $G = HK$ and $H \cap K = \{e\}$. Since $\gcd(m, n) = 1$, there are integers s and t such that $1 = sm + tn$. For any x in G, we have $x = x^1 = x^{sm+tn} = x^{sm}x^{tn}$ and, by Corollary 3 of Lagrange's Theorem, $x^{tn} \in H$ and $x^{sm} \in K$. Thus, $G = HK$. Now suppose some $x \in H \cap K$. Then $x^m = e = x^n$ and, by the corollary to Theorem 4.1, $|x|$ divides both m and n. Since $\gcd(m, n) = 1$, we have $|x| = 1$ and, therefore, $x = e$. ∎

Given an Abelian group G with $|G| = p_1^{n_1} p_2^{n_2} \ldots p_k^{n_k}$, where the p's are distinct primes, we let $G(p_i)$ denote the set $\{x \in G \mid x^{p_i^{n_i}} = e\}$. It then follows immediately from Lemma 1 and induction that $G = G(p_1) \times G(p_2) \times \cdots \times G(p_n)$. Hence, we turn our attention to groups of prime-power order.

Lemma 2 *Let G be an Abelian group of prime-power order and let a be an element of maximal order in G. Then G can be written in the form $\langle a \rangle \times K$.*

Proof. We denote $|G|$ by p^n and induct on n. If $n = 1$, then $G = \langle a \rangle \times \langle e \rangle$. Now assume the statement is true for all Abelian groups of order p^k where $k < n$. Among all the elements of G, choose a of maximal order p^m. Then $x^{p^m} = e$ for all x in G. We may assume that $G \neq \langle a \rangle$, otherwise there is nothing to prove. Now, among all the elements of G choose b of smallest order such that $b \notin \langle a \rangle$. We claim that $\langle a \rangle \cap \langle b \rangle = \{e\}$. Clearly, we may establish this claim by showing that $|b| = p$. Since $|b^p| = |b|/p$, we know $b^p \in \langle a \rangle$ by the manner in which b was chosen. Say, $b^p = a^i$. Notice that $e = b^{p^m} = (b^p)^{p^{m-1}} = (a^i)^{p^{m-1}}$, so $|a^i| \leq p^{m-1}$. Thus, a^i is not a generator of $\langle a \rangle$ and, therefore, by Theorem 4.2, $\gcd(p^m, i) \neq 1$. This proves that p divides i, so that we can write $i = pj$. Then $b^p = a^i = a^{pj}$. Consider the element $c = a^{-j}b$. Certainly, c is not in $\langle a \rangle$ for if it were, b would be too. Also, $c^p = a^{-jp}b^p = a^{-i}b^p = b^{-p}b^p = e$. Thus, we have found an element c of order p such that $c \notin \langle a \rangle$. Since b was chosen to have smallest order so that $b \notin \langle a \rangle$, we conclude that b also has order p and our claim is verified.

Now consider the factor group $\overline{\overline{G}} = G/\langle b \rangle$. To simplify the notation, we let $\overline{x}$ denote the coset $x\langle b \rangle$ in $\overline{G}$. If $|\overline{a}| < |a| = p^m$, then $\overline{a}^{p^{m-1}} = \overline{e}$. This means that $(a\langle b \rangle)^{p^{m-1}} = a^{p^{m-1}}\langle b \rangle = \langle b \rangle$, so that $a^{p^{m-1}} \in \langle a \rangle \cap \langle b \rangle = \{e\}$, contradicting the fact that $|a| = p^m$. Thus, $|\overline{a}| = |a| = p^m$, and therefore $\overline{a}$ is an element of maximal order in $\overline{G}$. By induction, we know $\overline{G}$ can be written in the form $\langle \overline{a} \rangle \times \overline{K}$ for some subgroups $\overline{K}$ of $\overline{G}$. Let K be the pullback of $\overline{K}$ under the natural homomorphism from G to $\overline{G}$ (that is, $K = \{x \in G \mid \overline{x} \in \overline{K}\}$). We claim that $\langle a \rangle \cap K = \{e\}$. For if $x \in \langle a \rangle \cap K$, then $\overline{x} \in \langle \overline{a} \rangle \cap \overline{K} = \{\overline{e}\} = \langle b \rangle$ and $x \in \langle a \rangle \cap \langle b \rangle = \{e\}$. It now follows from an order argument (see exercise 34) that $G = \langle a \rangle K$, and therefore $G = \langle a \rangle \times K$. ∎

Lemma 2 and induction now give the following.

Lemma 3   *A finite Abelian group of prime-power order is a direct product of cyclic groups.*

Let us pause to determine where we are in our effort to prove the Fundamental Theorem of Finite Abelian Groups. The remark following Lemma 1 shows that $G = G(p_1) \times G(p_2) \times \cdots \times G(p_n)$ where each $G(p_i)$ is a group of prime-power order, and Lemma 3 shows that each of these factors is a direct product of cyclic groups. Thus, we have proved G is a direct product of cyclic groups of prime-power order. All that remains to prove is the uniqueness of the factors. Certainly the groups $G(p_i)$ are uniquely determined by G since they comprise the elements of G of order a power of p_i. So we must prove there is only one way (up to isomorphism and rearrangement of factors) to write each $G(p_i)$ as a direct product of cyclic groups.

Lemma 4   *Suppose G is a finite Abelian group of prime-power order. If $G = H_1 \times H_2 \times \cdots \times H_m$ and $G = K_1 \times K_2 \times \cdots \times K_n$, where the H's and K's are nontrivial cyclic subgroups with $|H_1| \geqslant |H_2| \geqslant \cdots \geqslant |H_m|$ and $|K_1| \geqslant |K_2| \geqslant \cdots \geqslant |K_n|$, then $m = n$ and $|H_i| = |K_i|$ for all i.*

Proof.   We proceed by induction on $|G|$. Clearly, the case where $|G| = p$ is true. Now suppose the statement is true for all Abelian groups of order less than $|G|$. For any Abelian group L, the set $L^p = \{x^p \mid x \in L\}$ is a subgroup of L (see exercise 13 of Supplementary Exercises for Chapters 1–4). It follows that $G^p = H_1^p \times H_2^p \times \cdots \times H_{m'}^p$, and $G^p = K_1^p \times K_2^p \times \cdots \times K_{n'}^p$, where m' is the largest integer i such that $|H_i| > p$, and n' is the largest integer j such that $|K_j| > p$. (This ensures that our two direct products for G^p do not have trivial factors.) Since $|G^p| < |G|$, we have by induction that $m' = n'$ and $|H_i^p| = |K_i^p|$ for $i = 1, \ldots, m'$. Since $|H_i| = p|H_i^p|$, this proves $|H_i| = |K_i|$ for all $i = 1, \ldots, m'$. All that remains to prove is that the number of H_i of order p equals the number of K_i of order p; that is, we must prove $m - m' = n - n'$ (since $n' = m'$). This follows directly from the facts that $|H_1| |H_2| \cdots |H_{m'}| p^{m-m'} = |G| = |K_1| |K_2| \cdots |K_{n'}| p^{n-n'}$ and $|H_i| = |K_i|$ and $m' = n'$. ∎

Our proof of the Fundamental Theorem is now complete.

EXERCISES

Solving a problem is similar to building a house. We must collect the right material, but collecting the material is not enough; a heap of stones is not yet a house. To construct the house or the solution, we must put together the parts and organize them into a purposeful whole.

George Polya

1. What is the smallest positive integer n such that there are two nonisomorphic groups of order n?

2. What is the smallest positive integer n such that there are three nonisomorphic Abelian groups of order n?

3. What is the smallest positive integer n such that there are exactly four nonisomorphic Abelian groups of order n?

4. Calculate the number of elements of order 4 in each of Z_{16}, $Z_8 \oplus Z_2$, $Z_4 \oplus Z_4$, and $Z_4 \oplus Z_2 \oplus Z_2$.

5. Prove that any Abelian group of order 45 has an element of order 15. Does every Abelian group of order 45 have an element of order 9?

6. Show that there are two Abelian groups of order 108 that have exactly one subgroup of order 3.

7. Show that there are two Abelian groups of order 108 that have exactly four subgroups of order 3.

8. Show that there are two Abelian groups of order 108 that have exactly 13 subgroups of order 3.

9. Suppose G is an Abelian group of order 120 and G has exactly three elements of order 2. Determine the isomorphism class of G.

10. Prove that every finite Abelian group can be expressed as the (external) direct product of cyclic groups of orders $n_1, n_2, \ldots, n_t$ where n_{i+1} divides n_i for $i = 1, 2, \ldots, t - 1$. This exercise is referred to in Chapter 24.

11. Find all Abelian groups (up to isomorphism) of order 360.

12. Suppose the order of some finite Abelian group is divisible by 10. Prove that the group has a cyclic subgroup of order 10.

13. Show, by example, that if the order of a finite Abelian group is divisible by 4, the group need not have a cyclic subgroup of order 4.

14. On the basis of exercises 12 and 13, draw a general conclusion about the existence of cyclic subgroups of a finite Abelian group.

15. How many Abelian groups (up to isomorphism) are there
    a. of order 6?
    b. of order 15?
    c. of order 42?
    d. of order pq where p and q are distinct primes?
    e. of order pqr where p, q, and r are distinct primes?
    f. Generalize parts a, b, c, d, e.

16. How does the number (up to isomorphism) of Abelian groups of order n compare with the number (up to isomorphism) of Abelian groups of order m where
    a. $n = 3^2$ and $m = 5^2$?
    b. $n = 2^4$ and $m = 5^4$?
    c. $n = p^r$ and $m = q^r$ where p and q are prime?
    d. $n = p^r$ and $m = p^r q$ where p and q are distinct primes?
    e. $n = p^r$ and $m = p^r q^2$ where p and q are distinct primes?

17. The symmetry group of a nonsquare rectangle is an Abelian group of order 4. Is it isomorphic to Z_4 or $Z_2 \oplus Z_2$?

18. Verify the corollary to the Fundamental Theorem of Finite Abelian Groups in the case that the group has order 1080 and the divisor is 180.

19. The set $\{1, 9, 16, 22, 29, 53, 74, 79, 81\}$ is a group under multiplication modulo 91. Determine the isomorphism class of this group.

20. Determine the isomorphism class of the Nim group given in exercise 37 of Chapter 2.

21. Characterize those integers n such that the only Abelian groups of order n are cyclic.

22. Characterize those integers n such that any Abelian group of order n belongs to one of exactly four isomorphism classes.

23. Refer to Example 1 of this chapter and explain why it is unnecessary to compute the orders of the last five elements listed to determine the isomorphism class of G.

24. Let $G = \{1, 7, 17, 23, 49, 55, 65, 71\}$ under multiplication modulo 96. Express G as an external and internal direct product of cyclic groups.

25. Let $G = \{1, 7, 43, 49, 51, 57, 93, 99, 101, 107, 143, 149, 151, 157, 193, 199\}$ under multiplication modulo 200. Express G as an external and internal direct product of cyclic groups.

26. The set $G = \{1, 4, 11, 14, 16, 19, 26, 29, 31, 34, 41, 44\}$ is a group under multiplication modulo 45. Write G as an external and internal direct product of cyclic groups of prime-power order.

27. Suppose G is an Abelian group of order 9. What is the maximum number of elements (excluding the identity) one needs to compute the order of to determine the isomorphism class of G? What if G has order 18? What about 16?

28. Suppose that G is an Abelian group of order 16, and in computing the orders of its elements, you come across an element of order 8 and two elements of order 2. Explain why no further computations are needed to determine the isomorphism class of G.

29. Let G be an Abelian group of order 16. Suppose there are elements a and b in G such that $|a| = |b| = 4$ and $a^2 \neq b^2$. Determine the isomorphism class of G.

30. Prove that an Abelian group of order $2^n (n \geq 1)$ must have an odd number of elements of order 2.

31. Show that an Abelian group of odd order cannot have an element of even order.

32. Suppose G is an Abelian group with an odd number of elements. Show that the product of all of the elements of G is the identity.

33. Suppose G is a finite Abelian group. Prove that G has order p^n where p is prime if and only if the order of every element of G is a power of p.

34. Prove the assertion, made in Lemma 2, that $G = \langle a \rangle K$.

35. Dirichlet's Theorem says that, for every pair of relatively prime integers a and b, there are infinitely many primes of the form $at + b$. Use Dirichlet's Theorem to prove that every finite Abelian group is isomorphic to a subgroup of a U-group.

PROGRAMMING EXERCISES

It is a test of true theories not only to account for but to predict phenomena.
William Whewell, *Philosophy of the Inductive Sciences*

1. Write a program that lists the isomorphism classes of all finite Abelian groups of order n. Assume $n < 1,000,000$. Run your program for $n = 16, 24, 512, 2048, 441,000, 999,999$.

2. Write a program to determine how many integers in a given interval are the order of exactly one Abelian group, of exactly two Abelian groups, and so on, up to exactly nine Abelian groups. Run your program for the integers up to 1000.

3. Write a program that implements the algorithm given in this chapter for expressing an Abelian group as an internal direct product. Run your program for the groups $U(32)$, $U(80)$, and $U(65)$.

REFERENCE

1. R. Hirshon, "On Cancellations in Groups," *American Mathematical Monthly* 76 (1969): 1037–1039.

SUGGESTED READINGS

J. A. Gallian, "Computers in Group Theory," *Mathematics Magazine* 49 (1976): 69–73.

This paper discusses a number of computer-related projects in group theory done by undergraduate students.

J. Kane, "Distribution of Orders of Abelian Groups," *Mathematics Magazine* 49 (1976): 132–135.

In this note the author determines the percentages of integers k between 1 and n, for sufficiently large n, that have exactly one isomorphism class of Abelian groups of order k, exactly two isomorphism classes of Abelian groups of order k, and so on, up to 13 isomorphism classes.

SUPPLEMENTARY EXERCISES for Chapters 9–12

Every prospector drills many a dry hole, pulls out his rig, and moves on.

John L. Hess

1. Suppose H is a subgroup of G and each left coset of H in G is some right coset of H in G. Prove that H is normal in G.

2. Use a factor group-induction argument to prove that a finite Abelian group of order n has a subgroup of order m for every divisor m of n.

3. Suppose N is a subgroup of G of index 2. If x, $y \in G$ and $x \notin N$, $y \notin N$, prove that $xy \in N$.

4. Show that a group of order p^2, where p is prime, can be generated by two elements a and b such that every element can be expressed in the form $a^i b^j$.

5. Suppose H and K are subgroups of a group G and some left coset of H equals some left coset of K. Show that $H = K$. Show, by example, that H need not equal K if some left coset of H equals some right coset of K.

6. Show that a group of order 9 is Abelian.

7. Let H be a subgroup of G and let a, $b \in G$. Show that $aH = bH$ if and only if $Ha^{-1} = Hb^{-1}$.

8. Let $H \triangleleft G$. Show that $ab \in H$ implies $ba \in H$. Is this true when H is not normal?

9. Recall, $\text{diag}(G) = \{(g, g) \mid g \in G\}$. Prove that $\text{diag}(G) \triangleleft G \oplus G$ if and only if G is Abelian. When G is the set of real numbers, describe $\text{diag}(G)$ geometrically. When G is finite, what is the index of $\text{diag}(G)$ in G?

10. Let H be any group of rotations in D_n. Prove that H is normal in D_n.

11. Prove that $\text{Inn}(G) \triangleleft \text{Aut}(G)$.

12. Let G be a group, and let $a \in G$. Show that $\langle a \rangle Z(G)$ is Abelian.

13. The factor group $GL(2, \mathbf{R})/SL(2, \mathbf{R})$ is isomorphic to some very familiar group. What is this group?

14. Let k be a divisor of n. The factor group $(Z/\langle n \rangle)/(\langle k \rangle/\langle n \rangle)$ is isomorphic to some very familiar group. What is this group?

15. Let

$$H = \left\{ \begin{bmatrix} 1 & a & b \\ 0 & 1 & c \\ 0 & 0 & 1 \end{bmatrix} \middle| \; a, b, c \in Q \right\}.$$

Prove that

a. $Z(H)$ is isomorphic to Q under addition.

b. $H/Z(H)$ is isomorphic to $Q \oplus Q$.

c. Are your proofs for (a) and (b) valid when Q is replaced by $\mathbf{R}$? Are they valid when Q is replaced by $Z(\overline{p})$? $\mathbb{Z}_{\mathcal{Q}}$

16. Prove that $D_4/Z(D_4)$ is isomorphic to $Z_2 \oplus Z_2$.

17. Prove that Q/Z under addition is an infinite group in which every element has finite order.

18. Show that the intersection of any collection of normal subgroups of a group is a normal subgroup.

19. Let n be a fixed positive integer and G a group. If $H = \{x \in G \mid |x| = n\}$ is a subgroup of G, prove that it is normal in G. Give an example of a group G and an integer n where H is not a subgroup of G.

20. Suppose H and K are subgroups of a group and $|H|$ and $|K|$ are relatively prime. Show $H \cap K = \{e\}$.

21. How many subgroups of order 2 does $Z_2 \oplus Z_2$ have? How many of order 3 in $Z_3 \oplus Z_3$? How many subgroups of order p (p a prime) does $Z_p \oplus Z_p$ have? Prove that your answer is correct.

22. Without defining an isomorphism, explain why $D_6/Z(D_6)$ is isomorphic to D_3.

23. If H and K are normal Abelian subgroups of a group and if $H \cap K = \{e\}$, prove that HK is Abelian.

24. Let G be a group of permutations on the set $\{1, 2, \ldots, n\}$. Recall, $\text{stab}_G(1) = \{\alpha \in G \mid 1\alpha = 1\}$. If γ sends 1 to k, prove that $\text{stab}_G(1)\gamma = \{\beta \in G \mid 1\beta = k\}$.

25. Let G be an Abelian group and n be a positive integer. Let $G_n = \{g \mid g^n = e\}$ and $G^n = \{g^n \mid g \in G\}$. Prove that G/G_n is isomorphic to G^n.

26. Prove that the mapping $x \to x^2$ from a finite group to itself is one-to-one if the group has odd order.

27. Suppose G is a group of permutations on some set. If $|G| = 60$ and $\text{orb}_G(5) = \{1, 5\}$, prove that $\text{stab}_G(5)$ is normal in G.

28. Find a subgroup H and elements a and b of some group such that $aH = bH$ but $Ha \neq Hb$.

29. Let $n = 2m$ where m is odd. How many elements of order 2 does $D_n/Z(D_n)$ have? How many elements are in the subgroup $\langle R_{360/n} \rangle / Z(D_n)$? How do these compare with the number of elements of order 2 in D_m?

30. Suppose that G is isomorphic to $Z_2 \oplus Z_2 \oplus \cdots \oplus Z_2$ and $|G| > 2$. Show that the product of all the elements of G is the identity.

31. Let G be a finite Abelian group of order $2^n m$ where m is odd. If the subgroup of order 2^n is not cyclic, show that the product of all the elements of G is the identity. What is this product if the subgroup of order 2^n is cyclic?

32. Suppose $G = H \times K$ and N is a normal subgroup of H. Prove that N is normal in G.

RINGS

PART 3

Introduction to Rings

13

Example is the school of mankind, and they will learn at no other.

Edmund Burke, *On a Regicide Peace*

Motivation and Definition

Many sets are naturally endowed with two binary operations: addition and multiplication. Examples that quickly come to mind are the integers, the integers modulo n, the real numbers, matrices, and polynomials. When considering these sets as groups, we simply used addition and ignored multiplication. In many instances, however, one wishes to take into account both addition and multiplication. One abstract concept that does this is a ring.* This notion was originated in the mid–nineteenth century by Richard Dedekind.

> *DEFINITION*   Ring
>
> A *ring R* is a set with two binary operations, addition ($a + b$) and multiplication (ab) such that for all a, b, c in R
>
> 1. $a + b = b + a$.
> 2. $(a + b) + c = a + (b + c)$.
> 3. There is an element 0 in R such that $a + 0 = a$.
> 4. There is an element $-a$ in R such that $a + (-a) = 0$.
> 5. $a(bc) = (ab)c$.
> 6. $a(b + c) = ab + ac$ and $(b + c)a = ba + ca$.

---

*The term *ring* was coined by the German mathematician David Hilbert (1862–1943).

So, a ring is an Abelian group under addition, also having an associative multiplication that is left and right distributive over addition. Note that multiplication need not be commutative. When it is, we say the ring is *commutative*. Also, a ring need not have an identity under multiplication. When a ring other than {0} has an identity under multiplication, we say the ring has a *unity* (or *identity*). A nonzero element of a commutative ring with unity need not have a multiplicative inverse. When it does, we say it is a *unit* of the ring. Thus, a is a unit if a^{-1} exists.

The following terminology and notation is convenient. If a and b belong to a commutative ring R and a is nonzero, we say a *divides* b (or a is a *factor* of b) and write $a \mid b$, if there exists an element c in R such that $b = ac$. If a does not divide b, we write $a \nmid b$.

For an abstraction to be worthy of study, it must have many diverse concrete realizations. The following list of examples shows that the ring concept is pervasive.

Examples of Rings

Example 1   The set Z of integers under ordinary addition and multiplication is a commutative ring with unity 1. The units of Z are 1 and -1.   □

Example 2   The set $Z_n = \{0, 1, \ldots, n - 1\}$ under addition and multiplication modulo n is a commutative ring with unity 1. The set of units is $U(n)$.   □

Example 3   The set $Z[x]$ of all polynomials in the variable x with integer coefficients under ordinary addition and multiplication is a commutative ring with unity $f(x) = 1$.   □

Example 4   The set $M_2(Z)$ of 2×2 matrices with integer entries is a noncommutative ring with unity $\begin{bmatrix} 1 & 0 \\ 0 & 1 \end{bmatrix}$.   □

Example 5   The set $2Z$ of even integers under ordinary addition and multiplication is a commutative ring without unity.   □

Example 6   The set of all continuous real-valued functions of a real variable whose graphs pass through the point (1, 0) is a commutative ring without unity under the operations of pointwise addition and multiplication (that is, the operations $(f + g)(a) = f(a) + g(a)$ and $(f \cdot g)(a) = f(a) \cdot g(a)$).   □

Example 7   Let $R_1, R_2, \ldots, R_n$ be rings. We can use these to construct a new ring as follows. Let

$$R_1 \oplus R_2 \oplus \cdots \oplus R_n = \{(a_1, a_2, \ldots, a_n) \mid a_i \in R_i\}$$

and perform addition and multiplication componentwise; that is, define

$$(a_1, a_2, \ldots, a_n) + (b_1, b_2, \ldots, b_n) = (a_1 + b_1, a_2 + b_2, \ldots, a_n + b_n)$$

and

$$(a_1, a_2, \ldots, a_n)(b_1, b_2, \ldots, b_n) = (a_1b_1, a_2b_2, \ldots, a_nb_n).$$

This ring is called the *direct sum* of $R_1, R_2, \ldots, R_n$. ☐

All of the preceding examples are quite familiar to you, and the fact that they are rings is obvious. Our last example is included to show that rings, even those with only a handful of elements, can be unusual. It is one of the 11 different rings of order 4 (recall, there are only two groups of order 4).

Example 8 Let $R = \{0, a, b, c\}$. Define addition and multiplication by the Cayley tables:

+	0	a	b	c		·	0	a	b	c
0	0	a	b	c		0	0	0	0	0
a	a	0	c	b		a	0	a	b	c
b	b	c	0	a		b	0	a	b	c
c	c	b	a	0		c	0	0	0	0

Then R is a noncommutative ring without unity. ☐

In the next several chapters, we will see that many of the fundamental concepts of group theory can be naturally extended to rings. In particular, we will introduce rings homomorphisms and factor rings.

EXERCISES

1. The ring in Example 8 is finite and noncommutative. Give another example of a finite noncommutative ring. Give an example of an infinite noncommutative ring that does not have a unity.
2. The set $\{0, 2, 4\}$ under addition and multiplication modulo 6 has a unity. Find it.
3. Let $Z[\sqrt{2}] = \{a + b\sqrt{2} \mid a, b \in Z\}$. Prove that $Z[\sqrt{2}]$ is a ring under the ordinary addition and multiplication of real numbers.

4. Show that for fixed elements a and b in a ring, the equation $ax = b$ can have more than one solution. How does this compare with groups?

5. Prove that a ring can have at most one unity.

6. Find an integer n that shows the rings Z_n need not have the following properties that the ring of integers has.
   a. $a^2 = a$ implies $a = 0$ or $a = 1$.
   b. $ab = 0$ implies $a = 0$ or $b = 0$.
   c. $ab = ac$ and $a \neq 0$ implies $b = c$.
   Is the n you found prime?

7. Show that the three properties listed in exercise 6 are valid for Z_p where p is prime.

8. Show that the ring given in Example 8 has two left identities (i.e., an element r such that $rx = x$ for all x in the ring) but no right identity.

9. Show that a ring that is cyclic under addition is commutative.

10. Describe the elements of $M_2(Z)$ (see Example 4) that have multiplicative inverses.

11. Suppose that $R_1, R_2, \ldots, R_n$ are rings that contain nonzero elements. Show that $R_1 \oplus R_2 \oplus \cdots \oplus R_n$ has a unity if and only if each R_i has a unity.

12. Let R be a commutative ring with unity, and let $U(R)$ denote the set of units of R. Prove that $U(R)$ is a group under the multiplication of R. (This group is called the *group of units of R*.)

13. Describe the elements of $U(Z_n)$.

14. If $R_1, R_2, \ldots, R_n$ are commutative rings with unity show that
$$U(R_1 \oplus R_2 \oplus \cdots \oplus R_n) = U(R_1) \oplus U(R_2) \oplus \cdots \oplus U(R_n).$$

15. Determine $U(Z[x])$.

16. Determine $U(\mathbf{R}[x])$.

17. Show that a unit of a ring divides every element of the ring.

18. In Z_6, show that $4 \mid 2$; in Z_8, show that $3 \mid 7$; in Z_{15}, show that $9 \mid 12$.

19. Suppose a and b belong to a commutative ring R. If a is a unit of R and $b^2 = 0$, show that $a + b$ is a unit of R.

20. Suppose there is an integer $n > 1$ such that $a^n = a$ for all elements a of some ring. If m is a positive integer and $a^m = 0$ for some a, show that $a = 0$.

21. Give an example of ring elements a and b with the properties that $ab = 0$ but $ba \neq 0$.

22. In a ring in which $x^3 = x$ for all x, show that $ab = 0$ implies $ba = 0$.

23. For $p = 3, 7$, and 11, find all solutions of $a^2 + b^2 = 0$ in Z_p. Show that for $p = 5, 13$, and 17, there are nontrivial solutions (that is, $a \neq 0$, $b \neq 0$) of $a^2 + b^2 = 0$. Make a conjecture about the existence of nontrivial solutions of this equation in Z_p (p a prime) and the form of p.

PROGRAMMING EXERCISE

To make a start out of particulars and make them general. . . .

William Carlos Williams

1. Write a program to find all solutions of the equation $a^2 + b^2 = 0$ in Z_p. Run your program for all odd primes up to 37. Make a conjecture about the existence of nontrivial solutions in Z_p (p a prime) and the form of p.

Properties of Rings

14

In the beginning everything is self-evident, and it is hard to see whether one self-evident proposition follows from another or not. Obviousness is always the enemy of correctness.

Bertrand Russell

Properties of Rings

In this chapter we present some of the elementary properties of rings and introduce a few basic definitions needed in later chapters. Our first theorem shows how the operations of addition and multiplication intertwine.

> **Theorem 14.1** *Rules of Multiplication*
> *Let a, b, and c belong to a ring R. Then*
>
> 1. $a0 = 0a = 0$.
> 2. $a(-b) = (-a)b = -(ab)$.
> 3. $(-a)(-b) = ab.$*
> 4. $a(b - c) = ab - ac$ and $(b - c)a = ba - ca$.
>
> *Furthermore, if R has a unity element 1 then*
>
> 5. $(-1)a = -a$.
> 6. $(-1)(-1) = 1$.

---

*Minus times minus is plus,
The reason for this we need not discuss.*
                    W. H. Auden

Proof. We will prove (1) and (2) and leave the rest as easy exercises. To prove statements like those in Theorem 14.1, we need only "play off" the distributive property against the fact that R is a group under addition with additive identity 0. Consider (1). Clearly,

$$0 + a0 = a0 = a(0 + 0) = a0 + a0.$$

So by cancellation, $0 = a0$. Similarly, $0a = 0$.

To prove (2) we observe that

$$-(ab) + (ab) = 0 \quad \text{and} \quad a(-b) + ab = a(-b + b) = a0 = 0.$$

So $-(ab) + ab = a(-b) + ab$, and it follows from cancellation that $-(ab) = a(-b)$. The remainder of (2) is done analogously. ∎

Recall that in the case of groups, the identity and inverses are unique. What about rings? The same is true for rings, provided they exist. The proofs are identical to the ones given for groups and therefore are omitted.

Theorem 14.2 *Uniqueness of the Unity and Inverses*
If a ring has a unity, it is unique. If a ring element has an inverse, it is unique.

Many students have the mistaken tendency to treat a ring as if it were a group under *multiplication*. It is not. The two most common errors are to assume ring elements have multiplicative inverses—they need not, and that a ring has a multiplicative identity—it need not. For example, if a, b, and c belong to a ring, $a \neq 0$ and $ab = ac$, we *cannot* conclude that $b = c$. Similarly, if $a^2 = a$, we *cannot* conclude that $a = 0$ or 1 (as is the case with real numbers). In the first place, the ring need not have multiplicative cancellation, and in the second place, the ring need not have a multiplicative identity. There is an important class of rings where a multiplicative identity exists and multiplicative cancellation holds. This class is taken up in the next chapter.

Subrings

In our study of groups, subgroups played a crucial role. Subrings, the analogous idea in ring theory, play a much less important role than their counterparts in group theory. Nevertheless, subrings are important.

DEFINITION Subring
A subset S of a ring R is a *subring of R* if S is itself a ring with the operations of R.

Just as was the case for subgroups, there is a simple test for subrings.

Theorem 14.3 *Subring Test*
A nonempty subset S of a ring R is a subring if S is closed under subtraction and multiplication; that is, if a − b and ab are in S whenever a and b are in S.

Proof. Since addition in R is communicative and S is closed under subtraction, we know by the One-Step Subgroup Test (Theorem 3.1) that S is an Abelian group under addition. Also, since multiplication in R is associative as well as distributive over addition, the same is true for multiplication in S. Thus, the only condition remaining to check is that multiplication is a binary operation on S. But this is exactly what closure means. ∎

We leave it to the student to check that each of the following examples is a subring.

Example 1 $\{0\}$ and R are subrings of any ring R. $\{0\}$ is called the *trivial* subring of R. ☐

Example 2 $\{0, 2, 4\}$ is a subring of the ring Z_6, the integers modulo 6. ☐

Example 3 For each positive integer n, the set

$$nZ = \{0, \pm n, \pm 2n, \pm 3n, \ldots\}$$

is a subring of the integers Z. ☐

Example 4 The set of Gaussian integers

$$Z[i] = \{a + bi \mid a, b \in Z\}$$

is a subring of the complex numbers **C**. ☐

Example 5 Let R be the ring of all real-valued functions of a single real variable under pointwise addition and multiplication. The subset S of R of functions whose graphs pass through the origin form a subring of R. ☐

Example 6 The set

$$\left\{ \begin{bmatrix} a & 0 \\ 0 & b \end{bmatrix} \ \middle|\ a, b \in Z \right\}$$

of diagonal matrices is a subring of the ring of all 2×2 matrices over Z. ☐

We can picture the relationship between a ring and its various subrings by way of a subring lattice diagram. In such a diagram, any ring is a subring of all the rings that it is connected to by one or more upward lines. Figure 14.1 shows the relationship among some of the rings we have already discussed.

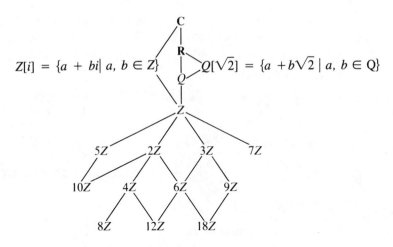

Figure 14.1   Partial subring lattice diagram of **C**.

EXERCISES

> Success and failure have much in common that is good. Both mean you're trying.
>
> Frank Tyger

1. What is wrong with the following "proof" that $(-a)(-b) = ab$. Observe that $(-a)(-b) = (-1)a(-1)b = (-1)(-1)ab = 1(ab) = ab$.

2. Verify that the six examples in this chapter are as stated.

3. Prove parts 3 through 6 of Theorem 14.1.

4. Give an example of a noncommutative ring that has exactly 16 elements.

5. Describe all the subrings of the ring of integers.

6. Show that if m and n are integers and a and b are elements from a ring, then $(ma)(nb) = (mn)(ab)$. (This exercise is referred to in Chapter 17.)

7. Show that if n is an integer and a is an element from a ring, then $n(-a) = -(na)$.

8. Let a and b belong to a ring R and let m be an integer. Prove that $m(ab) = (ma)b = a(mb)$.

9. Prove that the intersection of any collection of subrings of a ring R is a subring of R.

10. Let a belong to a ring R. Let $S = \{x \in R \mid ax = 0\}$. Show that S is a subring of R.

11. Let R be a ring. The *center of R* is the set $\{x \in R \mid ax = xa \text{ for all } a \text{ in } R\}$. Prove that the center of a ring is a subring.

12. Let m and n be positive integers and let k be the least common multiple of m and n. Show that $mZ \cap nZ = kZ$.

13. Explain why every subgroup of Z_n under addition is also a subring of Z_n.

14. Is Z_6 a subring of Z_{12}?

15. Suppose that R is a ring with unity 1 and a is an element of R such that $a^2 = 1$. Let $S = \{ara \mid r \in R\}$. Prove that S is a subring of R. Does S contain 1?

16. Let $M_2(Z)$ be the ring of all 2×2 matrices over the integers and $R = \left\{ \begin{bmatrix} a & a+b \\ a+b & b \end{bmatrix} \middle| a, b \in Z \right\}$. Prove or disprove that R is a subring of $M_2(Z)$.

17. Let $M_2(Z)$ be as in exercise 16 and $R = \left\{ \begin{bmatrix} a & a-b \\ a-b & b \end{bmatrix} \middle| a, b \in Z \right\}$. Prove or disprove that R is a subring of $M_2(Z)$.

18. Let $R = \left\{ \begin{bmatrix} a & a \\ b & b \end{bmatrix} \middle| a, b \in Z \right\}$. Prove or disprove that R is a subring of $M_2(Z)$.

19. Let $R = Z \oplus Z \oplus Z$ and $S = \{(a, b, c) \in R \mid a + b = c\}$. Prove or disprove that S is a subring of R.

20. Does the unity of a subring have to be the same as the unity of the whole ring? If yes, prove it; if no, give an example. What is the analogous situation for groups?

21. Show that $2Z \cup 3Z$ is not a subring of Z.

22. Let R be a ring with unity e. Show that $S = \{ne \mid n \in Z\}$ is a subring of R.

23. Determine the smallest subring of Q that contains $1/2$.

24. Determine the smallest subring of Q that contains $2/3$.

25. Let R be a ring. Prove that $a^2 - b^2 = (a + b)(a - b)$, for all a, b in R if and only if R is commutative.

26. Suppose R is a ring and $a^2 = a$ for all a in R. Show that R is commutative. (A ring in which $a^2 = a$ for all a is called a *Boolean* ring, in honor of the English mathematician George Boole [1815–1864].

27. Suppose that there is a positive even integer n such that $a^n = a$ for all elements a of some ring. Show that $-a = a$ for all a in the ring.

28. Give an example of a subset of a ring that is a subgroup under addition but not a subring.

PROGRAMMING EXERCISE

Theory is the general; experiments are the soldiers.

Leonardo da Vinci

1. Let $Z_n[i] = \{a + bi \mid a, b \in Z_n, i^2 = -1\}$ (the Gaussian integers modulo n). Write a program to find the group of units of this ring and the order of each element of the group. Run your program for $n = 3, 7, 11$, and 23. Is the group of units cyclic for these cases? Try to guess a formula for the order of the group of units of $Z_n[i]$ as a function of n when n is a prime and $n = 3$ modulo 4. Run your program for $n = 9$. Is this group cyclic? Does your formula predict the correct order for the group in this case? Run your program for $n = 5, 13$, and 17. Is the group cyclic for these cases? Try to guess a formula for the order of the group of units of $Z_n[i]$ as a function of n when n is a prime and $n = 1$ modulo 4. Run your program for $n = 25$. Is this group cyclic? Does your formula predict the correct order of the group in this case?

SUGGESTED READINGS

D. B. Erickson, "Orders for Finite Noncommutative Rings," *American Mathematical Monthly* 73 (1966): 376–377.

In this elementary paper, it is shown that there exists a noncommutative ring of order $m > 1$ if and only if m is divisible by the square of a prime.

Colin R. Fletcher, "Rings of Small Order," *The Mathematical Gazette* 64 (1980): 9–22.

This article gives a complete list of the 24 rings of order less than 8. The only hard case is the 11 rings of order 4.

I. N. Herstein

A whole generation of textbooks and an entire generation of mathematicians, myself included, have been profoundly influenced by that text [Herstein's Topics in Algebra.].

Georgia Benkart

I. N. HERSTEIN was born on March 28, 1923 in Poland. His family moved to Canada when he was seven. Herstein received a B.S. degree from the University of Manitoba, an M.A. from the University of Toronto and, in 1948, a Ph.D. degree from Indiana University under the supervision of Max Zorn. Before permanently settling at the University of Chicago in 1962, he held positions at the University of Kansas, Ohio State University, the University of Pennsylvania, and Cornell University.

Herstein wrote more than one hundred research papers and a dozen books. Although his principal interest was noncommutative ring theory, he also wrote papers on finite groups, linear algebra, and mathematical economics. His textbook, *Topics in Algebra,* first published in 1964, dominated the field for twenty years and has become a classic. Herstein had great influence through his teaching and his collaboration with colleagues. He had thirty Ph.D. students, and traveled and lectured widely. He died on February 9, 1988, after a long battle with cancer.

Integral Domains

$$15$$

The interplay between generality and individuality, deduction and construction, logic and imagination—this is the profound essence of live mathematics. Any one or another of these aspects of mathematics can be at the center of a given achievement. In a far reaching development all of them will be involved. Generally speaking, such a development will start from the "concrete" ground, then discard ballast by abstraction and rise to the lofty layers of thin air where navigation and observation are easy; after this flight comes the crucial test of landing and reaching specific goals in the newly surveyed low plains of individual "reality." In brief, the flight into abstract generality must start from and return again to the concrete and specific.

Richard Courant

Definition and Examples

To a certain degree, the notion of a ring was invented in an attempt to put the algebraic properties of the integers into an abstract setting. A ring is not the appropriate abstraction of the integers, however, for too much is lost in the process. Besides the two obvious properties of commutativity and existence of a unity, there is one further essential feature of the integers that rings in general do not enjoy. In this chapter, we introduce integral domains—a particular class of rings that have all three of these properties. Integral domains play a prominent role in number theory and algebraic geometry.

DEFINITION Zero-Divisors

A nonzero element a in a commutative ring R is called a *zero-divisor* if there is a nonzero element b in R such that $ab = 0$.

DEFINITION   Integral Domain

A commutative ring with a unity is said to be an *integral domain* if it has no zero-divisors.

Thus, in an integral domain, a product is zero only when one of the factors is zero; that is, $ab = 0$ only when $a = 0$ or $b = 0$. The following examples show that we have many friends that are integral domains and many friends that are not. For each example, the student should verify the assertion made.

Example 1   The ring of integers is an integral domain.   □

Example 2   The ring of Gaussian integers $Z[i] = \{a + bi \mid a, b \in Z\}$ is an integral domain.   □

Example 3   The ring $Z[x]$ of polynomials with integer coefficients is an integral domain.   □

Example 4   The ring $Z[\sqrt{2}] = \{a + b\sqrt{2} \mid a, b \in Z\}$ is an integral domain.   □

Example 5   The ring Z_p of integers modulo a prime p is an integral domain.   □

Example 6   The ring Z_n of integers modulo n is *not* an integral domain when n is not prime.   □

Example 7   The ring $M_2(Z)$ of 2×2 matrices over the integers is *not* an integral domain.   □

What makes integral domains particularly appealing is that they have an important multiplicative group-theoretic property, in spite of the fact that the nonzero elements need not form a group under multiplication. This property is cancellation.

Theorem 15.1   *Cancellation*
Let a, b, and c belong to an integral domain. If $a \neq 0$ and $ab = ac$, then $b = c$.

Proof.   From $ab = ac$ we have $a(b - c) = 0$. Since $a \neq 0$, we must have $b - c = 0$.   ■

Many authors prefer to define integral domains by the cancellation property, that is, as a commutative ring with unity in which the cancellation property holds. That definition is equivalent to ours.

Fields

In many applications, a particular kind of integral domain called a *field* is necessary.

DEFINITION Field

A commutative ring with a unity is called a *field* if every nonzero element is a unit.

It is often helpful to think of ab^{-1} as a divided by b. With this in mind, a field can be thought of as simply an algebraic system that is closed under addition, subtraction, multiplication, and division (except by 0). We have had numerous examples of fields: the complex numbers, the real numbers, the rational numbers. The abstract theory of fields was initiated by Heinrich Weber in 1893. Groups, rings, and fields are the three main branches of abstract algebra. Theorem 15.2 says that, in the finite case, fields and integral domains are the same.

Theorem 15.2 *Finite Integral Domains Are Fields*
A finite integral domain is a field.

Proof. Let D be a finite integral domain with unity 1. Let a be any nonzero element of D. We must show that a is a unit. If $a = 1$, a is its own inverse, so we may assume, $a \neq 1$. Now consider the following sequence of elements of D: $a, a^2, a^3, \ldots$. Since D is finite, there must be two positive integers i and j such that $i > j$ and $a^i = a^j$. Then, by cancellation, $a^{i-j} = 1$. Since $a \neq 1$, we know $i - j > 1$, and we have shown a^{i-j-1} is the inverse of a. ■

Corollary Z_p Is a Field
For every prime p, Z_p the ring of integers modulo p, is a field.

Proof. According to Theorem 15.2 we need only prove that Z_p has no zero-divisors. So, suppose a, $b \in Z_p$ and $ab = 0$. Then $ab = pk$ for some integer k. But, then, by Euclid's lemma (see Chapter 0), p divides a or p divides b. Thus, in Z_p, $a = 0$ or $b = 0$. ■

Putting the above corollary together with Example 6, we see that Z_n is a field if and only if n is prime. In a later chapter, we will describe how all finite fields can be constructed. For now, we give one example of a finite field not of the form Z_p.

Example 8 Field with Nine Elements
Let

$$Z_3[i] = \{a + bi \mid a, b \in Z_3\}$$
$$= \{0, 1, 2, i, 1 + i, 2 + i, 2i, 1 + 2i, 2 + 2i\}$$

Table 15.1   Multiplication Table for $Z_3[i]$

	1	2	i	$1 + i$	$2 + i$	$2i$	$1 + 2i$	$2 + 2i$
1	1	2	i	$1 + i$	$2 + i$	$2i$	$1 + 2i$	$2 + 2i$
2	2	1	$2i$	$2 + 2i$	$1 + 2i$	i	$2 + i$	$1 + i$
i	i	$2i$	2	$2 + i$	$2 + 2i$	1	$1 + i$	$1 + 2i$
$1 + i$	$1 + i$	$2 + 2i$	$2 + i$	$2i$	1	$1 + 2i$	2	i
$2 + i$	$2 + i$	$1 + 2i$	$2 + 2i$	1	i	$1 + i$	$2i$	2
$2i$	$2i$	i	1	$1 + 2i$	$1 + i$	2	$2 + 2i$	$2 + i$
$1 + 2i$	$1 + 2i$	$2 + i$	$1 + i$	2	$2i$	$2 + 2i$	i	1
$2 + 2i$	$2 + 2i$	$1 + i$	$1 + 2i$	i	2	$2 + i$	1	$2i$

where $i^2 = -1$. This is the ring of Gaussian integers modulo 3. Elements are added and multiplied as in the complex numbers, except that the coefficients are reduced modulo 3. In particular, $-1 = 2$. Table 15.1 is the multiplication table for the nonzero elements of $Z_3(i)$.   ☐

Example 9   Let $Q[\sqrt{2}] = \{a + b\sqrt{2} \mid a, b \in Q\}$. It is easy to see that $Q [\sqrt{2}]$ is a ring. To show that it is a field, we must exhibit an inverse for every nonzero element. To do this, we simply observe that $(a + b\sqrt{2})(a/(a^2 - 2b^2) - b\sqrt{2}/(a^2 - 2b^2)) = 1$. (Since $\sqrt{2}$ is irrational, $a^2 - 2b^2 = 0$ only when both a and b are 0.)   ☐

Characteristic of a Ring

Note that for any element x in $Z_3[i]$, we have $3x = x + x + x = 0$, since addition is done modulo 3. Similarly, in the subring $\{0, 3, 6, 9\}$ of Z_{12} we have $4x = x + x + x + x = 0$ for all x. This observation motivates the following definition.

> *DEFINITION*   Characteristic of a Ring
> The *characteristic* of a ring R is the least positive integer n such that $nx = 0$ for all x in R. If no such integer exists, we say R has characteristic 0.

Thus, the ring of integers has characteristic 0 and Z_n has characteristic n. An infinite ring can have nonzero characteristic. Indeed, the ring $Z_2[x]$ of all polynomials with coefficients in Z_2 has characteristic 2. (Addition and multiplication are done as with polynomials with ordinary integer coefficients except that the coefficients are reduced modulo 2.) When a ring has a unity, the task of determining the characteristic is simplified by Theorem 15.3.

> *Theorem 15.3*   *Characteristic of a Ring with Unity*
> Let R be a ring with unity 1. If 1 has infinite order under addition, then the

characteristic of R is 0. *If* 1 *has order n under addition, then the characteristic of R is n.*

Proof. If 1 has infinite order, then there is no positive integer n such that $n \cdot 1 = 0$, so R has characteristic 0. Now suppose that 1 has additive order n. Then $n \cdot 1 = 0$ and n is the least positive integer with this property. So for any x in R we have

$$nx = n(1x) = (n \cdot 1)x = 0x = 0.$$

Thus, R has characteristic n. ∎

In the case of an integral domain, the possibilities for the characteristic are severely limited.

Theorem 15.4 *Characteristic of an Integral Domain*
The characteristic of an integral domain is 0 *or prime.*

Proof. By Theorem 15.3, it suffices to show that if the additive order of 1 is finite, it must be prime. Suppose 1 has order n and $n = st$ where $1 < s, t < n$. Then

$$0 = n \cdot 1 = (st) \cdot 1 = (s \cdot 1)(t \cdot 1).$$

So, $s \cdot 1 = 0$ or $t \cdot 1 = 0$. But this contradicts the fact that 1 has order n. Hence, n cannot be factored as a product of two smaller integers. ∎

We conclude this chapter with a brief discussion of polynomials with coefficients from a ring—a topic we will consider in detail in later chapters. The existence of zero-divisors in a ring causes unusual results when finding roots of polynomials with coefficients in the ring. Consider, for example, the equation $x^2 - 4x + 3 = 0$. In the integers, we could find all solutions by factoring

$$x^2 - 4x + 3 = (x - 3)(x - 1) = 0$$

and setting each factor to 0. But notice, when we say we can find *all* solutions in this manner, we are using the fact that the only way for a product to equal zero is for one of the factors to be zero; that is, we are using the fact that Z is an integral domain. In Z_{12}, there are many pairs of nonzero elements whose product is zero: $2 \cdot 6 = 0$, $3 \cdot 4 = 0$, $4 \cdot 6 = 0$, $6 \cdot 8 = 0$, and so on. So, how do we find *all* solutions of $x^2 - 4x + 3 = 0$ in Z_{12}? The easiest way is to simply try every element! Upon doing so, we find four solutions: $x = 1$, $x = 3$, $x = 7$, and $x = 9$. Observe that we can find all four solutions of $x^2 - 4x + 3 = 0$ over Z_{11} or Z_{13}, say, by setting the two factors $x - 3$ and $x - 1$ to 0. Of course, the reason why this works for these rings is that they are integral domains. Perhaps this will convince you that integral domains are particularly nice rings. Table 15.2 gives a summary of some of the rings we have introduced and their properties.

Table 15.2 Summary of Rings and Their Properties

Ring	Form of Element	Unity	Commutative	Integral Domain	Field	Characteristic
Z	k	1	yes	yes	no	0
Z_n, n composite	k	1	yes	no	no	n
Z_p, p prime	k	1	yes	yes	yes	p
$Z[x]$	$a_n x^n + \cdots + a_1 x + a_0$	$f(x) = 1$	yes	yes	no	0
nZ, $n > 1$	nk	none	yes	no	no	0
$M_2(Z)$	$\begin{bmatrix} a & b \\ c & d \end{bmatrix}$	$\begin{bmatrix} 1 & 0 \\ 0 & 1 \end{bmatrix}$	no	no	no	0
$M_2(2Z)$	$\begin{bmatrix} 2a & 2b \\ 2c & 2d \end{bmatrix}$	none	no	no	no	0
$Z[i]$	$a + bi$	1	yes	yes	no	0
$Z_3[i]$	$a + bi$; $a, b \in Z_3$	1	yes	yes	yes	3
$Z_2[i]$	$a + bi$; $a, b \in Z_2$	1	yes	no	no	2
$Z[\sqrt{2}]$	$a + b\sqrt{2}$; $a, b \in Z$	1	yes	yes	no	0
$Q[\sqrt{2}]$	$a + b\sqrt{2}$; $a, b \in Q$	1	yes	yes	yes	0
$Z \oplus Z$	(a, b)	$(1, 1)$	yes	no	no	0

EXERCISES

It looked absolutely impossible. But it so happens that you go on worrying away at a problem in science and it seems to get tired, and lies down and lets you catch it.

William Lawrence Bragg*

1. Verify that Examples 1 through 7 are as claimed.

2. Which of Examples 1 through 5 are fields?

3. Show that a commutative ring with the cancellation property (under multiplication) has no zero-divisors.

4. List all zero-divisors in Z_{20}. Can you see a relationship between the zero-divisors of Z_{20} and the units of Z_{20}?

5. Show that every nonzero element of Z_n is a unit of a zero-divisor.

6. Find a nonzero element in a ring that is neither a zero-divisor nor a unit.

---

*Bragg, at age twenty-four, won the Nobel Prize for the invention of x-ray crystallography. He remains the youngest person ever to receive the Nobel Prize.

7. Let R be a finite commutative ring with unity. Prove that every nonzero element of R is either a zero divisor or a unit. What happens if we drop the "finite" condition on R?

8. Describe all zero-divisors and units of $Z \oplus Q \oplus Z$.

9. Let d be an integer that is not divisible by the square of any prime. Prove that $Z[\sqrt{d}] = \{a + b\sqrt{d} \mid a, b \in Z\}$ is an integral domain. (This exercise is referred to in Chapter 20.)

10. Show that a field is an integral domain.

11. Give an example of a commutative ring without zero-divisors that is not an integral domain.

12. Find two elements a and b in a ring such that both a and b are zero-divisors, but $a + b$ is not.

13. Let a belong to a ring R with unity and $a^n = 0$ for some positive integer n. (Such an element is called *nilpotent*.) Prove that $1 - a$ has a multiplicative inverse in R. (Hint: Consider $(1 - a)(1 + a + a^2 + \cdots + a^{n-1})$.)

14. Show that the nilpotent elements of a commutative ring form a subring.

15. Show that 0 is the only nilpotent element in an integral domain.

16. A ring element a is called an *idempotent* if $a^2 = a$. Prove that the only idempotents in an integral domain are 0 and 1.

17. Prove that the set of idempotents of a commutative ring is closed under multiplication.

18. Find a zero-divisor and a nonzero idempotent in $Z_5[i] = \{a + bi \mid a, b, \in Z_5, i^2 = -1\}$.

19. Determine the isomorphism class of the group of units of $Z_5[i]$.

20. Find all units, zero-divisors, idempotents, and nilpotent elements in $Z_3 \oplus Z_6$.

21. Determine all elements of a ring that are both units and idempotents.

22. Let R be the set of all real-valued functions of a single real variable under function addition and multiplication.
    a. Determine all zero-divisors of R.
    b. Determine all nilpotent elements of R.
    c. Show that every element is a zero-divisor or a unit.

23. (Subfield Test) Let F be a field and K a subset of F with at least two elements. Prove that K is a subfield of F if, for any a, b ($b \neq 0$) in K, $a - b$ and ab^{-1} belong to K.

24. Let d be a positive integer that is not a square. Prove that $Q[\sqrt{d}] = \{a + b\sqrt{d} \mid a, b \in Q\}$ is a field.

25. Let R be a ring with unity 1. If the product of any pair of nonzero elements of R is nonzero, prove that $ab = 1$ implies $ba = 1$.

26. Let $R = \{0, 2, 4, 6, 8\}$ under addition and multiplication modulo 10. Prove that R is a field.

27. Formulate the appropriate definition of a subdomain (that is, a "sub" integral domain). Let D be an integral domain with unity 1. Show that $P = \{n1 \mid n \in Z\}$ (that is, all integral multiples of 1) is a subdomain of D. Show that P is contained in every subdomain of D. What can we say about the order of P?

28. Prove that there is no integral domain with exactly six elements. Can your argument be adapted to show that there is no integral domain with exactly four elements? What about fifteen elements? Use these observations to guess a general result about the number of elements in a finite integral domain.

29. Is $Z \oplus Z$ an integral domain? Explain.

30. Determine all elements of an integral domain that are their own inverses under multiplication.

31. Suppose a and b belong to an integral domain.
    a. If $a^5 = b^5$ and $a^3 = b^3$, prove that $a = b$.
    b. If $a^m = b^m$ and $a^n = b^n$, where m and n are positive integers that are relatively prime, prove that $a = b$.

32. Find an example of an integral domain and distinct positive integers m and n such that $a^m = b^m$ and $a^n = b^n$, but $a \neq b$.

33. Verity Table 15.1. Look for short cuts. (For example, the sixth row can be obtained from the third one by doubling.)

34. Construct a multiplication table for $Z_2[i]$, the ring of Gaussian integers modulo 2. Is this ring a field? Is it an integral domain?

35. The nonzero elements of $Z_3[i]$ form an Abelian group of order 8 under multiplication. Is it isomorphic to Z_8, $Z_4 \oplus Z_2$, or $Z_2 \oplus Z_2 \oplus Z_2$?

36. Define $Z_n[i] = \{a + bi \mid a, b \in Z_n\}$, where $i^2 = -1$ in $Z_n[i]$. For some primes p, $Z_p[i]$ is a field, and for others it is not. In particular, when $p = 2$, 5, or 13, $Z_p[i]$ is *not* a field; but when $p = 3$, 7, or 11, $Z_p[i]$ is a field. From these examples, try to guess a condition that determines whether or not $Z_p[i]$ is a field. (Hint: It has to do with the equation $x^2 + 1 = 0$.)

37. Show that a finite commutative ring with no zero-divisors has a unity.

38. Suppose a and b belong to a commutative ring and ab is a zero-divisor. Show that either a is a zero-divisor or b is a zero-divisor.

39. Suppose R is a commutative ring without zero-divisors. Show that all the nonzero elements of R have the same additive order.

40. Suppose R is a commutative ring without zero-divisors. Show that the characteristic of R is 0 or prime.

41. Show that any finite field has order p^n where p is a prime. (Hint: Use facts about finite Abelian groups.)

42. Let x and y belong to an integral domain of prime characteristic p.
    a. Show that $(x + y)^p = x^p + y^p$
    b. Show that for all positive integers n, $(x + y)^{p^n} = x^{p^n} + y^{p^n}$.

c. Find elements x and y in a ring of characteristic 4 such that $(x + y)^4$ $\neq x^4 + y^4$.

43. Exhibit a finite field that contains two nonzero elements a and b such that $a^2 + b^2 = 0$.

44. Give an example of an infinite integral domain that has characteristic 3.

45. Let R be a ring and $M_2(R)$ the ring of 2×2 matrices with entries from R. Explain why these two rings have the same characteristic.

46. Let R be a ring with m elements. Show that the characteristic of R divides m.

47. Explain why a finite ring must have a nonzero characteristic.

48. Find all solutions of $x^2 - x + 2 = 0$ over $Z_3[i]$. (See Example 8.)

49. Consider the equation $x^2 - 5x + 6 = 0$.
    a. How many solutions does this equation have in Z_7.?
    b. Find all solutions of this equation in Z_8.
    c. Find all solutions of this equation in Z_{12}.
    d. Find all solutions of this equation in Z_{14}.

50. Find the characteristic of $Z_4 \oplus 4Z$.

51. Suppose R is an integral domain in which $20 \cdot 1 = 0$ and $12 \cdot 1 = 0$. (Recall $n \cdot 1$ means the sum $1 + 1 + \cdots + 1$ with n terms.) What is the characteristic of R?

52. In a commutative ring of characteristic 2, prove that the idempotents form a subring.

53. Describe the smallest subfield of the field of real numbers that contains $\sqrt{2}$.

54. Let F be a finite field with n elements. Prove that $x^{n-1} = 1$ for all nonzero x in F.

55. Let F be a field of prime characteristic p. Prove that $K = \{x \in F \mid x^p = x\}$ is a subfield of F.

56. Suppose a and b belong to a field of order 8 and $a^2 + ab + b^2 = 0$. Prove that $a = 0$ and $b = 0$. Do the same when the field has order 2^n with n odd.

57. Let F be a field of characteristic 2 with more than two elements. Show that $(x + y)^3 \neq x^3 + y^3$ for some x and y in F.

58. Suppose F is a field with characteristic not 2, and the nonzero elements of F form a cyclic group under multiplication. Prove that F is finite.

59. Suppose D is an integral domain and ϕ is a nonconstant function from D to the nonnegative integers such that $\phi(xy) = \phi(x)\phi(y)$. If x is a unit in D, show that $\phi(x) = 1$.

60. Let F be a field of order 32. Show that the only subfields of F are F itself and $\{0, 1\}$.

PROGRAMMING EXERCISES

"Data! data! data!" he cried impatiently. "I can't make bricks without clay."

Sherlock Holmes

1. Write a program to list all the idempotents (see exercise 16 for definition) in Z_n. Run your program for $1 \leq n \leq 200$. Use this output to make conjectures about the number of idempotents in Z_n as a function of n. For example, how many idempotents are there when n is a prime-power? What about when n is divisible by two distinct primes?

2. Write a program to list all the nilpotent elements (see exercise 13 for definition) in Z_n. Run your program for $1 \leq n \leq 200$. Use this output to make conjectures about nilpotent elements in Z_n.

3. Write a program to determine all the roots of the equation $x^3 - 2x^2 + x - 2 = (x - 2)(x^2 + 1)$ over Z_n. Run your program for $n = 5, 9, 25, 50$, and 100. How many roots does this polynomial have over Z?

SUGGESTED READINGS

R. A. Beaumont, "Equivalent Properties of a Ring," *American Mathematical Monthly* 57 (1950): 183.

In this brief note, it is proved that in any ring R, the following properties are equivalent.
1. R is commutative and contains no zero-divisors.
2. Every polynomial of degree n with coefficients in R has at most n zeros in R.
3. Every linear polynomial with coefficients in R has at most one zero in R.
4. If a, b, and c are in R, $b \neq 0$, then $ba = cb$ implies $a = c$.

N. A. Khan, "The Characteristic of a Ring," *American Mathematical Monthly* 70 (1963): 736.

Here it is shown that a ring has nonzero characteristic n if and only if n is the maximum of the orders of the elements of R.

K. Robin McLean, "Groups in Modular Arithmetic," *The Mathematical Gazette* 62 (1978): 94–104.

This article explores the interplay between various groups of integers under multiplication modulo n and the ring Z_n. It shows how to construct groups of integers in which the identity is not obvious, for example, 1977 is the identity of the group {1977, 5931} under multiplication modulo 7908.

Nathan Jacobson

Here, as in so many other parts of algebra, Jake's influence has been very noticeable.

Paul M. Cohn

NATHAN JACOBSON was born on September 8, 1910, in Warsaw Poland. Arriving in the United States in 1917, Jacobson grew up in Alabama, Mississippi, and Georgia, where his father owned small clothing stores. He received a B.A. degree from the University of Alabama in 1930 and a Ph.D. from Princeton in 1934. After brief periods as a professor at Bryn Mawr, the University of Chicago, the University of North Carolina, and Johns Hopkins, Jacobson accepted a position at Yale, where he remained until his retirement in 1981.

Jacobson's principal contributions to algebra lie in the fields of associative rings, Lie algebras, and Jordan algebras. In particular, he developed structure theories for these systems. He is the author of nine books and numerous articles.

Jacobson has held visiting positions in France, India, Italy, Israel, China, Australia, and Switzerland. Among his many honors are president of the American Mathematical Society, memberships in the National Academy of Sciences and the American Academy of Arts and Sciences, a Guggenheim Fellowship, and an honorary degree from the University of Chicago.

Ideals and Factor Rings

16

The secret of science is to ask the right questions, and it is the choice of problem more than anything else that marks the man of genius in the scientific world.

Sir Henry Tizard in C. P. Snow, *A Postscript to Science and Government*

Ideals

Normal subgroups play a special role in group theory—they permit us to construct factor groups. In this chapter, we introduce the analogous concepts for rings—ideals and factor rings.

> *DEFINITION*   Ideal
>
> A subring A of a ring R is called a (two-sided) *ideal* of R if for every $r \in R$ and every $a \in A$ both ra and ar are in A.

So, a subring A of a ring R is an ideal of R if A "absorbs" elements from R; that is, if $rA \subseteq A$ and $Ar \subseteq A$ for all $r \in R$.

An ideal A of R is called a *proper* ideal of A if A is a proper subset of R. In practice, one identifies ideals with the following test, which is an immediate consequence of the definition of ideal, and the subring test given in Chapter 14.

> ***Theorem 16.1**   Ideal Test*
>
> *A nonempty subset A of a ring R is an ideal of R if*
>
> 1. *$a - b \in A$ whenever $a, b \in A$.*
> 2. *ra and ar are in A whenever $a \in A$ and $r \in R$.*

Example 1   For any ring R, $\{0\}$ and R are ideals of R. The ideal $\{0\}$ is called the *trivial* ideal. $\qquad\square$

Example 2   For any positive integer n, the set $nZ = \{0, \pm n, \pm 2n, \ldots\}$ is an ideal of Z.   □

Example 3   Let R be a commutative ring with unity and let $a \in R$. The set $\langle a \rangle = \{ra \mid r \in R\}$ is an ideal of R called the *principal ideal generated by a*. (Notice that $\langle a \rangle$ is also the notation we used for the cyclic subgroup generated by a. However, the intended meaning will always be clear from the context.) The assumption that R is commutative is necessary in this example (see exercise 29 of the Supplementary Exercises for Chapters 13–16).   □

Example 4   Let $R[x]$ denote the set of all polynomials with real coefficients and A the subset of all polynomials with constant term zero. Then A is an ideal of $R[x]$.   □

Example 5   Let R be the ring of all real-valued functions of a real variable. The subset S of all differentiable functions is a subring of R but not an ideal of R.   □

Factor Rings

Let R be a ring and A an ideal of R. Since R is a group under addition and A is a normal subgroup of R, we may form the factor group $R/A = \{r + A \mid r \in R\}$. The natural question at this point is how may we form a ring of this group of cosets? The addition is already taken care of and, by analogy with groups of cosets, we define the product of two cosets $a + A$ and $b + A$ as $ab + A$. The next theorem shows that this definition works as long as A is an ideal and not just a subring of R.

Theorem 16.2   *Existence of Factor Rings*
Let R be a ring and A a subring of R. The set of cosets $\{r + A \mid r \in R\}$ is a ring under the operations $(s + A) + (t + A) = s + t + A$ and $(s + A)(t + A) = st + A$ if and only if A is an ideal of R.

Proof.   We do know that the set of cosets forms a group under addition, and it is trivial to check that the multiplication is associative and that multiplication is distributive over addition once we know that multiplication is indeed a binary operation on the set of cosets. Hence, the proof boils down to showing that multiplication is well defined if and only if A is an ideal of R. To do this, let us suppose A is an ideal and let $s + A = s' + A$ and $t + A = t' + A$. Then we must show that $st + A = s't' + A$. Well, by definition, $s = s' + a$ and $t = t' + b$, where a and b belong to A. Then,

$$st = (s' + a)(t' + b) = s't' + at' + s'b + ab,$$

and, so,

$$st + A = s't' + at' + s'b + ab + A = s't' + A,$$

since A absorbs the last three summands of the middle expression. Thus, multiplication is well defined when A is an ideal.

On the other hand, suppose A is not an ideal of R. Then there exist elements $a \in A$ and $r \in R$ such that $ar \notin A$ or $ra \notin A$. For convenience, say, $ar \notin A$. Consider the elements $a + A = 0 + A$ and $r + A$. Clearly, $(a + A)(r + A) = ar + A$ but $(0 + A)(r + A) = A$. Since $ar + A \neq A$, the multiplication is not well defined and the set of cosets is not a ring. ∎

Let's look at a few factor rings.

Example 6 $Z/4Z = \{0 + 4Z, 1 + 4Z, 2 + 4Z, 3 + 4Z\}$. To see how to add and multiply, consider $2 + 4Z$ and $3 + 4Z$.

$$(2 + 4Z) + (3 + 4Z) = 5 + 4Z = 1 + 4 + 4Z = 1 + 4Z,$$
$$(2 + 4Z)(3 + 4Z) = 6 + 4Z = 2 + 4 + 4Z = 2 + 4Z.$$

One can readily see that the two operations are essentially modulo 4 arithmetic. In the next chapter, we will see that $Z/4Z$ is isomorphic to Z_4. □

Example 7 $2Z/6Z = \{0 + 6Z, 2 + 6Z, 4 + 6Z\}$. Here the operations are essentially modulo 6 arithmetic and the ring is isomorphic to Z_3. □

Here is a noncommutative example of an ideal and factor ring.

Example 8   Let $R = \left\{ \begin{bmatrix} a_1 & a_2 \\ a_3 & a_4 \end{bmatrix} \middle| a_i \in Z \right\}$ and let I be the subset of R consisting of matrices with even entries. It is easy to show that I is indeed an ideal of R (exercise 16). Consider the factor ring R/I. The interesting question about this ring is its size. Just what is it? We claim it is 16; in fact, $R/I = \left\{ \begin{bmatrix} r_1 & r_2 \\ r_3 & r_4 \end{bmatrix} + I \middle| r_i \in \{0, 1\} \right\}$. An example illustrates the typical situation. Which of the 16 elements is $\begin{bmatrix} 7 & 8 \\ 5 & -3 \end{bmatrix} + I$? Well, observe that $\begin{bmatrix} 7 & 8 \\ 5 & -3 \end{bmatrix} + I = \begin{bmatrix} 1 & 0 \\ 1 & 1 \end{bmatrix} + \begin{bmatrix} 6 & 8 \\ 4 & -4 \end{bmatrix} + I = \begin{bmatrix} 1 & 0 \\ 1 & 1 \end{bmatrix} + I$, since an ideal absorbs its own elements. The general case is left to the reader. □

Example 9   Let $\mathbf{R}[x]$ denote the ring of polynomials with real coefficients and $\langle x^2 + 1 \rangle$ the principal ideal generated by $x^2 + 1$; that is,

$$\langle x^2 + 1 \rangle = \{f(x)(x^2 + 1) \mid f(x) \in \mathbf{R}[x]\}.$$

Then

$$\mathbf{R}[x]/\langle x^2 + 1 \rangle = \{g(x) + \langle x^2 + 1 \rangle\} = \{ax + b + \langle x^2 + 1 \rangle \mid a, b \in \mathbf{R}\}.$$

To see this last equality, note that if $g(x)$ is any member of $\mathbf{R}[x]$, then we may write $g(x)$ in the form $q(x)(x^2 + 1) + r(x)$, where $q(x)$ is the quotient and $r(x)$ is the remainder upon dividing $g(x)$ by $x^2 + 1$. In particular, $r(x) = 0$ or degree $r(x) < 2$ so that $r(x) = ax + b$ for some a and b in $\mathbf{R}$. Thus,

$$g(x) + \langle x^2 + 1 \rangle = q(x)(x^2 + 1) + r(x) + \langle x^2 + 1 \rangle = r(x) + \langle x^2 + 1 \rangle$$

since the ideal $\langle x^2 + 1 \rangle$ absorbs the term $q(x)(x^2 + 1)$.

How is multiplication done? Since

$$x^2 + 1 + \langle x^2 + 1 \rangle = 0 + \langle x^2 + 1 \rangle$$

one should think of $x^2 + 1$ as zero, or equivalently, as $x^2 = -1$. So, for example,

$$
\begin{aligned}
(x + 3 + \langle x^2 + 1 \rangle) \cdot (2x + 5 + \langle x^2 + 1 \rangle) \\
= 2x^2 + 11x + 15 + \langle x^2 + 1 \rangle \\
= 11x + 13 + \langle x^2 + 1 \rangle.
\end{aligned}
$$

In view of the fact that the elements of this ring have the form $ax + b + \langle x^2 + 1 \rangle$ where $x^2 + \langle x^2 + 1 \rangle = -1 + \langle x^2 + 1 \rangle$, it is perhaps not surprising that this ring turns out to be isomorphic to the ring of complex numbers. This observation was first made by Cauchy in 1847. $\qquad \square$

Example 9 illustrates one of the most important applications of factor rings— the construction of rings with highly desirable properties. In particular, we shall show how one may use factor rings to construct integral domains and fields.

Prime Ideals and Maximal Ideals

DEFINITION Prime Ideal, Maximal Ideal

A proper ideal A of a commutative ring R is said to be a *prime ideal* of R if $a, b \in R$ and $ab \in A$ implies $a \in A$ or $b \in A$. A proper ideal A of R is said to be a *maximal ideal* of R if, whenever B is an ideal of R and $A \subseteq B \subseteq R$, then $B = A$ or $B = R$.

So, the only ideal that properly contains a maximal ideal is the entire ring. The motivation for the definition of a prime ideal comes from the integers.

Example 10 Let n be a positive integer. Then, in the ring of integers, the ideal $n\mathbf{Z}$ is prime if and only if n is prime. $\qquad \square$

Example 11 The lattice of ideals of Z_{36} (Figure 16.1) shows that only $\langle 2 \rangle$ and $\langle 3 \rangle$ are maximal ideals. $\qquad \square$

Example 12 The ideal $\langle x^2 + 1 \rangle$ is maximal in $\mathbf{R}[x]$. To see this, assume that A is an ideal of $\mathbf{R}[x]$ that properly contains $\langle x^2 + 1 \rangle$. We will prove that $A = \mathbf{R}[x]$

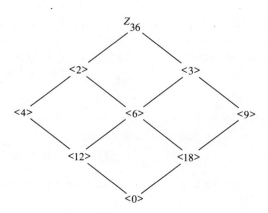

Figure 16.1

by showing that A contains some nonzero real number c. (This is the constant polynomial $h(x) = c$ for all x.) Then $1 = (1/c)c \in A$ and therefore, by exercise 14, $A = \mathbf{R}[x]$. To this end, let $f(x) \in A$, but $f(x) \notin \langle x^2 + 1 \rangle$. Then

$$f(x) = q(x)(x^2 + 1) + r(x)$$

where $r(x) \neq 0$ and degree $r(x) < 2$. It follows that $r(x) = ax + b$ where not both a and b are zero and

$$ax + b = r(x) = f(x) - q(x)(x^2 + 1) \in A.$$

Thus,

$$a^2x^2 - b^2 = (ax + b)(ax - b) \in A \qquad \text{and} \qquad a^2(x^2 + 1) \in A,$$

also. So,

$$a^2 + b^2 = (a^2x^2 + a^2) - (a^2x^2 - b^2) \in A. \qquad \square$$

Example 13 The ideal $\langle x^2 + 1 \rangle$ is not prime in $Z_2[x]$ since it contains $(x + 1)^2 = x^2 + 2x + 1 = x^2 + 1$ but does not contain $x + 1$. $\qquad \square$

Theorem 16.3 *R/A Is an Integral Domain if and Only if A Is Prime*
Let R be a commutative ring with unity and let A be an ideal of R. Then R/A is an integral domain if and only if A is prime.

Proof. Suppose R/A is an integral domain and $ab \in A$. Then $(a + A)(b + A) = ab + A = A$, the zero element of the ring R/A. So, either $a + A = A$ or $b + A = A$; that is, either $a \in A$ or $b \in A$. Hence, A is prime.

Now suppose that A is prime and $(a + A)(b + A) = 0 + A = A$. Then $ab \in A$ and therefore, $a \in A$ or $b \in A$. Thus, one of $a + A$ or $b + A$ is zero. ∎

For maximal ideals, we can do even better.

Theorem 16.4 *R/A Is a Field if and Only if A Is Maximal*
Let R be a commutative ring with unity and A an ideal of R. Then R/A is a field if and only if A is maximal.

Proof. Suppose R/A is a field and B is an ideal of R that properly contains A. Let $b \in B$ but $b \notin A$. Then $b + A$ is a nonzero element of R/A and, therefore, there exists an element $c + A$ such that $(b + A)(c + A) = 1 + A$, the multiplicative identity of R/A. Since $b \in B$, we have $bc \in B$. Because

$$1 + A = (b + A)(c + A) = bc + A,$$

we have $1 - bc \in A \subset B$. So, $1 = (1 - bc) + bc \in B$. By exercise 14, $B = R$. This proves that A is maximal.

Now suppose that A is maximal and let $b \in R$ but $b \notin A$. It suffices to show that $b + A$ has a multiplicative inverse. (All other properties for a field follow trivially.) Consider $B = \{br + a \mid r \in R, a \in A\}$. This is an ideal of R that properly contains A (exercise 19). Since A is maximal we must have $B = R$. Thus, $1 \in B$, say, $1 = bc + a'$ where $a' \in A$. Then

$$1 + A = bc + a' + A = bc + A = (b + A)(c + A). \qquad \blacksquare$$

EXERCISES

Problems worthy
of attack
prove their worth
by hitting back.
Piet Hein, "Problems," *Grooks* (1966)

1. Show that the set of rational numbers is a subring of the reals but is not an ideal.
2. Find a subring of $Z \oplus Z$ that is not an ideal of $Z \oplus Z$.
3. Let $S = \{a + bi \mid a, b \in Z, b \text{ is even}\}$. Show that S is a subring of $Z[i]$, but not an ideal of $Z[i]$.
4. Find all maximal ideals in
   a. Z_8, b. Z_{10}, c. Z_{12}, d. Z_n.
5. Let a belong to a commutative ring R. Show that $aR = \{ar \mid r \in R\}$ is an ideal of R. If R is the ring of even integers, list the elements of $4R$.
6. If n is a positive integer, show that $\langle n \rangle = nZ$ is a prime ideal of Z if and only if n is prime.
7. Prove that the intersection of any set of ideals of a ring is an ideal.

8. If A and B are ideals of a ring, show that the *sum* of A and B, $A + B = \{a + b \mid a \in A, b \in B\}$ is an ideal.

9. In the ring of integers, find a positive integer a such that
   a. $\langle a \rangle = \langle 2 \rangle + \langle 3 \rangle$
   b. $\langle a \rangle = \langle 3 \rangle + \langle 6 \rangle$
   c. $\langle a \rangle = \langle m \rangle + \langle n \rangle$

10. If A and B are ideals of a ring, show that the *product* of A and B,

    $$AB = \{a_1b_1 + a_2b_2 + \cdots + a_nb_n \mid a_i \in A, b_i \in B, n \text{ a positive integer}\}$$

    is an ideal.

11. Find a positive integer a such that
    a. $\langle a \rangle = \langle 3 \rangle\langle 4 \rangle$
    b. $\langle a \rangle = \langle 6 \rangle\langle 8 \rangle$
    c. $\langle a \rangle = \langle m \rangle\langle n \rangle$

12. Let A and B be ideals of a ring. Prove that $AB \subseteq A \cap B$.

13. If A and B are ideals of a commutative ring R with unity and $A + B = R$, show that $A \cap B = AB$.

14. If A is an ideal of a ring R and 1 belongs to A, prove that $A = R$. (This exercise is referred to in this chapter.)

15. If an ideal I of a ring R contains a unit, show that $I = R$.

16. Let R and I be as described in Example 8. Prove that I is an ideal of R.

17. Verify the claim made in Example 8 about the size of R/I.

18. Prove that the ideal $\langle x^2 + 1 \rangle$ is prime in $Z[x]$ but not maximal in $Z[x]$.

19. Show that the set B in the latter half of the proof of Theorem 16.4 is an ideal of R. (This exercise is referred to in this chapter.)

20. Construct a multiplication table for the ring $3Z/9Z$.

21. If R is a commutative ring with unity and A is an ideal of R, show that R/A is a commutative ring with unity.

22. Show that $\mathbf{R}[x]/\langle x^2 + 1 \rangle$ is a field.

23. Let R be a commutative ring with unity. Show that every maximal ideal of R is prime.

24. Show that $A = \{(3x, y) \mid x, y \in Z\}$ is a maximal ideal of $Z \oplus Z$.

25. Let R be the ring of continuous functions from $\mathbf{R}$ to $\mathbf{R}$. Show that $A = \{f \in R \mid f(0) = 0\}$ is a maximal ideal of R.

26. Let $R = Z_8 \oplus Z_{30}$. Find all maximal ideals of R, and for each maximal ideal I, identify the size of the field R/I.

27. How many elements are in $Z[i]/\langle 3 + i \rangle$? Give reasons for your answer.

28. In $Z[x]$, the ring of polynomials with integer coefficients, let $I = \{f \in Z[x] \mid f(0) = 0\}$. Prove that I is not a maximal ideal.

29. In $Z \oplus Z$, let $I = \{(a, 0) \mid a \in Z\}$. Show that I is a prime ideal but not a maximal ideal.

30. Let R be a ring and I an ideal in R. Prove that the factor ring R/I is commutative if and only if $rs - sr \in I$ for all r and s in R.

31. In $Z[x]$, let $I = \{f \in Z[x] \mid f(0) \text{ is an even integer}\}$. Prove that I is a prime ideal of $Z[x]$. Is it maximal?

32. Prove that $I = \langle 2 + 2i \rangle$ is not a prime ideal of $Z[i]$. How many elements are in $Z[i]/I$? What is the characteristic of $Z[i]/I$?

33. In $Z_5[x]$, let $I = \langle x^2 + x + 2 \rangle$. Find the multiplicative inverse of $2x + 3 + I$ in $Z_5[x]/I$.

34. An integral domain D is called a *principal ideal domain* if every ideal of D has the form $\langle a \rangle$ for some a in D. Show that Z is a principal ideal domain.

35. Let R be a ring and p be a fixed prime. Show that $I_p = \{r \in R \mid \text{additive order of } r \text{ is a power of } p\}$ is an ideal of R.

36. Let a and b belong to a commutative ring R. Prove that $\{x \in R \mid ax \in bR\}$ is an ideal.

37. Let R be a commutative ring and let A be any subset of R. Show that the *annihilator* of A, $\text{Ann}(A) = \{r \in R \mid ra = 0 \text{ for all } a \text{ in } A\}$, is an ideal.

38. Let R be a commutative ring and let A be an ideal of R. Show that the *nil radical* of A, $\sqrt{A} = \{r \in R \mid r^n \in A \text{ for some positive integer } n \text{ (n depends on r)}\}$, is an ideal. ($\sqrt{\langle 0 \rangle}$ is called the *nil radical* of R.)

39. Let $R = Z_{27}$. Find   a. $\sqrt{\langle 0 \rangle}$,   b. $\sqrt{\langle 3 \rangle}$,   c. $\sqrt{\langle 9 \rangle}$.

40. Let $R = Z_{36}$. Find   a. $\sqrt{\langle 0 \rangle}$,   b. $\sqrt{\langle 4 \rangle}$,   c. $\sqrt{\langle 6 \rangle}$.

41. Let R be a commutative ring. Show that $R/\sqrt{\langle 0 \rangle}$ has no nonzero nilpotent elements (that is, $\sqrt{R/\sqrt{\langle 0 \rangle}} = \langle 0 \rangle$).

42. Let A be an ideal in a commutative ring. Prove that $\sqrt{\sqrt{A}} = \sqrt{A}$.

43. Let $Z_2[x]$ be the ring of all polynomials with coefficients in Z_2 (that is, coefficients are 0 or 1, and addition and multiplication of coefficients are done modulo 2). Show that $Z_2[x]/\langle x^2 + x + 1 \rangle$ is a field.

44. List the elements of the field given in exercise 43, and make an addition and multiplication table for the field.

45. Show that $Z_3[x]/\langle x^2 + x + 1 \rangle$ is not a field.

46. Let R be a commutative ring without unity, and let $a \in R$. Describe the smallest ideal of R that contains a.

47. Let R be the ring of continuous functions from $\mathbf{R}$ to $\mathbf{R}$. Let $A = \{f \in R \mid f(0) \text{ is an even integer}\}$. Show that A is a subring of R, but not an ideal of R.

48. Show that $Z[i]/\langle 1 - i \rangle$ is a field. How many elements does this field have?

49. If R is a principal ideal domain and I is an ideal of R, prove that every ideal of R/I is principal (see exercise 34).

Richard Dedekind

Richard Dedekind was not only a mathematician, but one of the wholly great in the history of mathematics, now and in the past, the last hero of a great epoch, the last pupil of Gauss, for four decades himself a classic, from whose works not only we, but our teachers and the teachers of our teachers, have drawn.

Edmund Landau, Commemorative Address to the Royal Society of Göttingen

This stamp was issued by East Germany in 1981 to commemorate the 150th anniversary of Dedekind's birth. Notice that it features the representation of an ideal as the product of powers of prime ideals.

RICHARD DEDEKIND was born on October 6, 1831, in Brunswick, Germany, the birthplace of Gauss. Dedekind was the youngest of four children of a law professor. His early interests were in chemistry and physics, but he obtained a doctor's degree in mathematics at the age of twenty-one under Gauss at the University of Göttingen. Dedekind continued his studies at Göttingen for a few years and in 1854 he began to lecture there. When Gauss died the following year, Dedekind was selected to be one of the pallbearers. In 1857, Dedekind gave what is believed to be the first formal lectures on Galois theory. While at Göttingen, Dedekind developed a close relationship with two eminent mathematicians—Riemann and Dirichlet, Gauss's successor.

Dedekind spent the years 1858–1862 as a professor in Zurich. Then he accepted a position at an institute in Brunswick where he was once a student. Although this school was less than university level, Dedekind remained there for the next 50 years. He died in Brunswick in 1916.

During his career, Dedekind made numerous fundamental contributions to mathematics. His treatment of irrational numbers, "Dedekind cuts," put analysis on a firm, logical foundation. His work on unique factorization led to the modern theory of algebraic numbers. He was a pioneer in the theory of rings and fields. The notion of ideal as well as the term itself is due to Dedekind. Mathematics historian Morris Kline has called him "the effective founder of abstract algebra."

Emmy Noether

In the judgment of the most competent living mathematicians, Fräulein Noether was the most significant creative mathematical genius thus far produced since the higher education of women began. In the realm of algebra, in which the most gifted mathematicians have been busy for centuries, she discovered methods which have proved of enormous importance in the development of the present-day younger generation of mathematicians.

Albert Einstein, *The New York Times*

EMMY NOETHER was born on March 23, 1882, in Germany. When she entered the University of Erlangen, she was one of only two women among the 1000 students. Noether completed her doctorate in 1907.

In 1916. Noether went to Göttingen and, under the influence of David Hilbert and Felix Klein, became interested in general relativity. While there, she made a major contribution to physics with her theorem that, whenever there is a symmetry in nature, there is also a conservation law, and vice versa. Hilbert tried unsuccessfully to obtain a faculty appointment at Göttingen for Noether, saying "I do not see that the sex of the candidate is an argument against her admission as Privatdozent. After all, we are a university and not a bathing establishment."

It was not until she was thirty-eight that Noether's true genius revealed itself. Over the next 13 years, she used an axiomatic method to develop a general theory of ideals and noncommutative algebras. With this abstract theory, Noether was able to weld together many important concepts. Her approach was even more important than the individual results. Hermann Weyl said of Noether, "She originated above all a new and epoch-making style of thinking in algebra."

Noether was not good at lecturing, but she was an inspiring teacher. Her students were known as "the Noether boys," and many turned out to be important mathematicians. Weyl once said, "In my Göttingen years, 1930–1933, she was without doubt the strongest center of mathematical activity there, considering both the fertility of her scientific research program and her influence upon a large circle of pupils."

With the rise of Hitler in 1933, Noether, a Jew, fled to the United States and took a position at Bryn Mawr College. She died suddenly on April 14, 1935, following an operation.

SUPPLEMENTARY EXERCISES for Chapters 13–16

If at first you do succeed—try to hide your astonishment.

Harry F. Banks

1. Find all idempotent elements in Z_{10}, Z_{20}, Z_{30}. (Recall, a is idempotent if $a^2 = a$.)

2. If m and n are relatively prime integers greater than 1, prove that Z_{mn} has at least two idempotents besides 0 and 1.

3. Suppose R is a ring in which $a^2 = 0$ implies $a = 0$. Show that R has no nonzero nilpotent elements.

4. Let R be a commutative ring with more than one element. Prove that if for every nonzero element a of R, we have $aR = R$, then R is a field.

5. Let A, B, C be ideals of a ring R. If $AB \subseteq C$ and C is a prime ideal of R, show that $A \subseteq C$ or $B \subseteq C$.

6. Show, by example, that the intersection of two prime ideals need not be a prime ideal.

7. Show that the only ideals of a field F are $\{0\}$ and F.

8. Determine all factor rings of Z.

9. Let p be a prime. Show that $A = \{(px, y) \mid x, y \in Z\}$ is a maximal ideal of $Z \oplus Z$.

10. Let R be a commutative ring with unity. Suppose a is a unit and b is nilpotent. Show that $a + b$ is a unit. (Hint: See exercise 19 of Chapter 13.)

11. Let A, B, and C be subrings of a ring R. If $A \subseteq B \cup C$, show that $A \subseteq B$ or $A \subseteq C$.

12. For any element a in a ring R, define $\langle a \rangle$ to be the smallest ideal of R that contains a. If R is a commutative ring with unity, show that $\langle a \rangle = aR = \{ar \mid r \in R\}$. Show, by example, that if R is commutative but does not have a unity, then $\langle a \rangle$ and aR may be different.

13. Let R be a ring with unity. Show that $\langle a \rangle = \{s_1 a t_1 + s_2 a t_2 + \cdots + s_n a t_n \mid s_i, t_i \in R$ and n is a positive integer$\}$.

14. Let p be a prime. Show that $Z_p[x]$ has characteristic p.

15. Let A and B be ideals of a ring R. If $A \cap B = \{0\}$, show that $ab = 0$ when $a \in A$ and $b \in B$.

16. Show that the direct sum of two integral domains is not an integral domain.

17. Consider the ring $R = \{0, 2, 4, 6, 8, 10\}$ under addition and multiplication modulo 12. What is the characteristic of R?

18. What is the characteristic of $Z_m \oplus Z_n$?

19. Let R be a commutative ring with unity. Suppose the only ideals of R are $\{0\}$ and R. Show that R is a field.

20. Suppose I is an ideal of J and J is an ideal of R. Prove that if I has a unity, then I is an ideal of R. (Be careful not to assume that the unity of I is the unity of R. It need not be—see exercise 2 of Chapter 13.)

21. Recall an idempotent element b in a ring is one with the property that $b^2 = b$. Find a nontrivial idempotent (that is, not 0 and not 1) in $Q[x]/\langle x^4 + x^2 \rangle$.

22. In a principal ideal domain, show that every nontrivial prime ideal is a maximal ideal.

23. Find an example of a commutative ring R with unity such that $a, b \in R$, $a \neq b$, and $a^n = b^n$ and $a^m = b^m$ where n and m are positive integers that are relatively prime. (Compare with exercise 31, part b, of Chapter 15.)

24. Let $Q(\sqrt[3]{2})$ denote the smallest subfield of **R** that contains Q and $\sqrt[3]{2}$. Describe the elements of $Q(\sqrt[3]{2})$.

25. Let R be an integral domain with nonzero characteristic. If A is a proper ideal of R, show that R/A has the same characteristic as R.

26. Let F be a field of order p^n. Determine the group isomorphism class of F under the operation addition.

27. If R is a finite commutative ring with unity, prove that every prime ideal of R is a maximal ideal of R.

28. Let R be a noncommutative ring and $C(R)$ the center of R (see exercise 11 of Chapter 14). Prove that the additive group of $R/C(R)$ is not cyclic.

29. Let

$$R = \left\{ \begin{bmatrix} a & b \\ c & d \end{bmatrix} \mid a, b, c, d \in Z_2 \right\}$$

with ordinary matrix addition and multiplication modulo 2. Show that

$$\left\{ \begin{bmatrix} 1 & 0 \\ 0 & 0 \end{bmatrix} r \mid r \in R \right\}$$

is not an ideal of R. (Hence, in exercise 5 of Chapter 16, the commutativity assumption is necessary.)

30. If R is an integral domain and A is a proper ideal of R, must R/A be an integral domain?

31. Let $A = \{a + bi \mid a, b \in Z, a = b \bmod 2\}$. Show that A is a maximal ideal of $Z[i]$. How many elements does $Z[i]/A$ have?

Ring Homomorphisms

17

If there is one central idea which is common to all aspects of modern algebra it is the notion of homomorphism.

I. N. Herstein, *Topics in Algebra*

Definition and Examples

In our work with groups, we saw that one way to discover information about a group was to examine its interaction with other groups by way of homomorphisms. It should not be surprising to learn that this concept extends to rings with equally profitable results.

Just as a group homomorphism preserves the group operation, a ring homomorphism preserves the ring operations.

> *DEFINITIONS*   Ring Homomorphisms, Ring Isomophism
> A *ring homomorphism* ϕ from a ring R to a ring S is a mapping from R to S that preserves the two ring operations; that is,
>
> $$(a + b)\phi = a\phi + b\phi \quad \text{and} \quad (ab)\phi = (a\phi)(b\phi)$$
> $$\text{for all } a, b \text{ in } R.$$
>
> A one-to-one (ring) homomorphism that is also onto is called a (ring) *isomorphism*.

As was the case for groups, in the preceding definition the operations on the left of the equal signs are those of R, while the operations on the right of the equal signs are those of S.

As with group theory also, the roles of isomorphism and homomorphism

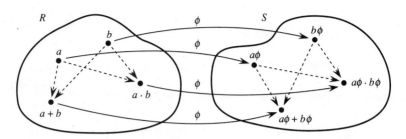

Figure 17.1

are entirely distinct. An isomorphism is used to show two rings are algebraically identical; a homomorphism is used to simplify a ring while retaining certain of its features.

A schematic representation of a ring homomorphism is given in Figure 17.1. The dashed arrows indicate the results of performing the ring operations.

The following examples illustrate homomorphisms. The reader should supply the missing details.

Example 1 For any positive integer n, the mapping $k \rightarrow k \bmod n$ is a ring homomorphism from Z onto Z_n. This mapping is called the *natural homomorphism* from Z to Z_n. $\qquad \square$

Example 2 The mapping $a + bi \rightarrow a - bi$ is a ring isomorphism from complex numbers onto the complex numbers. $\qquad \square$

Example 3 Let $\mathbf{R}[x]$ denote the ring of all polynomials with real coefficients. The mapping $f(x) \rightarrow f(1)$ is a ring homomorphism from $\mathbf{R}[x]$ onto $\mathbf{R}$. $\qquad \square$

Example 4 The mapping $\phi:x \rightarrow 5x$ from Z_4 to Z_{10} is a ring homomorphism. To check this, we first show that ϕ is well defined. To do this, suppose $x = y$ in Z_4. Then, $x - y = 4k$ for some integer k and $5x - 5y = 5(x - y) = 20k = 0$ in Z_{10}. So, $5x = 5y$ in Z_{10} and ϕ is well defined. Clearly,

$$(x + y)\phi = 5(x + y) = 5x + 5y = x\phi + y\phi.$$

Finally, using the fact that in Z_{10} we have $5 = 5 \cdot 5$, we see that $(xy)\phi = 5xy = 5 \cdot 5xy = x\phi y\phi$ and ϕ preserves multiplication as well. $\qquad \square$

Example 5 We determine all ring homomorphisms from Z_{12} to Z_{30}. By example 8 of Chapter 11, the only group homomorphisms from Z_{12} to Z_{30} are $x \rightarrow ax$ where $a = 0, 15, 10, 20, 5,$ or 25. But, since $1 \cdot 1 = 1$ in Z_{12} we must have $a \cdot a = a$ in Z_{30}. This requirement rules out 20 and 5 as possibilities for a. Finally, simple calculations show that each of the remaining four choices do yield ring homomorphisms. $\qquad \square$

Example 6   Let R be a commutative ring of characteristic 2. Then the mapping $a \rightarrow a^2$ is a ring homomorphism from R to R. $\square$

Example 7   Although $2Z$, the group of even integers under addition, is group-isomorphic to the group Z under addition, the ring $2Z$ is not ring-isomorphic to the ring Z. (Quick! What does Z have that $2Z$ doesn't?) $\square$

Our next three examples are applications of the natural homomorphism given in Example 1 to number theory.

Example 8   Test for Divisibility by 9
An integer n with decimal representation $a_k a_{k-1} \ldots a_0$ is divisible by 9 if and only if $a_k + a_{k-1} + \cdots + a_0$ is divisible by 9. To verify this, observe that $n = a_k 10^k + a_{k-1} 10^{k-1} + \cdots + a_0$. Then, letting α denote the natural homomorphism from Z to Z_9, we note that n is divisible by 9 if and only if

$$0\alpha = n\alpha = a_k \alpha 10\alpha + a_{k-1} \alpha 10\alpha + \cdots + a_0 \alpha$$
$$= a_k + a_{k-1} + \cdots + a_0.$$

$\square$

Example 9   Consider the sequence 3, 7, 11, 15, ... Is it possible that one of these integers is the sum of two squares? If so, the equation $3 + 4k = a^2 + b^2$ is satisfied for some integers $k, a,$ and b. Then, applying the natural homomorphism from Z to Z_4 to both sides of this equation, we see that the equation $3 = x^2 + y^2$ has a solution in Z_4. But, by direct substitution, one may verify that this equation has no solution. Thus, no integer in the sequence is the sum of two squares. $\square$

Example 10   We determine the positive integer solutions to the equation $x^3 + 15xy = 2192$. Let α be the natural homomorphism from Z to Z_3 and β be the natural homomorphism from Z to Z_5. If x and y are integer solutions to the equation, then $(x^3 + 15xy)\alpha = 2192\alpha$ and $(x^3 + 15xy)\beta = 2192\beta$. It follows then that $(x\alpha)^3 = 2$ and $(x\beta)^3 = 2$. Solving these two equations in Z_3 and Z_5, respectively, we see that $x\alpha = 2$ and $x\beta = 3$. Thus, $x \in \{2, 5, 8, \ldots\} \cap \{3, 8, 13, \ldots\}$. This mean that x has the form $8 + 15k$. If $k \geq 1$, then $x^3 > 2192$, so we conclude that $x = 8$. Substituting this into the original equation now gives $y = 14$. $\square$

Properties of Ring Homomorphisms

> **Theorem 17.1**   *Properties of Ring Homomorphisms*
> *Let ϕ be a homomorphism from a ring R to a ring S. Let A be a subring of R and B an ideal of S.*

1. *For any* $r \in R$ *and any positive integer* n, $(nr)\phi = n(r\phi)$ *and* $r^n\phi = (r\phi)^n$.
2. $A\phi = \{a\phi \mid a \in A\}$ *is a subring of* S.
3. *If* A *is an ideal and* ϕ *is onto* S, *then* $A\phi$ *is an ideal.*
4. $B\phi^{-1} = \{r \in R \mid r\phi \in B\}$ *is an ideal of* R.
5. *If* R *is commutative, then* $R\phi$ *is commutative.*
6. *If* R *has a unity* 1, $S \neq \{0\}$, *and* ϕ *is onto, then* 1ϕ *is the unity of* S.
7. ϕ *is an isomorphism if and only if* ϕ *is onto and* Ker $\phi = \{r \in R \mid r\phi = 0\} = \{0\}$.
8. *If* ϕ *is an isomorphism from* R *onto* S, *then* ϕ^{-1} *is an isomorphism from* S *onto* R.

Proof. The proofs of these properties are straightforward and are left as exercises. ∎

As with groups, the student should learn in words the various parts of Theorem 17.1 in addition to the symbols. Part (2) says that the homomorphic image of a subring is a subring. Part (4) says that the pullback of an ideal is an ideal, and so on.

The next three theorems parallel results we had for groups. The proofs are nearly identical to the group theory counterparts and are left as exercises.

Theorem 17.2 Kernels Are Ideals
Let ϕ *be a homomorphism from a ring* R *to a ring* S. *Then* Ker $\phi = \{r \in R \mid r\phi = 0\}$ *is an ideal of* R.

Theorem 17.3 First Isomorphism Theorem for Rings
Let ϕ *be a ring homomorphism from* R *to* S. *Then the mapping from* $R/\text{Ker } \phi$ *to* $R\phi$, *given by* $r + \text{Ker } \phi \rightarrow r\phi$ *is an isomorphism. In symbols,* $R/\text{Ker } \phi \approx R\phi$.

Theorem 17.4 Ideals Are Kernels
Every ideal of a ring R *is the kernel of a ring homomorphism of* R. *In particular, an ideal* A *is the kernel of the mapping* $r \rightarrow r + A$ *from* R *to* R/A.

The homomorphism from R to R/A given in Theorem 17.4 is called the *natural homomorphism* from R to R/A. Theorem 17.3 is often referred to as the Fundamental Theorem of Ring Homomorphisms.

As a corollary to the next theorem, we see that a ring with unity contains a copy of Z or Z_n.

Theorem 17.5 Homomorphism from Z to a Ring with Unity
Let R *be a ring with unity* e. *The mapping* $\phi : Z \rightarrow R$ *given by* $n \rightarrow ne$ *is a ring homomorphism.*

Proof. Let m, $n \in Z$. To show that addition is preserved, we consider three cases. First suppose that both m and n are nonnegative. Then

$$(m + n)\phi = (m + n)e = \underbrace{e + e + \cdots + e}_{(m + n) \text{ summands}}$$

$$= \underbrace{(e + e + \cdots + e)}_{m \text{ summands}} + \underbrace{(e + e + \cdots + e)}_{n \text{ summands}}$$

$$= me + ne = m\phi + n\phi.$$

Next, suppose that both m and n are negative. Then,

$$(m + n)\phi = (m + n)e = (-m - n)(-e)$$
$$= (-m)(-e) + (-n)(-e) = me + ne = m\phi + n\phi.$$

Third, suppose that one of m and n is nonnegative and the other is negative, say, $m \geq 0$, $n < 0$.

Then

$$(m + n)\phi = (m + n)e$$
$$= \underbrace{(e + e + \cdots + e)}_{m \text{ summands}} - \underbrace{(e + e + \cdots + e)}_{-n \text{ summands}}$$

$$= me + (-n)(-e) = me + ne = m\phi + n\phi.$$

So, ϕ preserves addition.

The multiplication can be handled in a single case with the aid of exercise 6 of Chapter 14, which says $(ma)(nb) = (mn)(ab)$ for all integers m and n. Thus, $(mn)\phi = (mn)e = (mn)(ee) = (me)(ne) = (m\phi)(n\phi)$. So, ϕ preserves multiplication as well. ∎

Corollary 1 *A Ring with Unity Contains Z_n or Z*
If R is a ring with unity and the characteristic of R is $n > 0$, then R contains a subring isomorphic to Z_n. If the characteristic of R is 0, then R contains a subring isomorphic to Z.

Proof. Let e by the unity element of R, and recall by exercise 22 of Chapter 14 that $S = \{ke \mid n \in Z\}$ is a subring of R. Now Theorem 17.5 shows that the mapping ϕ from Z onto S given by $k\phi = ke$ is a homomorphism, and by the First Isomorphism Theorem for rings, we have $Z/\text{Ker } \phi \approx S$. But, clearly, Ker $\phi = \langle n \rangle$ where n is the additive order of e and, by Theorem 15.3, n is the characteristic of R. So, $S \approx Z/\langle n \rangle \approx Z_n$. When R has characteristic 0, $S \approx Z/\langle 0 \rangle \approx Z$. ∎

Corollary 2 *Z_m Is a Homomorphic Image of Z*
For any positive integer m, the mapping $\phi : Z \to Z_m$ given by $x \to x \bmod m$ is a ring homomorphism.

Proof. This follows directly from the statement of Theorem 17.5, since in the ring Z_m, the integer x mod m is $x1$. (For example, in Z_3, if $x = 5$, we have $5(1) = 1 + 1 + 1 + 1 + 1 = 2$.) ∎

Corollary 3 *A Field Contains Z_p or Q (Steinitz—1910)*
If F is a field of characteristic p, then F contains a subfield isomorphic to Z_p. If F is a field of characteristic 0, then F contains a subfield isomorphic to the rational numbers.

Proof. By Corollary 1, F contains a subring isomorphic to Z_p if F has characteristic p, and F has a subring S isomorphic to Z if F has characteristic 0. In the latter case, let

$$T = \{ab^{-1} \mid a, b \in S, b \neq 0\}.$$

Then T is isomorphic to the rationals (exercise 47). ∎

Since the intersection of all subfields of a field is itself a subfield, every field has a smallest subfield (that is, a subfield that is contained in every subfield). This subfield is called the *prime subfield* of the field. It follows from Corollary 3 that the prime subfield of a field of characteristic p is isomorphic to Z_p, while the prime subfield of a field of characteristic 0 is isomorphic to Q. (See exercise 51.)

The Field of Quotients

Although the integral domain Z is not a field, it is at least contained in a field—the field of rational numbers. And notice that the field of rational numbers is nothing more than quotients of integers. Can we mimic the construction of the rationals from the integers for other integral domains? Yes.

Theorem 17.6 *Field of Quotients*
Let D be an integral domain. Then there exists a field F (called the field of quotients of D) that contains a subring isomorphic to D.

Proof. (Throughout this proof, you should keep in mind that we are using the construction of the rationals from the integers as a model for our construction of the field of quotients of D.)

Let S be the set of all formal symbols of the form a/b, where $a, b \in D$ and $b \neq 0$. Define an equivalence relation $\equiv$ on S by $a/b \equiv c/d$ if $ad = bc$ (just as we have $1/2 = 3/6$ for rationals). Now, let F be the set of equivalence classes of S under the relation $\equiv$. We define addition and multiplication on F by

$$[a/b] + [c/d] = [(ad + bc)/bd] \quad \text{and} \quad [a/b] \cdot [c/d] = [ac/bd].$$

(Notice that here we need the fact that D is an integral domain to ensure that multiplication is closed; that is, $bd \neq 0$ whenever $b \neq 0$ and $d \neq 0$.)

Since there are many representations of any particular element of F (just as in the rationals, we have $\frac{1}{2} = \frac{3}{6} = \frac{4}{8}$), we must show that these two operations are well defined. To do this, suppose $[a/b] = [a'/b']$ and $[c/d] = [c'/d']$ so that $ab' = a'b$ and $cd' = c'd$. It then follows that

$$(ad + bc)b'd' = adb'd' + bcb'd' = (ab')dd' + (cd')bb'$$
$$= (a'b)dd' + (c'd)bb' = a'd'bd + b'c'bd$$
$$= (a'd' + b'c')bd.$$

Thus, by definition, we have

$$[(ad + bc)/bd] = [(a'd' + b'c')/b'd'],$$

and, therefore, addition is well defined. We leave the verification that multiplication is well defined as an exercise. That F is a field is straightforward. Let 1 denote the unity of D. Then $[0/1]$ is the additive identity of F. The additive inverse of $[a/b]$ is $[-a/b]$; the multiplicative inverse of a nonzero element $[a/b]$ is $[b/a]$. The remaining field properties can be checked easily.

Finally, the mapping $\phi: D \rightarrow F$ given by $x \rightarrow [x/1]$ is a ring isomorphism from D to $D\phi$. ∎

In practice, it is customary to identify the equivalence class $[a/b]$ with the symbol a/b. This notation is less cumbersome and should not cause confusion. Again, the rational numbers provide a model, since we traditionally use $\frac{1}{2}$ in place of $[\frac{1}{2}]$, and so on. Just remember that a/b and c/d represent the same equivalence class if and only if $ad = cb$.

Example 11 Let $D = Z[x]$. Then the field of quotients of D is $\{f(x)/g(x) \mid f(x), g(x) \in D$, where $g(x)$ is not the zero polynomial$\}$. This set is usually called the set of *rational functions*. □

When F is a field, the field of quotients of $F[x]$ is traditionally denoted by $F(x)$. Thus, $Q(x) = \{f(x)/g(x) \mid f(x), g(x) \in Q[x]$, where $g(x)$ is not the zero polynomial$\}$.

EXERCISES

We can work it out.

Title of song by John Lennon and Paul McCartney, single, December 1965.

1. Show that the correspondence $x \rightarrow 5x$ from Z_5 to Z_{10} preserves both operations but is not well defined (that is, the correspondence is not a function).

2. Show that the correspondence $x \rightarrow 3x$ from Z_4 to Z_{12} is well defined and preserves addition but does not preserve multiplication.

3. a. Is the ring $2Z$ isomorphic to the ring $3Z$?
   b. Is the ring $2Z$ isomorphic to the ring $4Z$?

4. Let $Z_3[i] = \{a + bi \mid a, b \in Z_3\}$ (see Example 8 of Chapter 15). Show that $Z_3[i]$ is isomorphic to $Z_3[x]/\langle x^2 + 1 \rangle$ as fields.

5. Let

$$S = \left\{ \begin{bmatrix} a & b \\ -b & a \end{bmatrix} \mid a, b \in \mathbf{R} \right\}.$$

Show that $\phi:C \rightarrow S$ given by

$$(a + bi)\phi = \begin{bmatrix} a & b \\ -b & a \end{bmatrix}$$

is a ring isomorphism.

6. Let $Z[\sqrt{2}] = \{a + b\sqrt{2} \mid a, b \in Z\}$. Let

$$H = \left\{ \begin{bmatrix} a & 2b \\ b & a \end{bmatrix} \mid a, b \in Z \right\}.$$

Show that $Z[\sqrt{2}]$ and H are isomorphic as rings.

7. Consider the mapping from $M_2(Z)$ into Z given by $\begin{bmatrix} a & b \\ c & d \end{bmatrix} \rightarrow a$.
   Prove or disprove that this is a ring homomorphism.

8. Let $R = \left\{ \begin{bmatrix} a & b \\ 0 & c \end{bmatrix} \mid a, b, c \in Z \right\}$. Prove or disprove that the mapping $\begin{bmatrix} a & b \\ 0 & c \end{bmatrix} \rightarrow a$ is a ring homomorphism.

9. Is the mapping from Z_5 to Z_{30} given by $x \rightarrow 6x$ a ring homomorphism? Note that the image of the unity is the unity of the image but not the unity of Z_{30}.

10. Is the mapping from Z_{10} to Z_{10} given by $x \rightarrow 2x$ a ring homomorphism?

11. Describe the kernel of the homomorphism given in Example 3.

12. Let ϕ be a ring homomorphism from Z_m to Z_n. Prove that 1ϕ is an idempotent of Z_n. (Recall, an idempotent is an element a with the property that $a^2 = a$.)

13. In Z, let $A = \langle 2 \rangle$ and $B = \langle 8 \rangle$. Show that the group A/B is isomorphic to Z_4 but the ring A/B is not isomorphic to Z_4.

14. Determine all ring homomorphisms from Z_6 to Z_6. Determine all ring homomorphisms from Z_{20} to Z_{30}.

15. Determine all ring homomorphisms from Z to Z.

16. Determine all ring homomorphisms from Q to Q.

17. Prove that the sequence 2, 10, 18, 26, . . . contains no cube.

18. Prove that the sum of the squares of three consecutive integers cannot be a square.

19. Let m be a positive integer and let n be an integer obtained from m by rearranging the digits of m in some way. (For example, 72345 is a rearrangement of 35274.) Show that $m - n$ is divisible by 9.

20. (Test for divisibility by 11.) Let n be an integer with decimal representation $a_k a_{k-1} \ldots a_1 a_0$. Prove that n is divisible by 11 if and only if $a_0 - a_1 + a_2 - \cdots (-1)^k a_k$ is divisible by 11.

21. Show that the number 7,176,825,942,116,027,211 is divisible by 9 but not divisible by 11.

22. Show that the number 9,897,654,527,609,877 is divisible by 99.

23. (Test for divisibility by 3). Let n be an integer with decimal representation $a_k a_{k-1} \ldots a_1 a_0$. Prove that n is divisible by 3 if and only if $a_k + a_{k-1} + \cdots + a_1 + a_0$ is divisible by 3.

24. (Test for divisibility by 4.) Let n be an integer with decimal representation $a_k a_{k-1} \ldots a_1 a_0$. Prove that n is divisible by 4 if and only if $a_1 a_0$ is divisible by 4.

25. In your head, determine $(2 \cdot 10^{75} + 2)^{100} \bmod 3$ and $(10^{100} + 1)^{99} \bmod 3$.

26. Suppose ϕ is a ring homomorphism from $Z \oplus Z$ into $Z \oplus Z$. What are the possibilities for $(1, 0)\phi$?

27. Let R and S be commutative rings with unity. If ϕ is a homomorphism from R onto S and the characteristic of R is nonzero, prove that the characteristic of S divides the characteristic of R.

28. Let R be a commutative ring of prime characteristic p. Show that the *Frobenius* map $x \rightarrow x^p$ is a ring homomorphism from R to R.

29. Is there a ring homomorphism from the reals to some ring whose kernel is the integers?

30. Show that a homomorphism from a field onto a ring with more than one element must be an isomorphism.

31. Let ϕ be a ring homomorphism from R onto S and A be an ideal of S.
    a. If A is prime in S, show that $A\phi^{-1} = \{x \in R \mid x\phi \in A\}$ is prime in R.
    b. If A is maximal in S, show that $A\phi^{-1}$ is maximal in R.

32. Show that the homomorphic image of a principal ideal ring is a principal ideal ring.

33. Let R and S be rings.
    a. Show that the mapping from $R \oplus S$ onto R given by $(a,b) \rightarrow a$ is a ring homomorphism.
    b. Show that the mapping from R into $R \oplus S$ given by $a \rightarrow (a,0)$ is a one-to-one ring homomorphism.
    c. Show that $R \oplus S$ is isomorphic to $S \oplus R$.

34. Show that if m and n are distinct positive integers, then mZ is not ring isomorphic to nZ.

35. Prove or disprove that the field of real numbers is isomorphic to the field of complex numbers.

36. Show that the only automorphism of the real numbers is the identity mapping.

37. Determine all ring homomorphisms from **R** into **R**.

38. Show that the operation of multiplication defined in the proof of Theorem 17.6 is well defined.

39. Suppose n divides m and a is an idempotent of Z_n (that is, $a^2 = a$). Show that the mapping $x \rightarrow ax$ is a ring homomorphism from Z_m to Z_n. Show that the same correspondence need not yield a well-defined function if n does not divide m.

40. Let $Q[\sqrt{2}] = \{a + b\sqrt{2} \mid a, b \in Q\}$ and $Q[\sqrt{5}] = \{a + b\sqrt{5} \mid a, b \in Q\}$. Show that these two rings are not isomorphic.

41. Let $Z[i] = \{a + bi \mid a, b \in Z\}$. Show that the field of quotients of $Z[i]$ is isomorphic to $Q[i] = \{r + si \mid r, s \in Q\}$.

42. Let F be a field. Show that the field of quotients of F is isomorphic to F.

43. Let D be an integral domain and F the field of quotients of D. Show that if E is any field that contains D, then E contains a subfield isomorphic to F. (Thus, the field of quotients of an integral domain D is the smallest field containing D.)

44. Explain why a commutative ring with unity that is not an integral domain cannot be contained in a field.

45. Give an example of a ring without unity that is contained in a field.

46. Show that the relation $\equiv$ defined in the proof of Theorem 17.6 is an equivalence relation.

47. Prove that the set T in the proof of Corollary 3 of Theorem 17.5 is ring-isomorphic to the field of rational numbers.

48. Let $f(x) \in \mathbf{R}[x]$. If $a + bi$ is a complex zero of $f(x)$ (here $i = \sqrt{-1}$), show that $a - bi$ is a zero of $f(x)$. (This exercise is referred to in Chapter 33.)

49. Suppose $\phi : R \rightarrow S$ is a ring homomorphism and the image of ϕ is not $\{0\}$. If R has a unity and S is an integral domain, show that ϕ carries the unity of R to the unity of S. Give an example to show that the previous statement need not be true if S is not an integral domain.

50. Show that the mapping $\phi : D \rightarrow F$ in the proof of Theorem 17.6 is a ring homomorphism.

51. Show that the prime subfield of a field of characteristic p is isomorphic to Z_p and the prime subfield of a field of characteristic 0 is isomorphic to Q.

SUGGESTED READINGS

J. A. Gallian and D. S. Jungreis, "Homomorphisms from $Z_m[i]$ into $Z_n[i]$ and $Z_m[\rho]$ into $Z_n[\rho]$ where $i^2 + 1 = 0$ and $\rho^2 + \rho + 1 = 0$," *The American Mathematical Monthly* 95 (1988): 247–249.

This article gives formulas for counting the homomorphisms mentioned in the title.

J. A. Gallian and J. Van Buskirk, "The Number of Homomorphisms from Z_m into Z_n," *American Mathematical Monthly* 91 (1983): 196–197.

In this note, formulas are given for the number of group homomorphisms from Z_m into Z_n and the number of ring homomorphisms from Z_m into Z_n.

W. C. Waterhouse, "Rings with Cyclic Additive Group," *American Mathematical Monthly* 71 (1964): 449–450.

In this brief note, it is proved that, up to isomorphism, the number of rings whose additive group is cyclic of order m is the number of divisors of m.

Polynomial Rings

18

Very early in our mathematical education—in fact in junior high school or early in high school itself—we are introduced to polynomials. For a seemingly endless amount of time we are drilled, to the point of utter boredom, in factoring them, multiplying them, dividing them, simplifying them. Facility in factoring a quadratic becomes confused with genuine mathematical talent.

I. N. Herstein, *Topics in Algebra*

Notation and Terminology

One of the mathematical concepts that students are most familiar with and most comfortable with is that of a polynomial. In high school, students study polynomials with integer coefficients, rational coefficients, or real coefficients, and, perhaps, even complex coefficients. In earlier chapters of this book, we introduced something that was probably new—polynomials with coefficients from Z_n. Notice that all of these sets of polynomials are rings and, in each case, the set of coefficients is also a ring. In this chapter, we abstract all of these examples into one.

> *DEFINITION*   Ring of Polynomials over R
> Let R be a commutative ring. The set of formal symbols
>
> $$R[x] = \{a_n x^n + a_{n-1} x^{n-1} + \cdots + a_1 x + a_0 \mid a_i \in R,$$
> $$n \text{ is a nonnegative integer}\}$$
>
> is called the *ring of polynomials over R in the indeterminate x.* Two elements
>
> $$a_n x^n + a_{n-1} x^{n-1} + \cdots + a_1 x + a_0$$

and

$$b_m x^m + b_{m-1} x^{m-1} + \cdots + b_1 x + b_0$$

of $R[x]$ are considered equal if and only if $a_i = b_i$ for all nonnegative integers i. (Define $a_i = 0$ when $i > n$ and $b_i = 0$ when $i > m$.)

In this definition, the symbols $x, x^2, \ldots, x^n$ do not represent "unknown" elements or variables from the ring R. Rather, their purpose is to serve as convenient placeholders that separate the ring elements $a_n, a_{n-1}, \ldots, a_0$. We could have avoided the x's by defining a polynomial as an infinite sequence $a_0, a_1, a_2, \ldots, a_n, 0, 0, 0, \ldots$, but our method takes advantage of the student's experience in manipulating polynomials where x does represent a variable. The disadvantage of our method is that one must be careful not to confuse a polynomial with the function determined by a polynomial. For example, in $Z_3[x]$, the polynomials $f(x) = x^3 + 2x$ and $g(x) = x^5 + 2x$ determine the same function from Z_3 to Z_3 since $f(a) = g(a)$ for all a in Z_3.[*] But $f(x)$ and $g(x)$ are different elements of $Z_3[x]$. Also, in the ring $Z_n[x]$, be careful to reduce only the coefficients and not the exponents modulo n. For example, in $Z_3[x]$, $5x = 2x$, but $x^5 \neq x^2$.

To make $R[x]$ into a ring, we define addition and multiplication in the usual way.

DEFINITION   Addition and Multiplication in $R[x]$
Let R be a commutative ring and let

$$f(x) = a_n x^n + a_{n-1} x^{n-1} + \cdots + a_1 x + a_0$$

and

$$g(x) = b_m x^m + b_{m-1} x^{m-1} + \cdots + b_1 x + b_0$$

belong to $R[x]$. Then

$$f(x) + g(x) = (a_s + b_s) x^s + (a_{s-1} + b_{s-1}) x^{s-1} \\ + \cdots + (a_1 + b_1) x + a_0 + b_0$$

where $a_i = 0$ for $i > n$ and $b_i = 0$ for $i > m$. Also,

$$f(x)g(x) = c_{m+n} x^{m+n} + c_{m+n-1} x^{m+n-1} + \cdots + c_1 x + c_0$$

where

$$c_k = a_k b_0 + a_{k-1} b_1 + \cdots + a_1 b_{k-1} + a_0 b_k$$

for $k = 0, \ldots, m + n$.

---

[*]In general, given $f(x)$ in $R[x]$ and a in R, $f(a)$ means substitute a for x in the formula for $f(x)$. This substitution is a homomorphism from $R[x]$ to R.

Although the definition of multiplication might appear complicated, it is just a formalization of the familiar process of using the distributive property and collecting like terms. So, just multiply polynomials over a commutative ring R in the same way that polynomials are always multiplied. Here is an example.

Consider $f(x) = 2x^3 + x^2 + 2x + 2$ and $g(x) = 2x^2 + 2x + 1$ in $Z_3[x]$. Then in our preceding notation, $a_5 = 0$, $a_4 = 0$, $a_3 = 2$, $a_2 = 1$, $a_1 = 2$, $a_0 = 2$ and $b_5 = 0$, $b_4 = 0$, $b_3 = 0$, $b_2 = 2$, $b_0 = 1$. Now using the definitions and remembering that addition and multiplication of the coefficients is done modulo 3, we have

$$f(x) + g(x) = (2 + 0)x^3 + (1 + 2)x^2 + (2 + 2)x + (2 + 1)$$
$$= 2x^3 + 0x^2 + 1x + 0$$

and

$$f(x) \cdot g(x) = (0 \cdot 1 + 0 \cdot 2 + 2 \cdot 2 + 1 \cdot 0 + 2 \cdot 0 + 2 \cdot 0)x^5$$
$$+ (0 \cdot 1 + 2 \cdot 2 + 1 \cdot 2 + 2 \cdot 0 + 2 \cdot 0)x^4$$
$$+ (2 \cdot 1 + 1 \cdot 2 + 2 \cdot 2 + 2 \cdot 0)x^3$$
$$+ (1 \cdot 1 + 2 \cdot 2 + 2 \cdot 2)x^2 + (2 \cdot 1 + 2 \cdot 2)x + 2 \cdot 1$$
$$= x^5 + 0x^4 + 2x^3 + 0x^2 + 0x + 2.$$

Our definitions for addition and multiplication of polynomials were formulated so that $R[x]$ is commutative and associative, and so that multiplication is distributive over addition. After all, $R[x]$ is called the "ring of polynomials over R." We leave the verification that $R[x]$ is a ring as an exercise.

It is time to introduce some terminology for polynomials. If

$$f(x) = a_n x^n + a_{n-1}x^{n-1} + \cdots + a_1 x + a_0$$

where $a_n \neq 0$, we say $f(x)$ has *degree n*; the term a_n is called the *leading coefficient* of $f(x)$; and, if the leading coefficient is the multiplicative identity element of R, we say $f(x)$ is a *monic* polynomial. The polynomial $f(x) = 0$ has no degree. Polynomials of the form $f(x) = a_0$ are called *constant*. We often write deg $f(x) = n$ to indicate that $f(x)$ has degree n. In keeping with our experience with polynomials with real coefficients, we adopt the following notational conventions: we may insert or delete terms of the form $0x^k$; $1x^k$ will be denoted by x^k; $+(-a_k)x^k$ will be denoted by $-a_k x^k$.

Very often properties of R carry over to $R[x]$. Our first theorem is a case in point.

Theorem 18.1   *D an Integral Domain Implies D[x] Is*
If D is an integral domain, then D[x] is an integral domain.

Proof.   Since we already know that $D[x]$ is a ring, all we need to show is that $D[x]$ is commutative with a unity and has no zero-divisors. Clearly, $D[x]$ is commutative whenever D is. If 1 is the unity element of D, it is easy to check that $f(x) = 1$ is the unity element of $D[x]$. Finally, suppose

$$f(x) = a_n x^n + a_{n-1} x^{n-1} + \cdots + a_0$$

and

$$g(x) = b_m x^m + b_{m-1} x^{m-1} + \cdots + b_0$$

where $a_n \neq 0$ and $b_m \neq 0$. Then, by definition, $f(x)g(x)$ has leading coefficient $a_n b_m$ and since D is a integral domain, $a_n b_m \neq 0$. ∎

The Division Algorithm and Consequences

One of the properties of integers that we have used repeatedly is the division algorithm: If a and b are integers and $b \neq 0$, then there exist unique integers q and r such that $a = bq + r$ where $0 \leqslant r < |b|$. The next theorem is the analogous statement for polynomials over a field.

Theorem 18.2 *Division Algorithm for $F[x]$*
Let F be a field and let $f(x)$ and $g(x) \in F[x]$ with $g(x) \neq 0$. Then there exist unique polynomials $q(x)$ and $r(x)$ in $F[x]$ such that $f(x) = g(x)q(x) + r(x)$ and either $r(x) = 0$ or $\deg r(x) < \deg g(x)$.

Proof.   We begin by showing the existence of $q(x)$ and $r(x)$. If $f(x) = 0$ or $\deg f(x) < \deg g(x)$ we simply put $q(x) = 0$ and $r(x) = f(x)$. So, we may assume $n = \deg f(x) \geqslant \deg g(x) = m$ and let $f(x) = a_n x^n + \cdots + a_0$ and $g(x) = b_m x^m + \cdots + b_0$. The idea behind this proof is to begin just as if you were going to "long divide" $g(x)$ into $f(x)$, then use the Second Principle of Induction on $\deg f(x)$ to finish up. Thus, resorting to long division we let $f_1(x) = f(x) - a_n b_m^{-1} x^{n-m} g(x)$.* Then $f_1(x) = 0$ or $\deg$

---

*For example,

$$
\begin{array}{r}
3/2x^2 \\
2x^2 + 2\,\overline{\smash{\big)}\,3x^4 \qquad\quad + x + 1} \\
\underline{3x^4 \;\; + 3x^2} \\
-3x^2 + x + 1 \\
\end{array}
$$

$$-3x^2 + x + 1 = 3x^4 + x + 1 - 3/2x^2(2x^2 + 2)$$

In general,

$$
\begin{array}{r}
a_n b_m^{-1} x^{n-m} \\
b_m x^m + \cdots\,\overline{\smash{\big)}\,a_n x^n + \cdots} \\
\underline{a_n x^n + \cdots} \\
f_1(x) \\
\end{array}
$$

$$f_1(x) = (a_n x^n + \cdots) - a_n b_m^{-1} x^{n-m}(b_m x^m + \cdots)$$

$f_1(x) <$ deg $f(x)$; so, by our induction hypothesis, there exist $q_1(x)$ and $r_1(x)$ in $F[x]$ such that $f_1(x) = g(x)q_1(x) + r_1(x)$ where $r_1(x) = 0$ or deg $r_1(x) <$ deg $g(x)$. (Technically, we should get the induction started by proving the case deg $f(x) = 0$, but this is trivial.) Thus,

$$
\begin{aligned}
f(x) &= a_n b_m^{-1} x^{n-m} g(x) + f_1(x) \\
&= a_n b_m^{-1} x^{n-m} g(x) + q_1(x)g(x) + r_1(x) \\
&= [a_n b_m^{-1} x^{n-m} + q_1(x)]g(x) + r_1(x).
\end{aligned}
$$

So, the polynomials $q(x) = a_n b_m^{-1} x^{n-m} + q_1(x)$ and $r(x) = r_1(x)$ have the desired properties.

To prove uniqueness, suppose $f(x) = g(x)q(x) + r(x)$ and $f(x) = g(x)q\star(x) + r\star(x)$ where $r(x) = 0$ or deg $r(x) <$ deg $g(x)$ and $r\star(x) = 0$ or deg $r\star(x) <$ deg $g(x)$. Then, subtracting these two equations, we obtain

$$
0 = g(x)[q(x) - q\star(x)] + [r(x) - r\star(x)]
$$

or

$$
r\star(x) - r(x) = g(x)[q(x) - q\star(x)].
$$

Thus $r\star(x) - r(x)$ is zero or the degree of $[r\star(x) - r(x)]$ is at least that of $g(x)$. Since the latter is clearly impossible, we have $r\star(x) = r(x)$ and $q(x) = q\star(x)$ as well. ∎

The polynomials $q(x)$ and $r(x)$ in the Division Algorithm are called the *quotient* and *remainder* in the division of $f(x)$ by $g(x)$. When the ring of coefficients of a polynomial ring is a field, we can use the long division process to determine the quotient and remainder.

Example 1 To find the quotient and remainder upon dividing $f(x) = 3x^4 + x^3 + 2x^2 + 1$ by $g(x) = x^2 + 4x + 2$ where $f(x)$ and $g(x)$ belong to $Z_5[x]$, we may proceed by long division, provided we keep in mind that addition and multiplication are done modulo 5. Thus,

$$
\begin{array}{r}
3x^2 + 4x \\
x^2 + 4x + 2 \,\overline{)\, 3x^4 + x^3 + 2x^2 + 1} \\
\underline{3x^4 + 2x^3 + x^2 } \\
4x^3 + x^2 + 1 \\
\underline{4x^3 + x^2 + 3x } \\
2x + 1
\end{array}
$$

So, $3x^2 + 4x$ is the quotient and $2x + 1$ is the remainder. Also,

$$
3x^4 + x^3 + 2x^2 + 1 = (x^2 + 4x + 2)(3x^2 + 4x) + 2x + 1. \qquad \square
$$

Let D be an integral domain. If $f(x)$ and $g(x) \in D[x]$, we say $g(x)$ *divides* $f(x)$ in $D[x]$ (and write $g(x) \mid f(x)$) if there exists an $h(x) \in D[x]$ such that $f(x) = g(x)h(x)$. In this case, we also call $g(x)$ a *factor* of $f(x)$. An element a is a *zero* (or *root*) of a polynomial $f(x)$ if $f(a) = 0$. When F is a field, $a \in F$ and $f(x) \in F[x]$, we say a is a *zero of multiplicity* k if $(x - a)^k$ is a factor of $f(x)$, but $(x - a)^{k+1}$ is not a factor of $f(x)$. With these definitions, we may now give several important corollaries of the Division Algorithm. No doubt you have seen these for the special case where F is the field of real numbers.

Corollary 1   *The Remainder Theorem*
Let F be a field, $a \in F$, and $f(x) \in F[x]$. Then $f(a)$ is the remainder in the division of $f(x)$ by $x - a$.

Proof.   Exercise.   ■

Corollary 2   *The Factor Theorem*
Let F be a field, $a \in F$, and $f(x) \in F[x]$. Then a is a zero of $f(x)$ if and only if $x - a$ is a factor of $f(x)$.

Proof.   Exercise.   ■

Corollary 3   *Polynomials of Degree n Have at Most n Zeros*
A polynomial of degree n over a field has at most n zeros counting multiplicity.

Proof.   We proceed by induction on n. Clearly, a polynomial of degree 1 over a field has exactly one zero. Now suppose that $f(x)$ is a polynomial of degree n over a field and a is a zero of $f(x)$ of multiplicity k. Then, $f(x) = (x - a)^k q(x)$; and, since $n = \deg f(x) = \deg(x - a)^k q(x) = k + \deg q(x)$, we have $k \leqslant n$ (see exercise 16). If $f(x)$ has no zeros other than a, we are done. On the other hand, if $b \neq a$ and b is a zero of $f(x)$, then $0 = f(b) = (b - a)^k q(b)$ so that b is also a zero of $q(x)$. By the Second Principle of Mathematical Induction, we know that $q(x)$ has at most $\deg q(x) = n - k$ zeros, counting multiplicity. Thus, $f(x)$ has at most $k + n - k = n$ zeros, counting multiplicity.   ■

We remark that Corollary 3 is not true for arbitrary polynomial rings. For example, the polynomial $x^2 + 3x + 2$ has four zeros in Z_6. Lagrange was the first to prove Corollary 3 for polynomials in $Z_p[x]$.

Example 2   The Complex Zeros of $x^n - 1$
We find all complex zeros of $x^n - 1$. Let $\omega = \cos(360/n) + i \sin(360/n)$. It follows from DeMoivre's theorem (see Example 11 of Chapter 0) that $\omega^n = 1$ and $\omega^k \neq 1$ for $1 \leqslant k < n$. Thus, each of $1, \omega, \omega^2, \ldots, \omega^{n-1}$ is a zero of $x^n - 1$ and, by Corollary 3, there are no others.   □

The complex number ω in Example 2 is called a *primitive nth root of unity*. We conclude this chapter with an important theoretical application of the Division Algorithm, but first an important definition.

DEFINITION Principal Ideal Domain (PID)
A *principal ideal domain* is an integral domain R in which every ideal has the form $\langle a \rangle = \{ra \mid r \in R\}$ for some a in R.

Theorem 18.3 $F[x]$ Is a Principal Ideal Domain
Let F be a field. Then F[x] is a principal ideal domain.

Proof. By Theorem 18.1, we know $F[x]$ is an integral domain. Now, let I be an ideal in $F[x]$. If $I = \{0\}$, then $I = \langle 0 \rangle$. If $I \neq \{0\}$, then among all the elements of I, let $g(x)$ be one of minimum degree. We will show that $I = \langle g(x) \rangle$. To do this, let $f(x) \in I$. Then, by the Division Algorithm, we may write $f(x) = g(x)q(x) + r(x)$ where $r(x) = 0$ or deg $r(x) <$ deg $g(x)$. Since $r(x) = f(x) - g(x)q(x) \in I$, the minimality of deg $g(x)$ implies the latter condition cannot hold. So, $r(x) = 0$ and, therefore, $f(x) \in \langle g(x) \rangle$. ∎

The proof of Theorem 18.3 also establishes the following.

Theorem 18.4 Criterion for $I = \langle g(x) \rangle$
Let F be a field, I an ideal in F[x], and g(x) an element of F[x]. Then, $I = \langle g(x) \rangle$ if and only if g(x) is a nonzero polynomial of minimum degree in I.

As an application of the First Isomorphism Theorem for Rings (Theorem 17.3) and Theorem 18.4, we verify the remark we made in Example 9 of Chapter 16 that the ring $R[x]/\langle x^2 + 1 \rangle$ is isomorphic to the ring of complex numbers.

Example 3 Consider the homomorphism φ from $\mathbf{R}[x]$ onto $\mathbf{C}$ given by $f(x) \rightarrow f(i)$ (that is, evaluate every polynomial in $\mathbf{R}[x]$ at i). Then $x^2 + 1 \in$ Ker φ and is clearly a polynomial of minimum degree in Ker φ. Thus, Ker φ = $\langle x^2 + 1 \rangle$ and $\mathbf{R}[x]/\langle x^2 + 1 \rangle$ is isomorphic to $\mathbf{C}$. □

EXERCISES

The difference between a text without problems and a text with problems is like the difference between learning to read a language and learning to speak it.

Freeman Dyson, *Disturbing the Universe*

1. Let $f(x) = 4x^3 + 2x^2 + x + 3$ and $g(x) = 3x^4 + 3x^3 + 3x^2 + x + 4$ where $f(x), g(x) \in Z_5[x]$. Compute $f(x) + g(x)$ and $f(x) \cdot g(x)$.

2. Show that $x^2 + 3x + 2$ has 4 zeros in Z_6. (This exercise is referred to in this chapter.)

3. In $Z_3[x]$, show that $x^4 + x$ and $x^2 + x$ determine the same function from Z_3 to Z_3.

4. Prove Corollary 1 of Theorem 18.2.

5. Prove Corollary 2 of Theorem 18.2.

6. List all the polynomials of degree 2 in $Z_2[x]$.

7. If R is a commutative ring, show that the characteristic of $R[x]$ is the same as the characteristic of R.

8. If $\phi:R \rightarrow S$ is a ring homomorphism, define $\overline{\phi}:R[x] \rightarrow S[x]$ by $(a_n x^n + \cdots + a_0)\overline{\phi} = a_n\phi x^n + \cdots + a_0\phi$. Show that $\overline{\phi}$ is a ring homomorphism.

9. If the rings R and S are isomorphic, show that $R[x]$ and $S[x]$ are isomorphic.

10. Let R be a commutative ring. Show that $R[x]$ has a subring isomorphic to R.

11. Let $f(x) = x^3 + 2x + 4$ and $g(x) = 3x + 2$ in $Z_5[x]$. Determine the quotient and remainder upon dividing $f(x)$ by $g(x)$.

12. Let $f(x) = 5x^4 + 3x^3 + 1$ and $g(x) = 3x^2 + 2x + 1$ in $Z_7[x]$. Determine the quotient and remainder upon dividing $f(x)$ by $g(x)$.

13. Show that the polynomial $2x + 1$ in $Z_4[x]$ has a multiplicative inverse in $Z_4[x]$.

14. Are there any nonconstant polynomials in $Z[x]$ that have multiplicative inverses? Explain your answer.

15. Let p be a prime. Are there any nonconstant polynomials in $Z_p[x]$ that have multiplicative inverses? Explain your answer.

16. (Degree Rule) Let D be an integral domain and $f(x), g(x) \in D[x]$. Prove $\deg(f(x) \cdot g(x)) = \deg f(x) + \deg g(x)$.

17. Prove that the ideal $\langle x \rangle$ in $Z[x]$ is prime but not maximal.

18. Prove that the ideal $\langle x \rangle$ in $Q[x]$ is maximal.

19. Let F be an infinite field and $f(x) \in F[x]$. If $f(a) = 0$ for infinitely many elements a of F, show that $f(x) = 0$.

20. Let F be an infinite field and $f(x), g(x) \in F[x]$. If $f(a) = g(a)$ for infinitely many elements a of F, show that $f(x) = g(x)$.

21. Let F be a field and let $p(x) \in F[x]$. If $f(x), g(x) \in F[x]$ and $\deg f(x) < \deg p(x)$ and $\deg g(x) < \deg p(x)$, show that $f(x) + \langle p(x) \rangle = g(x) + \langle p(x) \rangle$ implies $f(x) = g(x)$. (This exercise is referred to in Chapter 22.)

22. Prove that $Z[x]$ is not a principal ideal domain. (Compare this with Theorem 18.3.)

23. Find a polynomial with integer coefficients that has $1/2$ and $-1/3$ as zeros.

24. Let $f(x) \in \mathbf{R}[x]$. Suppose $f(a) = 0$ but $f'(a) \neq 0$ where $f'(x)$ is the derivative of $f(x)$. Show that a is a zero of $f(x)$ of multiplicity 1.

25. Show that Corollary 2 of Theorem 18.2 is false for Z_m when m is not prime and $m > 1$.

26. Show that Corollary 3 of Theorem 18.2 is true for polynomials over integral domains.

27. Let F be a field and let

$$I = \{a_n x^n + a_{n-1} x^{n-1} + \cdots + a_0 \mid a_n, a_{n-1}, \ldots, a_0 \in F \text{ and}$$
$$a_n + a_{n-1} + \cdots + a_0 = 0\}.$$

Show that I is an ideal of $F[x]$ and find a generator for I.

28. Let m be a fixed positive integer. For any integer a, let $\bar{a}$ denote $a \bmod m$. Show that the mapping of $\phi: Z[x] \to Z_m[x]$ given by

$$(a_n x^n + a_{n-1} x^{n-1} + \cdots + a_0)\phi = \bar{a}_n x^n + \bar{a}_{n-1} x^{n-1} + \cdots + \bar{a}_0$$

is a ring homomorphism. (This exercise is referred to in Chapter 19.)

29. Let F be a field and $f(x) = a_n x^n + a_{n-1} x^{n-1} + \cdots + a_0 \in F[x]$. Prove that $x - 1$ is a factor of $f(x)$ if and only if $a_n + a_{n-1} + \cdots + a_0 = 0$.

30. Find infinitely many polynomials $f(x)$ in $Z_3[x]$ such that $f(a) = 0$ for all a in Z_3.

31. For every prime p, show that

$$x^{p-1} - 1 = (x - 1)(x - 2) \cdots [x - (p - 1)]$$

in $Z_p[x]$.

32. (Wilson's Theorem) For every integer $n > 1$, prove that $(n - 1)! = n - 1$ mod n if and only if n is prime.

33. For every prime p, show that $(p - 2)! = 1 \bmod p$.

34. Find the remainder upon dividing 98! by 101.

35. Prove that $(50!)^2 = -1 \bmod 101$.

36. If I is an ideal of a ring R, prove that $I[x]$ is an ideal of $R[x]$.

37. Give an example of a commutative ring R with unity and a maximal ideal I of R such that $I[x]$ is not a maximal ideal of $R[x]$.

38. Let R be a commutative ring with unity. If I is a prime ideal of R, prove that $I[x]$ is a prime ideal of $R[x]$.

39. Prove that $Q[x]/\langle x^2 - 2 \rangle$ is isomorphic to $Q[\sqrt{2}] = \{a + b\sqrt{2} \mid a, b \in Q\}$.

40. Let F be a field, and let $f(x)$ and $g(x)$ belong to $F[x]$. If there is no polynomial of positive degree in $F[x]$ that divides both $f(x)$ and $g(x)$ (in this case, $f(x)$ and $g(x)$ are said to be *relatively prime*), prove that there exist polynomials $h(x)$ and $k(x)$ in $F[x]$ with the property that $f(x)h(x) + g(x)k(x) = 1$. (This exercise is referred to in Chapter 23.)

41. Let $f(x) \in \mathbf{R}[x]$. If $f(a) = 0$ and $f'(a) = 0$ ($f'(a)$ is the derivative of $f(x)$ at a), show that $(x - a)^2$ divides $f(x)$.

Factorization of Polynomials

19

The value of a principle is the number of things it will explain.

Ralph Waldo Emerson

Reducibility Tests

In high school, students spend much time factoring polynomials and finding their roots. In this chapter, we consider the same problems in a more abstract setting.

To discuss factorization of polynomials, we must first introduce the polynomial analog of a prime integer.

DEFINITION Irreducible Polynomial, Reducible Polynomial
Let D be an integral domain. A polynomial $f(x)$ from $D[x]$ that is neither the zero polynomial nor a unit in $D[x]$ is said to be *irreducible over D* if, whenever $f(x)$ is expressed as a product $f(x) = g(x)h(x)$ with $g(x)$ and $h(x)$ from $D[x]$, then $g(x)$ or $h(x)$ is a unit in $D[x]$. A nonzero, nonunit element of $D[x]$ that is not irreducible over D is called *reducible over D*.

Notice that in the case that D is a field, a nonconstant $f(x) \in D[x]$ is irreducible over D if and only if $f(x)$ cannot be expressed as a product of two polynomials in $D[x]$ of lower degree (exercise).

Example 1 The polynomial $f(x) = 2x^2 + 4$ is irreducible over Q but reducible over Z. $\qquad \square$

Example 2 The polynomial $f(x) = 2x^2 + 4$ is irreducible over **R** but reducible over **C**. $\qquad \square$

Example 3 The polynomial $x^2 - 2$ is irreducible over Q but reducible over **R** ☐

Example 4 The polynomial $x^2 + 1$ is irreducible over Z_3 but reducible over Z_5. ☐

In general, it is a difficult problem to decide whether or not a particular polynomial is reducible over an integral domain but there are special cases when it is easy. Our first theorem is a case in point. It applies to the three preceding examples.

Theorem 19.1 *Reducibility Test for Degrees 2 and 3*
Let F be a field. If $f(x) \in F[x]$ and $\deg f(x) = 2$ or 3, then $f(x)$ is reducible over F if and only if $f(x)$ has a zero in F.

Proof. Suppose $f(x) = g(x)h(x)$ where both $g(x)$ and $h(x)$ belong to $F[x]$ and have degree less than that of $f(x)$. Since $\deg f(x) = \deg g(x) + \deg h(x)$ and $\deg f(x) = 2$ or 3, at least one of $g(x)$ or $h(x)$ has degree 1. Say, $g(x) = ax + b$. Then, clearly, $-a^{-1}b$ is a zero of $g(x)$ and therefore a zero of $f(x)$ as well.

Conversely, suppose $f(a) = 0$ where $a \in F$. Then, by the Factor Theorem, we know $x - a$ is a factor of $f(x)$ and, therefore, $f(x)$ is reducible over F. ∎

Theorem 19.1 is particularly easy to use when the field is Z_p for, in this case, we can check for reducibility of $f(x)$ by simply testing to see if $f(x) = 0$ for $x = 0, 1, \ldots, p - 1$.

Note that polynomials of degree larger than 3 may be reducible over a field, even though they do not have zeros in the field. For example, in $Q[x]$, the polynomial $x^4 + 2x^2 + 1 = (x^2 + 1)^2$, but has no zeros in Q.

Our next three tests deal with polynomials with integer coefficients. To simplify the proof of the first of these, we introduce some terminology and isolate a portion of the argument in the form of a lemma.

DEFINITION Content of Polynomial, Primitive Polynomial
The *content* of a polynomial $a_n x^n + a_{n-1} x^{n-1} + \cdots + a_0$ where the a's are integers is the greatest common divisor of the integers $a_n, a_{n-1}, \ldots, a_0$. A *primitive polynomial* is an element of $Z[x]$ with content 1.

Gauss's Lemma *The product of two primitive polynomials is primitive.*

Proof. [2] Let $f(x)$ and $g(x)$ be primitive polynomials, and suppose $f(x)g(x)$ is not primitive. Let p be a prime divisor of the content of $f(x)g(x)$, and let $\overline{f}(x)$, $\overline{g}(x)$ and $\overline{f(x)g(x)}$ be the polynomials obtained from $f(x)$, $g(x)$, and $f(x)g(x)$ by reducing the coefficients modulo p. Then, $\overline{f}(x)$ and $\overline{g}(x)$ belong to the integral domain $Z_p[x]$ and $\bar{f}(x)\bar{g}(x) = \overline{f(x)g(x)} = 0$, the zero element of $Z_p[x]$ (see exercise 28 of Chapter 18). Thus, $\bar{f}(x) = 0$ or $\bar{g}(x) = 0$. This means that either p divides every coefficient of $f(x)$ or p divides every coefficient of $g(x)$. Hence, either $f(x)$ is not primitive or $g(x)$ is not primitive. This contradiction completes the proof.   ∎

Remember that the question of reducibility depends on which ring of coefficients one permits. Thus, $x^2 - 2$ is irreducible over Z but reducible over $Q[\sqrt{2}]$. In a later chapter, we will prove every polynomial of degree greater than 1 with coefficients from an integral domain is reducible over some field. Theorem 19.2 shows that in the case of the integers, this field must be larger than the field of rational numbers.

Theorem 19.2 *Over Q Implies over Z*
Let $f(x) \in Z[x]$. If $f(x)$ is reducible over Q, then it is reducible over Z.

Proof. Suppose $f(x) = g(x)h(x)$ where $g(x)$ and $h(x) \in Q[x]$. Clearly, we may assume that $f(x)$ is primitive because we can divide both $f(x)$ and $g(x)h(x)$ by the content of $f(x)$. Let a be the least common multiple of the denominators of the coefficients of $g(x)$, and b the least common multiple of the denominators of the coefficients of $h(x)$. Then $abf(x) = ag(x) \cdot bh(x)$ where $ag(x)$ and $bh(x) \in Z[x]$. Let c_1 be the content of $ag(x)$ and c_2 be the content of $bh(x)$. Then $ag(x) = c_1g_1(x)$ and $bh(x) = c_2h_1(x)$ where both $g_1(x)$ and $h_1(x)$ are primitive and $abf(x) = c_1c_2g_1(x)h_1(x)$. Since $f(x)$ is primitive, the content of $abf(x)$ is ab. Also, since the product of two primitive polynomials is primitive, it follows that the content of $c_1c_2g_1(x)h_1(x)$ is c_1c_2. Thus, $ab = c_1c_2$ and $f(x) = g_1(x)h_1(x)$ where $g_1(x)$ and $h_1(x) \in Z[x]$.   ∎

Irreducibility Tests

Theorem 19.1 reduces the question of irreducibility of a polynomial of degree 2 or 3 to one of finding a zero. The next theorem often allows us to simplify the problem even further.

Theorem 19.3 *Mod p Irreducibility Test*
Let p be a prime and suppose $f(x) \in Z[x]$ with deg $f(x) \geq 1$. Let $\bar{f}(x)$ be the polynomial in $Z_p[x]$ obtained from $f(x)$ by reducing all the coefficients of $f(x)$ modulo p. If $\bar{f}(x)$ is irreducible over Z_p and deg $\bar{f}(x) = $ deg $f(x)$, then $f(x)$ is irreducible over Q.

Proof. It follows from the proof of Theorem 19.2 that if $f(x)$ is irreducible over Q then $f(x) = g(x)h(x)$ with $g(x)$, $h(x) \in Z[x]$ and both $g(x)$ and $h(x)$ have degree less than that of $f(x)$. Let $\bar{f}(x), \bar{g}(x),$ and $\bar{h}(x)$ be the polynomials obtained from $f(x)$, $g(x)$, and $h(x)$ by reducing all the coefficients modulo p. Since deg $f(x)$ = deg $\overline{f}(x)$, we have deg $\overline{g}(x) \leqslant$ deg $g(x) <$ deg $\overline{f}(x)$ and deg $\overline{h}(x) \leqslant$ deg $h(x) <$ deg $\overline{f}(x)$. But, $\overline{f}(x) = \overline{g}(x)\overline{h}(x)$, and this contradicts our assumption that $\overline{f}(x)$ is irreducible over Z_p. ∎

Example 5 Let $f(x) = 21x^3 - 3x^2 + 2x + 9$. Then, over $Z_2[x]$, we have $\overline{f}(x) = x^3 + x^2 + 1$ and, since $\overline{f}(0) = 1$ and $\overline{f}(1) = 1$, we see that $\overline{f}(x)$ is irreducible over Z_2. Thus, $f(x)$ is irreducible over Q. Notice that over Z_3, $\overline{f}(x) = 2x$ is irreducible but we may *not* apply Theorem 19.3 to conclude $f(x)$ is irreducible over Q. □

Be cautious not to use the converse of Theorem 19.3. If $f(x) \in Z[x]$ and $\overline{f}(x)$ is reducible over Z_p for some p, $f(x)$ may still be irreducible over Q. For example, consider $f(x) = 21x^3 - 3x^2 + 2x + 8$. Then over Z_2, $\overline{f}(x) = x^3 + x^2 = x^2(x + 1)$. But over Z_5, $\overline{f}(x)$ has no roots and, therefore is irreducible over Z_5. So, $f(x)$ is irreducible over Q. Note that this example shows that the Mod p Irreducibility Test may fail from some p and work for others. To conclude that a particular $f(x)$ in $Z[x]$ is irreducible over Q, all we need to do is find a single p for which the corresponding polynomial $\overline{f}(x)$ in Z_p is irreducible. However, this is not always possible since $f(x) = x^4 + 1$ is irreducible over Q but reducible over Z_p for *every* prime p. (See exercise 29.)

The Mod p Irreducibility Test can also be helpful in checking for irreducibility of polynomials of degree greater than 3 and polynomials with rational coefficients.

Example 6 Let $f(x) = (3/7)x^4 - (2/7)x^2 + (9/35)x + 3/5$. We will show that $f(x)$ is irreducible over Q. First, let $h(x) = 35f(x) = 15x^4 - 10x^2 + 9x + 21$. Then $f(x)$ is irreducible over Q if and only if $h(x)$ is irreducible over Z. Next, applying the Mod 2 Irreducibility Test to $h(x)$, we get $\overline{h}(x) = x^4 + x + 1$. Clearly, $\overline{h}(x)$ has no roots in Z_2. Furthermore, $\overline{h}(x)$ has no quadratic factor in $Z_2[x]$ either. (For if so, the factor would have to be either $x^2 + x + 1$ or $x^2 + 1$. Long division shows that $x^2 + x + 1$ is not a factor, and $x^2 + 1$ cannot be a factor because it has a root while $\overline{h}(x)$ does not.) Thus $\overline{h}(x)$ is irreducible over $Z_2[x]$. This guarantees that $h(x)$ is irreducible over Q. □

Example 7 Let $f(x) = x^5 + 2x + 4$. Obviously, neither Theorem 19.1 nor the Mod 2 Irreducibility Test helps here. Let's try mod 3. Substitution of 0, 1, and 2 into $f(x) = \overline{f}(x)$ does not yield 0, so there are no linear factors. But $\overline{f}(x)$ may have a quadratic factor. If so, we may assume it has the form $x^2 + ax + b$ (see exercise 5). This gives 9 possibilities to check. We can immediately rule out each of the nine that have a zero over Z_3, since $\overline{f}(x)$ does not have one. This leaves only $x^2 + 1$, $x^2 + x + 2$ and $x^2 + 2x + 2$ to check. These are eliminated by long division. So, since $\overline{f}(x)$ is irreducible over Z_3, $f(x)$ is irreducible over Q. (Why is it unnecessary to check for cubic or fourth degree factors?) □

Another important irreducibility test is the following one, credited to Ferdinand Eisenstein (1823–1852), a student of Gauss.

Theorem 19.4   Eisenstein Criterion (1850)
Let

$$f(x) = a_n x^n + a_{n-1} x^{n-1} + \cdots + a_0 \in Z[x].$$

If there is a prime p such that $p \nmid a_n$, $p \mid a_{n-1}, \ldots, p \mid a_0$ and $p^2 \nmid a_0$, then f(x) is irreducible over Q.

Proof.   If $f(x)$ is reducible over Q, we know by Theorem 19.2 that there exist elements $g(x)$ and $h(x)$ in $Z[x]$ such that $f(x) = g(x)h(x)$ and $1 \leqslant \deg g(x)$, $\deg h(x) < n$. Say, $g(x) = b_r x^r + \cdots + b_0$ and $h(x) = c_s x^s + \cdots + c_0$. Then, since $p \mid a_0 = b_0 c_0$ and $p^2 \nmid a_0$, it follows that p divides one of b_0 and c_0 but not the other. Let us say $p \mid b_0$ and $p \nmid c_0$. Also, since $p \nmid a_n = b_r c_s$, we know that $p \nmid b_r$. So, there is a least integer t such that $p \nmid b_t$. Now, consider $a_t = b_t c_0 + b_{t-1} c_1 + \cdots + b_0 c_t$. By assumption, p divides a_t and, by choice of t, every summand on the right after the first one is divisible by p. Clearly, this forces p to divide $b_t c_0$ as well. This is impossible, however, since p is prime and p divides neither b_t nor c_0.   ■

Corollary   *Irreducibility of pth Cyclotomic Polynomial*
For any prime p, the pth cyclotomic polynomial

$$\Phi_p(x) = \frac{x^p - 1}{x - 1} = x^{p-1} + x^{p-2} + \cdots + x + 1$$

is irreducible over Q.

Proof.   Let

$$f(x) = \Phi_p(x + 1) = \frac{(x + 1)^p - 1}{(x + 1) - 1} = x^{p-1} + px^{p-2} + \cdots + p.$$

Then, by Eisenstein's criterion, $f(x)$ is irreducible over Q. So, if $\Phi_p(x) = g(x)h(x)$ were a nontrivial factorization of $\Phi_p(x)$ over Q, then $f(x) = \Phi_p(x + 1) = g(x + 1)h(x + 1)$ would be a nontrivial factorization of $f(x)$ over Q. Since this is impossible, we conclude that $\Phi_p(x)$ is irreducible over Q.   ■

Example 8   The polynomial $3x^5 + 15x^4 - 20x^3 + 10x + 20$ is irreducible over Q because $5 \nmid 3$ and $25 \nmid 20$ but 5 does divide 15, -20, 10, and 20.   □

The principle reason for our interest in irreducible polynomials stems from the fact that there is an intimate connection between them, maximal ideals, and fields. This connection is revealed in the next theorem and its first corollary.

Theorem 19.5 *$p(x)$ Irreducible Implies $\langle p(x) \rangle$ Is Maximal*
Let F be a field and let $p(x) \in F[x]$. Then $\langle p(x) \rangle$ is a maximal ideal in $F[x]$ if and only if $p(x)$ is irreducible over F.

Proof. Suppose first that $\langle p(x) \rangle$ is a maximal ideal in $F[x]$. If $p(x) = g(x)h(x)$ is a factorization of $p(x)$ over F, then $\langle p(x) \rangle \subseteq \langle g(x) \rangle \subseteq F[x]$. Thus, $\langle p(x) \rangle = \langle g(x) \rangle$ or $F[x] = \langle g(x) \rangle$. In the first case, we must have $\deg p(x) = \deg g(x)$. In the second case, it follows that $\deg g(x) = 0$ and, consequently, $\deg h(x) = \deg p(x)$. Thus, $p(x)$ cannot be written as a product of two polynomials in $F[x]$ of lower degree.

Now, suppose that $p(x)$ is irreducible over F. Let I be any ideal of $F[x]$ such that $\langle p(x) \rangle \subseteq I \subseteq F[x]$. Because $F[x]$ is a principal ideal domain, we know that $I = \langle g(x) \rangle$ for some $g(x)$ in $F[x]$. So, $p(x) \in \langle g(x) \rangle$ and, therefore, $p(x) = g(x)h(x)$ where $h(x) \in F[x]$. Since $p(x)$ is irreducible over F, it follows that either $g(x)$ is a constant or $h(x)$ is a constant. In the first case, we have $I = F[x]$; in the second case, we have $\langle p(x) \rangle = \langle g(x) \rangle = I$. So, $\langle p(x) \rangle$ is maximal in $F[x]$. ∎

Corollary 1 *$F[x]/\langle p(x) \rangle$ Is a Field*
Let F be a field and $p(x)$ an irreducible polynomial over F. Then $F[x]/\langle p(x) \rangle$ is a field.

Proof. This follows directly from Theorems 19.5 and 16.4 ∎

The next corollary is a polynomial analog of Euclid's Lemma for primes (see Chapter 0).

Corollary 2 *$p(x) \mid a(x)b(x)$ Implies $p(x) \mid a(x)$ or $p(x) \mid b(x)$*
Let F be a field and let $p(x), a(x), b(x) \in F[x]$. If $p(x)$ is irreducible over F and $p(x) \mid a(x)b(x)$, then $p(x) \mid a(x)$ or $p(x) \mid b(x)$.

Proof. Since $p(x)$ is irreducible, $F[x]/\langle p(x) \rangle$ is a field and, therefore, an integral domain. Let $\bar{a}(x)$ and $\bar{b}(x)$ be the images of $a(x)$ and $b(x)$ under the natural homomorphism from $F[x]$ to $F[x]/\langle p(x) \rangle$. Since $p(x) \mid a(x)b(x)$, we have $\bar{a}(x)\bar{b}(x) = \bar{0}$, the zero element of $F[x]/\langle p(x) \rangle$. Thus, $\bar{a}(x) = \bar{0}$ or $\bar{b}(x) = \bar{0}$, and it follows that $p(x) \mid a(x)$ or $p(x) \mid b(x)$. ∎

The next two examples put the theory to work.

Example 9 We construct a field with 8 elements. By Theorem 19.1 and Corollary 1 of Theorem 19.5, it suffices to find a cubic polynomial over Z_2 that has no zero in Z_2. By inspection, $x^3 + x + 1$ fills the bill. Thus, $Z_2[x]/\langle x^3 + x + 1\rangle = \{ax^2 + bx + c + \langle x^3 + x + 1\rangle \mid a, b, c \in Z_2\}$ is a field with 8 elements. For practice, let us do a few calculations in this field. Since the sum of two polynomials of the form $ax^2 + bx + c$ is another one of the same form, addition is easy. Thus,

$$(x^2 + x + 1 + \langle x^3 + x + 1\rangle) + (x^2 + 1 + \langle x^3 + x + 1\rangle)$$
$$= x + \langle x^3 + x + 1\rangle.$$

On the other hand multiplication of two coset representatives need not yield one of the original 8 coset representatives:

$$(x^2 + x + 1 + \langle x^3 + x + 1\rangle) \cdot (x^2 + 1 + \langle x^3 + x + 1\rangle)$$
$$= x^4 + x^3 + x + 1 + \langle x^3 + x + 1\rangle$$
$$= x^4 + \langle x^3 + x + 1\rangle$$

(since the ideal absorbs the last three terms). How do we express this in the form $ax^2 + bx + c + \langle x^3 + x + 1\rangle$? One way is to long divide by $x^3 + x + 1$ to obtain the remainder of $x^2 + x$ (just as one reduces $12 + \langle 5\rangle$ to $2 + \langle 5\rangle$ by dividing by 5 to obtain the remainder). Another way is to observe that $x^3 + x + 1 + \langle x^3 + x + 1\rangle = 0 + \langle x^3 + x + 1\rangle$ implies $x^3 + \langle x^3 + x + 1\rangle = x + 1 + \langle x^3 + x + 1\rangle$. Thus, we may multiply both sides by x to obtain

$$x^4 + \langle x^3 + x + 1\rangle = x^2 + x + \langle x^3 + x + 1\rangle.$$

A partial multiplication table for this field is given in Table 19.1. To simplify the notation, we indicate a coset by its representative only. (Complete the table yourself. Keep in mind that x^3 can be replaced by $x + 1$ and x^4 by $x^2 + x$.) □

Example 10 Since $x^2 + 1$ has no root in Z_3, it is irreducible. Thus, $Z_3[x]/\langle x^2 + 1\rangle$ is a field. Analogous to Example 9 of Chapter 16, $Z_3[x]/\langle x^2 + 1\rangle = \{ax + b + \langle x^2 + 1\rangle \mid a, b \in Z_3\}$. Thus, this field has 9 elements. A multiplication table for this field can be obtained from Table 15.1 by replacing i by x. (Why does this work?) □

Table 19.1 A Partial Multiplication Table for Example 9.

	1	x	$x + 1$	x^2	$x^2 + 1$	$x^2 + x$	$x^2 + x + 1$
1							
x	x	x^2	$x^2 + x$	$x + 1$		1	$x^2 + x + 1$	$x^2 + 1$
$x + 1$							
x^2		x^2	$x + 1$	$x^2 + x + 1$	$x^2 + x$	x	$x^2 + 1$	1
$x^2 + 1$	$x^2 + 1$	1	x^2	x	$x^2 + x + 1$	$x + 1$	$x^2 + x$

Unique Factorization in Z[x]

As a further application of the ideas presented in this chapter, we next prove $Z[x]$ has an important factorization property. In Chapter 20, we will show that every principal ideal domain has the same factorization property. The case $Z[x]$ is handled separately because it is not a principal ideal domain. The first proof of Theorem 19.6 was given by Gauss. In the following theorem and its proof, keep in mind that the units in $Z[x]$ are precisely $f(x) = 1$ and $f(x) = -1$ (see exercise 15 of Chapter 13), and every polynomial from $Z[x]$ that is irreducible over Z is primitive (see exercise 2).

> **Theorem 19.6** *Unique Factorization in Z[x]*
> *Every polynomial in $Z[x]$ that is not the zero polynomial or a unit in $Z[x]$ can be written in the form $b_1 b_2 \ldots b_s p_1(x) p_2(x) \ldots p_m(x)$ where the b's are prime (that is, irreducible polynomials of degree 0), and the $p_i(x)$'s are irreducible polynomials of positive degree. Furthermore, if*
>
> $$b_1 b_2 \ldots b_s p_1(x) p_2(x) \ldots p_m(x) = c_1 c_2 \ldots c_t q_1(x) q_2(x) \ldots q_n(x)$$
>
> *where the b's and c's are irreducible polynomials of degree 0, and the p(x)'s and q(x)'s are irreducible polynomials of positive degree, then $s = t$, $m = n$, and, after renumbering the c's and q(x)'s, we have $b_i = \pm c_i$ for $i = 1, \ldots, s$; and $p_i(x) = \pm q_i(x)$ for $i = 1, \ldots, m$.*

Proof. Let $f(x)$ be a nonzero, nonunit polynomial from $Z[x]$. If $\deg f(x) = 0$, then $f(x)$ is constant and the result follows from the Fundamental Theorem of Arithmetic. If $\deg f(x) > 0$, let b denote the content of $f(x)$, and let $b_1 b_2 \ldots b_s$ be the factorization of b as a product of primes. Then, $f(x) = b_1 b_2 \ldots b_s f_1(x)$, where $f_1(x)$ belongs to $Z[x]$, is primitive and has positive degree. Thus, to prove the existence portion of the theorem, it suffices to show that a primitive polynomial $f(x)$ of positive degree can be written as a product of irreducible polynomials of positive degree. We proceed by induction of $\deg f(x)$. If $\deg f(x) = 1$, then $f(x)$ is already irreducible and we are done. Now suppose that every primitive polynomial of degree less than $\deg f(x)$ can be written as a product of irreducibles of positive degree. If $f(x)$ is irreducible, there is nothing to prove. Otherwise, $f(x) = g(x)h(x)$ where both $g(x)$ and $h(x)$ are primitive and have degree less than that of $f(x)$. Thus, by induction, both $g(x)$ and $h(x)$ can be written as a product of irreducibles of positive degree. Clearly, then, $f(x)$ is also such a product.

 To prove the uniqueness portion of the theorem, suppose $f(x) = b_1 b_2 \ldots b_s p_1(x) p_2(x) \ldots p_m(x) = c_1 c_2 \ldots c_t q_1(x) q_2(x) \ldots q_n(x)$ where the b's and c's are irreducible polynomials of degree 0, and the $p(x)$'s and $q(x)$'s are irreducible polynomials of positive degree. Let $b = b_1 b_2 \ldots b_s$ and $c = c_1 c_2 \ldots c_t$. Since the $p(x)$'s and $q(x)$'s are primitive, it follows from

Gauss's Lemma that $p_1(x)p_2(x) \ldots p_m(x)$ and $q_1(x)q_2(x) \ldots q_n(x)$ are primitive. Hence, both b and c must equal plus-or-minus the content of $f(x)$ and, therefore, are equal in absolute value. It then follows from the Fundamental Theorem of Arithmetic that $s = t$ and, after renumbering, $b_i = \pm c_i$ for $i = 1, 2, \ldots, s$. Thus, by cancelling the constant terms in the two factorizations for $f(x)$, we have $p_1(x)p_2(x) \cdots p_m(x) = \pm q_1(x) q_2(x) \cdots q_n(x)$. Now, viewing the $p(x)$'s and $q(x)$'s as elements of $Q[x]$ and noting that $p_1(x)$ divides $q_1(x) \cdots q_n(x)$, it follows from Corollary 2 of Theorem 19.5 and induction (see exercise 28) that $p_1(x) \mid q_i(x)$ for some i. By renumbering, we may assume $i = 1$. Then, since $q_1(x)$ is irreducible, we have $q_1(x) = (r/s) p_1(x)$, where $r, s \in Z$. However, because both $q_1(x)$ and $p_1(x)$ are primitive, we must have $r/s = \pm 1$. So, $q_1(x) = \pm p_1(x)$. Also, after cancelling, we have $p_2(x) \cdots p_m(x) = \pm q_2(x) \cdots q_n(x)$. Now, we may repeat the above argument with $p_2(x)$ in place of $p_1(x)$. If $m < n$, after m such steps we would have 1 on the left and a nonconstant polynomial on the right. Clearly, this is impossible. On the other hand, if $m > n$, after n steps we would have ± 1 on the right and a nonconstant polynomial on the left—another impossibility. So, $m = n$ and $p_i(x) = \pm q_i(x)$ after suitable renumbering of the $q(x)$'s.   ∎

Weird Dice: An Application of Unique Factorization

Example 11 Consider an ordinary pair of dice whose faces are labelled 1 through 6. The probability of rolling a sum of 7 is 6/36, the probability of rolling a sum of 6 is 5/36, and so on. In a 1978 issue of *Scientific American* [1], Martin Gardner remarked that if one were to label the six faces of one cube with integers 1, 2, 2, 3, 3, 4 and the six faces of another cube with the integers 1, 3, 4, 5, 6, 8, then the probability of obtaining any particular sum with these dice (called *Sicherman dice*) is the same as the probability of rolling that sum with ordinary dice (that is, 6/36 for a 7, 5/36 for a 6, and so on). See Figure 19.1. In this example, we show how the Sicherman labels can be derived, and that they are the only possible such labels besides 1 through 6. To do so, we utilize that fact that $Z[x]$ has the unique factorization property.

To begin with let us ask ourselves how we may obtain a sum of 6, say, with an ordinary pair of dice. Well, there are five possibilities for the two faces: (5, 1), (4, 2), (3, 3), (2, 4), and (1, 5). Next we consider the product of the two polynomials created by using the ordinary dice labels as exponents:

$$(x^6 + x^5 + x^4 + x^3 + x^2 + x)(x^6 + x^5 + x^4 + x^3 + x^2 + x).$$

Observe that we pick up the term x^6 in this product in precisely the following ways: $x^5 \cdot x^1, x^4 \cdot x^2, x^3 \cdot x^3, x^2 \cdot x^4, x^1 \cdot x^5$. Notice the correspondence between pairs of labels whose sum is 6 and pairs of terms whose product is x^6. This correspondence is one-to-one, and it is valid for all sums and all dice—including

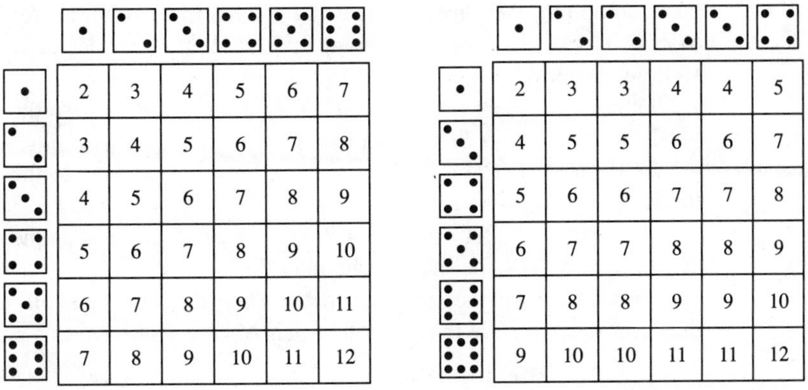

•	2	3	4	5	6	7
⠢	3	4	5	6	7	8
⠣	4	5	6	7	8	9
⠦	5	6	7	8	9	10
⠭	6	7	8	9	10	11
⠿	7	8	9	10	11	12

•	2	3	3	4	4	5
⠢	4	5	5	6	6	7
⠣	5	6	6	7	7	8
⠦	6	7	7	8	8	9
⠭	7	8	8	9	9	10
⠿	9	10	10	11	11	12

Figure 19.1

the Sicherman dice and any other dice that yield the desired probabilities. So, let $\{a_1, a_2, a_3, a_4, a_5, a_6\}$ and $\{b_1, b_2, b_3, b_4, b_5, b_6\}$ be any two sets of positive integer labels for a pair of cubes with the property that the probability of rolling any particular sum with these dice (let us call them *weird dice*) is the same as the probability of rolling that sum with ordinary dice labeled 1 through 6. Using our observation about products of polynomials, this means that

$$(*) \ (x^6 + x^5 + x^4 + x^3 + x^2 + x)(x^6 + x^5 + x^4 + x^3 + x^2 + x)$$
$$= (x^{a_1} + x^{a_2} + x^{a_3} + x^{a_4} + x^{a_5} + x^{a_6}).$$
$$(x^{b_1} + x^{b_2} + x^{b_3} + x^{b_4} + x^{b_5} + x^{b_6}).$$

Now all we have to do is solve this equation for the a's and b's. Here is where unique factorization in $Z[x]$ comes in. The polynomial $x^6 + x^5 + x^4 + x^3 + x^2 + x$ factors uniquely into irreducibles as

$$x(x + 1)(x^2 + x + 1)(x^2 - x + 1)$$

so that the left-hand side of equation (*) has the irreducible factorization

$$x^2(x + 1)^2(x^2 + x + 1)^2(x^2 - x + 1)^2.$$

So, by Theorem 19.6, this means that these factors are the only possible irreducible factors of $P(x) = x^{a_1} + x^{a_2} + x^{a_3} + x^{a_4} + x^{a_5} + x^{a_6}$. Thus, $P(x)$ has the form

$$x^q(x + 1)^r(x^2 + x + 1)^t(x^2 - x + 1)^u$$

where $0 \leq q, r, t, u \leq 2$.

To further restrict the possibilities for these four parameters, we evaluate $P(1)$ in two ways. $P(1) = 1^{a_1} + 1^{a_2} + \cdots + 1^{a_6} = 6$ and $P(1) = 1^q 2^r 3^t 1^u$. Clearly, this means that $r = 1$ and $t = 1$. What about q? Evaluating $P(0)$ in two ways shows $q \neq 0$. On the other hand, if $q = 2$, the smallest possible sum one could roll with the corresponding set of dice would be 3. Since this violates our assumption, we have now reduced our list of possibilities for q, r, t, and u to $q = 1$, $r = 1$, $t = 1$, and $u = 0$, 1, 2. Let's consider each of these possibilities in turn.

When $u = 0$, $P(x) = x^4 + x^3 + x^3 + x^2 + x^2 + x$ so the die labels are $\{4, 3, 3, 2, 2, 1\}$—a Sicherman die.

When $u = 1$, $P(x) = x^6 + x^5 + x^4 + x^3 + x^2 + x$ so the die labels are $\{6, 5, 4, 3, 2, 1\}$—an ordinary die.

When $u = 2$, $P(x) = x^8 + x^6 + x^5 + x^4 + x^3 + x$ so the die labels are $\{8, 6, 5, 4, 3, 1\}$—the other Sicherman die.

This proves that the Sicherman dice do give the same probabilities as ordinary dice *and* that they are the *only* other pair of dice that have this property. $\square$

EXERCISES

It is a great nuisance that knowledge can only be acquired by hard work.

W. Somerset Maugham

1. Suppose D is an integral domain and F is a field containing D. If $f(x) \in D[x]$ and $f(x)$ is irreducible over F but reducible over D, what can you say about the factorization of $f(x)$ over D?

2. Show that a polynomial from $Z[x]$ that is irreducible over Z is primitive.

3. Let p be a prime. Show that $x^n - p$ is irreducible over the rationals.

4. Suppose $f(x) = x^n + a_{n-1}x^{n-1} + \cdots + a_0 \in Z[x]$. If r is rational and $x - r$ divides $f(x)$, show that r is an integer.

5. Let F be a field and let a be a nonzero element of F.
   a. If $af(x)$ is irreducible over F, prove that $f(x)$ is irreducible over F.
   b. If $f(ax)$ is irreducible over F, prove that $f(x)$ is irreducible over F.
   c. If $f(x + a)$ is irreducible over F, prove that $f(x)$ is irreducible over F.

6. Show that $x^4 + 1$ is irreducible over Q but reducible over **R**.

7. Construct a field of order 25.

8. Construct a field order of 27.

9. Show that $x^3 + x^2 + x + 1$ is reducible over Q. Does this fact contradict the corollary to Theorem 19.4?

10. Determine which of the polynomials below are irreducible over Q.
    a. $x^5 + 9x^4 + 12x^2 + 6$
    b. $x^4 + x + 1$
    c. $x^4 + 3x^2 + 3$
    d. $x^5 + 5x^2 + 1$
    e. $5/2x^5 + 9/2x^4 + 15x^3 + 3/7x^2 + 6x + 3/14$.

11. Show that $x^2 + x + 4$ is irreducible over Z_{11}.

12. Suppose $f(x) \in Z_p[x]$ and is irreducible over Z_p. If $\deg f(x) = n$, prove that $Z_p[x]/\langle f(x) \rangle$ is a field with p^n elements.

13. Let $f(x) = x^3 + 6 \in Z_7[x]$. Write $f(x)$ as a product of irreducible polynomials over Z_7.

14. Let $f(x) = x^3 + x^2 + x + 1 \in Z_2[x]$. Write $f(x)$ as a product of irreducible polynomials over Z_2.

15. Let p be a prime.
    a. Show that the number of reducible polynomials over Z_p of the form $x^2 + ax + b$ is $p(p + 1)/2$.
    b. Determine the number of reducible quadratic polynomials over Z_p.

16. Let p be a prime.
    a. Determine the number of irreducible polynomials over Z_p of the form $x^2 + ax + b$.
    b. Determine the number of irreducible quadratic polynomials over Z_p.

17. Show that for every prime p there exists a field of order p^2.

18. Show that the field given in Example 10 of this chapter is isomorphic to the field given in Example 8 of Chapter 15.

19. Prove that, for every positive integer n, there are infinitely many polynomials of degree n in $Z[x]$ that are irreducible over Q.

20. Let $f(x) \in Z_p[x]$. Prove that if $f(x)$ has no factor of the form $x^2 + ax + b$, then it has no quadratic factor over Z_p.

21. Find all monic irreducible polynomials of degree 2 over Z_3.

22. Given that π is not the zero of a polynomial with rational coefficients, prove π^2 cannot be written in the form $a\pi + b$ where a and b are rational.

23. Find all the zeros and their multiplicity of $x^5 + 4x^4 + 4x^3 - x^2 - 4x + 1$ over Z_5.

24. Find all zeros of $f(x) = 2x^2 + 2x + 1$ over Z_5 by substitution. Find all zeros of $f(x)$ by using the quadratic formula $(-b \pm \sqrt{b^2 - 4ac})(2a)^{-1}$. Do your answers agree? Should they? Find all zeros of $g(x) = 2x^2 + x + 3$ over Z_5 by substitution. Try the quadratic formula on $g(x)$. Why doesn't it work? State necessary and sufficient conditions for the quadratic formula to yield the zeros of a quadratic from $Z_p[x]$ where p is prime.

25. Rational Root Theorem. Let

$$f(x) = a_n x^n + a_{n-1} x^{n-1} + \cdots + a_0 \in Z[x]$$

and $a_n \neq 0$. Prove that if r and s are relatively prime integers and $f(r/s) = 0$, then $r \mid a_0$ and $s \mid a_n$.

26. Let F be a field and $f(x) \in F[x]$. Show that, as far as deciding upon the irreducibility of $f(x)$ over F is concerned, we may assume that $f(x)$ is monic. (This assumption is useful when one uses a computer to check for irreducibility.)

27. Explain how the Mod p Irreducibility Test (Theorem 19.3) can be used to test members of $Q[x]$ for irreducibility.

28. Let F be a field and let $p(x)$, $a_1(x)$, $a_2(x)$, . . . , $a_k(x) \in F[x]$ where $p(x)$ is irreducible over F. If $p(x) \mid a_1(x)a_2(x) \ldots a_k(x)$, show that $p(x)$ divides some $a_i(x)$. (This exercise is referred to in the proof of Theorem 19.6.)

29. Show that $x^4 + 1$ is reducible over Z_p for every prime p.

30. Let F be a field and let $p(x)$ be irreducible over F. Show that $\{a + \langle p(x) \rangle \mid a \in F\}$ is a subfield of $F[x]/\langle p(x) \rangle$ isomorphic to F. (This exercise is referred to in Chapter 22.)

31. Let F be a field and let $p(x)$ be irreducible over F. If E is a field that contains F and there is an element a in E such that $p(a) = 0$, show that the mapping $\phi:F[x] \rightarrow E$ given by $f(x)\phi = f(a)$ is a ring homomorphism with kernel $\langle p(x) \rangle$. (This exercise is referred to in Chapter 22.)

32. Prove that the irreducible factorization of $x^6 + x^5 + x^4 + x^3 + x^2 + x$ over Z is $x(x + 1)(x^2 + x + 1)(x^2 - x + 1)$.

33. If p is a prime, prove that $x^{p-1} - x^{p-2} + x^{p-3} - \cdots - x + 1$ is irreducible over Q.

34. Carry out the analysis given in Example 11 for a pair of tetrahedrons instead of a pair of cubes. (Define ordinary tetrahedron dice as the ones labeled 1 through 4.)

35. Suppose in Example 11 that we begin with n ($n > 2$) ordinary dice each labeled 1 through 6 instead of just two. Show that the only possible labels that produce the same probabilities as n ordinary dice are the labels 1 through 6 and the two sets of Sicherman labels.

36. Show that one two-sided die labeled with 1 and 4 and another eighteen-sided die labeled with 1, 2, 2, 3, 3, 3, 4, 4, 4, 5, 5, 5, 6, 6, 6, 7, 7, 8 yield the same probabilities as an ordinary pair of cubes labeled 1 through 6. Carry out an analysis similar to that given in Example 11 to derive these sets of labels.

37. In the game of Monopoly, would the probabilities of landing on various properties be different if played with Sicherman dice instead of ordinary dice? Why?

38. (Algorithm for Factoring $x^n - 1$.) *Assume the following:*

   1. $\Phi_1(x) = x - 1$.
   2. For any prime p, define $\Phi_p(x) = x^{p-1} + x^{p-2} + \cdots + x + 1$.
   3. For k > 1, p a prime and m positive, $\Phi_{mp^k}(x) = \Phi_{mp}(x^{p^{k-1}})$.
   4. For n ≥ 3 odd, $\Phi_{2n}(x) = \Phi_n(-x)$.
   5. For p a prime and $p \nmid m$, $\Phi_{mp}(x) = \Phi_m(x^p)/\Phi_m(x)$. (For example,

$$\Phi_{15}(x) = \Phi_3(x^5)/\Phi_3(x) = \Phi_5(x^3)/\Phi_5(x)$$
$$= x^8 - x^7 + x^5 - x^4 + x^3 - x + 1.)$$

6. $x^n - 1 = \prod_{d|n} \Phi_d(x)$ is the irreducible factorization of $x^n - 1$ over Z. (For example,

$$x^{15} - 1 = \Phi_1(x)\,\Phi_3(x)\,\Phi_5(x)\,\Phi_{15}(x)(x - 1)(x^2 + x + 1) \cdot$$
$$(x^4 + x^3 + x^2 + x + 1)(x^8 - x^7 + x^5 - x^4 + x^3 - x + 1).)$$

Use this algorithm to find the irreducible factorization of each of the following:

a. $x^4 + x^3 + x^2 + x = \dfrac{x(x^4 - 1)}{x - 1}$

b. $x^8 + x^7 + \cdots + x = \dfrac{x(x^8 - 1)}{x - 1}$

c. $x^{20} + x^{19} + \cdots + x = \dfrac{x(x^{20} - 1)}{x - 1}$.

PROGRAMMING EXERCISES

Intelligence is not to make no mistakes but quickly to see how to make them good.

Bertolt Brecht, *The Measures Taken*

1. Write a program to test any polynomial in $Z_p[x]$ for zeros. Make reasonable assumptions about the size of p and the degree of the polynomial.

2. Write a program that will implement the "Mod p Irreducibility Test." Assume the degree of the polynomial is at most 5 and p is at most 11. Test your program with the examples given in this chapter and the polynomials given in exercise 10.

3. Write a program that will implement the Rational Root Theorem (see exercise 25) for polynomials with integer coefficients. Make reasonable assumptions about the size of the coefficients and the degree of the polynomial.

4. Program the algorithm given in exercise 38 to find the irreducible factorization over Z of all polynomials of the form $x^n - 1$ where n is between 2 and 100. On the basis of this information, make a conjecture about the coefficients of the irreducible factors of $x^n - 1$ for all n. Test your conjecture for $n = 105$.

5. Write a program that will implement Eisenstein's Criterion. Randomly generate polynomials of degree at most 5 with integer coefficients between -100 and 100. Keep track of how often Eisenstein's Criterion applies to these polynomials. If one of the generated polynomials $f(x)$ does not satisfy Eisenstein's Criterion, apply it to the polynomials $f(x + k)$ for $k = -10$, $-9, \ldots, -1, 1, \ldots, 10$. (From exercise 5c of this chapter, if $f(x + k)$

is irreducible, so is $f(x)$.) Keep track of how often Eisenstein's Criterion applies to $f(x)$ or $f(x + k)$. Does this give significantly better results than Eisenstein's Criterion alone?

REFERENCES

1. Martin Gardner, "Mathematical Games," *Scientific American* 238/2 (1978): 19–32.
2. Richard Singer, "Some Applications of a Morphism," *American Mathematical Monthly* 76 (1969): 1131–1132.

SUGGESTED READINGS

Duane Broline, "Renumbering the Faces of Dice," *Mathematics Magazine* 52 (1979): 312–315.

> In this article, the author extends the analysis we carried out in Example 11 to dice in the shape of Platonic solids.

J. A. Gallian and D. J. Rusin, "Cyclotonic Polynomials and Nonstandard Dice," *Discrete Mathematics* 27 (1979): 245–259.

> Here Example 11 is generalized to the case of n dice each with m labels for all n and m greater than 1.

M. A. Lee, "Some Irreducible Polynomials Which Are Reducible mod p for All p," *American Mathematical Monthly* 76 (1969): 1125.

> This brief note gives a class of polynomials that are reducible mod p for all primes p, but are irreducible over the integers.

Carl Friedrich Gauss

He [Gauss] lives everywhere in mathematics.

E. T. Bell, *Men of Mathematics*

This stamp was issued by East Germany in 1977. It commemorates Gauss's construction of a regular 17-sided polygon with a straightedge and compass.

CARL FRIEDRICH GAUSS, considered by many to be the greatest mathematician who has ever lived, was born in Brunswick, Germany, on April 30, 1777. By the age of three, he was able to perform long computations in his head; at ten, he studied algebra and analysis. While still a teenager, he made many fundamental discoveries. Among these were the method of "least squares" for handling statistical data, a proof that a 17-sided regular polygon can be constructed with a straightedge and compass (this result was the first of its kind since discoveries by the Greeks 2000 years earlier), and his quadratic reciprocity theorem. Gauss obtained his Ph.D. in 1799 from the University of Helmstedt, under the supervision of Pfaff. In his dissertation, he proved the Fundamental Theorem of Algebra.

In 1801, Gauss published his monumental book on number theory, *Disquisitiones Arithmeticae*, summarizing previous work in a systematic way and introducing many fundamental ideas of his own, including the notion of modular arithmetic. This book won Gauss great fame among mathematicians.

In 1801, Ceres (an asteroid) was observed by astronomers on three occasions before they lost track of it. In what seemed to be an almost superhuman feat, Gauss used these three observations to calculate the orbit of Ceres. In carrying out this work, he showed that the variation inherent in experimentally derived data follows a bell-shaped curve, now called the Gaussian distribution. Gauss also used the method of least squares in this problem. This achievement established Gauss's reputation as a scientific genius before he was twenty-five years old.

In 1807, Gauss became professor of astronomy and director of the new observatory at the University of Göttingen. During the decades to come, Gauss continued to make important contributions not only in nearly all branches of mathematics, but also in astronomy, mechanics, optics, geodesy, and magnetism. Gauss also invented, with the physicist Wilhelm Weber, the first practical electric telegraph.

The acceptance of complex numbers among mathematicians was brought about by Gauss's use of them. Gauss coined the term *complex number* and introduced the notation i for $\sqrt{-1}$. He proved that the ring $Z[i]$ is a unique factorization domain and a Euclidean domain.

Throughout his life, Gauss largely ignored the work of his contemporaries and, in fact, made enemies of many of them. Young mathematicians who sought encouragement from him were usually rebuffed. Despite this fact, Gauss had many outstanding students, including Eisenstein, Riemann, Kummer, Dirichlet, and Dedekind.

Gauss died in Göttingen at the age of seventy-eight on February 23, 1855. At Brunswick, there is a statue of him. Appropriately, the base is in the shape of a 17-point star.

Divisibility in Integral Domains

20

> Fundamental definitions do not arise at the start but at the end of the exploration, because in order to define a thing you must know what it is and what it is good for.
>
> Hans Freudenthal, *Developments in Mathematical Education*

Irreducibles, Primes

In the previous two chapters, we focused on factoring polynomials over the integers or a field. Several of those results, unique factorization in $Z[x]$ and the division algorithm for $F[x]$, for instance, are natural counterparts to theorems about the integers. In this chapter and the next, we examine factoring in a more abstract setting.

> **DEFINITION** Associates, Irreducibles, Primes
>
> Elements a and b of an integral domain D are called *associates* if $a = ub$ where u is a unit of D. A nonzero element a of an integral domain D is called *irreducible* if a is not a unit and, whenever $b, c \in D$ with $a = bc$, then b or c is a unit. A nonzero element a of an integral domain D is called *prime* if a is not a unit and $a \mid bc$ implies $a \mid b$ or $a \mid c$.

Roughly speaking, an irreducible is an element that can be factored only in a trivial way. Notice that an element a is prime if and only if $\langle a \rangle$ is a prime ideal.

Relating the above definitions to the integers may seem a bit confusing since in Chapter 0 we defined an integer to be prime if it satisfies our definition of an irreducible, and we proved a prime integer satisfies the definition of a prime in an integral domain (Euclid's Lemma). The source of the confusion is that, in the case of the integers, the concepts of irreducible and prime are equivalent but, as we will soon see, in general, they are not.

The distinction between primes and irreducibles is best illustrated by integral domains of the form $Z[\sqrt{d}] = \{a + b\sqrt{d} \mid a, b \in Z\}$ where d is not 1 and is not divisible by the square of a prime. (These rings are of fundamental importance in number theory.) To analyze these rings, we need a convenient method of determining their units, irreducibles, and primes. To do this, we define a function N, called the *norm*, from $Z[\sqrt{d}]$ into the nonnegative integers by $N(a + b\sqrt{d}) = |a^2 - db^2|$. We leave it to the reader to verify the following four properties: $N(x) = 0$ if and only if $x = 0$; $N(xy) = N(x)N(y)$ for all x and y; x is a unit if and only if $N(x) = 1$; and, if $N(x)$ is prime, then x is irreducible in $Z[\sqrt{d}]$.

Example 1 We exhibit an irreducible in $Z[\sqrt{-3}]$ that is not prime. Here $N(a + b\sqrt{-3}) = a^2 + 3b^2$. Consider $1 + \sqrt{-3}$. Suppose we can factor this as xy where neither x nor y is a unit. Then $N(xy) = N(x)N(y) = N(1 + \sqrt{-3}) = 4$, and it follows that $N(x) = 2$. But there are no integers a and b satisfying $a^2 + 3b^2 = 2$. Thus, x or y is a unit and $1 + \sqrt{-3}$ is an irreducible. To verify that it is not prime, we observe that $(1 + \sqrt{-3})(1 - \sqrt{-3}) = 4 = 2 \cdot 2$ so that $1 + \sqrt{-3}$ divides $2 \cdot 2$. On the other hand, for integers a and b to exist so that $2 = (1 + \sqrt{-3})(a + b\sqrt{-3}) = (a - 3b) + (a + b)\sqrt{-3}$, we must have $a - 3b = 2$ and $a + b = 0$, which is impossible. $\square$

Example 1 raises the question of whether or not there is an integral domain containing a prime that is not an irreducible. The answer: no.

Theorem 20.1 *Prime Implies Irreducible*
In an integral domain, every prime is irreducible.

Proof. Suppose a is a prime in an integral domain and $a = bc$. We must show that b or c is a unit. By definition of prime, we know $a \mid b$ or $a \mid c$. Say, $at = b$. Then $b \cdot 1 = b = at = (bc)t = b(ct)$ and, by cancellation, $1 = ct$. Thus, c is a unit. ■

Recall that a principal ideal domain is an integral domain in which every ideal has the form $\langle a \rangle$. The next theorem reveals a circumstance in which primes and irreducibles are equivalent.

Theorem 20.2 *PID Implies Irreducible Equals Prime*
In a principal ideal domain, an element is irreducible if and only if it is prime.

Proof. Theorem 20.1 shows primes are irreducible. To prove the converse, let a be an irreducible element of a principal ideal domain D and suppose $a \mid bc$. We must show $a \mid b$ or $a \mid c$. Consider the ideal $I = \{ax + by \mid x,$

$y \in D\}$ and let $\langle d \rangle = I$. Since $a \in I$, we can write $a = dr$, and because a is irreducible, d is a unit or r is a unit. If d is a unit, then $I = D$ and we may write $1 = ax + by$. Then $c = acx + bcy$, and since a divides both terms on the right, a also divides c.

On the other hand, if r is a unit, then $\langle a \rangle = \langle d \rangle = I$, and because $b \in I$, there is an element t in D such that $at = b$. Thus, a divides b. ∎

It is an easy consequence of the respective division algorithms for Z and $F[x]$, where F is a field, that Z and $F[x]$ are principal ideal domains (see exercise 34 of Chapter 16 and Theorem 18.3). Our next example shows, however, that one of the most familiar rings is not a principal ideal domain.

Example 2   We show $Z[x]$ is not a principal ideal domain. Consider the ideal $I = \{ax + 2b \mid a, b \in Z\}$. We claim I is not of the form $\langle h(x) \rangle$. If this were so, there would be $f(x)$ and $g(x)$ in $Z[x]$ such that $2 = h(x)f(x)$ and $x = h(x)g(x)$, since both 2 and x belong to I. By the degree rule (exercise 16 of Chapter 18), $0 = \deg 2 = \deg h(x) + \deg f(x)$ so that $h(x)$ is a constant polynomial. To determine which constant, we observe $2 = h(1)f(1)$. Thus, $h(1) = \pm 1$ or ± 2. Since 1 is not in I, we must have $h(x) = \pm 2$. But then $x = \pm 2g(x)$, which is nonsense.   □

We have previously proved that the integral domains Z and $Z[x]$ have important factorization properties: Every integer greater than 1 can be uniquely factored as a product of irreducibles (that is, primes), and every nonzero, nonunit polynomial can be uniquely factored as a product of irreducible polynomials. It is natural to ask whether all integral domains have this property. The question of unique factorization in integral domains first arose with the efforts to solve a famous problem in number theory that goes by the misnomer Fermat's Last Theorem.

Historical Discussion of Fermat's Last Theorem

There are infinitely many nonzero integers x, y, z that satisfy the equation $x^2 + y^2 = z^2$. But what about the equation $x^3 + y^3 = z^3$ or, more generally, $x^n + y^n = z^n$ where n is an integer greater than 2 and x, y, z are nonzero integers? Well, no one has ever found a single solution of this equation and no one probably ever will. The tremendous effort put forth by the likes of Euler, Legendre, Abel, Gauss, Dirichlet, Cauchy, Kummer, Kronecker, and Hilbert to prove that there are no solutions has greatly influenced the development of ring theory.

About a thousand years ago, Arab mathematicians gave an incorrect proof that there were no solutions when $n = 3$. The problem lay dormant until 1637, when the French mathematician Pierre de Fermat (1601–1665) wrote in the margin of a book, ". . . it is impossible to separate a cube into two cubes, a fourth power into two fourth powers, or, generally, any power above the second into two powers of the same degree: I have discovered a truly marvelous demonstration [of this general theorem] which this margin is too narrow to contain."

Because Fermat gave no proof, many mathematicians have tried to prove the result. The case where $n = 3$ was done by Euler in 1770, although his proof was incomplete. The case for $n = 4$ is elementary and was done by Fermat himself (see [1]). The case $n = 5$ was done in 1825 by Dirichlet, who had just turned 20, and by Legendre, who was past 70. Since the validity of the case for a particular integer implies the validity of all multiples of that integer, the next case of interest was $n = 7$. This case resisted the efforts of the best mathematicians until it was done by Gabriel Lamé in 1839. In 1847, Lamé stirred excitement by announcing that he had completely solved the problem. His approach was to factor the expression $x^p + y^p$ where p is prime into

$$(x + y)(x + \alpha y) \cdots (x + \alpha^{p-1}y)$$

where α is the complex number $\cos(2\pi/p) + i\sin(2\pi/p)$. Thus, his factorization took place in the ring $Z[\alpha] = \{a_0 + a_1\alpha + \cdots + a_{p-1}\alpha^{p-1} \mid a_i \in Z\}$. But Lamé made the mistake of assuming that in such a ring, factorization into the product of irreducibles is unique. In fact, three years earlier, Ernst Eduard Kummer had proved that this is not always the case. Undaunted by the failure of unique factorization, Kummer began developing a theory to "save" factorization by creating a new type of number. Within a few weeks of Lamé's announcement, Kummer had shown that Fermat's Theorem is true for all primes of a special type (see [1]). This proved that the theorem was true for all exponents less than 100, prime or not, except for 37, 59, 67, and 74. Kummer's work has led to the theory of ideals as we know it today.

Over the centuries many proofs have been proposed, but none has held up under scrutiny. The famous number theorist Edmund Laudau received so many of these that he had a form printed with "On page _____, lines _____ to _____ you will find there is a mistake." Martin Gardner, "Mathematical Games" columnist of *Scientific American,* had postcards printed to decline requests from readers asking him to examine their proof.

Aided by computers, contemporary mathematicians have made enormous advances. It is now known that Fermat's Theorem is true for all exponents up to 150,000 and all x up to $10^{1,800,000}$. The decimal representation for x^n in any counterexample would fill at least ten million printed pages! Recent discoveries have tied Fermat's Theorem closely to modern mathematical theories, giving hope that these theories might eventually lead to a proof. In March 1988, newspapers and scientific publications worldwide carried news of a proof by Yoichi Miyaoka (see Figure 20.1). Within weeks, however, cautious optimism about its validity had turned to pessimism (see Figure 20.2). The demise of Miyaoka's proof was charmingly lamented by a *Time* magazine essayist (see [2]). But the history book of Fermat's Theorem is replete with important developments arising from failed attempts. Miyaoka's work may prove to be another chapter.

Figure 20.1

In view of the fact that so many eminent mathematicians have been unable to prove Fermat's Theorem, despite the availability of vastly powerful theories that have been developed, it seems highly improbable that Fermat had a correct proof. Most likely, he made the error that his successors made by assuming that the properties of integers, such as unique factorization, carry over to integral domains in general.

Figure 20.2

Unique Factorization Domains

We now have the necessary terminology to formalize the idea of unique factorization.

> *DEFINITION*   Unique Factorization Domain (UFD)
> An integral domain D is a *unique factorization domain* if
>
> 1. every nonzero element of D that is not a unit can be written as a product of irreducibles of D, and
> 2. the factorization into irreducibles is unique up to associates and the order in which the factors appear.

Another way to formulate part 2 of this definition is the following. If $p_1^{n_1} p_2^{n_2} \cdots p_r^{n_r}$ and $q_1^{m_1} q_2^{m_2} \cdots q_s^{m_s}$ are two factorizations of some element as a product of irreducibles, where none of the p_i's are associates and none of the q_j's are associates, then $r = s$, and each $p_i^{n_i}$ is an associate of one and only one $q_j^{m_j}$.

Of course, the Fundamental Theorem of Arithmetic tells us that the ring of integers is a unique factorization domain and Theorem 19.6 says that $Z[x]$ is a unique factorization domain. In fact, as we shall soon see, most of the integral domains we have encountered are unique factorization domains.

Before proving our next theorem, we need the ascending chain condition for ideals.

Lemma   *Ascending Chain Condition for a PID*
In a principal ideal domain, any strictly increasing chain of ideals $I_1 \subset I_2 \subset \cdots$ must be finite in length.

Proof.   Let $I_1 \subset I_2 \subset \cdots$ be a chain of strictly increasing ideals in an integral domain D, and let I be the union of all the ideals in this chain. We leave it as an exercise to verify that I is an ideal of D.

Then, since D is a principal ideal domain, there is an element a in D such that $I = \langle a \rangle$. Because $a \in I$ and $I = \cup_k I_k$, a belongs to some member of the chain, say, $a \in I_n$. Clearly, then, for any member I_i of the chain, we have $I_i \subseteq I = \langle a \rangle \subseteq I_n$ so that I_n must be the last member of the chain.   ∎

Theorem 20.3   *PID Implies UFD*
Every principal ideal domain is a unique factorization domain.

Proof.   Let D be a principal ideal domain. We first show that any nonzero element of D that is not a unit is a product of irreducibles (the product could consist of only one factor). To do this, let a_0 be a nonzero, nonunit that is not irreducible. Then we may write $a_0 = b_1 a_1$ where neither b_1 nor a_1 is a unit. Now, if both b_1 and a_1 can be written as a product of irreducibles, then so can a_0. Thus, we may assume one of b_1 or a_1 cannot be written as a product of irreducibles, say a_1. Then, as before, we may write $a_1 = b_2 a_2$ where neither b_2 nor a_2 is a unit. Continuing in this fashion, we obtain an infinite sequence $b_1, b_2, \ldots$ of elements that are not units in D and an infinite sequence $a_0, a_1, a_2, \ldots$ of nonzero elements of D, with $a_n = b_{n+1} a_{n+1}$ for each n. Since b_{n+1} is not a unit, we have $\langle a_n \rangle \subset \langle a_{n+1} \rangle$ for each n (see exercise 3). Hence, $\langle a_0 \rangle \subset \langle a_1 \rangle \subset \cdots$ is an infinite strictly increasing chain of ideals. This contradicts the preceding lemma, so we conclude a_0 is, indeed, a product of irreducibles.

It remains to show that the factorization is unique up to associates and the order in which the factors appear. To do this, suppose some element a of D can be written

$$a = p_1 p_2 \ldots p_r = q_1 q_2 \ldots q_s,$$

where the p's and q's are irreducible and repetition is permitted. We induct on r. If $r = 1$, then a is irreducible and, clearly, $s = 1$ and $p_1 = q_1$. So we may assume that any element that can be expressed as a product of fewer than r irreducible factors can be done so in only one way (up to order and associates). Since p_1 divides $q_1 q_2 \ldots q_s$, it must divide some q_i (see exercise 24), say, $p_1 \mid q_1$. Then, $q_1 = u p_1$, where u is a unit of D. Thus,

$$ua = u p_1 p_2 \ldots p_r = q_1 (u q_2) \ldots q_s$$

and, by cancellation,

$$p_2 \ldots p_r = (u q_2) \ldots q_s.$$

The induction hypothesis now tells us that these two factorizations are identical up to associates and the order in which the factors appear. Hence, the same is true about the two factorizations of a.   ∎

In the existence portion of the proof of Theorem 20.3, the only way we used the fact that the integral domain D is a principal ideal domain was to say that D has the property that there is no infinite, strictly increasing chain of ideas in D. An integral domain with this property is called a *Noetherian domain*, in honor of Emmy Noether, who inaugurated the use of chain conditions in algebra. Noetherian domains are of the utmost importance in algebraic geometry. One reason for this is that, for many important rings R, the polynomial ring $R[x]$ is a Noetherian domain, but not a principal ideal domain. One such example is $Z[x]$. In particular, $Z[x]$ shows that a UFD need not be a PID.

As an immediate corollary of Theorem 20.3, we have the following fact.

Corollary *$F[x]$ Is a UFD*
Let F be a field. Then $F[x]$ is a unique factorization domain.

Proof. By Theorem 18.3, $F[x]$ is a principal ideal domain. So, $F[x]$ is a unique factorization domain, as well.   ∎

As an application of the preceding corollary, we give an elegant proof, due to Richard Singer, of Eisenstein's criterion (Theorem 19.4).

Example 3   Let

$$f(x) = a_n x^n + a_{n-1} x^{n-1} + \cdots + a_0 \in Z[x],$$

and suppose p is prime such that

$$p \nmid a_n,\ p \mid a_{n-1},\ \ldots,\ p \mid a_0,$$

and $p^2 \nmid a_0$. We will prove that $f(x)$ is irreducible over Q. If $f(x)$ is reducible over Q, we know there exist elements $g(x)$ and $h(x)$ in $Z[x]$ such that $f(x) = g(x)h(x)$ and $1 \le \deg g(x),\ \deg h(x) < n$. Let $\bar{f}(x)$, $\bar{g}(x)$, and $\bar{h}(x)$ be the polynomials in $Z_p[x]$ obtained from $f(x)$, $g(x)$, and $h(x)$ by reducing all coefficients modulo p. Then, since p divides all the coefficients of $f(x)$ except a_n, we have $\bar{a}_n x^n =$

$\bar{f}(x) = \bar{g}(x)\bar{h}(x)$. Since Z_p is a field, $Z_p[x]$ is a unique factorization domain. Thus, $x \mid \bar{g}(x)$ and $x \mid \bar{h}(x)$. So, $\bar{g}(0) = \bar{h}(0) = 0$ and, therefore, $p \mid g(0)$ and $p \mid h(0)$. But, then, $p^2 \mid g(0)h(0) = f(0) = a_0$, a contradiction. ☐

Euclidean Domains

Another important kind of integral domain is a Euclidean domain.

> *DEFINITION* Euclidean Domain
> An integral domain D is called a *Euclidean domain* if there is a function d from the nonzero elements of D to the nonnegative integers such that
>
> 1. $d(a) \leq d(ab)$ for all nonzero a, b in D; and
> 2. if a, $b \in D$, $b \neq 0$, then there exist elements q and r in D such that $a = bq + r$ where $r = 0$ or $d(r) < d(b)$.

Example 4 The ring Z is a Euclidean domain with $d(a) = |a|$ (the absolute value of a). ☐

Example 5 Let F be a field. Then $F[x]$ is a Euclidean domain with $d(f(x)) = \deg f(x)$ (see Theorem 18.2). ☐

The perceptive reader may have noticed the remarkable similarity between the ring of integers and the ring of polynomials over a field. These similarities are summarized in Table 20.1 on page 268.

Example 6 The ring of Gaussian integers

$$Z[i] = \{a + bi \mid a, b \in Z\}$$

is a Euclidean domain with $d(a + bi) = a^2 + b^2$. Unlike the previous two examples, the function d does not obviously satisfy the necessary conditions. That $d(x) \leq d(xy)$ for x, $y \in Z[i]$ follows directly from the fact that $d(xy) = d(x)d(y)$ (exercise 5). If x, $y \in Z[i]$ and $y \neq 0$ then $xy^{-1} \in Q[i]$, the field of quotients of $Z[i]$ (exercise 41 of Chapter 17). Say, $xy^{-1} = s + ti$ where $s, t \in Q$. Now let m be the integer nearest s, and let n be the integer nearest t. (These integers may not be uniquely determined but that does not matter.) Thus, $|m - s| \leq \frac{1}{2}$ and $|n - t| \leq \frac{1}{2}$. Then

$$xy^{-1} = s + ti = (m - m + s) + (n - n + t)i$$
$$= (m + ni) + [(s - m) + (t - n)i].$$

So,

$$x = (m + ni)y + [(s - m) + (t - n)i]y.$$

Table 20.1   Similarities Between Z and $F[x]$

Z		$F[x]$
Form of elements:	$\leftrightarrow$	Form of elements:
$a_n10^n + a_{n-1}10^{n-1} + \cdots + a_110 + a_0$		$a_nx^n + a_{n-1}x^{n-1} + \cdots + a_1x + a_0$
Euclidean domain:	$\leftrightarrow$	Euclidean domain:
$d(a) = |a|$		$d(f(x)) = \deg f(x)$
Units:		Units:
a is a unit if and only if $|a| = 1$		$f(x)$ is a unit if and only if $\deg f(x) = 0$
Division algorithm:	$\leftrightarrow$	Division algorithm:
For $a, b \in Z$, $b \neq 0$, there exists $q, r \in Z$ such that $a = bq + r$, $0 \leqslant r < |b|$		For $f(x), g(x) \in F[x]$, $g(x) \neq 0$, there exists $q(x), r(x) \in F[x]$ such that $f(x) = g(x)q(x) + r(x)$, $0 \leqslant \deg r(x) < \deg g(x)$ or $r(x) = 0$
PID:	$\leftrightarrow$	PID:
Every ideal $I = \langle a \rangle$ where $|a|$ is minimum		Every ideal $I = \langle f(x) \rangle$ where $\deg f(x)$ is minimum ($I \neq \{0\}$)
Prime:	$\leftrightarrow$	Irreducible:
No nontrivial factors		No nontrivial factors
UFD:	$\leftrightarrow$	
Every element is a "unique" product of primes		Every element is a "unique" product of irreducibles

We claim that the division condition of the definition of a Euclidean domain is satisfied with $q = m + ni$ and

$$r = [(s - m) + (t - n)i]y.$$

Clearly, q belongs to $Z[i]$, and since $r = x - qy$, so does r. Finally,

$$d(r) = d([(s - m) + (t - n)i])d(y)$$
$$= [(s - m)^2 + (t - n)^2]d(y)$$
$$\leqslant \left(\frac{1}{4} + \frac{1}{4}\right) d(y) < d(y). \qquad \square$$

Theorem 20.4   ED (Euclidean Domain) Implies PID
Every Euclidean domain is a principal ideal domain.

Proof.   Let D be a Euclidean domain and I a nonzero ideal of D. Among all the elements of I, let a be such that $d(a)$ is minimum. Then $I = \langle a \rangle$. For, if $b \in I$, there are elements q and r such that $b = aq + r$ where $r = 0$ or $d(r) < d(a)$. But $r = b - aq \in I$, so $d(r)$ cannot be less than $d(a)$. Thus, $r = 0$ and $b \in \langle a \rangle$. $\blacksquare$

For the curious, we remark that there are principal ideal domains that are not Euclidean domains. The first such example was given by T. Motzkin in 1949. A more accessible account of Motzkin's result can be found in [4].

As an immediate consequence of Theorems 20.3 and 20.4, we have the following important result.

Corollary ED Implies UFD
Every Euclidean domain is a unique factorization domain.

We may summarize our theorems and remarks as follows:

$$ED \Rightarrow PID \Rightarrow UFD$$
$$UFD \not\Rightarrow PID \not\Rightarrow ED$$

(You can remember these implications by listing the types alphabetically.)

In Chapter 19, we proved that $Z[x]$ is a unique factorization domain. Since Z is a unique factorization domain, the next theorem is a broad generalization of this fact. The proof is similar to that of the special case and we therefore omit it.

Theorem 20.5 D a UFD Implies D[x] a UFD
If D is a unique factorization domain, then D[x] is a unique factorization.

We conclude this chapter with an example of an integral domain that is not a unique factorization domain.

Example 7 The ring $Z[\sqrt{-5}] = \{a + b\sqrt{-5} \mid a, b \in Z\}$ is an integral domain but not a unique factorization domain. It is straightforward that $Z[\sqrt{-5}]$ is an integral domain (see exercise 9 of Chapter 15). To verify that unique factorization does not hold, we mimic the method used in Example 1 with $N(a + b\sqrt{-5}) = a^2 + 5b^2$. Since $N(xy) = N(x)N(y)$ and $N(x) = 1$ if and only if x is a unit (see exercise 14), it follows that the only units of $Z[\sqrt{-5}]$ are ± 1.

Now consider the following factorizations:

$$46 = 2 \cdot 23,$$
$$46 = (1 + 3\sqrt{-5})(1 - 3\sqrt{-5}).$$

We claim that each of these four factors is irreducible over $Z[\sqrt{-5}]$. Suppose, say, $2 = xy$ where $x, y \in Z[\sqrt{-5}]$ and neither is a unit. Then $4 = N(2) = N(x)N(y)$ and, therefore, $N(x) = N(y) = 2$, which is impossible. Likewise, if $23 = xy$ were a nontrivial factorization, then $N(x) = 23$. Thus, there would be integers a and b such that $a^2 + 5b^2 = 23$. Clearly, no such integers exist. The same argument applies to $1 \pm 3\sqrt{-5}$. □

In light of Examples 6 and 7, one can't help but wonder for which $d < 0$ is $Z[\sqrt{d}]$ a unique factorization domain. The answer is only when $d = -1$ or -2 (see [3, p. 297]). The case where $d = -1$ was first proved, naturally enough, by Gauss.

EXERCISES

To err is human, but when the eraser wears out ahead of the pencil, you're overdoing it.

Josh Jenkins

1. In an integral domain, show that the product of an irreducible and a unit is an irreducible.

2. Show that the union of a chain $I_1 \subset I_2 \subset \cdots$ of ideals of a ring R is an ideal of R.

3. Suppose a and b belong to an integral domain, $b \neq 0$, and a is not a unit. Show that $\langle ab \rangle$ is a proper subset of $\langle b \rangle$. (This exercise is referred to in the proof of Theorem 20.3.)

4. Let D be an integral domain. Define $a \sim b$ if a and b are associates. Show that this defines an equivalence relation on D.

5. In the notation of Example 6, show that $d(xy) = d(x)d(y)$.

6. Let D be a Euclidean domain and d the associated function. Prove that u is a unit in D if and only if $d(u) = d(1)$.

7. Let D be a Euclidean domain and d the associated function. Show that if a and b are associates in D then $d(a) = d(b)$.

8. Let D be a principal ideal domain. Show that every proper ideal of D is contained in a maximal ideal of D.

9. In $Z[\sqrt{-5}]$, show that 21 does not factor uniquely as a product of irreducibles.

10. Show that $1 - i$ is an irreducible in $Z[i]$.

11. Show that $Z[\sqrt{-6}]$ is not a unique factorization domain. (Hint: Factor 10 in two ways.) Why does this show that $Z[\sqrt{-6}]$ is not a principal ideal domain?

12. Give an example of a unique factorization domain with a subdomain that does not have unique factorization.

13. In $Z[i]$, show that 3 is irreducible but 2 and 5 are not.

14. For the ring $Z[\sqrt{d}] = \{a + b\sqrt{d} \mid a, b \in Z\}$ where $d \neq 1$ and d is not divisible by the square of a prime, prove that the norm $N(a + b\sqrt{d}) = |a^2 - db^2|$ satisfies the four assertions made preceding Example 1.

15. In an integral domain, show that a and b are associates if and only if $\langle a \rangle = \langle b \rangle$.

16. Prove that 7 is irreducible in $Z[\sqrt{6}]$, even though $N(7)$ is not prime. (Thus, the converse of the fourth part of exercise 14 is not true.)

17. Prove that if p is a prime in Z that can be written in the form $a^2 + b^2$, then $a + bi$ is irreducible in $Z[i]$. Find three primes that have this property and the corresponding irreducibles.

18. Prove that $Z[\sqrt{-3}]$ is not a principal ideal domain.

19. In $Z[\sqrt{-5}]$, prove that $1 + 3\sqrt{-5}$ is irreducible but not prime.

20. In $Z[\sqrt{5}]$, prove that both 2 and $1 + \sqrt{5}$ are irreducible but not prime.

21. Let d be an integer less than -1 that is not divisible by the square of a prime. Prove that the only units of $Z[\sqrt{d}]$ are $+1$ and -1.

22. If a and b belong to $Z[\sqrt{d}]$ where d is not divisible by the square of a prime and ab is a unit, prove that a and b are units.

23. Prove or disprove that if D is a principal ideal domain, then $D[x]$ is a principal ideal domain.

24. Let p be a prime in an integral domain. If $p \mid a_1a_2 \ldots a_n$ prove that p divides some a_i. This exercise is referred to in this chapter.

25. Determine the units in $Z[i]$.

26. Show that $3x^2 + 4x + 3 \in Z_5[x]$ factors as $(3x + 2)(x + 4)$ and $(4x + 1)(2x + 3)$. Explain why this does not contradict the corollary of Theorem 20.3.

27. Prove that $Z[\sqrt{5}]$ is not a unique factorization domain.

28. Let D be a principal ideal domain and let $p \in D$. Prove that $\langle p \rangle$ is a maximal ideal in D if and only if p is irreducible.

29. Let D be a principal ideal domain and p an irreducible element of D. Prove that $D/\langle p \rangle$ is a field.

30. Show that an integral domain with the property that every strictly decreasing chain of ideals $I_1 \supset I_2 \supset \cdots$ must be finite in length is a field.

31. An ideal A of a commutative ring R with unity is said to be *finitely generated* if there exist elements $a_1, a_2, \ldots, a_n$ of A such that every element of A can be written in the form $r_1a_1 + r_2a_2 + \cdots + r_na_n$ where $r_1, r_2, \ldots, r_n$ are in R. An integral domain R is said to satisfy the *ascending chain condition* if any strictly increasing chain of ideals $I_1 \subset I_2 \subset \cdots$ must be finite in length. Show that an integral domain R satisfies the ascending chain condition if and only if every ideal of R is finitely generated.

32. Prove or disprove that a subdomain of a Euclidean domain is a Euclidean domain.

33. Show that for any nontrivial ideal I of $Z[i]$, $Z[i]/I$ is finite.

REFERENCES

1. H. M. Edwards, *Fermat's Last Theorem: A Genetic Introduction to Algebraic Number Theory,* New York: Springer-Verlag, 1977.

2. C. Krauthhammer, "The Joy of Math, or Fermat's Revenge," *TIME,* April 18, 1988: 92.

3. H. M. Stark, *An Introduction to Number Theory,* Chicago, Ill.: Markham, 1970.

4. J. C. Wilson, "A Principal Ideal Ring That Is Not a Euclidean Ring," *Mathematics Magazine* 46 (1973): 74–78.

SUGGESTED READINGS

H. M. Edwards, "Fermat's Last Theorem," *Scientific American* 239/4 (1978): 104–122.

This well-written article traces the history of the efforts to prove Fermat's Last Theorem.

Steven Galovich, "Unique Factorization Rings with Zero Divisors," *Mathematics Magazine* 51 (1978): 277–283.

Here the concept of unique factorization is formulated for any commutative ring with unity. Rather complete structure theorems for unique factorization rings with zero divisors are given.

Sahib Singh, "Non-Euclidean Domains: An Example," *Mathematics Magazine* 49 (1976): 243.

This note gives a short proof that $Z[\sqrt{-n}] = \{a + b\sqrt{-n} \mid a, b \in Z\}$ is an integral domain that is not Euclidean when $n > 2$ and $-n = 2$ or 3 modulo 4.

Stan Wagon, "Fermat's Last Theorem," *The Mathematical Intelligence* 8 (1986): 59–61.

This article gives an overview of the work done on Fermat's Last Theorem.

Ernst Eduard Kummer

Modern arithmetic—after Gauss—began with Kummer.

E. T. Bell, *Men of Mathematics*

ERNST EDUARD KUMMER was born on January 29, 1810, in Sorau, Germany. He entered the University of Halle at the age of eighteen to study theology, but within three years, he had obtained a Ph.D. degree in mathematics. The next ten years of his life were spent doing research and teaching at the high-school level. One student greatly influenced by Kummer was Leopold Kronecker, who also became one of the nineteenth century's leading mathematicians (see Chapter 22 for a biography of Kronecker). Kummer received a professorship at the University of Breslau in 1842 and, 13 years later, moved on to the University of Berlin as Dirichlet's successor. In 1857, the French Academy of Sciences issued the following statement:

> Report on the competition for the grand prize in mathematical sciences. Already set in the competition for 1853 and prorogued to 1856. The committee, having found no work which seemed to it worthy of the prize among those submitted to it in competition, proposed to the Academy to award it to M. Kummer, for his beautiful researches on complex numbers composed of roots of unity and integers. The Academy adopted this proposal.

Throughout his career Kummer was a popular teacher because of the clarity of his lectures and his charm and sense of humor. At Berlin, he directed the dissertation of 39 students, several of whom became well-known mathematicians.

Although Kummer did his best work on unique factorization domains and Fermat's Last Theorem, he also made outstanding contributions to analysis, geometry, and physics. He died on May 14, 1893, at the age of eighty-three.

Sophie Germain

One of the very few women to overcome the prejudice and discrimination which have tended to exclude women from the pursuit of higher mathematics up to the present time was Sophie Germain.

Harold M. Edwards, *Fermat's Last Theorem*

SOPHIE GERMAIN was born in Paris on April 1, 1776. Although unable to attend a university because of discrimination against her sex, Germain educated herself by reading the works of Newton and Euler in Latin and the lecture notes of Lagrange. In 1804, Germain wrote to Gauss about her work in number theory but used the pseudonym Monsieur LeBlanc because she feared Gauss would not take seriously the efforts of a woman. Gauss gave Germain's results high praise and a few years later, upon learning her true identity, wrote to her [1, p. 61]:

> But how to describe to you my admiration and astonishment at seeing my esteemed correspondent Mr. LeBlanc metamorphose himself into this illustrious personage who gives such a brilliant example of what I would find it difficult to believe. A taste for the abstract sciences in general and above all the mysteries of numbers if excessively rare: it is not a subject which strikes everyone; the enchanting charms of this sublime science reveal themselves only to those who have the courage to go deeply into it. But when a person of the sex which, according to our customs and prejudices, must encounter infinitely more difficulties than men to familiarize herself with these thorny researches, succeeds nevertheless in surmounting these obstacles and penetrating the most obscure parts of them, then without doubt she must have the noblest courage, quite extraordinary talents, and a superior genius.

Germain is best known for her result on Fermat's Last Theorem. She proved that $x^n + y^n = z^n$ has no positive integer solutions if x, y, and z are relatively prime to each other and to n where $n < 100$.

Sophie Germain died on June 27, 1831, in Paris. A school and a street in Paris are named in her honor.

SUPPLEMENTARY EXERCISES for Chapters 17–20

The intelligence is proved not by ease of learning, but by understanding what we learn.

Joseph Whitney

1. Suppose F is a field and there is a ring homomorphism from Z onto F. Show that F is isomorphic to Z_p for some prime p.

2. Let $Q[\sqrt{2}] = \{r + s \sqrt{2} \mid r, s \in Q\}$. Determine all ring automorphisms of $Q[\sqrt{2}]$.

3. (Second Isomorphism Theorem for Rings) Let A be a subring of R and B an ideal of R. Show that $A \cap B$ is an ideal of A, and $A/(A \cap B)$ is isomorphic to $(A + B)/B$.

4. (Third Isomorphism Theorem for Rings) Let A and B be ideals of a ring R with $B \subseteq A$. Show that A/B is an ideal of R/B and $(R/B)/(A/B)$ is isomorphic to R/A.

5. Let $f(x)$ and $g(x)$ be irreducible polynomials over a field F. If $f(x)$ and $g(x)$ are not associates, prove that $F[x]/\langle f(x)g(x)\rangle$ is isomorphic to $F[x]/\langle f(x)\rangle \oplus F[x]/\langle g(x)\rangle$.

6. (Chinese Remainder Theorem for Rings) If R is a commutative ring and I and J are two proper ideals with $I + J = R$, prove that $R/(I \cap J)$ is isomorphic to $R/I \oplus R/J$. Explain why exercise 5 is a special case of this theorem.

7. Prove that the set of all polynomials with even coefficients is a prime ideal in $Z[x]$.

8. Let $R = Z[\sqrt{-5}]$ and let $I = \{a + b\sqrt{-5} \mid a, b \in Z, a - b \text{ is even}\}$. Show that I is a maximal ideal of R.

9. Let R be a ring with unity and let a be a unit in R. Show that the mapping from R into itself given by $x \rightarrow a^{-1}xa$ is a ring automorphism.

10. Let $a + b\sqrt{-5}$ belong to $Z[\sqrt{-5}]$ with $b \neq 0$. Show that 2 does not belong to $\langle a + b\sqrt{-5}\rangle$.

11. Show that $Z[i]/\langle 2 + i\rangle$ is a field. How many elements does it have?

12. Is the homomorphic image of a principal ideal domain a principal ideal domain?

13. In $Z[\sqrt{2}] = \{a + b\sqrt{2} \mid a, b \in Z\}$. Show that every element of the form $(3 + 2\sqrt{2})^n$ is a unit.

14. Let p be a prime. Show that there is exactly one ring homomorphism from Z_m to Z_{p^k} if p^k does not divide m, and exactly two ring homomorphisms from Z_m to Z_{p^k} if p^k does divide m.

15. Recall that a is an idempotent if $a^2 = a$. Show that if $1 + k$ is an idempotent in Z_n, then $n - k$ is an idempotent in Z_n.

16. Show that Z_n (where $n > 1$) always has an even number of idempotents. (The number is 2^d, where d is the number of distinct prime divisors of n.)

17. Show that if p is prime, the only idempotents in Z_{p^k} are 0 and 1.

18. Prove that if both k and $k + 1$ are idempotents in Z_n and $k \neq 0$, then $k = n/2$.

19. Prove that $x^4 + 15x^3 + 7$ is irreducible over Q.

20. For any integers m and n, prove that the polynomial $x^3 + (5m + 1)x + 5n + 1$ is irreducible over Z.

21. Prove that $\langle \sqrt{2} \rangle$ is a maximal ideal in $Z[\sqrt{2}]$.

22. Prove that $Z[\sqrt{-2}]$ and $Z[\sqrt{2}]$ are unique factorization domains. (Hint: Mimic Example 6 of Chapter 20.)

23. Is $\langle 3 \rangle$ a maximal ideal in $Z[i]$?

24. Express both 13 and $5 + i$ as a product of irreducibles from $Z[i]$.

25. Let $R = \{a/b \mid a, b \in Z, 3 \nmid b\}$. Prove that R is an integral domain. Find its field of quotients.

26. Give an example of a ring that contains a subring isomorphic to Z and a subring isomorphic to Z_3.

27. Show that $Z[i]/\langle 3 \rangle$ is not ring-isomorphic to $Z_3 \oplus Z_3$.

28. For any $n > 1$, prove that $R = \left\{ \begin{bmatrix} a & 0 \\ 0 & b \end{bmatrix} \middle| a, b \in Z_n \right\}$ is ring-isomorphic to $Z_n \oplus Z_n$.

FIELDS

PART 4

Vector Spaces

$$21$$

It is important to appreciate at the outset that the idea of a vector space is the algebraic abstraction and generalization of the cartesian coordinate system introduced into the euclidean plane—that is, a generalization of analytic geometry.

Richard A. Dean, *Elements of Abstract Algebra*

Definition and Examples

Abstract algebra has three basic components: groups, rings, and fields. Thus far we have covered groups and rings in some detail and we have touched on the notion of a field. To explore fields more deeply, we need the rudiments of vector space theory that are covered in a linear algebra course. In this chapter, we provide a concise review of this material.

DEFINITION Vector Space

A set V is said to be a *vector space* over a field F if V is an Abelian group under addition (denoted by $+$) and, if for each $a \in F$ and $v \in V$, there is an element av in V such that the following conditions hold for all a, b in F and all u, v in V.

1. $a(v + u) = av + au$.
2. $(a + b)v = av + bv$.
3. $a(bv) = (ab)v$.
4. $1v = v$.

The members of a vector space are called *vectors*. The members of the field are called *scalars*. The operation that combines a scalar a and a vector v to form the vector av is called *scalar multiplication*. In general, we will denote vectors by letters from the end of the alphabet, such as u, v, w, and scalars by letters from the beginning of the alphabet, such as a, b, c.

Example 1 Classic Vector Space $\mathbf{R}^n = \{(a_1, a_2, \ldots, a_n) \mid a_i \in \mathbf{R}\}$ is a vector space over $\mathbf{R}$. Here the operations are the obvious ones.

$$(a_1, a_2, \ldots, a_n) + (b_1, b_2, \ldots, b_n) = (a_1 + b_1, a_2 + b_2, \ldots, a_n + b_n)$$

and

$$b(a_1, a_2, \ldots, a_n) = (ba_1, ba_2, \ldots, ba_n). \qquad \square$$

Example 2 The set $M_2(Q)$ of 2×2 matrices with entries from Q is a vector space over Q. The operations are

$$\begin{bmatrix} a_1 & a_2 \\ a_3 & a_4 \end{bmatrix} + \begin{bmatrix} b_1 & b_2 \\ b_3 & b_4 \end{bmatrix} = \begin{bmatrix} a_1 + b_1 & a_2 + b_2 \\ a_3 + b_3 & a_4 + b_4 \end{bmatrix}$$

and

$$b\begin{bmatrix} a_1 & a_2 \\ a_3 & a_4 \end{bmatrix} = \begin{bmatrix} ba_1 & ba_3 \\ ba_2 & ba_4 \end{bmatrix}. \qquad \square$$

Example 3 The set $Z_p[x]$ of polynomials with coefficients from Z_p is a vector space over Z_p, where p is a prime. $\qquad \square$

Example 4 The complex numbers $\mathbf{C} = \{a + bi \mid a, b \in \mathbf{R}, i^2 = -1\}$ is a vector space over $\mathbf{R}$. The operations are the usual addition and multiplication of complex numbers. $\qquad \square$

The next example is a generalization of Example 4. Although it appears rather trivial, it is of the utmost importance in the theory of fields.

Example 5 Let E be a field and F a subfield of E. Then E is a vector space over F. The operations are the operations of E. $\qquad \square$

Subspaces

Of course, there is a natural analog of subgroup and subring.

DEFINITION Subspace
Let V be a vector space over a field F and let U be a subset of V. We say U is a *subspace* of V if U is also a vector space over F under the operations of V.

Example 6 The set $\{a_2x^2 + a_1x + a_0 \mid a_0, a_1, a_2 \in \mathbf{R}\}$ is a subspace of the set of all polynomials with real coefficients. □

Example 7 Let V be a vector space over F and let $v_1, v_2, \ldots, v_n$ be (not necessarily distinct) elements of V. Then the subset

$$\langle v_1, v_2, \ldots, v_n \rangle = \{a_1v_1 + a_2v_2 + \cdots + a_nv_n \mid a_1, a_2, \ldots, a_n \in F\}$$

is called the *subspace of V spanned by* $v_1, v_2, \ldots, v_n$. Any summand of the form $a_1v_1 + a_2v_2 + \cdots + a_nv_n$ is called a *linear combination* of $v_1, v_2, \ldots, v_n$. If $\langle v_1, v_2, \ldots, v_n \rangle = V$, we say that $v_1, v_2, \ldots, v_n$ *span V*. □

Linear Independence

The next definition is the heart of the theory.

> *DEFINITION* Linearly Dependent, Linearly Independent
> A set of vectors $\{v_1, v_2, \ldots, v_n\}$ is said to be *linearly dependent* over the field F if there are elements $a_1, a_2, \ldots, a_n$ from F, not all zero, such that $a_1v_1 + a_2v_2 + \cdots + a_nv_n = 0$. A set of vectors that is not linearly dependent over F is called *linearly independent* over F.

Example 8 In $\mathbf{R}^3$ the vectors $(1, 0, 0)$, $(1, 0, 1)$, and $(1, 1, 1)$ are linearly independent over $\mathbf{R}$. To verify this, assume there are real numbers a, b, and c such that $a(1, 0, 0) + b(1, 0, 1) + c(1, 1, 1) = (0, 0, 0)$. Then $(a + b + c, c, b + c) = (0, 0, 0)$. From this we see that $a = b = c = 0$. □

Certain kinds of linearly independent sets play a crucial role in the theory of vector spaces.

> *DEFINITION* Basis
> Let V be a vector space over F. A subset B of V is called a *basis* for V if B is linearly independent over F and B spans V.

The motivation for this definition is twofold. First, if B is a basis for a vector space V, then every member of V is a unique linear combination of the elements of B (see exercise 24). Second, with every vector space spanned by finitely many vectors, we can use the notion of basis to associate a unique integer that tells us much about the vector space. (In fact, this integer and the field completely determine the vector space up to isomorphism—see exercise 31.)

Example 9 The set $V = \left\{ \begin{bmatrix} a & a + b \\ a + b & b \end{bmatrix} \middle| a, b \in \mathbf{R} \right\}$ is a vector space.

over **R** (see exercise 17). We claim that $B = \left\{ \begin{bmatrix} 1 & 1 \\ 1 & 0 \end{bmatrix}, \begin{bmatrix} 0 & 1 \\ 1 & 1 \end{bmatrix} \right\}$ is a basis for V over **R**. To prove the set B is linearly independent, suppose that there are real numbers a and b such that

$$a \begin{bmatrix} 1 & 1 \\ 1 & 0 \end{bmatrix} + b \begin{bmatrix} 0 & 1 \\ 1 & 1 \end{bmatrix} = \begin{bmatrix} 0 & 0 \\ 0 & 0 \end{bmatrix}.$$

But this gives $\begin{bmatrix} a & a + b \\ a + b & b \end{bmatrix} = \begin{bmatrix} 0 & 0 \\ 0 & 0 \end{bmatrix}$ so that $a = b = 0$. On the other hand, since every member of V has the form

$$\begin{bmatrix} a & a + b \\ a + b & b \end{bmatrix} = a \begin{bmatrix} 1 & 1 \\ 1 & 0 \end{bmatrix} + b \begin{bmatrix} 0 & 1 \\ 1 & 1 \end{bmatrix},$$

we see that B spans V. □

We now come to the main result of this chapter.

Theorem 21.1 Invariance of Basis Size
If $\{u_1, u_2, \ldots, u_m\}$ and $\{w_1, w_2, \ldots, w_n\}$ are both bases of a vector space V, then $m = n$.

Proof.　Suppose $m \neq n$. To be specific, let us say $m < n$. Consider the set $\{w_1, u_1, u_2, \ldots, u_m\}$. Since the u's span V, we know that w_1 is a linear combination of the u's, say, $w_1 = a_1 u_1 + a_2 u_2 + \cdots + a_m u_m$. Clearly, not all the a's are zero. For convenience, say, $a_1 \neq 0$. Then $\{w_1, u_2, \ldots, u_m\}$ spans V (see exercise 22). Next consider the set $\{w_1, w_2, u_2, \ldots, u_m\}$. This time, w_2 is a linear combination of $w_1, u_2, \ldots, u_m$, say, $w_2 = b_1 w_1 + b_2 u_2 + \cdots + b_m u_m$. Then at least one of $b_2, \ldots, b_m$ is nonzero, for otherwise the w's are not linearly independent. Let us say $b_2 \neq 0$. Then $w_1, w_2, u_3, \ldots, u_m$ span V. Continuing in this fashion, we see that $\{w_1, w_2, \ldots, w_m\}$ spans V. But then w_{m+1} is a linear combination of $w_1, w_2, \ldots, w_m$ and, therefore, the set $\{w_1, \ldots, w_n\}$ is not linearly independent. This contradiction finishes the proof. ■

Theorem 21.1 shows that any two finite bases for a vector space have the same size. Of course, not all vector spaces have finite bases. However, there is no vector space that has a finite basis and an infinite basis (see exercise 26).

DEFINITION Dimension
A vector space that has a basis consisting of n elements is said to have *dimension n*. For completeness, the trivial vector space $\{0\}$ is said to be spanned by the empty set and to have dimension 0.

Although it requires a bit of set theory that is beyond the scope of this text, it can be shown that every vector space has a basis. A vector space that has a finite basis is called *finite dimensional*; otherwise it is called *infinite dimensional*.

EXERCISES

The good Lord made us with two ends—one to sit on and one to think with. How well you succeed in life depends on which one you use.

Isaac Dworetsky

1. Verify that each of the sets in Examples 1–4 satisfies the axioms for a vector space. Find a basis for each of the vector spaces in Examples 1–4.

2. (Subspace Test) Prove that a subset U of a vector space V over a field F is a subspace of V if for every u and u' in U and every a in F, $u - u' \in U$ and $au \in U$.

3. Verify that the set in Example 6 is a subspace. Find a basis for this subspace. Is $\{x^2 + x + 1, x + 5, 3\}$ a basis?

4. Verify that the set $\langle v_1, v_2, \ldots, v_n \rangle$ defined in Example 7 is a subspace.

5. Determine whether or not the set $\{(2, -1, 0), (1, 2, 5), (7, -1, 5)\}$ is linearly independent over $\mathbf{R}$.

6. Determine whether or not the set

$$\left\{ \begin{bmatrix} 2 & 1 \\ 1 & 0 \end{bmatrix}, \begin{bmatrix} 0 & 1 \\ 1 & 2 \end{bmatrix}, \begin{bmatrix} 1 & 1 \\ 1 & 1 \end{bmatrix} \right\}$$

is linearly independent over Z_5.

7. If $\{u, v, w\}$ is a linearly independent subset of a vector space, show that $\{u, u + v, u + v + w\}$ is also linearly independent.

8. If $\{v_1, v_2, \ldots, v_n\}$ is a linearly dependent set of vectors, prove that one of these vectors is a linear combination of the others.

9. (Every spanning collection contains a basis.) If $v_1, v_2, \ldots, v_n$ span a vector space V, prove that some subset of the v's is a basis for V.

10. (Every independent set is contained in a basis.) Let V be a finite dimensional vector space and let $\{v_1, v_2, \ldots, v_n\}$ be a linearly independent subset of V. Show that there are vectors $w_1, w_2, \ldots, w_m$ such that $\{v_1, v_2, \ldots, v_n, w_1, \ldots, w_m\}$ is a basis for V.

11. If V is a vector space over F of dimension 5 and U and W are subspaces of V of dimension 3, prove that $U \cap W \neq \{0\}$. Generalize.

12. Show that the solution set to the system of equations of the form

$$\begin{aligned} a_{11}x_1 + \cdots + a_{1n}x_n &= 0 \\ a_{21}x_1 + \cdots + a_{2n}x_n &= 0 \\ &\vdots \\ a_{m1}x_1 + \cdots + a_{mn}x_n &= 0 \end{aligned}$$

where the a's are real is a subspace of $\mathbf{R}^n$.

13. Let V be the set of all polynomials over Q of degree 2 together with the zero polynomial. Is V a vector space over Q?

14. Let $V = \mathbf{R} \oplus \mathbf{R} \oplus \mathbf{R}$ and $W = \{(a, b, c) \in V \mid a^2 + b^2 = c^2\}$. Is W a subspace of V? If so, what is its dimension?

15. Let $V = \mathbf{R} \oplus \mathbf{R} \oplus \mathbf{R}$ and $W = \{(a, b, c) \in V \mid a + b = c\}$. Is W a subspace of V? If so, what is its dimension?

16. Let $V = \left\{ \begin{bmatrix} a & b \\ b & c \end{bmatrix} \,\middle|\, a, b, c \in Q \right\}$. Prove that V is a vector space over Q and find a basis for V over Q.

17. Verify that the set V in Example 9 is a vector space over $\mathbf{R}$.

18. Let $P = \{(a, b, c) \mid a, b, c \in \mathbf{R}, a = 2b + 3c\}$. Prove that P is a subspace of $\mathbf{R}^3$. Find a basis for P. Give a geometric description of P.

19. Let U and W be subspaces of a vector space V. Show that $U \cap W$ is a subspace of V and that $U + W = \{u + w \mid u \in U, w \in W\}$ is a subspace of V.

20. If U is a proper subspace of a finite dimensional vector space V, show that the dimension of U is less than the dimension of V.

21. If V is a vector space of dimension n over the field Z_p, how many elements are in V?

22. Referring to the proof of Theorem 21.1, prove that $\{w_1, u_2, \ldots, u_n\}$ spans V.

23. Let $S = \{(a, b, c, d) \mid a, b, c, d \in \mathbf{R}, a = c, d = a + b\}$. What is the dimension of S?

24. Let B be a subset of a vector space V. Show that B is a basis for V if and only if every member of V is a unique linear combination of the elements of B. This exercise is referred to in this chapter and Chapter 22.

25. Let $u = (2, 3, 1)$, $v = (1, 3, 0)$, $w = (2, -3, 3)$. Since $\frac{1}{2}u - \frac{2}{3}v - \frac{1}{6}w = (0, 0, 0)$, can we conclude that u, v, and w are linearly dependent over Z_7?

26. If a vector space has one basis that contains infinitely many elements, prove that every basis contains infinitely many elements.

27. Define the vector space analog of a group and ring homomorphism. Such a mapping is called a *linear transformation*.

28. Define the vector space analog of a group and ring isomorphism.

29. Let T be a linear transformation of a vector space V. Prove that the *kernel* of $T = \{v \in V \mid vT = 0\}$ is a subspace of V.

30. Let T be a linear transformation of V onto W. If $\{v_1, v_2, \ldots, v_n\}$ spans V, show that $\{v_1T, v_2T, \ldots, v_nT\}$ spans W.

31. If V is a vector space over F of dimension n, prove that V is isomorphic as a vector space to $F^n = \{(a_1, a_2, \ldots, a_n) \mid a_i \in F\}$. (This exercise is referred to in Chapter 23.)

Emil Artin

For Artin, to be a mathematician meant to participate in a great common effort, to continue work begun thousands of years ago, to shed new light on old discoveries, to seek new ways to prepare the developments of the future. Whatever standards we use, he was a great mathematician.

Richard Brauer, *Bulletin of the American Mathematical Society*

EMIL ARTIN was one of the leading mathematicians of the twentieth century and a major contributor to linear algebra and abstract algebra. Artin was born on March 3, 1898, in Vienna, Austria, and grew up in what is now known as Czechoslovakia. After serving in the Austrian army during World War I, Artin enrolled at the University of Leipzig where he received a Ph.D. in 1921. From 1923 until he emigrated to America in 1937, he was a professor at the University of Hamburg. After one year at Notre Dame, Artin went to Indiana University. In 1946, he moved to Princeton where he stayed until 1958. The last four years of his career were spent where it began, at Hamburg.

Artin's mathematics is both deep and broad. He made contributions to number theory, group theory, ring theory (in fact, there is a class of rings named after him), field theory, Galois theory, geometric algebra, algebraic topology, and the theory of braids, a field he invented. Artin received the American Mathematical Society's Cole Prize in number theory, and he solved one of the twenty-three famous problems posed by the eminent mathematician David Hilbert in 1900. Besides mathematics, Artin had a deep interest in chemistry, astronomy, biology, and old music. He played the flute, the harpsicord, and the clavichord.

Artin was an outstanding teacher of mathematics at all levels, from freshman calculus to seminars for colleagues. Many of his Ph.D. students have become leading mathematicians. Through his research, teaching, and books, Artin exerted great influence among his contemporaries. He died of a heart attack, at the age of sixty-four, in 1962.

Extension Fields

22

The Fundamental Theorem of Field Theory

In our work on rings, we came across a number of fields both finite and infinite. Indeed, we saw that $Z_3[x]/\langle x^2 + 1\rangle$ is a field of order 9 while $\mathbf{R}[x]/\langle x^2 + 1\rangle$ is a field isomorphic to the complex numbers. In the next three chapters, we take up, in a systematic way, the subject of fields.

DEFINITION Extension Field
A field E is an *extension field* of a field F if $F \subseteq E$ and the operations of F are those of E restricted to F.

Cauchy's observation in 1847 that $\mathbf{R}[x]/\langle x^2 + 1\rangle$ is a field that contains a zero of $x^2 + 1$ prepared the way for the following sweeping generalization of that fact.

Theorem 22.1 *Fundamental Theorem of Field Theory (Kronecker's Theorem—1887)*
Let F be a field and $f(x)$ a nonconstant polynomial in $F[x]$. Then there is an extension field E of F in which $f(x)$ has a zero.

Proof. Since $F[x]$ is a unique factorization domain, $f(x)$ has an irreducible factor, say, $p(x)$. Clearly, it suffices to construct an extension field E of F

in which $p(x)$ has a zero. Our candidate for E is $F[x]/\langle p(x)\rangle$. We already know that this is a field from Corollary 1 of Theorem 19.5. Also, since the mapping of $\phi:F \to E$ given by $a\phi = a + \langle p(x)\rangle$ is one-to-one and preserves both operations, E has a subfield isomorphic to F. We may think of E as containing F if we simply identify the coset $a + \langle p(x)\rangle$ with its unique coset representative a (that is, think of $a + \langle p(x)\rangle$ as just a and vice versa; see exercise 30 of Chapter 19).

Finally, to show that $p(x)$ has a zero in E, write

$$p(x) = a_nx^n + a_{n-1}x^{n-1} + \cdots + a_0.$$

Then, in E, $x + \langle p(x)\rangle$ is a zero of $p(x)$. For

$$\begin{aligned} p(x + \langle p(x)\rangle) &= a_n(x + \langle p(x)\rangle)^n + a_{n-1}(x + \langle p(x)\rangle)^{n-1} + \cdots + a_0 \\ &= a_n(x^n + \langle p(x)\rangle) + a_{n-1}(x^{n-1} + \langle p(x)\rangle) + \cdots + a_0 \\ &= a_nx^n + a_{n-1}x^{n-1} + \cdots + a_0 + \langle p(x)\rangle \\ &= p(x) + \langle p(x)\rangle = 0 + \langle p(x)\rangle. \quad\blacksquare \end{aligned}$$

Example 1 Let $f(x) = x^2 + 1 \in Q[x]$. Then, in $E = Q[x]/\langle x^2 + 1\rangle$, we have

$$\begin{aligned} f(x + \langle x^2 + 1\rangle) &= (x + \langle x^2 + 1\rangle)^2 + 1 \\ &= x^2 + \langle x^2 + 1\rangle + 1 \\ &= x^2 + 1 + \langle x^2 + 1\rangle \\ &= 0 + \langle x^2 + 1\rangle. \end{aligned}$$

Of course, the polynomial $x^2 + 1$ has the complex number $\sqrt{-1}$ as a zero, but the point we wish to emphasize here is that we have constructed a field that contains the rational numbers and a zero for the polynomial $x^2 + 1$ by using only the rational numbers. No knowledge of complex numbers is necessary. Our method utilizes only the field we are given. □

Example 2 Let $f(x) = x^5 + 2x^2 + 2x + 2 \in Z_3[x]$. Then, the irreducible factorization of $f(x)$ over Z_3 is $(x^2 + 1)(x^3 + 2x + 2)$. So, we may take $E = Z_3[x]/\langle x^2 + 1\rangle$, a field with 9 elements, or $E = Z_3[x]/\langle x^3 + 2x + 2\rangle$, a field with 27 elements. □

Since every integral domain is contained in its field of quotients (Theorem 17.6), we see that every nonconstant polynomial with coefficients from an integral domain always has a zero in some field containing the ring of coefficients. The next example shows that this is not true for commutative rings in general.

Example 3 Let $f(x) = 2x + 1 \in Z_4[x]$. Then $f(x)$ has no zero in any ring containing Z_4. For if α were a zero, then $2\alpha + 1 = 0$ and $2(2\alpha + 1) = 2 = 0$ since in any ring containing Z_4, $4\alpha = 0$. □

Splitting Fields

To motivate the next definition and theorem, let's return to Example 1 for a moment. For notational convenience, in $Q[x]/\langle x^2 + 1 \rangle$, let $\alpha = x + \langle x^2 + 1 \rangle$. Then, since α and $-\alpha$ are both zeros of $x^2 + 1$, it should be the case that $x^2 + 1 = (x - \alpha)(x + \alpha)$. Let's check this out. First note that

$$(x - \alpha)(x + \alpha) = x^2 - \alpha^2 = x^2 - (x^2 + \langle x^2 + 1 \rangle).$$

At the same time,

$$x^2 + \langle x^2 + 1 \rangle = -1 + \langle x^2 + 1 \rangle$$

and we have agreed to identify -1 and $-1 + \langle x^2 + 1 \rangle$ so

$$(x - \alpha)(x + \alpha) = x^2 - (-1) = x^2 + 1.$$

This shows that $x^2 + 1$ can be written as a product of linear factors in some extension of Q. That was easy and you might argue coincidental. The polynomial given in Example 2 presents a greater challenge. Is there an extension of Z_3 in which that polynomial factors as a product of linear factors? Yes, there is. But first a definition.

> *DEFINITION*   Splitting Field
>
> Let E be an extension field of F and let $f(x) \in F[x]$. We say $f(x)$ *splits* in E if $f(x)$ can be factored as a product of linear factors in $E[x]$. We call E a *splitting field for $f(x)$ over F* if $f(x)$ splits in E but in no proper subfield of E.

Note that a splitting field of a polynomial over a field depends not only on the polynomial but the field as well. Indeed, a splitting field of $f(x)$ over F is just a smallest extension field of F in which $f(x)$ splits. The next example illustrates how a splitting field of a polynomial $f(x)$ over field F depends on F.

Example 4   Consider the polynomial $f(x) = x^2 + 1 \in Q[x]$. Since $x^2 + 1 = (x + \sqrt{-1})(x - \sqrt{-1})$, we see that $f(x)$ splits in **C**, but a splitting field over Q is $Q(i) = \{r + si \mid r, s \in Q\}$. A splitting field for $x^2 + 1$ over **R** is **C**. Likewise, $x^2 - 2 \in Q[x]$ splits in **R**, but a splitting field over Q is $Q(\sqrt{2}) = \{r + s\sqrt{2} \mid r, s \in Q\}$. $\square$

There is a useful analogy between the definition of splitting field and the definition of an irreducible polynomial. Just as it makes no sense to say "$f(x)$ is irreducible" it makes no sense to say "E is a splitting field for $f(x)$." In each case, the underlying field must be specified; that is, one must say "$f(x)$ is irreducible over F" and "E is a splitting field for $f(x)$ over F."

The following notation is convenient. Let F be a field and let $a_1, a_2, \ldots, a_n$ be elements of some extension E of F. We use $F(a_1, a_2, \ldots, a_n)$ to denote the smallest subfield of E that contains F and the set $\{a_1, a_2, \ldots, a_n\}$. It is an easy exercise to show that $F(a_1, a_2, \ldots, a_n)$ is the intersection of all subfields of E that contain F and the set $\{a_1, a_2, \ldots, a_n\}$.

Notice that if $f(x) \in F[x]$ and $f(x)$ factors as

$$b(x - a_1)(x - a_2) \cdots (x - a_n)$$

over some extension E of F, then $F(a_1, \ldots, a_n)$ is the splitting field for $f(x)$ over F.

This notation appears to be inconsistent with the notation used in earlier chapters. For example, we denoted the set $\{a + b\sqrt{2} \mid a, b \in Z\}$ by $Z[\sqrt{2}]$ and the set $\{a + b\sqrt{2} \mid a, b \in Q\}$ by $Q[\sqrt{2}]$. The difference is $Z[\sqrt{2}]$ is merely a ring while $Q[\sqrt{2}]$ is a field. In general, parentheses are used when one wishes to indicate that the set is a field, although no harm would be done by using, say, $Q[\sqrt{2}]$ to denote $\{a + b\sqrt{2} \mid a, b \in Q\}$ if we were concerned with its ring properties only. After all, notation is just a convenient way to convey information. Using parentheses rather than brackets simply conveys a bit more information about the set.

Theorem 22.2 *Existence of Splitting Fields*
Let F be a field and let $f(x)$ be a nonconstant element of $F[x]$. Then there exists a splitting field E for $f(x)$ over F.

Proof. We proceed by induction on deg $f(x)$. If deg $f(x) = 1$, then $f(x)$ is already linear and $E = F$. Now suppose the statement is true for all fields and all polynomials of degree less than that of $f(x)$. By Theorem 22.1, there is an extension E of F in which $f(x)$ has a zero, say, a_1. Then we may write $f(x) = (x - a_1)g(x)$ where $g(x) \in E[x]$. Since deg $g(x) <$ deg $f(x)$, by induction, there is a field K that contains E and all the zeros of $g(x)$, say, $a_2, \ldots, a_n$. Clearly, then, the splitting field for $f(x)$ over F is $F(a_1, a_2, \ldots, a_n)$. ∎

Example 5 Consider

$$f(x) = x^4 - x^2 - 2 = (x^2 - 2)(x^2 + 1)$$

over Q. Obviously, the zeros of $f(x)$ are $\pm\sqrt{2}$ and $\pm i$. So a splitting field for $f(x)$ over Q is

$$Q(\sqrt{2}, i) = Q(\sqrt{2})(i) = \{\alpha + \beta i \mid \alpha, \beta \in Q(\sqrt{2})\}$$
$$= \{(a + b\sqrt{2}) + (c + d\sqrt{2})i \mid a, b, c, d \in Q\} \square$$

Example 6 Consider $f(x) = x^2 + x + 2$ over Z_3. Then $Z_3(i) = \{a + bi \mid a, b \in Z_3\}$ (see Example 8 of Chapter 15) is a splitting field for $f(x)$ over Z_3 because

$$f(x) = [x - (1 + i)][x - (1 - i)].$$

At the same time, we know by the proof of Kronecker's Theorem that the element $x + \langle x^2 + x + 2 \rangle$ of

$$F = Z_3[x]/\langle x^2 + x + 2 \rangle$$

is a zero of $f(x)$. Since $f(x)$ has degree 2, it follows from the Factor Theorem (Corollary 2 of Theorem 18.2) that the other zero of $f(x)$ must also be in F. Thus, $f(x)$ splits in F; and because F has only nine elements, it is obvious that F is also a splitting field of $f(x)$ over Z_3. But how do we factor $f(x)$ in F? Factoring $f(x)$ in F is confusing because we are using the symbol x in two distinct ways. It is used as a placeholder to write the polynomial $f(x)$, and it is used to create the coset representatives of the elements of F. This confusion can be avoided by simply identifying the coset $1 + \langle x^2 + x + 2 \rangle$ with the element 1 in Z_3 and denoting the coset $x + \langle x^2 + x + 2 \rangle$ by β. To obtain the factorization of $f(x)$ in F, we observe that

$$(x - \beta)[x - (2\beta + 2)] = x^2 + (-\beta - 2\beta - 2)x + 2(\beta^2 + \beta)$$
$$= x^2 + x + 2 = f(x).$$

Here we have used the facts that F has characteristic 3 and $\beta^2 + \beta = 1$. Thus, we have found two splitting fields for $x^2 + x + 2$ over Z_3, one of the form $F(a)$ and one of the form $F[x]/\langle p(x) \rangle$ (where $F = Z_3$ and $p(x) = x^2 + x + 2$). □

The next theorem shows how the fields $F(a)$ and $F[x]/\langle p(x) \rangle$ are related in the case that $p(x)$ is irreducible over F and a is a zero of $p(x)$ in some extension of F.

Theorem 22.3 $F(a) \approx F[x]/\langle p(x) \rangle$
Let F be a field and let $p(x) \in F[x]$ be irreducible over F. If a is a zero of $p(x)$ in some extension E of F then $F(a)$ is isomorphic to $F[x]/\langle p(x) \rangle$. Furthermore, if $\deg p(x) = n$, then every member of $F(a)$ can be uniquely expressed in the form

$$c_{n-1}a^{n-1} + c_{n-2}a^{n-2} + \cdots + c_1a + c_0$$
where $c_0, c_1, \ldots, c_{n-1} \in F$.

Proof. Consider the function ϕ from $F[x]$ to $F(a)$ given by $f(x)\phi = f(a)$. Clearly, ϕ is a ring homomorphism. We claim Ker $\phi = \langle p(x) \rangle$. (This is exercise 31 of Chapter 19.) Since $p(a) = 0$, we have $\langle p(x) \rangle \subseteq$ Ker ϕ. Now, suppose $f(x) \in$ Ker ϕ and $f(x) \notin \langle p(x) \rangle$. Then as in the proof of Theorem 16.4 (see also exercise 40 of Chapter 18), there are elements $h(x)$ and $g(x)$ in $F[x]$ such that $1 = h(x)f(x) + g(x)p(x)$. But this is impossible, since $f(a) = 0 = p(a)$. Thus, Ker $\phi = \langle p(x) \rangle$. At this point, it follows from the First Isomorphism Theorem for Rings and Corollary 1 of Theorem 19.5 that $F[x]\phi$ is a subfield of $F(a)$. Noting that $F[x]\phi$ contains both F

and *a* and recalling that $F(a)$ is the smallest such field, we have $F[x]/\langle p(x)\rangle \approx F[x]\phi = F(a)$.

The final assertion of the theorem follows from the fact that every element of $F[x]/\langle p(x)\rangle$ can be expressed uniquely in the form

$$c_{n-1}x^{n-1} + \cdots + c_0 + \langle p(x)\rangle$$

where $c_0, \ldots, c_{n-1} \in F$ (see exercise 21 of Chapter 18) and the natural isomorphism from $F[x]/\langle p(x)\rangle$ to $F(a)$ carries $c_k x^k + \langle p(x)\rangle$ to $c_k a^k$. ∎

Recall that a basis for an *n*-dimensional vector space over a field *F* is a set of *n* vectors $v_1, v_2, \ldots, v_n$ with the property that every member of the vector space can be expressed uniquely in the form $a_1 v_1 + a_2 v_2 + \cdots + a_n v_n$, where the *a*'s belong to *F* (exercise 24 of Chapter 21). So, in the language of vector spaces, the latter portion of Theorem 22.3 says that if *a* is a zero of an irreducible polynomial over *F* of degree *n*, then the set $\{1, a, \ldots, a^{n-1}\}$ is a basis for $F(a)$ over *F*.

Theorem 22.3 often gives a convenient way to describe the elements of a field.

Example 7   Consider the irreducible polynomial $f(x) = x^6 - 2$ over *Q*. Since $\sqrt[6]{2}$ is a zero of $f(x)$, we know from Theorem 22.3 that the set $\{1, 2^{1/6}, 2^{2/6}, 2^{3/6}, 2^{4/6}, 2^{5/6}\}$ is a basis for $Q(\sqrt[6]{2})$ over *Q*. Thus,

$$Q(\sqrt[6]{2}) = \{a_0 + a_1 2^{1/6} + a_2 2^{2/6} + a_3 2^{3/6} + a_4 2^{4/6} + a_5 2^{5/6} \mid a_i \in Q\}.$$

This field is isomorphic to $Q[x]/\langle x^6 - 2\rangle$.   □

Notice that Theorem 22.3 does not apply to $Q(\pi)$ since π is not the zero of a polynomial in $Q[x]$. (This important fact was first proved by Ferdinand Lindemann (1852–1939) in 1882.)

In Example 6, we produced two splitting fields for the polynomial $x^2 + x + 2$ over Z_3. Likewise, it is an easy exercise to show that both $Q[x]/\langle x^2 + 1\rangle$ and $Q(i) = \{r + si \mid r, s \in Q\}$ are splitting fields of the polynomial $x^2 + 1$ over *Q*. But are these different-looking splitting fields algebraically different? Not really. We conclude our discussion of splitting fields by proving splitting fields are unique up to isomorphism. For technical reasons, it is easier to prove a more general result.

We begin by observing first that any field isomorphism ϕ from *F* to *F'* has a natural extension from $F[x]$ to $F'[x]$ given by $c_n x^n + c_{n-1}x^{n-1} + \cdots + c_1 x + c_0 \rightarrow (c_n\phi)x^n + (c_{n-1}\phi)x^{n-1} + \cdots + (c_1\phi)x + c_0\phi$. Since this mapping agrees with ϕ on *F*, it is convenient and natural to use ϕ to denote this mapping as well.

Lemma   *Let F be a field, let $p(x) \in F[x]$ be irreducible over F, and let a be a zero of $p(x)$ in some extension of F. If ϕ is a field isomorphism from*

F to F' and b is a zero of $p(x)\phi$ in some extension of F', then there is an isomorphism from F(a) to F'(b) that agrees with ϕ on F and carries a to b.

Proof. First observe that since $p(x)$ is irreducible over F, $p(x)\phi$ is irreducible over $F'[x]$. It is straightforward to check that the mapping from $F[x]/\langle p(x)\rangle$ to $F'[x]/\langle p(x)\phi\rangle$ given by

$$f(x) + \langle p(x)\rangle \rightarrow f(x)\phi + \langle p(x)\phi\rangle$$

is a field isomorphism. By a slight abuse of notation, we denote this mapping by ϕ also. (If you object, put a bar over the ϕ.) From the proof of Theorem 22.3, we know there is an isomorphism α from $F(a)$ to $F[x]/\langle p(x)\rangle$ that is the identity on F and carries a to $x + \langle p(x)\rangle$. Similarly, there is an isomorphism β from $F'[x]/\langle p(x)\phi\rangle$ to $F'(b)$ that is the identity on F' and carries $x + \langle p(x)\phi\rangle$ to b. Thus, $\alpha\phi\beta$ is the desired mapping. See Figure 22.1.∎

Theorem 22.4 *Let ϕ be an isomorphism from a field F to a field F' and let $f(x) \in F[x]$. If E is a splitting field for $f(x)$ over F and E' is a splitting field for $f(x)\phi$ over F', then there is an isomorphism from E to E' that agrees with ϕ on F.*

Proof. We induct on $\deg f(x)$. If $\deg f(x) = 1$, then $E = F$ and $E' = F'$ so that ϕ itself is the desired mapping. If $\deg f(x) > 1$, let $p(x)$ be an irreducible factor of $f(x)$, let a be a zero of $p(x)$ in E, and let b be a zero of $p(x)\phi$ in E'. By the preceding lemma, there is an isomorphism α from $F(a)$ to $F'(b)$ that agrees with ϕ on F and carries a to b. Now write $f(x) = (x - a)g(x)$ where $g(x) \in F(a)[x]$. Then E is a splitting field for $g(x)$ over $F(a)$ and E' is a splitting field for $g(x)\alpha$ over $F'(b)$. Since $\deg g(x) < \deg f(x)$, there is an isomorphism from E to E' that agrees with α on $F(a)$ and therefore with ϕ on F. ∎

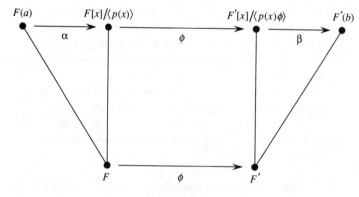

Figure 22.1

Corollary   *Splitting Fields Are Unique*
Let F be a field and let $f(x) \in F[x]$. Then any two splitting fields of $f(x)$ over F are isomorphic.

Proof.   Suppose E and E' are splitting fields of $f(x)$ over F. The result follows immediately from Theorem 22.4 by letting ϕ be the identity from F to F.   ∎

In light of the above corollary, we may refer to "the" splitting field of a polynomial without ambiguity.

Even though $x^6 - 2$ has a zero in $Q(\sqrt[6]{2})$, it does not split in $Q(\sqrt[6]{2})$. The splitting field is easy to obtain, however.

Example 8   The Splitting Field of $x^n - a$ over Q
Let a be a positive rational number and let ω be a primitive nth root of unity (see Example 2 of Chapter 18). Then each of

$$a^{1/n}, \ \omega a^{1/n}, \ \omega^2 a^{1/n}, \ \ldots, \ \omega^{n-1} a^{1/n}$$

is a zero of $x^n - a$ in $Q(\sqrt[n]{a}, \omega)$.   □

Zeros of an Irreducible Polynomial

Now that we know that every nonconstant polynomial over a field splits in some extension, we ask whether irreducible polynomials must split in some special way. Yes, they do. To discover how, we borrow something whose origins are in calculus.

> *DEFINITION*   Let $f(x) = a_n x^n + a_{n-1} x^{n-1} + \cdots + a_1 x + a_0$ belong to $F[x]$. The *derivative* of $f(x)$, denoted $f'(x)$, is the polynomial $na_n x^{n-1} + (n-1)a_{n-1} x^{n-2} + \cdots + a_1$ in $F[x]$.

Notice that our definition does not involve the notion of a limit. The standard rules for handling sums and products of functions in calculus carry over to arbitrary fields as well.

Lemma   *Let $f(x)$ and $g(x) \in F[x]$ and let $a \in F$. Then*

1. $(f(x) + g(x))' = f'(x) + g'(x)$
2. $(af(x))' = af'(x)$
3. $(f(x)g(x))' = f(x)g'(x) + g(x)f'(x)$.

Proof.   Parts 1 and 2 follow from straightforward applications of the definition. Using part 1 and induction on $\deg f(x)$, part 3 reduces to the special case that $f(x) = a_n x^n$. This also follows directly from the definition.   ∎

Before addressing the question of the nature of the zeros of an irreducible polynomial, we establish a general result concerning zeros of multiplicity greater than 1. Such zeros are called *multiple* zeros.

Theorem 22.5   Criterion for Multiple Zeros

A polynomial f(x) over a field F has a multiple zero if and only if f(x) and f'(x) have a common factor of positive degree in F[x].

Proof.   If a is a multiple zero of $f(x)$, then there is a $g(x)$ in some extension E of F such that $f(x) = (x - a)^2 g(x)$. Since $f'(x) = (x - a)^2 g'(x) + 2(x - a)g(x)$, we see that $f'(a) = 0$. Thus $x - a$ is a factor of both $f(x)$ and $f'(x)$ in the extension E of F. Now if $f(x)$ and $f'(x)$ have no common divisor of positive degree in $F[x]$, there are polynomials $h(x)$ and $k(x)$ in $F[x]$ such that $f(x)h(x) + f'(x)k(x) = 1$ (see exercise 40 of Chapter 18). Viewing $f(x)h(x) + f'(x)k(x)$ as an element of $E[x]$, we see also that $x - a$ is a factor of 1. Since this is nonsense, $f(x)$ and $f'(x)$ must have a common divisor of positive degree in $F[x]$.

Conversely, suppose $f(x)$ and $f'(x)$ have a common factor of positive degree, but $f(x)$ has no multiple zeros. We will obtain a contradiction. Let a be a zero of the common factor. Then a is a zero of $f(x)$ and $f'(x)$. Since a is a zero of $f(x)$ of multiplicity 1, there is a polynomial $q(x)$ such that $f(x) = (x - a)q(x)$ where $q(a) \neq 0$. Then $f'(x) = (x - a)q'(x) + q(x)$ and $0 = f'(a) = q(a)$, a contradiction.   ∎

Theorem 22.6   Zeros of an Irreducible

Let f(x) be an irreducible polynomial over a field F. If F has characteristic 0, then f(x) has no multiple zeros. If F has characteristic $p \neq 0$, then f(x) has a multiple zero only if it is of the form $f(x) = g(x^p)$ for some g(x) in F[x].

Proof.   If $f(x)$ has a multiple zero, then, by Theorem 22.5, $f(x)$ and $f'(x)$ have a common divisor of positive degree in $F[x]$. Since the only divisor of positive degree of $f(x)$ in $F[x]$ is $f(x)$ itself (up to associates) we see that $f(x)$ divides $f'(x)$. Because a polynomial over a field cannot divide a polynomial of smaller degree, we must have $f'(x) = 0$.

Now what does it mean to say that $f'(x) = 0$? If we write $f(x) = a_n x^n + a_{n-1}x^{n-1} + \cdots + a_1 x + a_0$, then $f'(x) = na_n x^{n-1} + (n - 1)a_{n-1}x^{n-2} + \cdots + a_1$. Thus, $f'(x) = 0$ only when $ka_k = 0$ for $k = 1, \ldots, n$.

So, when char $F = 0$, we have $f(x) = a_0$ and when char $F = p \neq 0$, we have $a_k = 0$ when p does not divide k. Thus, the only powers of x that appear in the sum $a_n x^n + \cdots + a_1 x + a_0$ are those of the form $x^{pi} = (x^p)^i$. It follows that $f(x) = g(x^p)$ for some $g(x) \in F[x]$. (For example, if $f(x) = x^{4p} + 3x^{2p} + x^p$, then $g(x) = x^4 + 3x^2 + x$.)   ∎

EXERCISES

I have yet to see any problem, however complicated, which, when you looked at it in the right way, did not become still more complicated.

Paul Anderson, New Scientist

1. Describe the elements of $Q(\sqrt[3]{5})$.
2. Show $Q(\sqrt{2}, \sqrt{3}) = Q(\sqrt{2} + \sqrt{3})$.
3. Find the splitting field of $x^3 - 1$ over Q. Express your answer in the form $Q(a)$.
4. Find the splitting field of $x^4 + 1$ over Q.
5. Find the splitting field of

$$x^4 + x^2 + 1 = (x^2 + x + 1)(x^2 - x + 1)$$

   over Q.
6. Let $a, b \in \mathbf{R}$ with $b \neq 0$. Show that $\mathbf{R}(a + bi) = \mathbf{C}$.
7. Find a polynomial $p(x)$ in $Q[x]$ so that $Q(\sqrt{1 + \sqrt{5}})$ is isomorphic to $Q[x]/\langle p(x)\rangle$.
8. Let $F = Z_2$ and let $f(x) = x^3 + x + 1 \in F[x]$. Suppose a is a zero of $f(x)$ in some extension of F. How many elements does $F(a)$ have? Express each member of $F(a)$ in terms of a. Write out a complete multiplication table for $F(a)$.
9. Let $F(a)$ be the field described in exercise 8. Express each of a^5, a^{-2} and a^{100} in the form $c_2a^2 + c_1a + c_0$.
10. Let $F(a)$ be the field described in exercise 8. Show that $a + 1$, $a^2 + 1$ and $a^2 + a + 1$ are all zeros of $x^3 + x^2 + 1$.
11. Describe the elements in $Q(\pi)$.
12. Let $F = Q(\pi^3)$. Find a basis for $F(\pi)$ over F.
13. Show that $Q(\sqrt{2})$ is not isomorphic to $Q(\sqrt{3})$.
14. Find all automorphisms of $Q(\sqrt[3]{5})$.
15. Let F be a field of characteristic p and let $f(x) = x^p - a \in F[x]$. Show that $f(x)$ is irreducible over F or $f(x)$ splits in F.
16. Let F be a field and let $f(x) \in F[x]$ be irreducible over F. Suppose a and b are zeros of $f(x)$ in some extension of F. Show that there is an isomorphism from $F(a)$ onto $F(b)$ that acts as the identity on F and carries a to b.
17. Find a, b, c in Q so that

$$(1 + \sqrt[3]{4})\,(2 - \sqrt[3]{2}) = a + b\sqrt[3]{2} + c\sqrt[3]{4}.$$

Note that such a, b, c exist since

$$(1 + \sqrt[3]{4})/(2 - \sqrt[3]{2}) \in Q(\sqrt[3]{2}) = \{a + b\sqrt[3]{2} + c\sqrt[3]{4} \mid a, b, c \in Q\}.$$

18. Express $(3 + 4\sqrt{2})^{-1}$ in the form $a + b\sqrt{2}$ where $a, b \in Q$.

19. Show that $Q(4 - i) = Q(1 + i)$ where $i = \sqrt{-1}$.

20. Let F be a field, and let a and b belong to F with $a \neq 0$. If c belongs to some extension of F, prove that $F(c) = F(ac + b)$. (F "absorbs" its own elements.)

21. Let $f(x) \in F[x]$ and let $a \in F$. Show that $f(x)$ and $f(x + a)$ have the same splitting field over F.

22. Recall, two polynomials $f(x)$ and $g(x)$ from $F[x]$ are said to be relatively prime if there is no polynomial of positive degree in $F[x]$ that divides both $f(x)$ and $g(x)$. Show that if $f(x)$ and $g(x)$ are relatively prime in $F[x]$, they are relatively prime in $K[x]$ where K is any extension of F.

23. Determine all of the subfields of $Q(\sqrt{2})$.

Leopold Kronecker

But the worst of it is that Kronecker uses his authority to proclaim that all those who up to now have labored to establish the theory of functions are sinners before the Lord ... such a verdict from a man whose eminent talent and distinguished performance in mathematical research I admire as sincerely and with as much pleasure as all his colleagues [is humiliating].

Karl Weierstrass in a letter to Sonja Kowalewski

LEOPOLD KRONECKER, the son of a businessman, was born on December 7, 1823 in Liegnitz, Prussia. As a young schoolboy, he excelled in all subjects and received special instruction from the great algebraist Kummer. Kronecker entered the University of Berlin in 1841 and completed his Ph.D. dissertation in 1845 on the units in a certain ring.

Kronecker devoted the years from 1845 to 1853 to business affairs, relegating mathematics to a hobby. Thereafter, being well-off financially, he spent most of his time doing research in algebra and number theory. Kronecker was one of the early advocates of the abstract approach to algebra. He innovatively applied rings and fields in his investigations of algebraic numbers and was the first mathematician to master Galois's theory of fields.

Kronecker advocated constructive methods for all proofs and definitions. He believed all mathematics should be based on relationships among integers. He went so far as to say to Lindemann, who proved that π is transcendental, that irrational numbers do not exist. His most famous remark on the matter is "God made the integers, all the rest is the work of man." And Kronecker believed all other numbers, being the work of man, were to be avoided. Although Kronecker's mathematical philosophy gained no supporters among his contemporaries, his beliefs about constructive proofs and definitions were advocated many years later by the so-called intuitionists. Kronecker died on December 29, 1891, at the age of sixty-eight.

Algebraic Extensions

Banach once told me, "Good mathematicians see analogies between theorems or theories, the very best ones see analogies between analogies."

S. M. Ulam, *Adventures of a Mathematician*

Characterization of Extensions

In Chapter 22, we saw that every element in the field $Q(\sqrt{2})$ has the particularly simple form $a + b\sqrt{2}$ where a and b are rational. On the other hand, the elements of $Q(\pi)$ have the more complicated form

$$(a_n\pi^n + a_{n-1}\pi^{n-1} + \cdots + a_1\pi + a_0)/(b_m\pi^m + b_{m-1}\pi^{m-1} + \cdots + b_1\pi + b_0)$$

where the a's and b's are rational. The fields of the first type have a great deal of structure. This structure is the subject of this chapter.

> *DEFINITION* Types of Extensions
>
> Let E be an extension field of a field F and let $a \in E$. We call a *algebraic over F* if a is the zero of some polynomial in $F[x]$. If a is not algebraic over F, it is called *transcendental over F*. An extension E of F is called an *algebraic* extension of F if every element of E is algebraic over F. If E is not an algebraic extension of F, it is called a *transcendental* extension of F. An extension of F of the form $F(a)$ is called a *simple* extension of F.

Leonhard Euler used the term *transcendental* for numbers that are not algebraic because "they transcended the power of algebraic methods." Although

Euler made this distinction in 1744, it wasn't until 1844 that the existence of transcendental numbers over Q was proved by Joseph Liouville. Charles Hermite proved e is transcendental over Q in 1873, and Lindemann showed that π is transcendental over Q in 1882. To this day, it is not known whether $\pi + e$ is transcendental over Q. With a precise definition of "almost all" it can be shown that almost all real numbers are transcendental over Q.

Theorem 23.1 shows why we make the distinction between elements that are algebraic over a field and elements that are transcendental over a field. Recall, $F(x)$ is the field of quotients of $F[x]$; that is,

$$F(x) = \{f(x)/g(x) \mid f(x), g(x) \in F[x], g(x) \neq 0\}.$$

Theorem 23.1   Characterization of Extensions
Let E be an extension field of the field F and let $a \in E$. If a is transcendental over F, then $F(a) \approx F(x)$. If a is algebraic over F, then $F(a) \approx F[x]/\langle p(x)\rangle$ where $p(x)$ is a polynomial in $F[x]$ of minimum degree such that $p(a) = 0$. Moreover, $p(x)$ is irreducible over F.

Proof.   Consider the homomorphism $\phi:F[x] \to F(a)$ given by $f(x)\phi = f(a)$. If a is transcendental over F, then Ker $\phi = \{0\}$, and so we may extend ϕ to an isomorphism $\overline{\phi}:F(x) \to F(a)$ by defining $[f(x)/g(x)]\overline{\phi} = f(a)/g(a)$.

If a is algebraic over F, then Ker $\phi \neq \{0\}$; and by the corollary to Theorem 18.3, there is a polynomial $p(x)$ in $F[x]$ such that Ker $\phi = \langle p(x)\rangle$ and $p(x)$ has minimum degree among all nonzero elements of Ker ϕ. Thus, $p(a) = 0$ and, since $p(x)$ is a polynomial of minimum degree with this property, it is irreducible over F.   ∎

The proof of Theorem 23.1 can readily be adapted to yield the next two results also. The details are left to the reader (see exercise 1).

Theorem 23.2   Uniqueness Property
If a is algebraic over a field F, then there is a unique monic irreducible polynomial $p(x)$ in $F[x]$ such that $p(a) = 0$.

The polynomial with the property specified in Theorem 23.2 is called the *minimal polynomial for a over F.*

Theorem 23.3   Divisibility Property
Let a be algebraic over F, and let $p(x)$ be the minimal polynomial for a over F. If $f(x) \in F[x]$ and $f(a) = 0$, then $p(x)$ divides $f(x)$ in $F[x]$.

If E is an extension field of F we may view E as a vector space over F (that is, the elements of E are the vectors and the elements of F are the scalars). We are then able to use such notions as dimension and basis in our discussion.

Finite Extensions

DEFINITION Degree of an Extension

Let E be a field extension of a field F. We say E *has degree n over F* and write $[E:F] = n$, if E has dimension n as a vector space over F. If $[E:F]$ is finite, E is called a *finite extension* of F; otherwise, we say E is an *infinite extension* of F.

Figure 23.1 illustrates a convenient method of depicting the degree of a field extension over a field.

Example 1 The field of complex numbers has degree 2 over the reals since $\{1, i\}$ is a basis. The field of complex numbers is an infinite extension of the rationals. □

Example 2 If a is algebraic over F and its minimal polynomial over F has degree n, then, by Theorem 22.3, we know $\{1, a, \ldots, a^{n-1}\}$ is a basis for $F(a)$ over F; and, therefore, $[F(a):F] = n$. In this case, we say a has *degree n over F*. □

Theorem 23.4 Finite Implies Algebraic
If E is a finite extension of F, then E is an algebraic extension of F.

Proof. Suppose $[E:F] = n$ and $a \in E$. Then the set $\{1, a, \ldots, a^n\}$ is linearly dependent over F; that is, there are elements $c_0, c_1, \ldots, c_n$ in F, not all zero, such that

$$c_n a^n + c_{n-1} a^{n-1} + \cdots + c_1 a + c_0 = 0.$$

Clearly, then, a is a zero of the nonzero polynomial

$$f(x) = c_n x^n + c_{n-1} x^{n-1} + \cdots + c_1 x + c_0. \qquad \blacksquare$$

The converse of Theorem 23.4 is not true. For otherwise, the degrees of the elements of every algebraic extension of E over F would be bounded. But

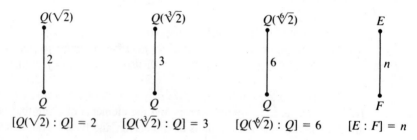

Figure 23.1

$Q(\sqrt{2}, \sqrt[3]{2}, \sqrt[4]{2}, \ldots)$ is an algebraic extension of Q that contains elements of every degree over Q.

The next theorem is the field theory counterpart of Lagrange's Theorem for finite groups. Like all counting theorems, it has far-reaching consequences.

Theorem 23.5 $[K:F] = [K:E][E:F]$
Let K be a finite extension field of the field E and let E be a finite extension field of the field F. Then K is a finite extension field of F and $[K:F] = [K:E][E:F]$.

Proof. Let $X = \{x_1, x_2, \ldots, x_n\}$ be a basis for K over E, and $Y = \{y_1, y_2, \ldots, y_m\}$ a basis for E over F. It suffices to prove that

$$YX = \{y_j x_i \mid 1 \leq j \leq m, \, 1 \leq i \leq n\}$$

is a basis for K over F. To do this, let $a \in K$. Then there are elements b_1, $b_2, \ldots, b_n \in E$ such that

$$a = b_1 x_1 + b_2 x_2 + \cdots + b_n x_n.$$

And, for each $i = 1, \ldots, n$, there are elements $c_{i1}, c_{i2}, \ldots, c_{im} \in F$ such that

$$b_i = c_{i1} y_1 + c_{i2} y_2 + \cdots + c_{im} y_m.$$

Thus,

$$a = \sum_{i=1}^{n} b_i x_i = \sum_{i=1}^{n} \left(\sum_{j=1}^{m} c_{ij} y_j \right) x_i = \sum_{i,j} c_{ij}(y_j x_i).$$

This proves that YX spans K over F.

Now suppose there are elements c_{ij} in F such that

$$0 = \sum_{i,j} c_{ij}(y_j x_i) = \sum_i \sum_j (c_{ij} y_j) x_i.$$

Then, since each $c_{ij} y_j \in E$ and X is a basis for K over E, we have

$$\sum_j c_{ij} y_j = 0$$

for each i. But each $c_{ij} \in F$ and Y is a basis for E over F so each $c_{ij} = 0$. This proves that the set YX is linearly independent over F. ∎

Using the fact that for any field extension L of a field J, $[L:J] = n$ if and only if L is isomorphic to J^n as vector spaces (see exercise 28), we may give a concise conceptual proof of Theorem 23.5 as follows. Let $[K:E] = m$ and $[E:F] = n$. Then $K \approx E^m$ and $E \approx F^n$ so that $K \approx E^m \approx (F^n)^m \approx F^{nm}$. Thus, $[K:F] = nm$.

The content of Theorem 23.5 can be pictured as in Figure 23.2. Examples 3, 4, and 5 show how Theorem 23.5 is often utilized.

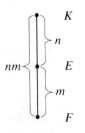

$$[K : F] = [K : E][E : F]$$

Figure 23.2

Example 3 Consider $Q(\sqrt{2}, \sqrt[3]{2})$. Then $[Q(\sqrt{2}, \sqrt[3]{2}):Q] = 6$. For, clearly, $Q(\sqrt{2}, \sqrt[3]{2}) \subseteq Q(\sqrt[6]{2})$ so that

$$[Q(\sqrt{2}, \sqrt[3]{2}):Q] \leq [Q(\sqrt[6]{2}):Q] = 6.$$

On the other hand, since

$$Q(\sqrt{2}) \subseteq Q(\sqrt{2}, \sqrt[3]{2}) \quad \text{and} \quad Q(\sqrt[3]{2}) \subseteq Q(\sqrt{2}, \sqrt[3]{2}),$$

we have $2 = [Q(\sqrt{2}):Q]$ divides $[Q(\sqrt{2}, \sqrt[3]{2}):Q]$ and $3 = [Q(\sqrt[3]{2}:Q]$ divides $[Q(\sqrt{2}, \sqrt[3]{2}):Q]$. It now follows that $Q(\sqrt{2}, \sqrt[3]{2}) = Q(\sqrt[6]{2})$. (See Figure 23.3.) $\qquad\square$

Example 4 Since $\{1, \sqrt{3}\}$ is a basis for $Q(\sqrt{3}, \sqrt{5})$ over $Q(\sqrt{5})$ and $\{1, \sqrt{5}\}$ is a basis for $Q(\sqrt{5})$ over Q, the proof of Theorem 23.5 shows that $\{1, \sqrt{3}, \sqrt{5}, \sqrt{15}\}$ is a basis for $Q(\sqrt{3}, \sqrt{5})$ over Q. (See Figure 23.4.) $\qquad\square$

Example 5 Consider $Q(\sqrt{3}, \sqrt{5})$. We claim $Q(\sqrt{3}, \sqrt{5}) = Q(\sqrt{3} + \sqrt{5})$. The inclusion $Q(\sqrt{3} + \sqrt{5}) \subseteq Q(\sqrt{3}, \sqrt{5})$ is clear. Now note that since $(\sqrt{3} + \sqrt{5})^3 = 18\sqrt{3} + 14\sqrt{5}$ and $-14\sqrt{3} - 14\sqrt{5}$, both belong to

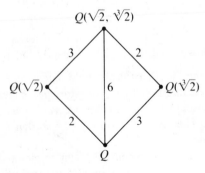

Figure 23.3

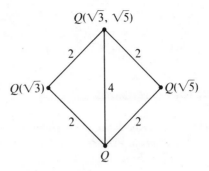

Figure 23.4

$Q(\sqrt{3} + \sqrt{5})$, so does their sum $4\sqrt{3}$. Therefore, $1/4(4\sqrt{3}) = \sqrt{3} \in Q(\sqrt{3} + \sqrt{5})$. Of course, $\sqrt{5} = \sqrt{3} + \sqrt{5} - \sqrt{3} \in Q(\sqrt{3} + \sqrt{5})$ as well. Thus, $Q(\sqrt{3}, \sqrt{5}) \subseteq Q(\sqrt{3} + \sqrt{5})$. $\qquad\square$

The preceding example shows that an extension obtained by adjoining two elements to a field can sometimes be obtained by adjoining a single element to the field. Our next theorem shows that under certain conditions, that can always be done.

Theorem 23.6 *Primitive Element Theorem (Steinitz, 1910)*
If F is a field of characteristic 0, and a and b are algebraic over F, then there is an element c in F(a, b) such that F(a, b) = F(c).

Proof. Let $p(x)$ and $q(x)$ be the minimum polynomials over F for a and b, respectively. In some extension K of F, let $a = a_1, a_2, \ldots, a_m$ and $b = b_1, b_2, \ldots, b_n$ be the distinct zeros of $p(x)$ and $q(x)$, respectively. Among the infinitely many elements of F, choose an element d not equal to $(a - a_i)/(b - b_j)$ for all i and all $j > 1$. In particular, $a_i \neq a + d(b - b_j)$ for $j > 1$.

We shall show that $c = a + db$ has the property that $F(a, b) = F(c)$. Certainly, $F(c) \subseteq F(a, b)$. To verify that $F(a, b) \subseteq F(c)$, it suffices to prove that $b \in F(c)$ for then b, c, and d belong to $F(c)$ and $a = c - bd$. Consider the polynomials $q(x)$ and $r(x) = p(c - dx)$ over $F(c)$. Since both $q(b) = 0$ and $r(b) = p(c - db) = p(a) = 0$, both $q(x)$ and $r(x)$ are divisible by the minimum polynomial $s(x)$ for b over $F(c)$ (see Theorem 23.3). Because $s(x) \in F(c)[x]$, we may complete the proof by proving $s(x) = x - b$. Since $s(x)$ is a common divisor of $q(x)$ and $r(x)$, the only possible zeros of $s(x)$ in K are the zeros of $q(x)$ that are also zeros of $r(x)$. But $r(b_j) = p(c - db_j) = p(a + db - db_j) = p(a + d(b - b_j))$ and d was chosen so that $a + d(b - b_j) \neq a_i$ for $j > 1$. It follows that b is the only zero of $s(x)$ in $K[x]$ and, therefore, $s(x) = (x - b)^u$. Since $s(x)$ is irreducible and F has characteristic 0, Theorem 22.6 guarantees that $u = 1$. ∎

In the terminology introduced earlier, Theorem 23.6 says that any finite extension of a field of characteristic 0 is a simple extension. An element a with the property that $E = F(a)$ is called a *primitive element* of E.

Properties of Algebraic Extensions

Theorem 23.7 *Algebraic over Algebraic Is Algebraic*
If K is an algebraic extension of E and E is an algebraic extension of F, then K is an algebraic extension of F.

Proof. Let $a \in K$. It suffices to show that a belongs to some finite extension of F. Since a is algebraic over E, we know a is the zero of some irreducible polynomial in $E[x]$, say, $p(x) = b_n x^n + \cdots + b_0$. Now we construct a tower of field extensions of F as follows:

$$F_0 = F(b_0),$$
$$F_1 = F_0(b_1), \ldots, F_n = F_{n-1}(b_n).$$

In particular,

$$F_n = F(b_0, b_1, \ldots, b_n)$$

so that $p(x) \in F_n[x]$. Thus, $[F_n(a):F_n] = n$; and because each b_i is algebraic over F, we know each $[F_{i+1}:F_i]$ is finite. So,

$$[F_n(a):F] = [F_n(a):F_n][F_n:F_{n-1}] \cdots [F_1:F_0][F_0:F]$$

is finite. (See Figure 23.5.) ∎

Corollary *Subfield of Algebraic Elements*
Let E be an extension field of the field F. Then the set of all elements of E that are algebraic over F is a subfield of E.

Proof. Suppose $a, b \in E$ are algebraic over F and $b \neq 0$. To show that $a + b$, $a - b$, ab, and a/b are algebraic over F, it suffices to show that $[F(a, b):F]$ is finite, since each of these four elements belongs to $F(a, b)$. But note that

$$[F(a, b):F] = [F(a, b):F(b)][F(b):F].$$

Also, since a is algebraic over F, it is certainly algebraic over $F(b)$. Thus, both $[F(a, b):F(b)]$ and $[F(b):F]$ are finite. ∎

For any extension E of a field F, the subfield of E of the elements that are algebraic over F is called the *algebraic closure of F over E*.

One might wonder if there is such a thing as a maximal algebraic extension of a field F; that is, whether there is an algebraic extension E of F that has no proper algebraic extensions. For such an E to exist, it is necessary that every

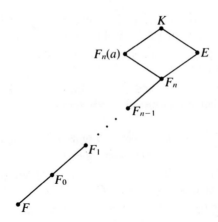

Figure 23.5

polynomial in $E[x]$ split in E. Otherwise, it follows from Kronecker's Theorem that E would have a proper algebraic extension. This condition is also sufficient. If every member of $E[x]$ splits in E, and K were an algebraic extension of E, then every member of K is a zero of some element of $E[x]$. But the zeros of elements of $E[x]$ are in E. A field that has no proper algebraic extension is called *algebraically closed*. In 1910, Ernst Steinitz proved that every field has a unique (up to isomorphism) algebraic extension that is algebraically closed. A proof of this result requires a sophisticated set theory background. The reader can find a proof in [1, p. 344–345].

In 1799, Gauss, at the age of 22, proved that **C** is algebraically closed. That fact was considered so important at the time that it was called "The Fundamental Theorem of Algebra." Over a fifty-year period, Gauss found three additional proofs of the Fundamental Theorem. Today more than 100 proofs exist. In view of the ascendancy of abstract algebra in this century, a more appropriate phrase for Gauss's result would be "The Fundamental Theorem of Classical Algebra."

EXERCISES

The more we do, the more we can do.
William Hazlitt

1. Prove Theorem 23.2 and Theorem 23.3.
2. Prove that $Q(\sqrt{2}, \sqrt[3]{2}, \sqrt[4]{2}, \ldots)$ is an algebraic extension of Q but not a finite extension of Q.
3. Let E be the algebraic closure of F. Show that every polynomial in $F[x]$ splits in E.

4. Let E be an algebraic extension of F. If every polynomial in $F[x]$ splits in E, show that E is algebraically closed.

5. Suppose F is a field and every irreducible polynomial in $F[x]$ is linear. Show that F is algebraically closed.

6. Suppose $f(x)$ and $g(x)$ are irreducible over F and deg $f(x)$ and deg $g(x)$ are relatively prime. If a is a zero of $f(x)$ in some extension of F, show that $g(x)$ is irreducible over $F(a)$.

7. Let a and b belong to Q with $b \neq 0$. Show that $Q(\sqrt{a}) = Q(\sqrt{b})$ if and only if there exists some $c \in Q$ such that $a = bc^2$.

8. Find the degree and a basis for $Q(\sqrt{3} + \sqrt{5})$ over $Q(\sqrt{15})$. Find the degree and a basis for $Q(\sqrt{2}, \sqrt[3]{2}, \sqrt[4]{2})$ over Q.

9. Suppose E is an extension of F of prime degree. Show that for every a in E, $F(a) = F$ or $F(a) = E$.

10. Suppose α is algebraic over Q. Show that $\sqrt{\alpha}$ is algebraic over Q.

11. Suppose E is an extension of F and $a, b \in E$. If a is algebraic over F of degree m and b is algebraic over F of degree n where m and n are relatively prime, show that $[F(a, b):F] = mn$.

12. Show, by example, that if a is algebraic over F of degree m and b is algebraic over F of degree n, then $[F(a, b):F]$ need not be mn.

13. Let K be a field extension of F and let $a \in K$. Show that $[F(a):F(a^3)] = 1$ or 3.

14. Find the minimal polynomial for $\sqrt{-3} + \sqrt{2}$ over Q.

15. Let K be an extension of F. Suppose E_1 and E_2 are contained in K and are extensions of F. If $[E_1:F]$ and $[E_2:F]$ are both prime, show that $E_1 = E_2$ or $E_1 \cap E_2 = F$.

16. Find the minimal polynomial for $\sqrt[3]{2} + \sqrt[3]{4}$ over Q.

17. Let E be a finite extension of $\mathbf{R}$. Prove that $E = \mathbf{C}$ or $E = \mathbf{R}$.

18. Suppose $[E:Q] = 2$. Show that there is an integer d such that $E = Q(\sqrt{d})$ and d is not divisible by the square of any prime.

19. Suppose $p(x) \in F[x]$ and E is a finite extension of F. If $p(x)$ is irreducible over F and deg $p(x)$ and $[E:F]$ are relatively prime, show that $p(x)$ is irreducible over E.

20. Let E be a field extension of F. Show that $[E:F]$ is finite if and only if $E = F(a_1, a_2, \ldots, a_n)$ where $a_1, a_2, \ldots, a_n$ are algebraic over F.

21. If α and β are transcendental over Q, show that either $\alpha\beta$ or $\alpha + \beta$ is also transcendental over Q.

22. Let $f(x) \in F[x]$. If a belongs to some extension of F and $f(a)$ is algebraic over F, prove that a is algebraic over F.

23. Let $f(x) = ax^2 + bx + c \in Q[x]$. Find a primitive element for the splitting field for $f(x)$ over Q.

24. Find the splitting field for $x^4 - x^2 - 2$ over Z_3.

25. Let $f(x) \in F[x]$. If $\deg f(x) = 2$ and a is a zero of $f(x)$ in some extension of F, prove that $F(a)$ is the splitting field for $f(x)$ over F.

26. Find the splitting field for $x^3 + x + 1$ over Z_2. Express $x^3 + x + 1$ as a product of linear factors over the splitting field.

27. If F is a field and the multiplicative group of nonzero elements of F is cyclic, prove that F is finite.

28. Prove that if K is a field extension of F, then $[K{:}F] = n$ if and only if K is isomorphic to F^n as vector spaces. (See exercise 21 of Chapter 21 for the appropriate definition.)

REFERENCE

1. E. A. Walker, *Introduction to Abstract Algebra,* New York: Random House, 1987.

SUGGESTED READINGS

Charles J. Parry and David Perin, "Equivalence of Extension Fields," *Mathematics Magazine* 50 (1977): 36–42.

Let K be a field, a and b nonzero elements of K, and n a positive integer relatively prime to the characteristic of K. This paper gives necessary and sufficient conditions for $K(a^{1/n}) = K(b^{1/n})$.

R. L. Roth, "On Extensions of Q by Square Roots," *American Mathematical Monthly* 78 (1971): 392–393.

In this paper, it is proved that if $p_1, p_2, \ldots, p_n$ are distinct primes, then $[Q(\sqrt{p_1}, \sqrt{p_2}, \ldots, \sqrt{p_n}){:}Q] = 2^n$.

Paul B. Yale, "Automorphisms of the Complex Numbers," *Mathematics Magazine* 39 (1966): 135–141.

This award-winning expository paper is devoted to various results on automorphisms of the complex numbers.

Irving Kaplansky

He got to the top of the heap by being a first-rate
doer and expositor of algebra.

Paul R. Halmos, *I Have a Photographic Memory*

IRVING KAPLANSKY was born on March 22, 1917, in Toronto, Canada, a few
years after his parents emigrated from Poland. As a young boy, he demonstrated a
talent for music and took piano lessons for eleven years. Although his parents thought
he would pursue a career in music, Kaplansky knew early on that mathematics was
what he wanted to do. (To this day, however, music remains a hobby.) As an undergraduate at the University of Toronto, Kaplansky was a member of the winning
team of the first William Lowell Putnam competition, a mathematical contest for
United States and Canadian college students. Kaplansky received a B.A. degree from
Toronto in 1938 and an M.A. in 1939. In 1939 he entered Harvard University for his
doctorate as the first recipient of a Putnam Fellowship. After receiving his Ph.D. from
Harvard in 1941, Kaplansky stayed on as Benjamin Pierce instructor until 1944. After
one year at Columbia University, he went to the University of Chicago, where he
remained until his retirement in 1984. He then became the first director of the Mathematical Sciences Research Institute at the University of California, Berkeley.

Kaplansky is the author of numerous books and research papers. His interests are
broad, including areas such as ring theory, group theory, field theory, Galois theory,
ergodic theory, algebras, metric spaces, number theory, statistics, and probability. In
much of his mathematics there is a recurring theme that might be described as "algebra
with an infinite flavor." He is also regarded as a first-rate mathematical stylist.

Among the many honors received by Kaplansky are a Guggenheim Fellowship,
election to both the National Academy of Sciences and the American Academy of
Arts and Sciences, appointment as president of the American Mathematical Society,
and honorary degrees from the University of Waterloo and Queens University. Kaplansky also received the University of Chicago Quantrell Prize for excellence in
undergraduate teaching.

Finite Fields

24

This theory [of finite fields] is of considerable interest in its own right and it provides a particularly beautiful example of how the general theory of the preceding chapters fits together to provide a rather detailed description of all finite fields.

Richard A. Dean, *Elements of Abstract Algebra*

Classification of Finite Fields

In this, our final chapter on field theory, we take up one of the most beautiful and important areas of abstract algebra—finite fields. Finite fields were first introduced by Galois in 1830 in his proof of the unsolvability of the general quintic equation. When Cayley invented matrices a few decades later, it was natural to investigate groups of matrices over finite fields. To this day, matrix groups over finite fields are among the most important classes of groups. In the past thirty years, there have been important applications of finite fields to the design of computers and to the problem of reliably transmitting information. But besides the many uses of finite fields in pure and applied mathematics, there is yet another good reason for studying them. They are just plain fun!

The most striking fact about finite fields is the restricted nature of their order and structure. We have already seen that every finite field has prime-power order (exercise 41 of Chapter 15). A converse of sorts is also true.

Theorem 24.1. *Classification of Finite Fields*
For each prime p and each positive integer n there is, up to isomorphism, a unique finite field of order p^n.

Proof. Consider the splitting field E of $f(x) = x^{p^n} - x$ over Z_p. We will show that $|E| = p^n$. Since $f(x)$ splits in E, we know that $f(x)$ has exactly

p^n zeros in E, counting multiplicity. Note that $f(x) = x(x^{p^n-1} - 1)$ so that 0 is zero of $f(x)$ of multiplicity 1. Now, if a were some zero of $f(x)$ of multiplicity greater than 1, then $f(x)$ could be written in the form $x(x - a)g(x)$, where $g(x) \in E[x]$ and $g(a) = 0$. But long division shows that

$$g(x) = x^{p^n-2} + ax^{p^n-3} + a^2x^{p^n-4} + \cdots + a^{p^n-2}$$

so that

$$0 = g(a) = (p^n - 1) a^{p^n-2} = -a^{p^n-2}$$

since E has characteristic p. This implies that $a = 0$ and gives us a contradiction. Thus, $f(x)$ has p^n distinct zeros in E. On the other hand, the set of zeros of $f(x)$ in E is closed under addition, subtraction, multiplication, and division by nonzero elements (see exercise 27) so that the set of zeros $f(x)$ is itself a field extension of Z_p in which $f(x)$ splits. Thus, the set of zeros of $f(x)$ is E and, therefore, $|E| = p^n$.

To show that there is a unique field for each prime-power, suppose K is any field of order p^n. Then K has a subfield isomorphic to Z_p and, because the nonzero elements of K form a multiplicative group of order $p^n - 1$, every element of K is a zero of $f(x) = x^{p^n} - x$ (see exercise 18). So, K must be a splitting field for $f(x)$ over Z_p. By the corollary to Theorem 22.4, there is only one such field up to isomorphism. ∎

The existence portion of Theorem 24.1 appeared in the works of Galois and Gauss in the first third of the nineteenth century. Rigorous proofs were given by Dedekind in 1857, and by Jordan in 1870 in his classic book on group theory. The uniqueness half of the theorem was proved by E. H. Moore in an 1893 paper concerning finite simple groups. The mathematics historian E. T. Bell once said that this paper by Moore marked the beginning of abstract algebra in America.

Because there is only one field for each prime-power p^n, we may unambiguously denote it by GF(p^n), in honor of Galois, and call it the *Galois field of order* p^n.

Structure of Finite Fields

The next theorem tells us the additive and multiplicative structure of a field of order p^n.

> ***Theorem 24.2*** *Structure of Finite Fields*
> *As a group under addition,* GF(p^n) *is isomorphic to*
>
> $$\underbrace{Z_p \oplus Z_p \oplus \cdots \oplus Z_p}_{n \text{ factors}};$$

as a group under multiplication, the set of nonzero elements of $GF(p^n)$ is isomorphic to Z_{p^n-1} (and is, therefore, cyclic).

Proof. Since $GF(p^n)$ has characteristic p, we know $px = 0$ for all x in $GF(p^n)$. Thus, every nonzero element of $GF(p^n)$ has additive order p. Clearly then, under addition, $GF(p^n)$ is isomorphic to a direct product of n copies of Z_p.

To see that the multiplicative group $GF(p^n)^\#$ of nonzero elements of $GF(p^n)$ is cyclic, we first note, by exercise 10 of Chapter 12 that it is isomorphic to a direct product of the form $Z_{n_1} \oplus Z_{n_2} \oplus \cdots \oplus Z_{n_k}$ where each n_{i+1} divides n_i. So, for any element $a = (a_1, a_2, \ldots, a_k)$ in this product, we have

$$a^{n_1} = (n_1 a_1, n_1 a_2, \ldots, n_1 a_k) = (0, 0, \ldots, 0).$$

(Remember, the operation is componentwise addition.) Thus, the polynomial $x^{n_1} - 1$ has $p^n - 1$ zeros in $GF(p^n)$. Since the number of zeros of a polynomial over a field cannot exceed the degree of the polynomial (Corollary 3 of Theorem 18.2), we know $p^n - 1 \leq n_1$. On the other hand, since $GF(p^n)^\#$ has a subgroup isomorphic to Z_{n_1}, we also have $n_1 \leq p^n - 1$. It follows then that $GF(p^n)^\#$ is isomorphic to Z_{p^n-1}. ∎

Corollary 1 *GF(p^n) Contains an Element of Degree n*
Let a be a generator of the group of nonzero elements of $GF(p^n)$ under multiplication. Then a is algebraic over $GF(p)$ of degree n.

Proof. Let $f(x) = x^{p^n} - x$ and let $F = GF(p)$. Since $f(x)$ can be written as a product of irreducibles over F and, by the proof of Theorem 24.1, $f(a) = 0$, we know a is a zero of some irreducible over F, say, $g(x)$. Then, $F(a) \approx F[x]/\langle g(x) \rangle$, and since $F(a) = GF(p^n)$ has degree n over F, $g(x)$ has degree n. ∎

Corollary 1, together with Example 2 of Chapter 23, gives us the following useful and aesthetically appealing formula.

Corollary 2 $[GF(p^n): GF(p)] = n$

Example 1 Let's examine the field GF(16) in detail. Since $x^4 + x + 1$ is irreducible over Z_2 we know

$$GF(16) \approx \{ax^3 + bx^2 + cx + d + \langle x^4 + x + 1 \rangle \mid a, b, c, d \in Z_2\}.$$

Thus, we may think of GF(16) as the set

$$F = \{ax^3 + bx^2 + cx + d \mid a, b, c, d \in Z_2\}$$

where addition is done as in $Z_2[x]$, but multiplication is done modulo $x^4 + x + 1$.

For example,

$$(x^3 + x^2 + x + 1)(x^3 + x) = x^3 + x^2$$

since the remainder upon dividing

$$(x^3 + x^2 + x + 1)(x^3 + x) = x^6 + x^5 + x^2 + x$$

by $x^4 + x + 1$ in Z_2 is $x^3 + x^2$. An easier way to perform the same calculation is to observe that in this context $x^4 + x + 1$ *is* 0 so

$$x^4 = -x - 1 = x + 1,$$
$$x^5 = x^2 + x,$$

and

$$x^6 = x^3 + x^2.$$

Thus,

$$x^6 + x^5 + x^2 + x = (x^3 + x^2) + (x^2 + x) + x^2 + x = x^3 + x^2.$$

Another way to simplify the multiplication process is to make use of the fact that the nonzero elements of GF(16) form a cyclic group of order 15. To take advantage of this, we must first find a generator of this group. Since any element $F^\#$ must have multiplicative order that divides 15, all we need do is find an element α in $F^\#$ so that $\alpha^3 \neq 1$ and $\alpha^5 \neq 1$. Obviously, x has these properties. So, we may think of GF(16) as the set $\{0, 1, x, x^2, \ldots, x^{14}\}$ where $x^{15} = 1$. This makes multiplication in F trivial, but, unfortunately, it makes addition more difficult. For example, $x^{10} \cdot x^7 = x^{17} = x^2$, but what is $x^{10} + x^7$? So, we face a dilemma. If we write the elements of $F^\#$ in the additive form $ax^3 + bx^2 + cx + d,$ then addition is easy and multiplication is hard. On the other hand, if we write the elements of $F^\#$ in the multiplicative form x^i, then multiplication is easy and addition is hard. Can we have the best of both? Yes, we can. All we need do is use the relation $x^4 = x + 1$ to make a two-way conversion table, as in Table 24.1.

So, we see from Table 24.1 that

$$x^{10} + x^7 = (x^2 + x + 1) + (x^3 + x + 1)$$
$$= x^3 + x^2 = x^6$$

and

$$(x^3 + x^2 + 1)(x^3 + x^2 + x + 1) = x^{13} \cdot x^{12}$$
$$= x^{25} = x^{10} = x^2 + x + 1. \qquad \square$$

Don't be misled by the preceding example into believing that the element x is always a generator for the cyclic multiplicative group of nonzero elements. It is not. (See exercise 10.) Although any two irreducible polynomials of the same degree over $Z_p[x]$ yield isomorphic fields, some are better than others for computation purposes.

Table 24.1 Conversion Table for Addition and Multiplication in GF(16)

Multiplicative form to additive form		Additive form to multiplicative form	
1	1	1	1
x	x	x	x
x^2	x^2	$x + 1$	x^4
x^3	x^3	x^2	x^2
x^4	$x + 1$	$x^2 + x$	x^5
x^5	$x^2 + x$	$x^2 + 1$	x^8
x^6	$x^3 + x^2$	$x^2 + x + 1$	x^{10}
x^7	$x^3 + x + 1$	x^3	x^3
x^8	$x^2 + 1$	$x^3 + x^2$	x^6
x^9	$x^3 + x$	$x^3 + x$	x^9
x^{10}	$x^2 + x + 1$	$x^3 + 1$	x^{14}
x^{11}	$x^3 + x^2 + x$	$x^3 + x^2 + x$	x^{11}
x^{12}	$x^3 + x^2 + x + 1$	$x^3 + x^2 + 1$	x^{13}
x^{13}	$x^3 + x^2 + 1$	$x^3 + x + 1$	x^7
x^{14}	$x^3 + 1$	$x^3 + x^2 + x + 1$	x^{12}

Subfields of a Finite Field

Theorem 24.1 gave us a complete description of all finite fields. The following theorem gives us a complete description of all the subfields of a finite field. Notice the close analogy between this theorem and Theorem 4.3, which describes the subgroups of a finite cyclic group.

Theorem 24.3 *Subfields of a Finite Field*
For each divisor m of n, $GF(p^n)$ has a unique subfield of order p^m. Moreover, these are the only subfields of $GF(p^n)$.

Proof. To show the existence portion of the theorem, suppose m divides n. Then, since

$$p^n - 1 = (p^m - 1)(p^{n-m} + p^{n-2m} + \cdots + p^m + 1),$$

we see that $p^m - 1$ divides $p^n - 1$. This implies that $x^{p^m-1} - 1$ divides $x^{p^n-1} - 1$ in $Z_p[x]$. Thus, every zero of $x(x^{p^m-1} - 1)$ is also a zero of $x(x^{p^n-1} - 1)$. But the proof of Theorem 24.1 shows that the set of zeros of $x(x^{p^m-1} - 1)$ is $GF(p^m)$ and the set of zeros of $x(x^{p^n-1} - 1)$ in $GF(p^n)$ is $GF(p^n)$. Hence, $GF(p^m)$ is a subfield of $GF(p^n)$ whenever m divides n. The uniqueness portion of the theorem follows from the observation that if $GF(p^n)$ had two distinct subfields of order p^m, then the polynomial $x^{p^m} - x$ would have more than p^m zeros in $GF(p^n)$. This contradicts Corollary 3 of Theorem 18.2.

Finally suppose F is a subfield of $GF(p^n)$. Then F is isomorphic to $GF(p^m)$ for some m and, by Theorem 23.3,

$$
\begin{aligned}
n &= [GF(p^n): GF(p)] \\
&= [GF(p^n): GF(p^m)][GF(p^m): GF(p)] \\
&= [GF(p^n): GF(p^m)]m.
\end{aligned}
$$

Thus, m divides n. ∎

Theorems 24.2 and 24.3, together with Theorem 4.3, make the task of finding the subfields of a field a simple exercise in arithmetic.

Example 2 Let F be the field of order 16 given in Example 1. Then there are exactly three subfields of F and their orders are 2, 4, and 16. Obviously, the subfield of order 2 is $\{0, 1\}$ and the subfield of order 16 if F itself. To find the subfield of order 4, we merely observe that the three nonzero elements of this subfield must be the cyclic subgroup of $F^{\#} = \langle x \rangle$ of order 3. So the subfield of order 4 is

$$
\{0, 1, x^5, x^{10}\} = \{0, 1, x^2 + x, x^2 + x + 1\}. \qquad \square
$$

Example 3 If F is a field of order $3^6 = 729$ and α is a generator of $F^{\#}$, then the subfields of F are

1. $GF(3) = \{0\} \cup \langle \alpha^{364} \rangle = \{0, 1, 2\}$.
2. $GF(9) = \{0\} \cup \langle \alpha^{91} \rangle$.
3. $GF(27) = \{0\} \cup \langle \alpha^{28} \rangle$.
4. $GF(729) = \{0\} \cup \langle \alpha \rangle$. $\qquad \square$

The subfield lattice of $GF(2^{24})$ is illustrated in Figure 24.1.

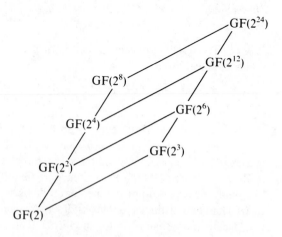

Figure 24.1 Subfield lattice of $GF(2^{24})$.

EXERCISES

We are an intelligent species and the use of our intelligence quite properly gives us pleasure. In this respect the brain is like a muscle. When it is used we feel very good. Understanding is joyous.

Carl Sagan, *Broca's Brain*

1. Find [GF(729):GF(9)] and [GF(64):GF(8)].

2. If n divides m, show that $[\mathrm{GF}(p^m):\mathrm{GF}(p^n)] = m/n$.

3. Let K be a finite extension field of a finite field F. Show that there is an element a in K such that $K = F(a)$.

4. Let F be as in Example 1. Use the generator $z = x^2 + 1$ for $F^{\#}$ and construct a table that converts polynomials in $F^{\#}$ to powers of z, and vice versa.

5. Let $f(x)$ be a cubic irreducible over Z_2. Prove that the splitting field of $f(x)$ over Z_2 has order 8 or 64.

6. Prove that the rings $Z_3[x]/\langle x^2 + x + 2 \rangle$ and $Z_3[x]/\langle x^2 + 2x + 2 \rangle$ are isomorphic.

7. Prove that the maximum degree of any irreducible factor of $x^8 - x$ over Z_2 is 3.

8. Prove that the maximum degree of any irreducible factor of $x^{p^n} - x$ over Z_p is n.

9. Show that x is a generator of the cyclic group $(Z_3[x]/\langle x^3 + 2x + 1 \rangle)^{\#}$.

10. Show that x is not a generator of the cyclic group $(Z_3[x]/\langle x^3 + 2x + 2 \rangle)^{\#}$. Find one.

11. Without actually calculating $|x|$ explain why x is a generator of the cyclic group $(Z_2[x]/\langle x^5 + x^3 + 1 \rangle)^{\#}$.

12. Suppose α and β belong to $\mathrm{GF}(81)^{\#}$ with $|\alpha| = 5$ and $|\beta| = 16$. Show that $\alpha\beta$ is a generator of $\mathrm{GF}(81)^{\#}$.

13. Construct a field of order 27 and carry out the analysis done in Example 1, including the conversion table.

14. Show that any finite subgroup of the multiplicative group of a field is cyclic.

15. Suppose m and n are positive integers and m divides n. If F is any field, show that $x^m - 1$ divides $x^n - 1$ in $F[x]$.

16. If $g(x)$ is irreducible over $\mathrm{GF}(p)$ and $g(x)$ divides $x^{p^n} - x$, prove that deg $g(x)$ divides n.

17. Draw the subfield lattice of $\mathrm{GF}(3^{18})$ and $\mathrm{GF}(2^{30})$.

18. Use a purely group-theoretical argument to show that if F is a field of order p^n, then every element of $F^{\#}$ is a zero of $x^{p^n} - x$. (This exercise is referred to in the proof of Theorem 24.1.)

19. How does the subfield lattice of $\mathrm{GF}(2^{30})$ compare with the subfield lattice of $\mathrm{GF}(3^{30})$?

20. If $p(x)$ is an irreducible polynomial in $Z_p[x]$ with no multiple zeros, show that $p(x)$ divides $x^{p^n} - x$ for some n.

21. Suppose p is a prime and $p \neq 2$. Let a be a nonsquare in GF(p) (that is, a does not have the form b^2 for any b in GF(p)). Show that a is a nonsquare in GF(p^n) if n is odd and a is a square in GF(p^n) if n is even.

22. Show that the *Frobenius mapping* ϕ: GF(p^n) $\rightarrow$ GF(p^n), given by $a \rightarrow a^p$ is an automorphism of order n (that is, ϕ^n is the identity mapping).

23. Show that every element of GF(p^n) can be written in the form a^p for some unique a in GF(p^n).

24. Suppose F is a field of order 1024 and $F^{\#} = \langle\alpha\rangle$. List the elements of each subfield of F.

25. Suppose F is a field of order 125 and $F^{\#} = \langle\alpha\rangle$. Show that $\alpha^{62} = -1$.

26. Show that no finite field is algebraically closed.

27. Let E be the splitting field of $f(x) = x^{p^n} - x$ over Z_p. Show that the set of zeros of $f(x)$ in E is closed under addition, subtraction, multiplication, and division (by nonzero elements). (This exercise is referred to in the proof of Theorem 24.1.)

28. Suppose L and K are subfields of GF(p^n). If L has p^s elements and K has p^t elements, how many elements does $L \cap K$ have?

PROGRAMMING EXERCISES

He who labors diligently need never despair; for all things are accomplished by diligence and labor.

Menander of Athens

1. The number of monic irreducible polynomials of degree d over GF(p) is denoted by $I_p(d)$. Use the equation $p^n = \Sigma\, dI_p(d)$, where the summation is over all divisors d of n, to compute $I_p(d)$. (For example, $3^1 = I_3(1)$; $3^2 = I_3(1) + 2I_3(2)$; $3^3 = I_3(1) + 3I_3(3)$. Thus, $I_3(1) = 3, I_3(2) = 3$, and $I_3(3) = 8$.) Make reasonable assumptions on p and d. Run your program for $p = 2, 3, 5, 7$ and $d = 1, 2, 3, 4, 5$.

SUGGESTED READINGS

Shalom Feigelstock, "Mersenne Primes and Group Theory," *Mathematics Magazine* 49 (1976): 198–199.

A group G is called a *hereditary field group* if G and each of its subgroups is the multiplicative group of nonzero elements of a finite field. This paper gives a theorem that describes all hereditary field groups.

Judy L. Smith and J. A. Gallian, "Factoring Finite Factor Rings," *Mathematics Magazine* 58 (1985): 93–95.

This paper gives an algorithm for finding the group of units of the ring $F[x]/\langle g(x)^m\rangle$.

L. E. Dickson

One of the books [written by L. E. Dickson] is his major, three-volume *History of the Theory of Numbers* which would be a life's work by itself for a more ordinary man.

A. A. Albert, *Bulletin of the American Mathematical Society*

LEONARD EUGENE DICKSON was born in Independence, Iowa, on January 22, 1874. Dickson was the valedictorian of the 1893 class at the University of Texas. In 1894, he went to the University of Chicago and studied under E. H. Moore. Two years later he received a Ph.D., the first to be awarded in mathematics at Chicago. After spending a few years at the University of California and University of Texas, he was appointed to the faculty at Chicago and remained there until his retirement in 1939.

Dickson was one of the most prolific mathematicians of this century, writing 267 research papers and 18 books. His three-volume *History of the Theory of Numbers* took nine years to write. His principal interests were matrix groups, finite fields, algebra, and number theory. He published the first extensive exposition of the theory of finite fields.

Although Dickson supervised the Ph.D. theses of 64 students, he was not a good teacher in the conventional sense. Rather, he inspired his students to emulate him as a research mathematician.

Dickson had a disdainful attitude toward applicable mathematics; he would often say, "Thank God that number theory is unsullied by any applications." He also had a sense of humor. Dickson would often mention his honeymoon: "It was a great success," he said, "except that I only got two research papers written."

Dickson received many honors in his career. He was the first to be awarded the prize from the American Association for the Advancement of Science for the most notable contribution to the advancement of science, and the first to receive the Cole Prize in algebra from the American Mathematical Society; he was president of the American Mathematical Society, and was given honorary degrees by Harvard and Princeton. The University of Chicago has research instructorships named after him. Dickson died on January 17, 1954.

Geometric Constructions

25

Failure properly to understand the theoretical character of the question of geometrical construction and stubbornness in refusing to take cognizance of well-
established scientific facts are responsible for the persistence of an unending
line of angle-trisectors and circle-squarers. Those among them who are able to
understand elementary mathematics might profit by studying this chapter.

Richard Courant and Herbert Robbins, *What Is Mathematics?*

Historical Discussion of Geometric Constructions

The ancient Greeks were fond of geometric constructions. They were especially
interested in constructions that could be achieved using only a straightedge
without markings and a compass. They knew, for example, that any angle can
be bisected and they knew how to construct an equilateral triangle, a square, a
regular pentagon, and a regular hexagon. But they did not know how to trisect
every angle or how to construct a regular seven-sided polygon (heptagon). Another problem that they attempted was the duplication of the cube—that is, given
any cube, they tried to construct a new cube having twice the volume of the
given one by means of a straightedge and compass. Legend has it that the ancient
Athenians were told by the oracle at Delos that a plague would end if they
constructed a new altar to Apollo in the shape of a cube with double the volume
of the old altar, which was also a cube. Besides "doubling the cube," the Greeks
also attempted to "square the circle"—to construct a square with area equal to
that of a given circle. They knew how to solve all these problems using other
means, such as a compass and a straightedge with two marks, or a straightedge
and a spiral, but they could not achieve any of the constructions with a compass

and straightedge alone. These problems vexed mathematicians for over 2000 years.

The resolution of the perplexities was made possible when they were transferred from questions of geometry to questions of algebra in the nineteenth century. With the introduction of coordinate geometry, it was obvious that the existence of straightedge constructions corresponds to solving linear equations, while the use of a compass corresponds to solving quadratic equations. Thus, a straightedge and compass construction is possible if and only if linear and quadratic equations can be solved by a series of additions, subtractions, multiplications, divisions, and extractions of square roots.

The first of the famous problems of antiquity to be solved was that of the construction of regular polygons. It was known since Euclid that regular polygons with a number of sides of the form 2^k, $2^k \cdot 3$, $2^k \cdot 5$ and $2^k \cdot 3 \cdot 5$ could be constructed, and it was believed that no others were possible. In 1796, while still a teenager, Gauss conceived a proof that the seventeen-sided regular polygon is constructible. In 1801 Gauss asserted that a regular polygon of n sides is constructible if and only if n has the form $2^k p_1 p_2 \ldots p_t$ where the p's are distinct primes of the form $2^{2^s} + 1$ and $k \geq 0$.

Thus, regular polygons with 3, 4, 5, 6, 8, 10, 12, 15, 16, 17, and 20 sides are possible to construct, while those with 7, 9, 11, 13, 14, 18, and 19 are not. How these constructions can be effected is another matter. One person spent ten years trying to determine a way to construct the 65,537-sided polygon.

Gauss's result on the constructibility of regular n-gons eliminated another of the famous unsolved problems—the ability to trisect a 60° angle enables one to construct a regular 9-gon. Thus, there is no method for trisecting a 60° angle with a straightedge and compass. In 1837 Wantzel proved that it was not possible to double the cube. The last of the four problems, the squaring of the circle, resisted attempts until 1882, when Ferdinand Lindemann proved that π is transcendental since, as we will show, all constructible numbers are algebraic.

Constructible Numbers

With the field theory we now have, it is an easy matter to solve the following problem: Given an unmarked straightedge, compass, and a unit length, what other lengths can be constructed? To begin, we call a real number α *constructible* if by means of a straightedge, a compass, and a line segment of length 1, we can construct a line segment of length $|\alpha|$ in a finite number of steps. It follows from plane geometry that if α and β ($\beta \neq 0$) are constructible numbers, then so are $\alpha + \beta$, $\alpha - \beta$, $\alpha \cdot \beta$, and α/β. (See the exercises for hints.) Thus, the set of constructible numbers contains Q and is a subfield of the real numbers. What we desire is an algebraic characterization of this field. To do this, let F be any subfield of the reals. Call the subset $F \times F$ of the real plane the *plane of F*, any line joining two points in the plane of F a *line in F*, and any circle whose center and radius are in the plane of F a *circle in F*. Then a line in F has an equation of the form $ax + by + c = 0$ where a, b, c, $\in F$, and a circle in F has an equation of the form

$$x^2 + y^2 + ax + by + c = 0 \qquad \text{where} \qquad a, b, c \in F.$$

In particular, note that to find the point of intersection of a pair of lines in F or the points of intersection of a line in F and a circle in F, one need only solve a linear or quadratic equation in F. We now come to the crucial question. Starting with points in the plane of some field F, which points in the real plane can be obtained with a straightedge and compass? Well, there are only three ways to construct points, starting with points in the plane of F.

1. Intersect two lines in F.
2. Intersect a circle in F and a line in F.
3. Intersect two circles in F.

In case 1, we do not obtain any new points because two lines in F intersect in a point in the plane of F. In case 2, the point of intersection is either the solution to a linear equation in F or a quadratic equation in F. So, the point lies in the plane of F or the plane of $F(\sqrt{\alpha})$, where $\alpha \in F$ and α is positive. Case 3 can be reduced to case 2 by choosing one of the circles and the line joining the two points of intersection of the circles. (We leave it as an exercise to prove that this line is in F or in $F(\sqrt{\alpha})$, where $\alpha \in F$.)

It follows then that the only points in the real plane, which can be constructed from the plane of a field F, are those whose coordinates lie in fields of the form $F(\sqrt{\alpha})$, where $\alpha \in F$ and α is positive. Of course, we can start over with $F_1 = F(\sqrt{\alpha})$ and construct points whose coordinates lie in fields of the form $F_2 = F_1(\sqrt{\beta})$, where $\beta \in F_1$ and β is positive. Continuing in this fashion, we see that a real number c is constructible if and only if there is a series of fields $Q = F_1 \subseteq F_2 \subseteq \cdots \subseteq F_n \subseteq \mathbf{R}$ such that $F_{i+1} = F_i(\sqrt{\alpha_i})$, where $\alpha_i \in F_i$ and $c \in F_n$. Since $[F_{i+1}:F_i] = 1$ or 2, we see by Theorem 23.5 that if c is constructible, then $[Q(c):Q] = 2^k$ for some nonnegative integer k.

We now dispatch the problems that plagued the Greeks. Consider doubling the cube of volume 1. The enlarged cube would have an edge of length $\sqrt[3]{2}$. But $[Q(\sqrt[3]{2}):Q] = 3$, so such a cube cannot be constructed.

Next consider the possibility of trisecting a 60° angle. If it were possible to trisect an angle of 60°, then cos 20° would be constructible. (See Figure 25.1.)

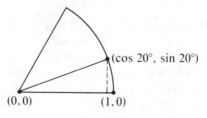

$(0,0)$ $(1,0)$ (cos 20°, sin 20°)

Figure 25.1

In particular, $[Q(\cos 20°):Q] = 2^k$ for some k. Now, using the trigonometric identity $\cos 3\theta = 4 \cos^3 \theta - 3 \cos \theta$ with $\theta = 20°$, we see that $1/2 = 4 \cos^3 20° - 3 \cos 20°$ so that $\cos 20°$ is a root of $8x^3 - 6x - 1$. But, since $8x^3 - 6x - 1$ is irreducible over Q, we must also have $[Q(\cos 20°):Q] = 3$. This contradiction shows that trisecting a 60° angle is impossible.

The remaining problems are relegated to the exercises.

Angle-Trisectors and Circle-Squarers

Down through the centuries, hundreds of people have claimed to have achieved one or more of the impossible constructions. In 1775, the Paris Academy, so overwhelmed with these, passed a resolution no longer to examine them or machines purported to exhibit perpetual motion. Although it has been more than 100 years since the last of the constructions has been shown to be impossible, there continues to be a steady parade of people who claim to have done one or more of them. Most of these people have heard that this is impossible but refused to believe it. One person insisted he could trisect any angle with a straightedge alone [2, p. 158]. Another found his trisection in 1973 after 12,000 hours of work [2, p. 80]. One got his from God [2, p. 73]. In 1971 a person with a Ph.D. in mathematics asserted that he had a valid trisection method [2, p. 127]. Many people have claimed the hat trick: trisecting the angle, doubling the cube, and squaring the circle. Two men who did this in 1961 succeeded in having notice of their accomplishment in the *Congressional Record* [2, p. 110]. Occasionally, newspapers and magazines have run stories about "doing the impossible," often giving the impression that the construction may be valid. Many angle-trisectors and circle-squarers have had their work published at their own expense and distributed to college and universities. One had his printed in four languages! There are two delightful books written by mathematicians about their encounters with these people. The books are full of wit, charm, and humor ([1] and [2]).

EXERCISES

Only prove to me that it is impossible, and I will set about it this very evening.

Spoken by a member of the audience after DeMorgan
gave a lecture on the impossibility of squaring the circle.

1. If a and b are constructible numbers, give a geometrical proof that $a + b$ and $a - b$ are constructible.

2. If a and b are constructible, give a geometrical proof that ab is constructible. (Hint: Consider the following figure. Notice that all segments in the figure can be made with a straightedge and compass.)

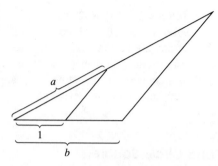

3. If a and b ($b \neq 0$) are constructible numbers, give a geometrical proof that a/b is constructible. (Hint: Consider the following figure.)

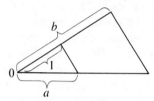

4. Prove that if c is a constructible number, then so is $\sqrt{|c|}$. (Hint: Consider the following semicircle with diameter $1 + |c|$.)

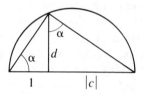

5. Prove that $\sin \theta$ is constructible if and only if $\cos \theta$ is constructible.

6. Prove that an angle θ is constructible if and only if $\sin \theta$ is constructible.

7. Prove that $\cos 2\theta$ is constructible if and only if $\cos \theta$ is constructible.

8. Prove that $30°$ is a constructible angle.

9. Prove that a $45°$ angle can be trisected with a straightedge and compass.

10. Prove that a $40°$ angle is not constructible.

11. Show that the point of intersection of two lines in the plane of a field F lie in the plane of F.

12. Show that the points of intersection of a circle in the plane of a field F and a line in the plane of F are points in the plane of F or the plane of $F(\sqrt{\alpha})$ where $\alpha \in F$ and α is positive. Give an example of a circle and a line in the plane of Q whose points of intersection are not in the plane of Q.

13. Prove that $8x^3 - 2x - 1$ is irreducible over Q.

14. Many of the proposed methods of trisection are equivalent to the assertion that $\sin(\theta/3) = \sin\theta/(2 + \cos\theta)$ is an identity. Show that it is not.

15. Many of the proposed trisections are equivalent to the assertion that $\tan(\theta/3) = 2\sin(\theta/2)/(2 + \cos(\theta/2))$ is an identity. Show that it is not.

16. Use the fact that $8\cos^3(2\pi/7) + 4\cos^2(2\pi/7) - 4\cos(2\pi/7) - 1 = 0$ to prove that a regular seven-sided polygon is not constructible with a straightedge and a compass.

17. Show that a regular 9-gon cannot be constructed with a straightedge and a compass.

18. (Squaring the Circle.) Show that it is impossible to construct with a straightedge and a compass, a square whose area equals that of a circle of radius 1. You may use the fact that π is transcendental over Q.

19. Can the cube be "tripled"?

20. Can the cube be "quadrupled"?

21. Can the circle be "cubed"?

REFERENCES

1. Augustus DeMorgan, *A Budget of Paradoxes,* 2nd ed., Salem, N.H.: Ayer, 1915.
2. Underwood Dudley, *A Budget of Trisections,* New York: Springer-Verlag, 1987.

SUGGESTED READING

Underwood Dudley, *A Budget of Trisections,* New York: Springer-Verlag, 1987.

This highly entertaining book includes detailed information about the personalities of trisectors and their constructions. There is also a chapter on methods for trisecting an angle with tools other than a straightedge and a compass. Some of these will give you a chuckle: one uses a tomahawk, another a watch. According to Dudley, "no one has yet shown how to accomplish the trisection with a digital watch." No doubt there are people working on this right now.

SUPPLEMENTARY EXERCISES for Chapters 21–25

Difficulties strengthen the mind, as labor does the body.

Seneca

1. Show that $x^{50} - 1$ has no multiple zeros in any extension of Z_3.

2. Suppose $p(x)$ is a quadratic polynomial with rational coefficients and is irreducible over Q. Show that $p(x)$ has two zeros in $Q[x]/\langle p(x)\rangle$ (counting multiplicity).

3. Let F be a finite field of order q and let $a \in F$. If n divides $q - 1$, prove that the equation $x^n = a$ has either no solutions or n distinct solutions.

4. Without using the Primitive Element Theorem, prove that if $[K:F]$ is prime, then K has a primitive element.

5. Let a be a complex zero of $x^2 + x + 1$. Express $(5a^2 + 2)/a$ in the form $c + ba$, where c and b are rational.

6. Describe the elements of the extension $Q(\sqrt[4]{2}.)$ over the field $Q(\sqrt{2})$.

7. If $[F(a):F] = 5$, fine $[F(a^3):F]$.

8. If $p(x) \in F[x]$ and deg $p(x) = n$, show that the splitting field for $p(x)$ over F has degree at most $n!$.

9. Let a be a nonzero algebraic element over F of degree n. Show that a^{-1} is also algebraic over F of degree n.

10. Prove that $\pi^2 - 1$ is algebraic over $Q(\pi^3)$.

11. If ab is algebraic over F, prove that a is algebraic over $F(b)$.

12. Let E be an algebraic extension of a field F. If R is a ring and $E \supseteq R \supseteq F$, show that R must be a field.

13. If a is transcendental over F, show that every element of $F(a)$ that is not in F is transcendental over F.

SPECIAL TOPICS

PART 5

Sylow Theorems

26

Generally these three results are implied by the expression "Sylow's Theorem." All of them are of fundamental importance. In fact, if the theorems of group theory were arranged in order of their importance Sylow's Theorem might reasonably occupy the second place—coming next to Lagrange's Theorem in such an arrangement.

G. A. Miller, *Theory and Application of Finite Groups*

Conjugacy Classes

In this chapter, we derive a number of important arithmetical relationships between a group and certain of its subgroups. Recall from Chapter 9 that Lagrange's Theorem was proved by showing that the cosets of a subgroup partition the group. Another fruitful method of partitioning the elements of a group is by way of conjugacy classes.

> *DEFINITION* Conjugacy Class of a
>
> Let a and b be elements of a group G. We say a and b are *conjugate* in G (and call b a *conjugate* of a) if $x^{-1} ax = b$ for some x in G. The *conjugacy class of a* is the set cl$(a) = \{x^{-1}ax \mid x \in G\}$.

We leave it to the reader (exercise 1) to prove that conjugacy is an equivalence relation on G, and that the conjugacy class of a is the equivalence class of a under conjugacy. Thus, we may partition any group into disjoint conjugacy classes. Let's look at one example. In D_4 we have

$$\text{cl}(H) = \{R_0^{-1}HR_0, R_{90}^{-1}HR_{90}, R_{180}^{-1}HR_{180}, R_{270}^{-1}HR_{270},$$
$$H^{-1}HH, V^{-1}HV, D^{-1}HD, D'^{-1}HD'\} = \{H, V\}.$$

Similarly, one may verify that

$$cl(R_0) = \{R_0\},$$
$$cl(R_{90}) = \{R_{90}, R_{270}\} = cl(R_{270}),$$
$$cl(R_{180}) = \{R_{180}\},$$
$$cl(V) = \{V, H\} = cl(H),$$
$$cl(D) = \{D, D'\} = cl(D').$$

Theorem 26.1 gives an arithmetical relationship between the size of the conjugacy class of a and the size of the centralizer of a.

Theorem 26.1   The Number of Conjugates of a
Let G be a group and a be an element of G. Then, $|cl(a)| = |G{:}C(a)|$.

Proof.   Consider the function T that sends the coset $C(a)x$ to the conjugate $x^{-1}ax$ of a. A routine calculation shows that T is well defined, one-to-one, and maps the set of right cosets onto the conjugacy class of a(exercise). Thus, the number of conjugates of a is the index of the centralizer of a. ∎

The Class Equation

Since the conjugacy classes partition a group, the following important counting principle is a corollary to Theorem 26.1.

Corollary   The Class Equation
For any finite group G,

$$|G| = \sum |G{:}C(a)|$$

where the sum runs over one element a from each conjugacy class of G.

In finite group theory, counting principles such as this corollary are powerful tools.* Theorem 26.2, which is the single most important fact about finite p-groups, is a vivid illustration of this.

Theorem 26.2   p-Groups Have Nontrivial Centers
Let G be a finite group whose order is a power of a prime p. Then $Z(G)$ has more than one element.

Proof.   First observe that $cl(a) = \{a\}$ if and only if $a \in Z(G)$. Thus, by culling out these elements, we may write the class equation in the form

$$|G| = |Z(G)| + \sum |G{:}C(a)|$$

---

*"Never underestimate a theorem that counts something." John Fraleigh, *A First Course in Abstract Algebra.*

where the sum runs over representatives of all conjugacy classes with more than one element (this set may be empty). But $|G{:}C(a)| = |G|/|C(a)|$ so each term in $\Sigma\ |G{:}C(a)|$ has the form p^k with $k \geq 1$. Hence,

$$|G| - \sum |G{:}C(a)| = |Z(G)|$$

where each term on the left is divisible by p. It follows then that p also divides $|Z(G)|$ and, hence, $|Z(G)| \neq 1$. ∎

Corollary *Groups of Order p^2 Are Abelian*
If $|G| = p^2$ where p is prime, then G is Abelian.

Proof. By Theorem 26.2 and Lagrange's Theorem, $|Z(G)| = p$ or p^2. If $|Z(G)| = p^2$, then $G = Z(G)$ and G is Abelian. If $|Z(G)| = p$, then $|G/Z(G)| = p$ so that $G/Z(G)$ is cyclic. But, then, by Theorem 10.3 G is Abelian. ∎

The Probability That Two Elements Commute

Before proceeding to the main goal of this chapter, we pause for an interesting application of Theorem 26.1 and the class equation. (Our discussion is based on [1] and [2].) Suppose we select two elements at random (with replacement) from a finite group. What is the probability that these two elements commute? Well, suppose G is a finite group of order n. Then the probability, $\Pr(G)$, that two elements selected at random from G commute is $|K|/n^2$ where $K = \{(x, y) \in G \oplus G \mid xy = yx\}$. Now notice that for each $x \in G$, we have $(x, y) \in K$ if and only if $y \in C(x)$. Thus,

$$|K| = \sum_{x \in G} |C(x)|.$$

Also, exercise 44 states that if x and y are in the same conjugacy class, then $|C(x)| = |C(y)|$. If, for example, $\text{cl}(a) = \{a_1, a_2, \ldots, a_t\}$, then

$$|C(a_1)| + |C(a_2)| + \cdots + |C(a_t)| = t|C(a)| = |G{:}C(a)|\,|C(a)| = |G| = n.$$

So, by choosing one representative from each conjugacy class, say, $x_1, x_2, \ldots, x_m$, we have

$$|K| = \sum_{i \in G} |C(x)| = \sum_{i=1}^{m} |G{:}C(x_i)|\,|C(x_i)| = m \cdot n.$$

Thus, the answer to our question is m/n where m is the number of conjugacy classes in G and n is the number of elements of G.

Obviously, when G is non-Abelian, $\Pr(G)$ is less than 1. But how much less than 1? Clearly, the more conjugacy classes there are, the larger $\Pr(G)$ is. Consequently, $\Pr(G)$ is large when the sizes of the conjugacy classes are small. Noting that $|\text{cl}(a)| = 1$ if and only if $a \in Z(G)$, we obtain the maximum number of conjugacy classes when $|Z(G)|$ is as large as possible and all other conjugacy classes have exactly two elements in each. Since G is non-Abelian, it follows from Theorem 10.3 that $|G/Z(G)| \geq 4$ and, therefore, $|Z(G)| \leq |G|/4$. Thus, in the extreme case, we would have $|Z(G)| = |G|/4$, and the remaining $\frac{3}{4}|G|$ elements would be distributed in conjugacy classes with two elements each. So, in a non-Abelian group, the number of conjugacy classes is no more then $|G|/4 + \frac{1}{2} \cdot \frac{3}{4}|G|$, and $\Pr(G)$ is less than or equal to $\frac{5}{8}$. The dihedral group D_4 is an example of a group that has probability equal to $\frac{5}{8}$.

The Sylow Theorems

Now back to Sylow's Theorems. Recall that the converse of Lagrange's Theorem is false; that is, if G is a group of order m and n divides m, G need *not* have a subgroup of order n. Our next theorem is a partial converse to Lagrange's Theorem. It, as well as Theorem 26.2, was first proved by the Norwegian mathematician Ludwig Sylow (1832–1918). Sylow's Theorem and Lagrange's Theorem are the two most important results in finite group theory. The first gives a sufficient condition for the existence of subgroups, while the second gives a necessary condition.

> **Theorem 26.3**   *Existence of Subgroups of Prime-Power Order (Sylow's First Theorem, 1872)*
> *Let G be a finite group and p a prime. If p^k divides $|G|$, then G has at least one subgroup of order p^k.*

Proof.   We proceed by induction on $|G|$. If $|G| = 1$, Theorem 26.3 is trivially true. Now assume the statement is true for all groups of order less than $|G|$. If G has a proper subgroup H such that p^k divides $|H|$, then, by our inductive assumption, H has a subgroup of order p^k and we are done. Thus, we may henceforth assume that p^k does not divide the order of any proper subgroup of G. Next, consider the class equation for G in the form

$$|G| = |Z(G)| + \sum |G:C(a)|$$

where we sum over a representative of each conjugacy class cl(a) where $a \notin Z(G)$. Since p^k divides $|G| = |G:C(a)| \, |C(a)|$ and p^k does not divide $|C(a)|$, we know p must divide $|G:C(a)|$ for all $a \notin Z(G)$. It then follows from the class equation that p divides $|Z(G)|$. The Fundamental Theorem of Finite Abelian Groups then guarantees that $Z(G)$ contains an element of order p, say, x. Since x is in the center of G, $\langle x \rangle$ is a normal subgroup of G, and we may form the factor group $G/\langle x \rangle$. Now observe that p^{k-1} divides $|G/\langle x \rangle|$. Thus, by the induction hypothesis, $G/\langle x \rangle$ has a subgroup of order p^{k-1} and, by exercise 43 of Chapter 11, this subgroup has the form $H/\langle x \rangle$ where H is a subgroup of G. Finally, note that $|H/\langle x \rangle| = p^{k-1}$ and $|\langle x \rangle| = p$ imply that $|H| = p^k$, and this completes the proof. ∎

Let's be sure we understand exactly what Sylow's Theorem means. Say we have a group G of order $2^3 \cdot 3^2 \cdot 5^4 \cdot 7$. Then Sylow's Theorem says G must have at least one subgroup of each of the following orders: 2, 4, 8, 3, 9, 5, 25, 125, 625, and 7. On the other hand, Sylow's Theorem tells us nothing about the possible existence of subgroups of orders 6, 10, 15, 30, or any other divisor of $|G|$ that has two or more distinct prime factors. Because certain subgroups guaranteed by Sylow's Theorem play a central role in the theory of finite groups, they are given a special name.

DEFINITION Sylow *p*-Subgroup

Let G be a finite group and let p be a prime divisor of $|G|$. If p^k divides $|G|$ and p^{k+1} does not divide $|G|$, then any subgroup of G or order p^k is called a *Sylow p-subgroup of G*.

So, returning to our group G of order $2^3 \cdot 3^2 \cdot 5^4 \cdot 7$, we call any subgroup of order 8 a Sylow 2-subgroup of G, any subgroup of order 625 a Sylow 5-subgroup of G, and so on. Notice that a Sylow p-subgroup of G is a subgroup whose order is the largest power of p consistent with Lagrange's Theorem.

Since any subgroup of order p is cyclic, we have the following corollary, first proved by Cauchy in 1844.

Corollary *Cauchy's Theorem*

Let G be a finite group and p a prime that divides the order of G. Then G has an element of order p.

Sylow's Theorem is so fundamental to finite group theory that many different proofs of it have been published over the years (our proof is essentially the one given by Georg Frobenius [1849–1917] in 1887). Likewise, there are scores of generalizations of Sylow's Theorem.

Observe that the corollary to the Fundamental Theorem of Finite Abelian Groups and Sylow's Theorem shows that the converse of Lagrange's Theorem is true for all finite Abelian groups and all finite Abelian groups of prime-power order.

There are two more Sylow theorems that are extremely valuable tools in finite group theory. But first we introduce a new term.

DEFINITION Conjugate Subgroups

Let H and K be subgroups of a group G. We say H and K are *conjugate* in G if there is an element g in G such that $H = g^{-1}Kg$.

Recall from Chapter 9 that if G is a group of permutations on a set S and $i \in S$, then $\text{orb}_G(i) = \{i\phi \mid \phi \in G\}$ and $|\text{orb}_G(i)|$ divides $|G|$.

Theorem 26.4 *Sylow's Second Theorem*

If H is a subgroup of a finite group G and $|H|$ is a power of a prime p, then H is contained in some Sylow p-subgroup of G.

Proof. Let K be a Sylow p-subgroup of G and let $C = \{K = K_1, K_2, \ldots, K_n\}$ be the set of all conjugates of K in G. Since conjugation is an automorphism, each element of C is a Sylow p-subgroup of G. Let S_C denote the group of all permutations of C. For each $g \in G$, define $\phi_g : C \to C$ by $K_i \phi_g = g^{-1} K_i g$. It is an easy exercise to show that each $\phi_g \in S_C$.

Now define a mapping $T : G \to S_C$ by $gT = \phi_g$. Since $K_i \phi_{gh} = (gh)^{-1} K_i (gh) = h^{-1}(g^{-1} K_i g)h = h^{-1}(K_i \phi_g)h = (K_i \phi_g)\phi_h = K_i \phi_g \phi_h$, we have $\phi_{gh} = \phi_g \phi_h$ and, therefore, T is a homomorphism from G to S_C.

Next consider HT, the image of H under T. Since $|H|$ is a power of p, so is $|HT|$ (see part 9 of Theorem 11.1). Thus, by the Orbit-Stabilizer Theorem (Theorem 9.2), for each i, $|\mathrm{orb}_{HT}(K_i)|$ divides $|HT|$ so that $|\mathrm{orb}_{HT}(K_i)|$ is a power of p. Now we ask: Under what conditions does $|\mathrm{orb}_{HT}(K_i)| = 1$? Well, $|\mathrm{orb}_{HT}(K_i)| = 1$ means that $K_i \phi_g = g^{-1} K_i g = K_i$ for all $g \in H$; that is, $|\mathrm{orb}_{HT}(K_i)| = 1$ if and only if $H \leq N(K_i)$. But the only elements of order a power of p that belong to $N(K_i)$ are those of K_i (see exercise 9). Thus, $|\mathrm{orb}_{HT}(K_i)| = 1$ if and only if $H \leq K_i$.

So, to complete the proof, all we need do is show that for some i, $|\mathrm{orb}_{HT}(K_i)| = 1$. Analogous to Theorem 26.1, we have $|C| = |G{:}N(K)|$ (see exercise 38). And since $|G{:}K| = |G{:}N(K)| \; |N(K){:}K|$ is not divisible by p, neither is $|C|$. Because the orbits partition C, $|C|$ is the sum of powers of p. If no orbit has size 1, then p divides each summand and, therefore, p divides $|C|$. Thus, there is an orbit of size 1 and the proof is complete. ∎

Theorem 26.5 *Sylow's Third Theorem*
The number n_p of Sylow p-subgroups of G is equal to 1 modulo p and divides $|G|$. Furthermore, any two Sylow p-subgroups of G are conjugate.

Proof. Let K be any Sylow p-subgroup of G and let $C = \{K = K_1, K_2, \ldots, K_n\}$ be the set of all conjugates of K in G. We first prove that $n = 1 \bmod p$.

Let S_C and T be as in the proof of Theorem 26.4. This time we consider KT, the image of K under T. As before, we have $|\mathrm{orb}_{KT}(K_i)|$ is a power of p for each i and $|\mathrm{orb}_{KT}(K_i)| = 1$ if and only if $K \leq K_i$. Thus, $|\mathrm{orb}_{KT}(K_1)| = 1$ and $|\mathrm{orb}_{KT}(K_i)|$ is a power of p greater than 1 for all $i \neq 1$. Since the orbits partition C, it follows that $n = |C| = 1 \bmod p$.

Next we show that every Sylow p-subgroup of G belongs to C. To do this, suppose that H is a Sylow p-subgroup of G that is not in C. Let S_C and T be as in the proof of Theorem 26.4 and this time consider HT. As in the previous paragraph, $|C|$ is the sum of the orbits sizes under the action of HT. However, no orbit has size 1 since H is not in C. Thus, $|C|$ is a sum of terms each divisible by p so that $n = |C| = 0 \bmod p$. This contradiction proves that H belongs to C and n is the number of Sylow p-subgroups of G.

Finally, that n divides $|G|$ follows directly from the fact that $n = |G{:}N(K)|$ (see exercise 38). ∎

Observe that the first portion of Sylow's Third Theorem is a counting principle.* As an important consequence of Sylow's Third Theorem, we have the following corollary.

Corollary *A Unique Sylow p-Subgroup Is Normal*
A Sylow p-subgroup of a finite group G is a normal subgroup of G if and only if it is the only Sylow p-subgroup of G.

We illustrate Sylow's Third Theorem with two examples.

Example 1 Consider the Sylow 2-subgroups of S_3. They are $\{(1), (12)\}$, $\{(1), (23)\}$, and $\{(1), (13)\}$. According to Sylow's Third Theorem, we should be able to obtain the latter two of these from the first by conjugation. Indeed,

$$(13)^{-1}\{(1), (12)\}(13) = \{(1), (23)\}$$
$$(23)^{-1}\{(1), (12)\}(13) = \{(1), (13)\}. \qquad \square$$

Example 2 Consider the Sylow 3-subgroups of A_4. They are $\{\alpha_1, \alpha_5, \alpha_9\}$, $\{\alpha_1, \alpha_6, \alpha_{11}\}$, $\{\alpha_1, \alpha_7, \alpha_{12}\}$, and $\{\alpha_1, \alpha_8, \alpha_{10}\}$. (See the table on page 92.) Then,

$$\alpha_2^{-1}\{\alpha_1, \alpha_5, \alpha_9\} \alpha_2 = \{\alpha_1, \alpha_7, \alpha_{12}\}$$
$$\alpha_3^{-1}\{\alpha_1, \alpha_5, \alpha_9\} \alpha_3 = \{\alpha_1, \alpha_8, \alpha_{10}\}$$
$$\alpha_4^{-1}\{\alpha_1, \alpha_5, \alpha_9\} \alpha_4 = \{\alpha_1, \alpha_6, \alpha_{11}\}$$

Thus, the number of Sylow 3-subgroups is 1 modulo 3 and the four Sylow 3-subgroups are conjugate. $\qquad \square$

Figure 26.1 shows the subgroup lattice for S_3 and A_4. We have connected the Sylow p-groups with a dashed circle to indicate that they belong to one orbit under conjugation. Notice that the three subgroups of order 2 in A_4 are contained in a Sylow 2-group as required by Sylow's Second Theorem. As it happens, these three subgroups also belong to one orbit under conjugation, but this is not a consequence of Sylow's Third Theorem.

In contrast to the two preceding examples, observe that the dihedral group of order 12 has 7 subgroups of order 2, but conjugating $\{R_0, R_{180}\}$ does not yield any of the other six. (Why?)

Applications of Sylow's Theorems

A few numerical examples will make the Sylow theorems come to life. Say, G is a group of order 40. What do the Sylow theorems tell us about G? A great deal! Since 1 is the only divisor of 40 that is congruent to 1 modulo 5, we know

---

*"Whenever you can, count." Sir Francis Galton (1822–1911), *The World of Mathematics*.

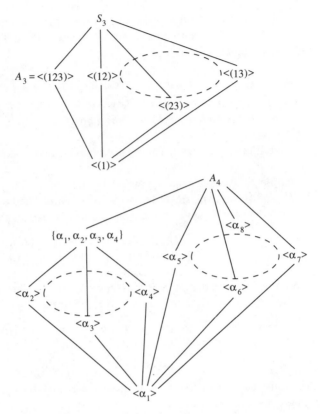

Figure 26.1 Lattice of subgroups for S_3 and A_4.

G has exactly one subgroup of order 5 and it is normal. Similarly, G has either one or five subgroups of order 8. If there is only one subgroup of order 8, it is normal. If there are five subgroups of order 8, none are normal and all five can be obtained by starting with any particular one, say, H, by computing $x^{-1}Hx$ for various x's. Finally, if we let K denote the normal subgroup of order 5 and H any subgroup of order 8, then $G = HK$. (See exercise 40 of Chapter 10 and exercise 7, Supplementary Exercises for Chapters 5–8.) If H happens to be normal, we can say even more: $G = H \times K$.

What about a group G of order 30? It must have either one or six subgroups of order 5 and one or ten subgroups of order 3. However, G cannot have both six subgroups of order 5 *and* ten subgroups of order 3 (then G would have more than 30 elements). Thus, G has a normal subgroup of order 3 or 5. It follows then that the product of these two subgroups is a group of order 15 that is both cyclic (exercise 24) and normal (exercise 7 of Chapter 10) in G. (This, in turn, implies that *both* the subgroup of order 3 and the subgroup of order 5 are normal in G (exercise 44 of Chapter 10).) So, if we let y be a generator of the cyclic subgroup of order 15 and x by an element of order 2 (the existence of which is guaranteed by Cauchy's Theorem), we see that

$$G = \{x^i y^j \mid 0 \leqslant i \leqslant 1, 0 \leqslant j \leqslant 14\}$$

Note that in these two examples we were able to deduce all of this information from knowing only the order of the group. So many conclusions from one assumption! This is the beauty of finite group theory.

In Chapter 9, we classified all groups of order 6. As a further illustration of the power of the Sylow theorems, we now determine all groups of order $2p$ where p is prime.

Theorem 26.6 *Classification of Groups of Order $2p$*
Let $|G| = 2p$ where p is an odd prime. Then G is isomorphic to Z_{2p} or D_p.

Proof. By Sylow's Theorem, G has subgroups of order 2 and p. So, G has an element a of order 2 and an element b of order p. Note that G is generated by a and b. Since $\langle b \rangle$ has index 2, it is normal and therefore $a^{-1}ba = b^k$ for some positive integer k less than p. Thus, $b^{k^2} = (b^k)^k = (a^{-1}ba)^k = a^{-1}b^k a = a^{-1}(a^{-1}ba)a = a^{-2}ba^2$. Because $|a| = 2$, we must have $b = b^{k^2}$. So $b^{k^2-1} = e$, which implies p divides $k^2 - 1 = (k - 1)(k + 1)$. Since $1 \leqslant k < p$, this implies $k - 1 = 0$ or $k + 1 = p$. In the first case, we have $k = 1$ and $ab = ba$. Thus, $|ab| = 2p$ and G is isomorphic to Z_{2p}. In the second case, the $2p$ elements of G can be written in the form $a^i b^j$ where $0 \leqslant i \leqslant 1$ and $0 \leqslant j \leqslant p - 1$. Furthermore,

$$a^i b^{j_1} a b^{j_2} = a^{i+1} b^{j_2 - j_1}$$

and we leave it as an exercise to show that this guarantees that G is isomorphic to D_p. ∎

The argument used to prove Theorem 26.6 can be extended to show that there are at most two groups of order pq for any primes p and q. (This was first proved by E. Netto in 1882—a proof is given in [3, p. 204].) One special case of this result of particular interest is the following.

Theorem 26.7 *Cyclic Groups of Order pq*
If G is a group of order pq where p and q are primes, $p < q$ and p does not divide $q - 1$, then G is cyclic. In particular, G is isomorphic to Z_{pq}.

Proof. Let H be a Sylow p-subgroup of G and K a Sylow q-subgroup of G. Sylow's Third Theorem states that the number of Sylow p-subgroups of G (which we called n_p) is of the form $1 + kp$ and divides pq. So $1 + kp = 1$, p, q, or pq. From this and the fact that $p \nmid q - 1$, it follows that $k = 0$ and, therefore, H is the only Sylow p-subgroup of G.

Similarly, there is only one Sylow q-subgroup of G. Thus, by the corollary to Theorem 26.5, H and K are normal subgroups of G. Let $H = \langle x \rangle$

and $K = \langle y \rangle$. To show that G is cyclic, it suffices to show that x and y commute, then $|xy| = |x| \, |y| = pq$. But observe, since H and K are normal, we have

$$x^{-1}y^{-1}xy = (x^{-1}y^{-1}x)y \in Ky = K$$

and

$$x^{-1}y^{-1}xy = x^{-1}(y^{-1}xy) \in x^{-1}H = H$$

Thus, $x^{-1}y^{-1}xy \in K \cap H = \{e\}$ and, hence, $xy = yx$. (See also exercise 43 of Chapter 10.) ∎

Theorems 26.6 and 26.7 demonstrate the power of the Sylow theorems in classifying the finite groups that have a small number of prime factors. Similar results exist for groups of order p^2q, p^2q^2, p^3, p^4 where p and q are prime.

For your amusement, Figure 26.2 gives a list of the number of nonisomorphic groups with orders at most 100. Note, in particular, the large number of groups of order 64. Also, observe that, generally speaking, it is not the size of the group that gives rise to a large number of groups of that size but the number of prime factors involved. In all, there are 1047 nonisomorphic groups with 100 or less elements. Contrast this with the fact reported in 1978 that there are 2358 group of order 128 alone [4].

As a final application of the Sylow theorems, you might enjoy seeing a determination of the groups of orders 99, 66, and 255. In fact, our arguments serve as a good review of much of our work in group theory.

Order	1	2	3	4	5	6	7	8	9	10	11	12	13	14	15	16	17	18	19	20
Number	1	1	1	2	1	2	1	5	2	2	1	5	1	2	1	14	1	5	1	5

Order	21	22	23	24	25	26	27	28	29	30	31	32	33	34	35	36	37	38	39	40
Number	2	2	1	15	2	2	5	4	1	4	1	51	1	2	1	14	1	2	2	14

Order	41	42	43	44	45	46	47	48	49	50	51	52	53	54	55	56	57	58	59	60
Number	1	6	1	4	2	2	1	52	2	5	1	5	1	15	2	13	2	2	1	13

Order	61	62	63	64	65	66	67	68	69	70	71	72	73	74	75	76	77	78	79	80
Number	1	2	4	267	1	4	1	5	1	4	1	50	1	2	3	4	1	6	1	52

Order	81	82	83	84	85	86	87	88	89	90	91	92	93	94	95	96	97	98	99	100
Number	15	2	1	15	1	2	1	12	1	10	1	4	2	2	1	230	1	5	2	16

Figure 26.2 The number of groups of a given order ≤ 100.

Example 3 Determination of the Groups of Order 99

Suppose G is a group of order 99. Let H be a Sylow 3-subgroup of G and K a Sylow 11-subgroup of G. Since 1 is the only positive divisor of 99 that is equal to 1 mod 11, we know from Sylow's Third Theorem and its corollary that K is normal in G. Similarly, H is normal in G. It follows that elements from H and K commute (see exercise 43 of Chapter 10) and, therefore, $G = H \times K$. Since both H and K are Abelian, G is also Abelian. Thus, G is isomorphic to Z_{99} or $Z_3 \oplus Z_{33}$. $\square$

Example 4 Determination of the Groups of Order 66

Suppose G is a group of order 66. Let H be a Sylow 3-subgroup of G and K a Sylow 11-subgroup of G. Since 1 is the only positive divisor of 66 that is equal to 1 mod 11, we know K is normal in G. Thus, HK is a subgroup of G of order 33 (exercise 40 of Chapter 10 and exercise 7, Supplementary Exercises for Chapters 5–8). Since any group of order 33 is cyclic (Theorem 26.7), we may write $HK = \langle x \rangle$. Next, let $y \in G$ and $|y| = 2$. Since $\langle x \rangle$ has index 2 in G, we know it is normal. So $y^{-1}xy = x^i$ for some i from 1 to 32. Then, $xy = yx^i$ and, since every member of G is of the form $x^s y^t$, the structure of G is completely determined by the value of i. We claim that there are only four possibilities for i. To prove this, observe that $|x^i| = |x|$ (exercise 6, Supplementary Exercises for Chapters 1–4). Thus, i and 33 are relatively prime. But also, since y has order 2,

$$x = y(y^{-1}xy)y^{-1} = yx^iy^{-1} = y^{-1}x^iy$$
$$= (y^{-1}xy)^i = (x^i)^i = x^{i^2}.$$

So $x^{i^2-1} = e$ and, therefore, 33 divides $i^2 - 1$. From this it follows that 11 divides $i \pm 1$ and, therefore, $i = 0 \pm 1$, $i = 11 \pm 1$, $i = 22 \pm 1$, or $i = 33 \pm 1$. Putting this together with the other information we have about i, we see that $i = 1, 10, 23,$ or 32. This proves that there are at most four groups of order 66.

To prove that there are exactly four, we simply observe that Z_{66}, D_{33}, $D_{11} \oplus Z_3$, and $D_3 \oplus Z_{11}$ each has order 66 and no two are isomorphic. For example, $D_{11} \oplus Z_3$ has 11 elements of order 2, while $D_3 \oplus Z_{11}$ has only 3 elements of order 2. $\square$

Example 5 The Only Group of Order 255 Is Z_{255}

Let G be a subgroup of order $255 = 3 \cdot 5 \cdot 17$, and let H be a Sylow 17-subgroup of G. By Sylow's Third Theorem, H is the only Sylow 17-subgroup of G, so $N(H) = G$. By Example 13 of Chapter 11, $|N(H)/C(H)|$ divides $|\text{Aut}(H)| = |\text{Aut}(Z_{17})|$. By Theorem 6.3, $|\text{Aut}(Z_{17})| = |U(17)| = 16$. Since $|N(H)/C(H)|$ must divide 255 and 16, we have $|N(H)/C(H)| = 1$. Thus, $C(H) = G$. This means that every element of G commutes with every element of H and, therefore, $H \subseteq Z(G)$. Thus, 17 divides $|Z(G)|$, which in turn divides 255. So $|Z(G)| = 17, 51, 85,$ or 255 and $|G/Z(G)| = 15, 5, 3,$ or 1. But the only groups of orders 15, 5, 3, or 1 are the cyclic ones, so we know $G/Z(G)$ is cyclic. Now the "G/Z theorem" (Theorem 10.3) shows that G is Abelian, and the Fundamental Theorem of Finite Abelian Groups tell us that G is cyclic. □

EXERCISES

If I rest, I rust.

Martin Luther

1. Show that conjugacy is an equivalence relation on a group.
2. Calculate all conjugacy classes for the quaternions (see exercise 4, Supplementary Exercises for Chapters 1–4).
3. Show that $cl(a) = \{a\}$ if and only if $a \in Z(G)$.
4. Describe the conjugacy classes of an Abelian group.
5. Exhibit a Sylow 2-subgroup of S_4. Describe an isomorphism from this group to D_4.
6. If $|G| = 36$ and is non-Abelian, prove that the Sylow 2-subgroups or the Sylow 3-subgroups are not normal.
7. Suppose G is a group of order 48. Show that the intersection of any two distinct Sylow 2-subgroups of G has order 8.
8. Find all the Sylow 3-subgroups of A_4.
9. Let K be a Sylow p-subgroup of a finite group G. Prove that if $x \in N(K)$ and the order of x is a power of p, then $x \in K$. (This exercise is referred to in this chapter.)
10. Let H be a Sylow p-subgroup of G. Prove that H is the only Sylow p-subgroup of G contained in $N(H)$.
11. Suppose G is a group of order 168. If G has more than one Sylow 7-subgroup, exactly how many does it have?
12. Show that every group of order 56 has a proper nontrivial normal subgroup.
13. What is the smallest composite (that is, nonprime and greater than 1) integer n such that there is a unique group of order n?
14. Let G be a noncyclic group of order 21. How many Sylow 3-subgroups does G have?
15. Prove that a noncyclic group of order 21 must have 14 elements of order 3.
16. How many Sylow 5-subgroups of S_5 are there? Exhibit two.
17. How many Sylow 3-subgroups of S_5 are there? Exhibit five.
18. Prove that a group of order 175 is Abelian.
19. Generalize the argument given in Example 3 to obtain a theorem about groups of order p^2q where p and q are distinct primes.

20. Let G be a group of order p^2q^2 where p and q are distinct primes and $q \nmid p^2 - 1$ and $p \nmid q^2 - 1$. Prove that G is Abelian. List three pairs of primes that satisfy these conditions.

21. What is the smallest possible odd-order non-Abelian group?

22. Prove that a group of order 375 has a subgroup of order 15.

23. Prove that a group of order 105 contains a subgroup of order 35.

24. Without using Theorem 26.8, prove that a group of order 15 is cyclic. (This exercise is referred to in the discussion about groups of order 30.)

25. Prove that a group of order 595 has a normal Sylow 17-subgroup.

26. Let G be a group of order 60. Show that G has exactly four elements of order 5 or exactly 24 elements of order 5. Which of these cases hold for A_5?

27. Show that the center of a group of order 60 cannot have order 4.

28. Suppose G is a group of order 60 and G has a normal subgroup N of order 2. Show that
    a. G has normal subgroups of order 6, 10, and 30
    b. G has subgroups of order 12 and 20
    c. G has a cyclic subgroup of order 30

29. Let G be a group of order 60. If the Sylow 3-subgroup is normal, show that the Sylow 5-subgroup is normal.

30. Show that if G is a group of order 168 that has a normal subgroup of order 4, then G has a normal subgroup of order 28.

31. Suppose p is prime and $|G| = p^n$. Show that G has normal subgroups of order p^k for all k between 1 and n.

32. Suppose p is a prime and $|G| = p^n$. If H is a subgroup of G, prove that $N(H) > H$. (This exercise is referred to in Chapter 27.)

33. Suppose p is a prime and $|G| = p^n$. If H is a subgroup of G of index p, prove that H is normal in G.

34. Suppose G is a finite group and all its Sylow subgroups are normal. Show that G is a direct product of its Sylow subgroups.

35. Let G be a finite group and H a normal Sylow p-subgroup of G. Show that $H\alpha = H$ for all automorphisms α of G.

36. If H is a normal subgroup of a finite group G and $|H| = p^k$ for some prime p, show that it is contained in every Sylow p-subgroup of G.

37. Let H and K denote a Sylow 3-subgroup and Sylow 5-subgroup of a group, respectively. Suppose $|H| = 3$ and $|K| = 5$. If 3 divides $|N(K)|$, show that 5 divides $|N(H)|$.

38. Let H be a subgroup of a group G. Prove that the number of conjugates of H in G is $|G{:}N(H)|$. (Hint: Mimic the proof of Theorem 26.1.)

39. Let H be a normal subgroup of a group G. Show that H is the union of the conjugacy classes of the elements of H. Is that true when H is not normal in G?

40. Let p be a prime. If each element of a finite group G has order a power of p, prove that G has order a power of p. (Such a group is called a *p-group*.)

41. Prove that all Sylow p-subgroups of a finite group are isomorphic.

42. What is the probability that a randomly selected element from D_4 commutes with V?

43. Let G be a finite group and let $a \in G$. Express the probability that a randomly selected element from G commutes with a in terms of orders of subgroups of G.

44. Prove that if x and y are in the same conjugacy class of a group, then $|C(x)| = |C(y)|$. (This exercise is referred to in the discussion on the probability that two elements from a group commute.)

45. Find $\Pr(D_4)$, $\Pr(S_3)$, and $\Pr(A_4)$.

46. Prove that $\Pr(G \oplus H) = \Pr(G) \cdot \Pr(H)$.

47. Let R be a finite noncommutative ring. Show that the probability that two randomly chosen elements from **R** commute is at most $\frac{5}{8}$. (Hint: Mimic the group case and use the fact that the additive group $R/C(R)$ is not cyclic.)

REFERENCES

1. W. H. Gustafson, "What Is the Probability that Two Group Elements Commute?" *American Mathematical Monthly* 80 (1973): 1031–1034.

2. Desmond MacHale, "How Commutative Can a Non-Commutative Group Be?" *The Mathematical Gazette* 58 (1974): 199–202.

3. H. Paley and P. Weichsel, *A First Course in Abstract Algebra*, New York: Holt, Rinehart & Winston, 1966.

4. Eugene Rodemich, "The Groups of Order 128," *Notices of the American Mathematical Society* 25 (1978): A-71.

SUGGESTED READINGS

W. H. Gustafson, "What Is the Probability that Two Group Elements Commute?" *American Mathematical Monthly* 80 (1973): 1031–1034.

This paper is concerned with the problem posed in the title. It is shown that for all finite non-Abelian groups and certain infinite non-Abelian groups, the probability that two elements from a group commute is at most $\frac{5}{8}$. The paper concludes with a number of exercises.

Desmond MacHale, "Commutativity in Finite Rings," *American Mathematical Monthly* 83 (1976): 30–32.

In this easy-to-read paper, it is shown that the probability that two elements from a finite noncommutative ring commute is at most $\frac{5}{8}$. A number of properties of $\Pr(G)$ when G is a finite group are stated. For example, if $H \leq G$, then $\Pr(G) \leq \Pr(H)$. Also, there is no group G such that $7/16 < \Pr(G) < 1/2$.

Ludvig Sylow

Sylow's Theorem is 100 years old. In the course of a century this remarkable theorem has been the basis for the construction of numerous theories.

L. A. Shemetkov

LUDVIG SYLOW (pronounced "SEE-loe") was born on December 12, 1832, in Christiania (now Oslo), Norway. While a student at Christiania University, Sylow won a gold medal for competitive problem-solving. In 1855, he became a high school teacher, and despite the long hours required by his teaching duties, Sylow found time to study the papers of Abel. During the school year 1862–1863, Sylow received a temporary appointment at Christiania University and gave lectures on Galois's theory and permutation groups. Among his students that year was the great mathematician Sophus Lie (pronounced "Lee"), after whom Lie algebras and Lie groups are named. From 1873 to 1881, Sylow, with some help from Lie, prepared a new edition of Abel's works. In 1902, Sylow and Elling Holst published Abel's correspondence.

Sylow's great discovery, Sylow's Theorem, came in 1872. Upon learning of Sylow's result, C. Jordan called it "one of the essential points in the theory of permutations." The result took on greater importance when the theory of abstract groups flowered in the late nineteenth century and early twentieth century.

In 1869, Sylow was offered a professorship at Christiania University, but turned it down. Upon Sylow's retirement at age sixty-five from high school teaching, Lie mounted a successful campaign to establish a chair for Sylow at Christiania University. Sylow held this position until his death on September 7, 1918.

Finite Simple Groups

27

It is a widely held opinion that the problem of classifying finite simple groups is close to a complete solution. This will certainly be one of the great achievements of mathematics of this century.

Nathan Jacobson

Historical Background

We now come to the El Dorado of finite group theory—the simple groups. Simple group theory is a vast and difficult subject; we call it the El Dorado of group theory because of the enormous effort put forth by hundreds of mathematicians during recent years to discover and classify all finite simple groups. Let's begin our discussion with the definition of a simple group and some historical background.

DEFINITION Simple Group

A group is *simple* if its only normal subgroups are the identity subgroup and the group itself.

The notion of a simple group was introduced by Galois about 150 years ago. The simplicity of A_5, the group of even permutations on five symbols, played a crucial role in his proof that there is not a solution by radicals of the general fifth degree polynomial (that is, there is no "quintic formula"). But what makes simple groups important in the theory of groups? They are important because they play a role in group theory somewhat analogous to that of primes in number theory or the elements in chemistry; that is, they serve as the building blocks for all groups. These building blocks may be determined in the following way. Given a finite group G, choose a normal subgroup G_1 of $G = G_0$ of largest order. Then the factor group G_0/G_1 is simple, and we next choose a normal subgroup G_2 of G_1 of largest order. Then G_1/G_2 is also simple, and we continue in this fashion until we arrive at $G_n = \{e\}$. The simple groups G_0/G_1, $G_1/G_2, \ldots, G_{n-1}/G_n$ are called the *composition factors* of G. More than 100 years ago Jordan and Hölder proved that these factors are independent of the choices of the normal subgroups made in the process described. In a certain sense, a group can be reconstructed from its composition factors and many of the properties of a group are determined by the nature of its composition factors. This and the fact that many questions about finite groups can be reduced (by induction) to questions about simple groups make clear the importance of determining all finite simple groups.

Just which groups are the simple ones? The Abelian simple groups are precisely Z_n where $n = 1$ or n is prime. This follows directly from the corollary in Chapter 12. Unfortunately, it is not at all easy to describe the known non-Abelian simple groups. The best we can do here is to give a few examples and mention a few words about their discovery. It was Galois who first observed that A_n is simple for all $n \geq 5$. The next discoveries were made by Jordan in 1870 when he found four infinite families of simple matrix groups over the field Z_p, where p is prime. Between the years 1892 and 1905, the American mathematician Dickson (see Chapter 24 for a biography) generalized Jordan's results to arbitrary finite fields and discovered several new infinite families of simple groups.

The next important discoveries came in the 1950s. In that decade, many new infinite families of simple groups were found and the initial steps down the long and winding road that led to the complete classification of all finite simple groups were taken. The first step was Brauer's observation that the centralizer of an element of order 2 was an important tool for studying simple groups. A few years later, Thompson, in his Ph.D. thesis, introduced the crucial idea of studying the normalizers of various subgroups of prime-power order.

In the early 1960s came the momentous Feit-Thompson Theorem, which says that a non-Abelian simple group must have even order. This property was first conjectured around 1900 by one of the pioneers of modern group-theoretical methods, the Englishman Burnside (see Chapter 28 for a biography). The proof of the Feit-Thompson Theorem filled an entire issue of a journal, 255 pages in all (see Figure 27.1, [2]). This result provided the impetus to classify the finite simple groups, that is, a program to discover all finite simple groups and *prove* that there are no more to be found. Throughout the 1960s, the methods introduced in the Feit-Thompson proof were generalized and improved with great success by a number of mathematicians. At the same time, a small number of simple groups—called *sporadic simple groups*—were constructed by ad hoc methods that did not yield infinitely many possibilities. Despite many spectacular achieve-

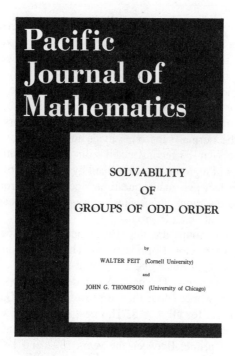

Pacific Journal of Mathematics

SOLVABILITY

OF

GROUPS OF ODD ORDER

by

WALTER FEIT   (Cornell University)

and

JOHN G. THOMPSON   (University of Chicago)

Oh, what are the orders of all simple groups?
I speak of the honest ones, not of the loops.
It seems that old Burnside their orders has
  guessed
Except for the cyclic ones, even the rest.

CHORUS: Finding all groups that are simple
          is no simple task.

Groups made up with permutes will produce
  some more:
For A_n is simple, if n exceeds 4.
Then, there was Sir Matthew who came into
  view
Exhibiting groups of an order quite new.

Still others have come on to study this thing.
Of Artin and Chevalley now we shall sing.
With matrices finite they made quite a list
The question is: Could there be others they've
  missed?

Suzuki and Ree then maintained it's the case
That these methods had not reached the end of
  the chase.
They wrote down some matrices, just four by
  four.
That made up a simple group. Why not make
  more?

And then came the opus of Thompson and
  Feit
Which shed on the problem remarkable light.
A group, when the order won't factor by two
Is cyclic or solvable. That's what is true.

Suzuki and Ree had caused eyebrows to raise,
But the theoreticians they just couldn't faze.
Their groups were not new: if you added a
  twist,
You could get them from old ones with a flick
  of the wrist.

Figure 27.1

Still, some hardy souls felt a thorn in their side.
For the five groups of Mathieu all reason defied;
Not A_n, not twisted, and not Chevalley,
They called them sporadic and filed them away.

Are Mathieu groups creatures of heaven or hell?
Zvonimir Janko determined to tell.
He found out that nobody wanted to know:
The masters had missed 1 7 5 5 6 0.

The floodgates were opened! New groups were the rage!
(And twelve or more sprouted, to greet the new age.)
By Janko and Conway and Fischer and Held
McLaughlin, Suzuki, and Higman, and Sims.

No doubt you noted the last lines don't rhyme.
Well, that is, quite simply, a sign of the time.
There's chaos, not order, among simple groups;
And maybe we'd better go back to the loops.

Figure 27.1 (continued)

ments, research in simple group theory in the 1960s was haphazard and the decade ended with many people believing that the classification would never be completed. (The anonymously written "song" in Figure 27.1 captures the spirit of the times.) Others, more optimistic, were predicting it would be accomplished in the 1990s.

The 1970s began with Thompson receiving the Fields Medal for his fundamental contributions to simple group theory. This honor is the highest recognition a mathematician can receive (more information about the Fields Medal is given at the end of this chapter). Within a few years three major events took place that ultimately led to the classification. First, Thompson published what is regarded as the single most important paper in simple group theory—the N-group paper. Here, Thompson introduced many fundamental techniques and supplied a model for the classification of a broad family of simple groups. Second was Gorenstein's elaborate outline for the classification delivered in a series of lectures at the University of Chicago in 1972. Here a broad plan for the overall proof was laid out. The army of researchers now had a battle plan and a commander-in-chief. But this army still needed more and better weapons. Thus, came the third critical development: the involvement of Michael Aschbacher. In a dazzling series of papers, Aschbacher combined his own insight with the methods of Thompson, which had been generalized throughout the 1960s, and a geometric approach pioneered by Bernd Fischer to achieve one brilliant result after another in rapid succession. In fact, so much progress was made by Aschbacher and others that by 1976, it was clear to nearly everyone involved that enough techniques had been developed to complete the classification. Only details remained.

The 1980s were ushered in with Aschbacher following in the footsteps of Feit and Thompson by winning the American Mathematical Society's Cole Prize in algebra (see the note at the end of this chapter). A week later, Robert L.

Griess made the spectacular announcement that he had constructed the "Monster."* The Monster is the largest of the sporadic simple groups. In fact, it has vastly more elements than there are atoms on the Earth! Its order is 808,017,424,794,512,875,886,459,904,961,710,757,005,754,368,000,000,000 (hence, the name). This is approximately 8×10^{53}. The Monster is a group of rotations in 196,883 dimensions. Thus, each element can be expressed as a $196,883 \times 196,883$ matrix.

The "Twenty-Five Years' War" came to an end in January 1981, when Gorenstein, speaking for "The Team," announced at an American Mathematical Society meeting that all finite simple groups had been classified; that is, there is now a complete list of all the finite simple groups. The proof that this list is complete runs over 10,000 journal pages. Many of the mathematicians involved in this effort are now searching for ways to simplify this proof.

Nonsimplicity Tests

In view of the fact that simple groups are the building blocks for all groups, it is surprising how scarce the non-Abelian simple groups are. For example, A_5 is the only one whose order is less than 168; there are only five non-Abelian simple groups of order less than 1000 and only 56 of order less than 1,000,000. In this chapter, we give a few theorems that are useful in proving that a particular integer is not the order of a non-Abelian simple group. Our first such result is an easy arithmetical test that comes from combining Sylow's Third Theorem and the fact that groups of prime-power order have nontrivial centers.

Theorem 27.1 *Sylow Test for Nonsimplicity*
Let n be a positive integer that is not prime, and let p be a prime divisor of n. If 1 is the only divisor of n that is congruent to 1 modulo p, then there does not exist a simple group of order n.

Proof. If n is a prime-power, then a group of order n has a nontrivial center and, therefore, is not simple. If n is not a prime-power, then every Sylow subgroup is proper and, by Sylow's Third Theorem, we know that the number of Sylow p-subgroups of a group of order n is congruent to 1 modulo p and divides n. Since 1 is the only such number, the Sylow p-subgroup is unique and, therefore, by the corollary to Sylow's Third Theorem, it is normal. ∎

How good is this test? Well, if we were to program a computer to apply this criterion to all the nonprime integers between 1 and 200 and eliminate any that satisfy it, only the following would be left as possible orders of finite simple groups: 12, 24, 30, 36, 48, 56, 60, 72, 80, 90, 96, 105, 108, 112, 120, 132,

---

*The name was coined by the English mathematician John H. Conway.

144, 150, 160, 168, 180, and 192. (In fact, computer experiments have revealed that for large intervals, say, 500 or more, this test eliminates over 90% of the nonprime integers as possible orders of simple groups. See [3] for more on this.)

Our next test rules out 30, 90, and 150.

Theorem 27.2   2 · Odd Test
An integer of the form 2 · n where n is an odd number greater than 1, is not the order of a simple group.

Proof.   Let G be a group of order $2n$ where n is odd and greater than 1. Recall from the proof of Cayley's Theorem (Example 8 of Chapter 6) that the mapping $g \to T_g$ is an isomorphism from G to a permutation group on the elements of G (where $xT_g = xg$ for all x in G). Since $|G| = 2n$, Cauchy's Theorem guarantees that there is an element g in G of order 2. Then, when the permutation T_g is written in disjoint cycle form, each cycle must have length 1 or 2, otherwise, $|g| \neq 2$. But T_g can contain no 1-cycles, because the 1-cycle (x) would mean $x = xT_g = xg$, so $g = e$. Thus, in cycle form, T_g consists of exactly n transpositions, where n is odd. Therefore, T_g is an odd permutation. This means that the set of even permutations in G is a normal subgroup of index 2. (See exercise 19 of Chapter 5 and exercise 7 of Chapter 10.) Hence, G is not simple.   ■

The next theorem is a broad generalization of Cayley's Theorem. We will make heavy use of its two corollaries.

Theorem 27.3   Generalized Cayley Theorem
Let G be a group and H a subgroup of G. Let S be the group of all permutations of the right cosets of H in G. Then there is a homomorphism from G into S whose kernel lies in H and contains every normal subgroup of G that is contained in H.

Proof.   For each $g \in G$, define a permutation T_g of the right cosets of H by $(Hx)T_g = Hxg$. As in the proof of Cayley's Theorem, it is easy to verify that the mapping of $\alpha: g \to T_g$ is a homomorphism from G into S.

Now if $g \in \text{Ker } \alpha$, then T_g is the identity map, so $H = HT_g = Hg$, and, therefore, g belongs to H. Thus, $\text{Ker } \alpha \subseteq H$. On the other hand, if K is normal in G and $K \subseteq H$, then for any $k \in K$ and any x in G, there is an element k' in K so that $xk = k'x$. Thus,

$$HxT_k = Hxk = Hk'x = Hx$$

and, therefore, T_k is the identity permutation. This means $k \in \text{Ker } \alpha$. We have proved then that every normal subgroup of G contained in H is also contained in $\text{Ker } \alpha$.   ■

As a consequence of Theorem 27.3, we obtain the following very powerful arithmetical test for nonsimplicity.

Corollary 1 *Index Theorem*

If G is a finite group and H is a proper subgroup of G such that $|G|$ does not divide $|G{:}H|!$, then H contains a nontrivial normal subgroup of G. In particular, G is not simple.

Proof. Let α be the homomorphism given in Theorem 27.3. Then $G/\text{Ker } \alpha$ is isomorphic to a subgroup of S and $\text{Ker } \alpha \subseteq H$. Thus, $|G/\text{Ker } \alpha|$ divides $|S| = |G{:}H|!$. Since $|G|$ does not divide $|G{:}H|!$, the order of $\text{Ker } \alpha$ must be greater than 1. ∎

Corollary 2 *Embedding Theorem*

If a finite non-Abelian simple group G has a subgroup of index n, then G is isomorphic to a subgroup of A_n.

Proof. Let H be the subgroup of index n and S_n the group of all permutations of the n right cosets of H in G. By the Generalized Cayley Theorem, there is a nontrivial homomorphism from G into S_n. Since G is simple and the kernel of a homomorphism is a normal subgroup of G, we see that the mapping from G into S_n is one-to-one so that G is isomorphic to some subgroup of S_n. Recall from exercise 19 of Chapter 5 that any subgroup of S_n consists of even permutations only or half even and half odd. If G were isomorphic to a subgroup of the latter type, the even permutations would be a normal subgroup of index 2 (see exercise 7 of Chapter 10), which would contradict that fact that G is simple. Thus, G is isomorphic to a subgroup of A_n. ∎

Using the Index Theorem with the largest Sylow subgroup for H reduces our list of possible orders of non-Abelian simple groups still further. For example, let G be any group of order $80 = 16 \cdot 5$. We may choose H to be a subgroup of order 16. Since 80 is not a divisor of 5!, there is no simple group of order 80. The same argument applies to 12, 24, 36, 48, 72, 96, 108, 160, and 192, thereby leaving only 56, 60, 105, 112, 120, 132, 144, 168, and 180 as possible orders of non-Abelian simple groups up to 200. Let's consider these. Quite often we may use a counting argument to eliminate an integer. Consider 56. By Sylow's Theorem, we know that a simple group of order $56 = 8 \cdot 7$ would contain eight Sylow 7-subgroups and seven Sylow 2-subgroups. Now any two Sylow p-subgroups that have order p must intersect in only the identity. So the union of the eight Sylow 7-subgroups yields 48 elements of order 7, while the union of any two Sylow 2-subgroups gives at least $8 + 8 - 4 = 12$ new elements. But there are only 56 elements in all. This contradiction shows that there is not a simple group of order 56. An analogous argument also eliminates the integers 105 and 132.

So, our list of possible orders of non-Abelian simple groups up to 200 is down to 60, 112, 120, 144, 168, and 180. Of these, 60 and 168 do correspond to simple groups. The others can be eliminated with a bit of razzle dazzle.

The easiest case to handle is $112 = 2^4 \cdot 7$. Suppose there were a simple group G of order 112. A Sylow 2-subgroup of G must have index 7. So, by the Embedding Theorem, G is isomorphic to a subgroup of A_7. But 112 does not divide $|A_7|$, a contradiction.

Next consider the possibility of a simple group G of order $144 = 9 \cdot 16$. By the Sylow theorems, we know that $n_3 = 4$ or 16 and $n_2 \geq 3$. (Recall n_p denotes the number of Sylow p-subgroups of G.) The Index Theorem rules out the case where $n_3 = 4$, so we know there are 16 Sylow 3-subgroups. Now, if every pair of Sylow 3-subgroups had only the identity in common, a straightforward counting argument would produce more than 144 elements. So, let H and H' be a pair of Sylow 3-subgroups whose intersection has order 3. Then $H \cap H'$ is a subgroup of both H and H' and by exercise 32 of Chapter 26, we see that $N(H \cap H')$ must contain both H and H' and, therefore, the set HH'. (HH' need not be a subgroup.) Thus,

$$|N(H \cap H')| \geq |HH'| = \frac{|H| \, |H'|}{|H \cap H'|} = \frac{9 \cdot 9}{3} = 27.$$

At this stage, we have three arithmetical conditions on $k = |N(H \cap H')|$. We know 9 divides k; k divides 144; and, $k \geq 27$. Clearly then $k \geq 36$ and so $|G{:}N(H \cap H')| \leq 4$. The Index Theorem now gives us the desired contradiction.

Finally, suppose G is a non-Abelian simple group of order $180 = 2^2 \cdot 3^2 \cdot 5$. Then $n_5 = 6$ or 36 and $n_3 = 10$. First, assume that $n_5 = 36$. Then G has $36 \cdot 4 = 144$ elements of order 5. Now if each pair of the Sylow 3-subgroups intersects in only the identity, then there are 80 more elements in the group, a contradiction. So, we may assume there are two Sylow 3-subgroups L_3 and L_3' whose intersection has order 3. Then, as was the case for the integer 144, we have

$$|N(L_3 \cap L_3')| \geq |L_3 L_3'| = \frac{9 \cdot 9}{3} = 27.$$

Thus,

$$|N(L_3 \cap L_3')| = 9 \cdot k$$

where $k \geq 3$ and k divides 20. Clearly then,

$$|N(L_3 \cap L_3')| \geq 36$$

and

$$|G{:}N(L_3 \cap L_3')| \leq 5.$$

The Index Theorem now gives us another contradiction. Hence, we may assume that $n_5 = 6$. In this case, we let H be the normalizer of a Sylow 5-subgroup of G. By Sylow's Third Theorem, we have $6 = |G{:}H|$, so that $|H| = 30$. In Chapter 26, we proved that every group of order 30 has an element of order 15. On the other hand, since $n_5 = 6$, G has a subgroup of index 6 and the Embedding

Theorem tells us that G is isomorphic to a subgroup of A_6. But A_6 has no element of order 15. (See exercise 6 of Chapter 5.)

Unfortunately, the argument for 120 is fairly long and complicated. However, no new techniques are required to do it. We leave this as an exercise. Some hints are given in the answer section.

The Simplicity of A_5

Once 120 is disposed of, we will have shown that the only integers between 1 and 200 that can be the orders of simple groups are 60 and 168. For completeness, we will now prove that A_5, which has order 60, is a simple group. A similar argument can be used to show that the factor group $SL(2, 7)/Z(SL(2, 7))$ is a simple group of order 168. (This group is denoted by $PSL(2, 7)$.)

If A_5 had a nontrivial proper normal subgroup H, then $|H| = 2, 3, 4, 5, 6, 10, 12, 15, 20$, or 30. By exercise 34 of Chapter 5, A_5 has 24 elements of order 5, 20 elements of order 3, and no elements of order 15. Now, if $|H| = 3, 6, 12$, or 15, then $|A_5/H|$ is relatively prime to 3, and by exercise 45 of Chapter 10, H would have to contain all 20 elements of order 3. If $|H| = 5, 10$, or 20, then $|A_5/H|$ is relatively prime to 5 and, therefore, H would have to contain the 24 elements of order 5. If $|H| = 30$, then $|A_5/H|$ is relatively prime to both 3 and 5, and so H would have to contain all the elements of order 3 and 5. Finally, if $|H| = 2$ or 4, then $|A_5/H| = 30$ or 15. But we know from our results in Chapter 26 that any group of order 30 or 15 has an element of order 15. However, since A_5 contains no such element, neither does A_5/H. This proves that A_5 is simple.

The simplicity of A_5 was known to Galois in 1830, although the first formal proof was done by Jordan in 1870. A few years later, Felix Klein showed that the group of rotations of a regular icosahedron is simple and, therefore, isomorphic to A_5 (see exercise 28). Since then it has frequently been called the *icosahedral group*. Klein was the first to prove that there is a simple group of order 168.

The problem of determining which integers in a certain interval are possible orders for finite simple groups goes back to 1892, when Hölder went up to 200. His arguments for the integers 144 and 180 alone used up ten pages. By 1975, this problem had been pushed to well beyond 1,000,000. See [4] for a detailed account of this endeavor. Of course, now that all finite simple groups have been classified, this problem is merely a historical curiosity.

The Fields Medal

The highest award for mathematical achievement is the Fields Medal. Two to four such awards are bestowed at the opening session of the International Congress of Mathematicians, held once every four years. Although the Fields Medal is considered by most mathematicians as the equivalent of the Nobel Prize, there are great differences between these awards. Besides the huge disparity in publicity and monetary value associated with the two honors, the Fields Medal is restricted to those under forty years of age.* This tradition stems from John Charles Fields's stipulation, in his will establishing the medal, that the awards should be "an encouragement for further achievement."

More details about the Fields Medal and a list of the recipients can be found in [1] and [5]. The latter article also includes photographs of the medal.

The Cole Prize

Approximately every five years since 1928, the American Mathematical Society awards one or two Cole Prizes for research in algebra and one or two Cole Prizes for research in algebraic number theory. The prize was founded in honor of Frank Nelson Cole on the occasion of his retirement as secretary of the American Mathematical Society. In view of the fact that Cole was one of the first people interested in simple groups, it is interesting to note that no fewer than six recipients of the prize—Dickson, Chevalley, Brauer, Feit, Thompson, and Aschbacher—have made fundamental contributions to simple groups at some time in their careers. In 1980, Aschbacher received an award of $1500.

EXERCISES

If you don't learn from your mistakes, there's no sense making them.

Herbert V. Prochnow

1. Prove that there is no simple group of order $210 = 2 \cdot 3 \cdot 5 \cdot 7$.
2. Prove that there is no simple group of order $280 = 2^3 \cdot 5 \cdot 7$.
3. Prove that there is no simple group of order $216 = 2^3 \cdot 3^3$.
4. Prove that there is no simple group of order $300 = 2^2 \cdot 3 \cdot 5^2$.
5. Prove that there is no simple group of order $525 = 3 \cdot 5^2 \cdot 7$.
6. Prove that there is no simple group of order $540 = 2^2 \cdot 3^3 \cdot 5$.
7. Prove that there is no simple group of order $528 = 2^4 \cdot 3 \cdot 11$.

---

*"Take the sum of human achievement in action, in science, in art, in literature—subtract the work of the men above forty, and while we should miss great treasures, even priceless treasures, we would practically be where we are to-day. . . . The effective, moving, vitalizing work of the world is done between the ages of twenty-five and forty." Sir William Osler (1849–1919), *Life of Sir William Osler*, vol. I, chap. 24 (The Fixed Period).

8. Prove that there is no simple group of order $315 = 3^2 \cdot 5 \cdot 7$.

9. Prove that there is no simple group of order $396 = 2^2 \cdot 3^2 \cdot 11$.

10. Prove that there is no simple group of order n where $201 \leqslant n \leqslant 235$.

11. Without using the Generalized Cayley Theorem or its corollaries, prove that there is no simple group of order 112.

12. Without using the "$2 \cdot$ odd" test, prove that there is no simple group of order 210.

13. You may have noticed that all the "hard" integers are even. Choose three odd integers between 200 and 1000. Show that none of these is the order of a simple group unless it is prime.

14. Show that there is no simple group of order pqr where p, q, and r are primes (p, q, and r need not be distinct).

15. Show that A_5 cannot contain a subgroup of order 30, 20, or 15.

16. Show that S_5 cannot contain a subgroup of order 40 or 30. (This exercise is referred to in Chapter 33.)

17. Prove that a simple group order 60 has a subgroup of order 6 and a subgroup of order 10.

18. Prove that if G is a finite group and H a proper normal subgroup of largest order, then G/H is simple.

19. Suppose H is a subgroup of a finite group G and $|H|$ and $(|G{:}H| - 1)!$ are relatively prime. Prove that H is normal in G. What does this tell you about a subgroup of index 2 in a finite group?

20. Suppose p is the smallest prime that divides $|G|$. Show that any subgroup of index p in G is normal in G.

21. Prove that there is no simple group of order $120 = 2^3 \cdot 3 \cdot 5$.

22. Show that the group of rotations of a regular dodecahedron is simple.

23. Show that the group of rotations of a regular icosahedron is simple.

24. Prove that the only nontrivial proper normal subgroup of S_5 is A_5. (This exercise is referred to in Chapter 33.)

25. Show that $PSL(2, 7) = SL(2, 7)/Z(SL(2, 7))$ is simple. (This exercise is referred to in this chapter.)

26. Show that the permutations (12) and (12345) generate S_5.

27. Suppose a subgroup H of S_5 contains a 5-cycle and a 2-cycle. Show that $H = S_5$. (This exercise is referred to in Chapter 33.)

28. Show that (up to isomorphism) A_5 is the only simple group of order 60.

29. Suppose G is a finite simple group and G contains subgroups H and K such that $|G{:}H|$ and $|G{:}K|$ are prime. Show that $|H| = |K|$.

PROGRAMMING EXERCISES

One machine can do the work of fifty ordinary men. No machine can do the work of one extraordinary man.

Elbert Hubbard, *Roycraft Dictionary and Book of Epigrams*

1. Program Theorem 27.1. Use a counter M to keep track of how many integers the theorem eliminates in any given interval. Run your program for the following intervals: 1–100; 501–600; 5001–5100; 10,001–10,100. How does M seem to behave as the size of the integers grow?

2. Program the Index Theorem. Use a counter M to keep track of how many integers it eliminates in any given interval. Run your program for the same intervals as in exercise 1. How does M seem to behave as the size of the integers grow?

REFERENCES

1. H. Edwards, "A Short History of the Fields Medal," *The Mathematical Intelligencer* 1 (1978): 127–129.

2. W. Feit and J. G. Thompson, "Solvability of Groups of Odd Order," *Pacific Journal of Mathematics* 13 (1963): 775–1029.

3. J. A. Gallian, "Computers in Group Theory," *Mathematics Magazine* 49 (1976): 69–73.

4. J. A. Gallian, "The Search for Finite Simple Groups," *Mathematics Magazine* 49 (1976): 163–179.

5. H. S. Tropp, "The Origins and History of the Fields Medal," *Historia Mathematica* 3 (1976): 167–181.

SUGGESTED READINGS

G. Cornell, N. Pele, and M. Wage, "Simple Groups of Orders Less Than 1000," *Journal of Undergraduate Research* 5 (1973): 77–86.

In this charming article, three undergraduate students use slightly more theory than was given in this chapter to show that the only integers less than 1000 that could be orders of simple groups are 60, 168, 320, 504, 660, and 720. All but the last one are orders of simple groups. The proof that there is no simple group of order 720 is omitted because it is significantly beyond most undergraduates.

Karl David, "Using Commutators to Prove A_5 Is Simple," *The American Mathematical Monthly* 94 (1987): 775–776.

This note gives an elementary proof that A_5 is simple using commutators.

J. A. Gallian, "The Search for Finite Simple Groups," *Mathematics Magazine* 49 (1976): 163–179.

A historical account of the search for finite simple groups is given.

Anthony Gardiner, "Groups of Monsters," *New Scientist* April 5 (1979): 34.

In this article, the author briefly discusses the construction of the sporadic simple groups. He mentions that Charles Sims constructed the "Baby Monster" (order 4,154,781,481,226,426,191,177,580,544,000,000) using $8,000 of computer time, and that Sims has a technique to construct the "Monster" at a cost of $3,000,000 and occupy the entire Rutgers University Computer Complex for a year! (Incidentally, Griess's construction was entirely by hand.)

Martin Gardner, "The Capture of the Monster: A Mathematical Group with a Ridiculous Number of Elements," *Scientific American* 242 (6) (1980): 20–32.

This article gives an elementary introduction to groups and a discussion of simple groups, including the "Monster."

Daniel Gorenstein, "The Enormous Theorem," *Scientific American* 253(6) 1985: 104–115.

You won't find an article on a complex subject better written for the layperson than this one. Gorenstein, the driving force behind the classification, uses concrete examples, analogies, and nontechnical terms to make the difficult subject matter of simple groups accessible.

A. L. Hammond, "Sporadic Groups: Exceptions, or Part of a Pattern?" *Science* 181 (1973): 146–148.

This article gives a brief discussion of sporadic simple groups and their connection with error-correcting codes such as those used to transmit data from spacecrafts to earth.

Richard Silvestri, "Simple Groups of Finite Order," *Archive for the History of Exact Sciences* 20 (1979): 313–356.

This article contains a plethora of historical information about the work on simple groups in the nineteenth century.

Michael Aschbacher

Fresh out of graduate school, he [Aschbacher] had just entered the field, and from that moment he became the driving force behind my program. In rapid succession he proved one astonishing theorem after another. Although there were many other major contributors to this final assault, Aschbacher alone was responsible for shrinking my projected 30-year timetable to a mere 10 years.

Daniel Gorenstein, *Scientific American*

MICHAEL ASCHBACHER was born on April 8, 1944, in Little Rock, Arkansas. Shortly after his birth, his family moved to Illinois, where his father was a professor of accounting and his mother was a high school English teacher. When he was nine years old, his family moved to East Lansing, Michigan; six years later, they moved to Los Angeles.

After high school, Aschbacher enrolled in the California Institute of Technology. In addition to his schoolwork, he passed the first four actuary exams and was employed for a few years as an actuary, full-time in the summers and part-time during the academic year. Two of the Cal Tech mathematicians who influenced him were Marshall Hall and Donald Knuth. In his senior year, Aschbacher took abstract algebra but showed little interest in the course. Accordingly, he received a grade of C.

In 1966, Aschbacher went to the University of Wisconsin for a Ph.D. degree. He completed his thesis in 1969 and, after spending one year as an assistant professor at the University of Illinois, he returned to Cal Tech and quickly moved up to the rank of professor.

Aschbacher's thesis work in the area of combinatorial geometries had led him to consider certain group-theoretical questions. Gradually, he turned his attention more and more to purely group-theoretical problems, particularly those bearing on the classification of finite simple groups. The 1980 Cole Prize Selection Committee said of one of his papers, "[It] *lifted the subject to a new plateau and brought the classification within reach.*"

Daniel Gorenstein

The techniques of these three papers [by Daniel Gorenstein and John Walter, classifying finite simple groups with dihedral Sylow 2-subgroups] cover the spectrum of finite group theory more thoroughly than any single paper known to the reviewer

John G. Thompson, *Mathematical Reviews*

DANIEL GORENSTEIN was born in Boston on January 1, 1923. He became interested in mathematics at the age of twelve, when he taught himself calculus. After graduating from the Boston Latin School, he entered Harvard University. His senior thesis was done, under the direction of Saunders MacLane, on finite groups. Upon graduating in 1943, Gorenstein was offered an instructorship at Harvard to teach mathematics to army personnel. After the war ended, he began graduate work at Harvard. He received his Ph.D. degree in 1951 working in algebraic geometry under Oscar Zariski. It was in his thesis that he introduced the class of rings that is now named after him. In 1951, Gorenstein was offered a position at Clark University in Worcester, Massachusetts, where he stayed until moving to Northeastern University in 1964. He accepted his present position at Rutgers in 1969.

A milestone in Gorenstein's development as a group theorist came in 1960–1961 when he was invited to participate in a "Group Theory Year" at the University of Chicago. It was there that Gorenstein, assimilating the revolutionary techniques then being developed by John Thompson, began his fundamental work that contributed to the classification of finite simple groups.

Through his pioneering research papers, dynamic lectures, numerous personal contacts, and his influential book on finite groups, Gorenstein became the leader in the twenty-five-year effort, by hundreds of mathematicians, that led to the classification of the finite simple groups.

John Thompson

There seemed to be no limit to his power.

Daniel Gorenstein

JOHN G. THOMPSON was born on October 13, 1932, in Ottawa, Kansas. In 1951, he entered Yale University as a divinity student, but switched to mathematics in his sophomore year. In 1955, he began graduate school at the University of Chicago and obtained his Ph.D. degree four years later. After one year on the faculty at Harvard, Thompson returned to Chicago. He remained there until 1968, when he moved to Cambridge University in England. At Cambridge, he is the Rouse Ball Professor of Mathematics.

Thompson's brilliance was evident early. In his dissertation, he verified a fifty-year-old conjecture about finite groups possessing a certain kind of automorphism. (An article about his achievement appeared in the *New York Times*!) The novel methods Thompson used in his thesis foreshadowed the revolutionary ideas he would later introduce in the Feit-Thompson paper and the classification of minimal simple groups (i.e., simple groups that contain no proper non-Abelian simple subgroups). The assimilation and extension of Thompson's methods by others throughout the 1960s and 1970s ultimately led to the classification of finite simple groups.

John Thompson received the Cole Prize in algebra from the American Mathematical Society in 1965 and the Fields Medal in 1970. In 1987, Thompson was presented an honorary degree of Doctor of Science by the University of Oxford.

Generators and Relations

28

One cannot escape the feeling that these mathematical formulae have an independent existence and an intelligence of their own, that they are wiser than we are, wiser even than their discoverers, that we get more out of them than we originally put into them.

Heinrich Hertz

Motivation

In this chapter, we give a convenient way to define a group with certain prescribed properties. Simply put, we begin with a set of elements that we want to generate the group, and a set of equations (called *relations*) that specify conditions these generators are to satisfy. Among all such possible groups, we will select one as large as possible. This will uniquely determine the group up to isomorphism.

To motivate the theory involved, we begin with a concrete example. Consider D_4, the group of symmetries of a square. Recall that $R = R_{90}$ and H, a reflection across a horizontal axis, generate the group. Observe that R and H are related in the following ways:

$$(1) \qquad R^4 = H^2 = (RH)^2 = R_0 \qquad \text{(the identity)}.$$

Other relations between R and H such as $HR = R^3H$ and $RHR = H$ also exist, but they can be derived from those given in (1). For example, $(RH)^2 = R_0$ yields $HR = R^{-1}H^{-1}$ and $R^4 = H^2 = R_0$ give $R^{-1} = R^3$ and $H^{-1} = H$. So, $HR = R^3H$. In fact, every relation between R and H can be derived from those given in (1).

Thus, D_4 is a group that is generated by a pair of elements a and b subject to the relations $a^4 = b^2 = (ab)^2 = e$ and such that all other relations between *a* and *b* can be derived from these. This last stipulation is necessary because the subgroup $\{R_0, R_{180}, H, V\}$ of D_4 is generated by two elements satisfying the relations in (1) with $a = R_{180}$ and $b = H$. However, the "extra" relation $a^2 = e$ satisfied by this subgroup cannot be derived from the original ones (since $R_{90}^2 \neq R_0$). It is natural to ask whether this description of D_4 applies to some other group as well. The answer is no. Any other group generated by two elements α and β satisfying only the relations $\alpha^4 = \beta^2 = (\alpha\beta)^2 = e$, and those that can be derived from these, is isomorphic to D_4.

Similarly, one can show that the group $Z_4 \oplus Z_2$ is generated by two elements a and b such that $a^4 = b^2 = e$ and $ab = ba$, and any other relation between a and b can be derived from these. The purpose of this chapter is to show that this procedure can be reversed; that is, we can begin with any set of generators and relations among the generators and construct a group that is uniquely described by these generators and relations, subject to the stipulation that all other relations among the generators can be derived from the original ones.

Definitions and Notation

We begin with some definitions and notation. For any set $S = \{a, b, c, \ldots\}$ of distinct symbols, we create a new set $\overline{S} = \{a^{-1}, b^{-1}, c^{-1}, \ldots\}$ by replacing each x in S by x^{-1}. Define the set $W(S)$ to be the collection of all formal finite strings of the form $x_1 x_2 \cdots x_k$, where each $x_i \in S \cup \overline{S}$. The elements of $W(S)$ are called *words from S*. We also permit the string with no elements to be in $W(S)$. This word is called the *empty word* and is denoted by e.

We may define a binary operation on the set $W(S)$ by juxtaposition; that is, if $x_1 x_2 \cdots x_k$ and $y_1 y_2 \cdots y_t$ belong to $W(S)$, then so does $x_1 x_2 \cdots x_k y_1 y_2 \cdots y_t$. Observe that this operation is associative and the empty word is the identity. Also, notice that a word such as aa^{-1} is not the identity because we are treating the elements of $W(S)$ as formal symbols with no implied meaning.

At this stage we have everything we need to make a group out of $W(S)$ except inverses. Here a difficulty arises since it seems reasonable that the inverse of the word ab, say, should be $b^{-1}a^{-1}$. But $abb^{-1}a^{-1}$ is not the empty word! You may recall that we faced a similar obstacle long ago when we carried out the construction of the field of quotients of an integral domain. There we had formal symbols of the form a/b and we wanted the inverse of a/b to be b/a. But their product, ab/ba, was a formal symbol not the same as the formal symbol $1/1$, the identity. So, we proceed here as we did there—by way of equivalence classes.

DEFINITION Equivalence Classes of Words

For any pair of elements u and v of $W(S)$, we say u is related to v if v can be obtained from u by a finite sequence of insertions or deletions of words of the form xx^{-1} or $x^{-1}x$ where $x \in S$.

We leave it as an exercise to show that this relation is an equivalence relation on $W(S)$.

Example 1 Let $S = \{a, b, c\}$. Then $acc^{-1}b$ is equivalent to ab; $aab^{-1}bbaccc^{-1}$ is equivalent to $aabac$; the word $a^{-1}aabb^{-1}a^{-1}$ is equivalent to the empty word; the word $ca^{-1}b$ is equivalent to $cc^{-1}caa^{-1}a^{-1}bbca^{-1}ac^{-1}b^{-1}$. Note however that $cac^{-1}b$ is not equivalent to ab. $\qquad\qquad\square$

Free Group

Theorem 28.1 *Equivalence Classes Form a Group*
Let S be a set of distinct symbols. For any word u in $W(S)$ let $\overline{u}$ denote the set of all words in $W(S)$ equivalent to u (that is, $\overline{u}$ is the equivalence class containing u). Then the set of all equivalence classes of elements of $W(S)$ is a group under the operation $\overline{u} \cdot \overline{v} = \overline{uv}$.

Proof. This proof is left as an exercise. ∎

The group defined in Theorem 28.1 is called a *free group on S*. Theorem 28.2 shows why free groups are important.

Theorem 28.2 *The Universal Mapping Property*
Every group is a homomorphic image of a free group.

Proof. Let G be a group and let S be a set of generators for G. (Such a set exists because we may take S to be G itself.) Now let F be the free group on S. Unfortunately, since any word in $W(S)$ is also an element of G, we have created a notational problem for ourselves. So, to distinguish between these two cases we will denote the word $x_1x_2 \cdots x_n$ in $W(S)$ by $(x_1x_2 \cdots x_n)_F$ and the product $x_1x_2 \cdots x_n$ in G by $(x_1x_2 \cdots x_n)_G$. As before $\overline{x_1x_2 \cdots x_n}$ denotes the equivalence class in F containing the word $x_1x_2 \cdots x_n$ in $W(S)$. Notice that $(x_1x_2 \cdots x_n)_F$ and $(x_1x_2 \cdots x_n)_G$ may be entirely different elements since the operations on $W(S)$ and G are different.
 Now consider the mapping from F into G given by

$$(\overline{x_1x_2 \cdots x_n})_F \phi = (x_1x_2 \cdots x_n)_G.$$

Clearly, ϕ is well defined, for inserting or deleting expressions of the form xx^{-1} or $x^{-1}x$ in elements of $W(S)$ corresponds to inserting or deleting the identity in G. To check that ϕ is operation-preserving, observe that

$$(\overline{x_1x_2 \cdots x_n})_F(\overline{y_1y_2 \cdots y_m})_F \phi = (\overline{x_1x_2 \cdots x_ny_1y_2 \cdots y_m})_F \phi$$
$$= (x_1x_2 \cdots x_ny_1y_2 \cdots y_m)_G$$
$$= (x_1x_2 \cdots x_n)_G(y_1y_2 \cdots y_m)_G.$$

(All we are doing is taking a product in F and viewing it as a product in G. For example, if G is the cyclic group of order 4 generated by a, then

$$(\overline{aaaaa})_F \phi = (aaaaa)_G = a.)$$

Finally, ϕ is onto G because S generates G.   ■

The following corollary is an immediate consequence of Theorem 28.2 and the First Isomorphism Theorem for Groups.

Corollary   *Universal Factor Group Property*
Every group is isomorphic to a factor group of a free group.

Generators and Relations

We have now laid the foundation for defining a group by way of generators and relations. Before doing so, we will illustrate the basic idea with an example.

Example 2   Let F be the free group on the set $\{a, b\}$ and let N be the smallest normal subgroup of F containing the set $\{a^4, b^2, (ab)^2\}$. We will show that F/N is isomorphic to D_4. We begin by observing that the mapping ϕ from F onto D_4, which takes a to R_{90} and b to H (horizontal reflection), is a homomorphism whose kernel contains N. Thus, $F/\text{Ker } \phi$ is isomorphic to D_4. On the other hand, we claim that the set

$$K = \{N, aN, a^2N, a^3N, bN, abN, a^2bN, a^3bN\}$$

of left cosets of F/N is F/N itself. To see this, it suffices to show that K is closed under multiplication on the left by a and b. Clearly, every member of F/N can be generated by starting with N and successively multiplying on the left by various combinations of a's and b's. So, once we have generated K, the closure property implies that no further elements of F/N can be generated. It is trivial that K is closed under left multiplication by a. For b, we will do only one of the eight cases. The others can be done in a similar fashion. Consider $b(aN)$. From $(ab)^2N = N$ and $a^4N = N$, we deduce $babN = a^{-1}N = a^3N$. From the normality of N, we obtain $babN = baNb$. So, $baNb = a^3N$ and, therefore, $baNb^2 = a^3Nb$. Finally, since $b^2N = N$ and N is normal, this last relation yields $b(aN) = a^3bN$. Upon completion of the other cases, we know F/N has at most eight elements. At the same time, we know $F/\text{Ker } \phi$ has exactly eight elements. Since $F/\text{Ker } \phi$ is a factor group of F/N (indeed, $F/\text{Ker } \phi \approx (F/N)/(\text{Ker } \phi/N)$), it follows that F/N also has eight elements and $F/N = F/\text{Ker } \phi \approx D_4$.   □

DEFINITION   Generators and Relations
Let G be a group generated by some set $A = \{a_1, a_2, \ldots, a_n\}$ and let F be the free group on A. Let $W = \{w_1, w_2, \ldots, w_t\}$ be a subset of F and let N be the smallest normal subgroup of F containing W. We say G is *given by the generators* $a_1, a_2, \ldots, a_n$ *and the relations* $w_1 = w_2 = \cdots = w_t = e$ if there is an isomorphism from F/N onto G that carries a_iN to a_i.

The notation for this situation is

$$G = \langle a_1, a_2, \ldots, a_n \mid w_1 = w_2 = \cdots = w_t = e \rangle.$$

As a matter of convenience, we have restricted the number of generators and relations in our definition to be finite. This restriction is not necessary, however. Also, it is often more convenient to write a relation in implicit form. For example, the relation $a^{-1}b^{-3}ab = e$ is often written as $ab = b^3a$. In practice, one does not bother writing down the normal subgroup N that contains the relations. Instead, one just thinks of anything in N as the identity, as our notation suggests. Rather than say G is given by

$$\langle a_1, a_2, \ldots, a_n \mid w_1 = w_2 = \cdots = w_t = e \rangle,$$

many authors prefer to say that G has the *presentation*

$$\langle a_1, a_2, \ldots, a_n \mid w_1 = w_2 = \cdots = w_t = e \rangle.$$

Notice that a free group is "free" of relations; that is, the equivalence class containing the empty word is the only relation. We mention in passing the fact that a subgroup of a free group is also a free group. (See [3, p. 242] for a proof.) Free groups are of fundamental importance in a branch of algebra known as combinatorial group theory.

Example 3   The discussion in Example 2 can now be summed up by writing

$$D_4 = \langle a, b \mid a^4 = b^2 = (ab)^2 = e \rangle. \qquad \square$$

Example 4   The group of integers is the free group on one letter; that is, $Z \approx \langle a \rangle$. (This is the only nontrivial Abelian group that is free.) $\qquad \square$

The next theorem formalizes the argument used in Example 2 to prove that the group defined there had eight elements.

Theorem 28.3   *(Dyck, 1882)*
Let
$$G = \langle a_1, a_2, \ldots, a_n \mid w_1 = w_2 = \cdots = w_t = e \rangle$$
and let
$$\overline{G} = \langle a_1, a_2, \ldots, a_n \mid w_1 = w_2 = \cdots = w_t$$
$$= w_{t+1} = \cdots = w_{t+k} = e \rangle.$$
Then $\overline{G}$ is a homomorphic image of G.

Proof.   Exercise. ∎

Corollary   *Largest Group Satisfying Defining Relations*
If K is a group satisfying the defining relations of a finite group G and $|K| \geq |G|$, then K is isomorphic to G.

Proof.   Exercise.   ■

Example 5   Quaternions

Consider the group $G = \langle a, b \mid a^2 = b^2 = (ab)^2 \rangle$. What does G look like?
Formally, G is of course isomorphic to F/N where F is free on $\{a, b\}$ and N is
the smallest normal subgroup of F containing $b^{-2}a^2$ and $(ab)^{-2}a^2$. But as we
have already said, we need not use the N. We just think of the elements of G
as words in a and b where $a^2 = b^2 = (ab)^2$. Now let $H = \langle b \rangle$ and $S = \{H,$
$aH\}$. Then, just as in Example 2, it follows that S is closed under multiplication
by a and b from the left. So, as in Example 2, we have $G = H \cup aH$. Thus,
we can determine the elements of G once we know exactly how many elements
there are in H. (Here again, the three relations come in.) To do this, first observe
that $b^2 = (ab)^2 = abab$ implies $b = aba$. Then $a^2 = b^2 = (aba)(aba) =$
$aba^2ba = ab^4a$ and, therefore, $b^4 = e$. Hence, H has at most four elements and
therefore G has at most eight; namely, $e, b, b^2, b^3, a, ab, ab^2,$ and ab^3. It is
conceivable, however, that not all of these eight are distinct. For example,
$Z_2 \oplus Z_2$ satisfies the defining relations and has only four elements. Perhaps it
is the largest group satisfying the relations. How can we show that the eight
elements listed above are distinct? Well, consider the group $\overline{G}$ generated by the
matrices

$$ A = \begin{bmatrix} 0 & 1 \\ -1 & 0 \end{bmatrix} \quad \text{and} \quad B = \begin{bmatrix} 0 & i \\ i & 0 \end{bmatrix} $$

where $i = \sqrt{-1}$. Direct calculations show that in $\overline{G}$ the elements $e, B, B^2, B^3,$
A, AB, AB^2, AB^3 are distinct and $\overline{G}$ satisfies the relations $A^2 = B^2 = (AB)^2$.
So, it follows from the corollary to Dyck's Theorem, that G has order 8.   □

The next example illustrates why in Examples 2 and 5, it is necessary to
show that the eight elements listed for the group are distinct.

Example 6   Let

$$ G = \langle a, b \mid a^3 = b^9 = e, a^{-1}ba = b^{-1} \rangle. $$

Once again, we let $H = \langle b \rangle$ and observe that $G = H \cup aH \cup a^2H$. Thus,

$$ G = \{a^i b^j \mid 0 \leqslant i \leqslant 2, 0 \leqslant j \leqslant 8\} $$

and, therefore, G has at most 27 elements. But this time we will not be able to
find some concrete group of order 27 satisfying the same relations that G does,
for notice that $b^{-1} = a^{-1}ba$ implies

$$ b = (a^{-1}ba)^{-1} = a^{-1}b^{-1}a. $$

Hence,

$$ b = ebe = a^{-3}ba^3 = a^{-2}(a^{-1}ba)a^2 = a^{-2}b^{-1}a^2 $$
$$ = a^{-1}(a^{-1}b^{-1}a)a = a^{-1}ba = b^{-1}. $$

So, the original three relations imply the additional relation $b^2 = e$. But $b^2 = e = b^9$ further implies $b = e$. It follows then that G has only three distinct elements, namely, e, a, and a^2. $\qquad\square$

We hope Example 6 convinces you of the fact that once a list of the elements of the group given by a set of generators and relations is obtained, one must further verify that this list has no duplications. Typically, this is accomplished by exhibiting a specific group that satisfies the given set of generators and relations that has the same size as the list. Obviously, experience plays a role here.

Classification of Groups of Order up to 15

The next theorem illustrates the utility of the ideas presented in this chapter.

Theorem 28.4 *(Cayley, 1859) Classification of Groups of Order 8*
Up to isomorphism, there are only five groups of order 8: Z_8, $Z_4 \oplus Z_2$, $Z_2 \oplus Z_2 \oplus Z_2$, D_4, and the quaternions.

Proof. The Fundamental Theorem of Finite Abelian Groups takes care of the Abelian cases. Now, let G be a non-Abelian group of order 8. Also, let $G_1 = \langle a, b \mid a^4 = b^2 = (ab)^2 = e \rangle$ and let $G_2 = \langle a, b \mid a^2 = b^2 = (ab)^2 \rangle$. We know from the preceding examples that G_1 is isomorphic to D_4 and G_2 is isomorphic to the quaternions. Thus, it suffices to show that G must satisfy the defining relations for G_1 or G_2. It follows from exercise 38 of Chapter 2 and Lagrange's Theorem that G has an element of order 4; call it a. Then if b is any element of G not in $\langle a \rangle$, we know

$$G = \langle a \rangle \cup \langle a \rangle b = \{e, a, a^2, a^3, b, ab, a^2b, a^3b\}.$$

Consider the element b^2 of G. Which of the eight elements of G can it be? Not b, ab, a^2b, or a^3b, by cancellation. Not a, for b^2 commutes with b and a does not. Not a^3, for the same reason. Thus, $b^2 = e$ or $b^2 = a^2$. Suppose $b^2 = e$. Since $\langle a \rangle$ is a normal subgroup of G, we know $b^{-1}ab \in \langle a \rangle$. From this and the fact that $|b^{-1}ab| = |a|$, we then conclude that $b^{-1}ab = a$ or $b^{-1}ab = a^{-1}$. The first relation would mean that G is Abelian, so we know $b^{-1}ab = a^{-1}$. But then, since $b^2 = e$, we have $(ab)^2 = e$ and, therefore, G satisfies the defining relations for G_1.

Finally, if $b^2 = a^2$ holds instead of $b^2 = e$, we can show by an argument like that already given that $(ab)^2 = a^2$, and, therefore, G satisfies the defining relations for G_2. $\qquad\blacksquare$

The classification of the groups of order 8, together with our results on groups of order p^2, $2p$, and pq from Chapter 28, allow us to classify the groups of order up to 15 with the exception of those of order 12. We already know four

Table 28.1 Classification of Groups of Order Up to 15

Order	Abelian Groups	Non-Abelian Groups
1	Z_1	
2	Z_2	
3	Z_3	
4	$Z_4, Z_2 \oplus Z_2$	
5	Z_5	
6	Z_6	D_3
7	Z_7	
8	$Z_8, Z_4 \oplus Z_2, Z_2 \oplus Z_2 \oplus Z_2$	D_4, Q_4
9	$Z_9, Z_3 \oplus Z_3$	
10	Z_{10}	D_5
11	Z_{11}	
12	$Z_{12}, Z_6 \oplus Z_2$	D_6, A_4, Q_6
13	Z_{13}	
14	Z_{14}	D_7
15	Z_{15}	

groups of order 12, namely, $Z_{12}, Z_6 \oplus Z_2, D_6,$ and A_4. An argument along the lines of Theorem 28.4 can be given to show there is only one more group of order 12. This group, called the *dicyclic group of order 12* and denoted by Q_3, has presentation $\langle a, b \mid a^6 = e, a^3 = b^2, b^{-1}ab = a^{-1} \rangle$. Table 28.1 lists the groups of order at most 15. We use Q_4 to denote the quaternions (see Example 5 of this chapter).

Characterization of Dihedral Groups

As another nice application of generators and relations, we may now characterize the dihedral groups. For $n \geq 3$, we have used D_n to denote the group of symmetries of a regular n-gon. Imitating Example 2, one can show that $D_n \approx \langle a, b \mid a^n = b^2 = (ab)^2 = e \rangle$ (see exercise 8). By analogy, these generators and relations serve to define D_1 and D_2 also. (These are also called dihedral groups.) Finally, we define the infinite dihedral group D_∞ as $\langle a, b \mid a^2 = b^2 = e \rangle$. The elements of D_∞ can be listed as $e, a, b, ab, ba, (ab)a, (ba)b, (ab)^2, (ba)^2, (ab)^2a, (ba)^2b, (ab)^3, (ba)^3, \ldots$.

> **Theorem 28.5** *Characterization of Dihedral Groups*
> *Any group generated by a pair of elements of order 2 is dihedral.*

Proof. Let G be a group generated by a pair of elements of order 2, say, a and b. We consider the order of ab. If $|ab| = \infty$, then G is infinite and satisfies the relations of D_∞. We will show that G is isomorphic to D_∞. By Dyck's Theorem, G is isomorphic to some factor group of D_∞, say, D_∞/H.

Now suppose $x \in H$ and $x \neq e$. Since every element of D_∞ has one of the forms $(ab)^i$, $(ba)^i$, $(ab)^i a$, or $(ba)^i b$, by symmetry, we may assume $x = (ab)^i$ or $x = (ab)^i a$. If $x = (ab)^i$, then

$$H = (ab)^i H = (abH)^i$$

so that $(abH)^{-1} = (abH)^{i-1}$. But

$$(abH)^{-1} H = (abH)^{-1} = b^{-1} a^{-1} H = baH$$

and it follows that

$$aHabHaH = baH = (abH)^{-1}.$$

Thus,

$$D_\infty/H = \langle aH, bH \rangle = \langle aH, abH \rangle$$

(see exercise 7) and D_∞/H satisfies the defining relations for D_i (use exercise 8 with $x = aH$ and $y = abH$). In particular, G is finite—an impossibility.

If $x = (ab)^i a$, then

$$H = (ab)^i aH = (ab)^i HaH$$

and, therefore,

$$(abH)^i = (ab)^i H = (aH)^{-1} = a^{-1} H = aH.$$

It follows that

$$\langle aH, bH \rangle = \langle aH, abH \rangle \subseteq \langle abH \rangle.$$

However,

$$(abH)^{2i} = (aH)^2 = a^2 H = H$$

so that D_∞/H is again finite. This contradiction forces $H = \{e\}$ and G to be isomorphic to D_∞.

Finally, suppose $|ab| = n$. Since $G = \langle a, b \rangle = \langle a, ab \rangle$, we can show that G is isomorphic to D_n by proving that $b(ab)b = (ab)^{-1}$, which is the same as $ba = (ab)^{-1}$ (see exercise 8). But $(ab)^{-1} = b^{-1} a^{-1} = ba$, since a and b have order 2. ∎

The preceding characterization of the dihedral groups has been known for over 100 years.

Realizing the Dihedral Groups with Mirrors

A geometric realization of D_∞ can be obtained by placing two mirrors in a parallel position, as shown in Figure 28.1.* If we let a and b denote reflections in mirrors

---

*Perhaps the most spectacular illustration of this is the photograph in *Time* magazine of Ann-Margaret in a dancing costume between a pair of parallel mirrors. The result is infinitely many images of Ann-Margaret! *Time,* Sept. 18 (1978):96.

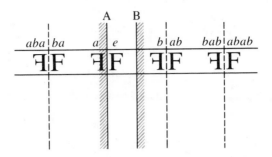

Figure 28.1 The group D_∞—reflections in parallel mirrors.

A and B, respectively, then ab (that is, a followed by b) represents a translation through twice the distance between the two mirrors to the right, and ba is the translation through the same distance to the left.

The finite dihedral groups can also be realized with a pair of mirrors. For example, if we place a pair of mirrors at a 45° angle, we obtain the group D_4. Notice that in Figure 28.2, the effect of reflecting an object in mirror A, then mirror B, is a rotation of twice the angle between the two mirrors (that is, 90°).

In Figure 28.3, we see a portion of the pattern produced by reflections in a pair of mirrors set at a 1° angle. The corresponding group is D_{180}. In general, reflections in a pair of mirrors set at the angle 180°/n correspond to the group D_n. As n becomes larger and larger, the mirrors approach a parallel position. In the limiting case, we have the group D_∞.

The ideas discussed in this section are relevant to the design of kaleidoscopes.* These are mechanical devices that utilize mirrors to produce pleasing images. The reader may find information about building various kaleidoscopes in [2] and [4, pp. 68–69, 171–172, 200–201].

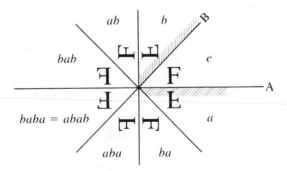

Figure 28.2 The group D_4—reflections in mirrors at a 45° angle.

---

*The word *kaleidoscope* is derived from three Greek words meaning "beautiful," "form," and "to see." The term was coined by Sir David Brewster, who wrote a treatise on its theory and history.

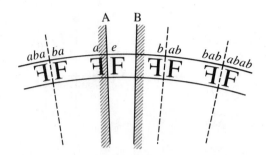

Figure 28.3 The group D_{180}—reflections in mirrors at a 1° angle.

We conclude this chapter by commenting on the advantages and disadvantages of using generators and relations to define groups. The principal advantage is that in many situations, particularly in knot theory, algebraic topology, and geometry, groups defined by way of generators and relations arise in a natural way. Within group theory itself, it is often convenient to construct examples and counterexamples with generators and relations. Among the disadvantages in defining a group by generators and relations is the fact that it is often difficult to decide whether or not the group is finite, or even whether or not a particular element is the identity. Furthermore, the same group can be defined with entirely different sets of generators and relations, and, given two groups defined by generators and relations, it is often extremely difficult to decide whether or not these two groups are isomorphic. Nowadays, these questions are frequently tackled with the aid of a powerful computer.

Generators and relations for many well-known groups can be found in [1].

EXERCISES

It don't come easy.

Title of a song by Ringo Starr, May 1971

1. Let n be an even integer. Prove that $D_n/Z(D_n)$ is isomorphic to $D_{n/2}$.
2. Let S be a set of distinct symbols. Show that the relation defined on $W(S)$ in this chapter is an equivalence relation.
3. Show that $\langle a, b \mid a^5 = b^2 = e, ba = a^2b \rangle$ is isomorphic to Z_2.
4. Verify that the set K in Example 2 is closed under multiplication on the left by b.
5. Prove Theorem 28.3 and its corollary.
6. Let G be the group $\{\pm 1, \pm i, \pm j, \pm k\}$ with multiplication defined as in exercise 38 of Chapter 10. Show that G is isomorphic to $\langle a, b \mid a^2 = b^2 = (ab) \rangle$. (Hence, the name quaternions.)

7. In any group, show that $\langle a, b \rangle = \langle a, ab \rangle$. (This exercise is referred to in the proof of Theorem 28.5.)

8. Prove that $G = \langle x, y \mid x^2 = y^n = e, xyx = y^{-1} \rangle$ is isomorphic to D_n. (This exercise is referred to in the proof of Theorem 28.5.)

9. Let

$$M = \tfrac{1}{3}\begin{bmatrix} 0 & -2 & 1 \\ 2 & 0 & -2 \\ -1 & 2 & 0 \end{bmatrix} \quad \text{and} \quad N = \tfrac{1}{3}\begin{bmatrix} 1 & -2 & 0 \\ -2 & 0 & 2 \\ 0 & 2 & -1 \end{bmatrix}.$$

Show that the group generated by M and N is isomorphic to D_4.

10. What is the minimum number of generators needed for $Z_2 \oplus Z_2 \oplus Z_2$? Find a set of generators and relations for this group.

11. Suppose $x^2 = y^2 = e$ and $yz = zxy$. Show that $xy = yx$.

12. Let $G = \langle a, b \mid a^2 = b^4 = e, ab = b^3a \rangle$.
    a. Express $a^3b^2abab^3$ in the form b^ia^j.
    b. Express b^3abab^3a in the form b^ia^j.

13. Let $G = \langle a, b \mid a^2 = b^2 = (ab)^2 \rangle$.
    a. Express $a^3b^2abab^3$ in the form b^ia^j.
    b. Express b^3abab^3a in the form b^ia^j.

14. Let G be the group defined by the following table. Show that G is isomorphic to D_n.

	1	2	3	4	5	6	$\cdots$	$2n$
1	1	2	3	4	5	6	$\cdots$	$2n$
2	2	1	$2n$	$2n-1$	$2n-2$	$2n-3$	$\cdots$	3
3	3	4	5	6	7	8	$\cdots$	2
4	4	3	2	1	$2n$	$2n-1$	$\cdots$	5
5	5	6	7	8	9	10	$\cdots$	4
6	6	5	4	3	2	1	$\cdots$	7
.	.	.	.	.	.	.	.	.
.	.	.	.	.	.	.	.	.
.	.	.	.	.	.	.	.	.
$2n$	$2n$	$2n-1$	$2n-2$	$2n-3$	$2n-4$	$2n-5$	$\cdots$	1

15. Let $G = \langle x, y \mid x^8 = y^2 = e, yxyx^3 = e \rangle$. Show that $|G| \leq 16$. Find the center of G. Assuming G does have 16 elements, find the order of xy.

16. Classify all groups of order ≤ 11.

17. Let G be defined by some set of generators and relations. Show that every factor group of G satisfies the generators and relations defining G.

18. Let $G = \langle s, t \mid sts = tst \rangle$. Show that the permutations (23) and (13) satisfy the defining relations of G. Explain why this proves that G is non-Abelian.

19. Let G be generated by a and b and suppose $\langle b \rangle$ is normal in G. Show that every element of G can be written in the form $a^i b^j$.

20. Let $G = \langle x, y \mid x^{2n} = e, x^n = y^2, y^{-1}xy = x^{-1} \rangle$. Show that $Z(G) = \{e, x^n\}$. Assuming that $|G| = 4n$, show that $G/Z(G)$ is isomorphic to D_n. (The group G is called the *dicyclic* group of order $4n$.)

21. Let $G = \langle a, b \mid a^6 = b^3 = e, b^{-1}ab = a^3 \rangle$. How many elements does G have? What familiar group is G isomorphic to?

22. Let $G = \langle x, y \mid x^4 = y^4 = e, xyxy^{-1} = e \rangle$. Show that $|G| \leq 16$. Find the center of G. Assuming $|G| = 16$, show that $G\langle y^2 \rangle$ is isomorphic to D_4.

23. The set $H_n = \{0, 1, 2, \ldots, 2n - 1\}$ with the binary operation $x * y = [x + (-1)^x y] \bmod 2n$ is a group. Show that H_n is isomorphic to the dihedral group of order $2n$.

24. The group G, defined by the following table, is isomorphic to one of the groups listed in Table 28.1. Which one?

1	2	3	4	5	6	7	8	9	10	11	12
2	3	4	5	6	1	8	9	10	11	12	7
3	4	5	6	1	2	9	10	11	12	7	8
4	5	6	1	2	3	10	11	12	7	8	9
5	6	1	2	3	4	11	12	7	8	9	10
6	1	2	3	4	5	12	7	8	9	10	11
7	12	11	10	9	8	4	3	2	1	6	5
8	7	12	11	10	9	5	4	3	2	1	6
9	8	7	12	11	10	6	5	4	3	2	1
10	9	8	7	12	11	1	6	5	4	3	2
11	10	9	8	7	12	2	1	6	5	4	3
12	11	10	9	8	7	3	2	1	6	5	4

REFERENCES

1. H. S. M. Coxeter and W. O. J. Moser, *Generators and Relations for Discrete Groups*, 4th ed., Berlin: Springer-Verlag, 1980.

2. J. Kennedy and D. Thomas, *Kaleidoscope Math*, Palo Alto: Creative Publications, 1978.

3. Joseph J. Rotman, *The Theory of Groups: An Introduction*, Boston: Allyn and Bacon, 1965.

4. A. V. Shubnikov and V. A. Koptsik, *Symmetry in Science*, New York: Plenum Press, 1974.

SUGGESTED READINGS

Alexander H. Fran, Jr. and David Singmaster, *Handbook of Cubik Math*, Hillside, New Jersey: Enslow, 1982.

This book is replete with the group-theoretical aspects of **the Magic** Cube. It uses permutation group theory and generators and relations to **discuss the** solutions to the cube and related results. The book has numerous **challenging** exercises stated in group-theoretical terms.

I. Kleiner, "The Evolution of Group Theory: A Brief Survey," *Mathematics Magazine* 59 (1986): 195–215.

This award-winning paper outlines the origins of the main concepts in group theory.

Lee Neuwirth, "The Theory of Knots," *Scientific American* 240 (1979): 110–124.

This article shows how a unique group can be associated with a knotted string. Mathematically, a knot is just a one-dimensional curve situated in three-dimensional space. The theory of knots—a branch of topology—seeks to classify and analyze the different ways of tracing such a curve. Around the turn of this century, Henri

The cloverleaf knot.

Poincaré observed that important geometrical characteristics of knots could be described in terms of group generators and relations—the so-called knot group. Among others, Neuwirth describes the construction of the knot group for the cloverleaf knot pictured. One set of generators and relations for this group is $\langle x, y, z \mid xy = yz, zx = yz \rangle$.

B. L. van der Waerden, "Hamilton's Discovery of Quaternions," *Mathematics Magazine* 49 (1976): 227–234.

This award-winning paper uses Hamilton's papers and letters to describe how he came to discover the quaternions. (van der Waerden is the author of a classic text on modern algebra published in 1930. It is one of the most influential mathematics books written in this century.)

William Burnside

Burnside, during a life of steadfast devotion to his science, has contributed to many an issue. In one of the most abstract domains of thought, he has systematized and amplified its range so that, there, his work stands as a landmark in the widening expanse of knowledge. Whatever be the estimate of Burnside made by posterity, contemporaries salute him as a Master among the mathematicians of his own generation.

A. R. Forsyth, *Journal of the London Mathematical Society*

WILLIAM BURNSIDE was born on July 2, 1852, in London. In 1871, he entered Cambridge University and was considered the best of his college class. After graduation in 1875, Burnside was appointed lecturer at Cambridge, where he stayed until 1885. He then accepted a position at the Royal Naval College at Greenwich and spent the rest of his career in this post.

Burnside wrote more than 150 research papers in many fields. Most of his early papers were devoted to applied mathematics, principally hydrodynamics. He also published papers on differential geometry, elliptic functions, and probability theory. He is best remembered, however, for his pioneering work in group theory, which appeared in some fifty papers, and his classic book entitled *Theory of Groups*. Because of his emphasis on the abstract approach, many consider Burnside as the first pure group theorist.

One mark of greatness in a mathematician is the ability to pose important and challenging problems—problems that open up new areas of research for future generations. Here, Burnside excelled. It was he who first conjectured that a group of odd order is solvable (i.e., the group G has a series of normal subgroups, $G = G_0 \geq G_1 \geq G_2 \geq \cdots \geq G_n = \{e\}$, such that $|G_i/G_{i+1}|$ is prime). This extremely important conjecture was finally proved over fifty years later by Feit and Thompson in a 255-page prize-winning paper (see Chapter 27 for more on this).

Burnside was elected a Fellow of the Royal Society and awarded two Royal medals. He served as president of the Council of the London Mathematical Society and received their DeMorgan medal. Burnside died on August 21, 1927.

Symmetry Groups

29

It [group theory] provides a sensitive instrument for investigating symmetry, one of the most pervasive and elemental phenomena of the real world.

M. I. Kargapolov and Ju. I. Marzljakov, *Fundamentals of the Theory of Groups*

Isometries

In the early chapters of this book, we briefly discussed symmetry groups. In this chapter and the next, we examine this fundamentally important concept in some detail. It is convenient to begin such a discussion with the definition of an isometry (from the Greek *isometros,* meaning "equal measure") in $\mathbf{R}^n$.

DEFINITION Isometry
An *isometry* of n-dimensional space $\mathbf{R}^n$ is a one-to-one function from $\mathbf{R}^n$ into $\mathbf{R}^n$ that preserves distance.

In other words, a one-to-one function T from $\mathbf{R}^n$ onto $\mathbf{R}^n$ is an isometry if, for every pair of points p and q in $\mathbf{R}^n$, the distance from pT to qT is the same as the distance from p to q. With this definition, we may now make precise the definition of the symmetry group of an n-dimensional figure.

DEFINITION   Symmetry Group of a Figure in $\mathbf{R}^n$
Let F be a set of points in $\mathbf{R}^n$. The *symmetry group of F in $\mathbf{R}^n$* is the set of all isometries of $\mathbf{R}^n$ that carry F onto itself.

It is important to realize that the symmetry group of an object depends not only upon the object, but also on the space in which we view it. For example, the symmetry group of a line segment in $\mathbf{R}^1$ has order 2; the symmetry group

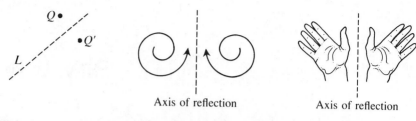

Axis of reflection Axis of reflection

Figure 29.1 Reflected images.

of a line segment considered as a set of points in $\mathbf{R}^2$ has order 4; the symmetry group of a line segment viewed as a set of points in $\mathbf{R}^3$ has infinite order (see exercise 9).

Although we have formulated our definitions for all finite dimensions, our chief interest will be the 2-dimensional case. It can be shown [1, p. 46] that every isometry of $\mathbf{R}^2$ is one of four types: rotation, reflection, translation, and glide-reflection. Rotation about a point in a plane needs no explanation. A *reflection across a line L* is that transformation that leaves every point of L fixed and takes every point Q, not on L, to the point Q' so that L is the perpendicular bisector of the line segment from Q to Q' (see Figure 29.1). The line L is called the *axis of reflection*. In an x, y coordinate plane, the transformation $(x, y) \to (x, -y)$ is a reflection across the x-axis, while $(x, y) \to (y, x)$ is a reflection across the line $y = x$. Some authors call an axis of reflective symmetry L a *mirror* because L acts like a two-sided mirror; that is, the image of a point Q in a mirror placed on the line L is, in fact, the image of Q under the reflection across the line L. Reflections are called *opposite* isometries because they reverse orientation. For example, the reflected image of a clockwise spiral is a counterclockwise spiral. Similarly, the reflected image of a right hand is a left hand. (See Figure 29.1.)

A *translation* is simply a function that carries all points the same distance in the same direction. For example, if p and q are points in a plane and T is a translation, then the two vectors joining p to pT and q to qT have the same length and direction. A *glide-reflection* is the product of a translation and a reflection across the line containing the translation vector. In Figure 29.2, the vector gives the direction and length of the translation, and is contained in the axis of reflection. A glide-reflection is also an opposite isometry. Successive footprints in wet sand are related by a glide-reflection.

$p \bullet$ $\bullet pT$

Figure 29.2 Glide-reflection.

Classification of Finite Plane Symmetry Groups

Our first goal in this chapter is to classify all finite plane symmetry groups. As we have already seen in earlier chapters, the dihedral group D_n is the plane symmetry group of a regular n-gon. (For convenience, call D_2 the plane symmetry group of a nonsquare rectangle and D_1 the plane symmetry group of the letter "V." In particular, $D_2 \approx Z_2 \oplus Z_2$ and $D_1 \approx Z_2$.) The cyclic groups Z_n are easily seen to be plane symmetry groups also. Figure 29.3 gives an illustration of an organism whose plane symmetry group consists of four rotations and is isomorphic to Z_4. The surprising fact is that the cyclic groups and dihedral groups are the only finite plane symmetry groups. The famous mathematician Hermann Weyl attributes this theorem to Leonardo da Vinci (1452–1519).

Theorem 29.1 *Finite Symmetry Groups in the Plane*
The only finite plane symmetry groups are Z_n and D_n.

Proof. Let G be a finite plane symmetry group of some figure. We first observe that G cannot contain a translation or a glide-reflection, otherwise G would be infinite. Next, we show that there is some point in the plane that is left fixed by every member of G. To do this, let us suppose the plane is coordinatized, and let $S = \{(x_1, y_1), (x_2, y_2), \ldots, (x_m, y_m)\}$ be the orbit of $(0, 0)$ under G (that is, $S = (0, 0)G = \{(0, 0)\phi \mid \phi \in G\}$). Then,

$$(\bar{x}, \bar{y}) = \left(\frac{1}{m} \sum_{i=1}^{m} x_i, \frac{1}{m} \sum_{i=1}^{m} y_i \right)$$

Figure 29.3 *Aurelia insulinda.* An organism whose plane symmetry group is Z_4.

is the centroid of the system of points in S. Since every member of G preserves distances, it follows that for any ϕ in G, the point $(\bar{x}, \bar{y})\phi$ is the centroid of $S\phi$. But, because $S\phi = ((0, 0)G)\phi = (0, 0)G\phi = (0, 0)G = S$, we must have $(\bar{x}, \bar{y})\phi = (\bar{x}, \bar{y})$. So, $(\bar{x}, \bar{y})$ is left fixed by every element of G. In particular, for any reflection in G, the axis of reflection must contain $(\bar{x}, \bar{y})$, and, for any rotation in G, the center of rotation is $(\bar{x}, \bar{y})$.

For convenience, let us denote a rotation about $(\bar{x}, \bar{y})$ of σ degrees by R_σ. Now, among all rotations in G, let β be the smallest positive angle of rotation. (Such an angle exists, since G is finite and R_{360} belong to G.) We claim that every rotation in G is some power of R_β. To see this, suppose R_σ is in G. We may assume $0° < \sigma \le 360°$. Then, $\beta \le \sigma$ and there is some integer t such that $t\beta \le \sigma < (t + 1)\beta$. But, then, $R_{\sigma-t\beta} = R_\sigma \circ (R_\beta)^{-t}$ is in G and $0 \le \sigma - t\beta < \beta$. Since β represents the smallest positive angle of rotation among the elements of G, we must have $\sigma - t\beta = 0$, and therefore, $R_\sigma = (R_\beta)^t$. This verifies the claim.

For convenience, let us say that $|R_\beta| = n$. Now if G has no reflections, we have proved $G = \langle R_\beta \rangle \approx Z_n$. If G has at least one reflection, say, α, then

$$\alpha, \ \alpha R_\beta, \ \alpha(R_\beta)^2, \ \ldots, \ \alpha(R_\beta)^{n-1}$$

are also reflections. Furthermore, this is the entire set of reflections of G. For if γ is any reflection in G, then $\alpha\gamma$ is a rotation, and so $\alpha\gamma = (R_\beta)^k$ for some k. Thus, $\gamma = \alpha^{-1}(R_\beta)^k = \alpha(R_\beta)^k$. So

$$G = \{R_0, R_\beta, (R_\beta)^2, \ \ldots, \ (R_\beta)^{n-1}, \alpha, \ \alpha R_\beta, \ \alpha(R_\beta)^2, \ \ldots, \ \alpha(R_\beta)^{n-1}\},$$

and G is generated by the pair of reflections α and αR_β. So, by our characterization of the dihedral groups (Theorem 28.5), G is the dihedral group D_n. ∎

Classification of Finite Groups of Rotations in R³

One might think that the set of all possible finite symmetry groups in three dimensions is much more diverse. Surprisingly, this is not the case. For example, moving to three dimensions only introduces three new groups of rotations. This observation was first made by the physicist and mineralogist Auguste Bravais in 1849, in his study of possible structures of crystals.

Theorem 29.2 *Finite Groups of Rotations in R³.*
Up to isomorphism, the finite groups of rotations in R³ are Z_n, D_n, A_4, S_4, and A_5.

Theorem 29.2, together with the Orbit-Stabilizer Theorem (Theorem 9.2), make easy work of determining the group of rotations of an object in R³.

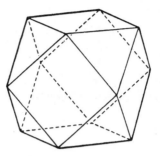

Figure 29.4

Example 1 We determine the group G of rotations of the solid in Figure 29.4. We begin by singling out one of the squares. Obviously, there are four rotations that fix this square, and the designated square can be rotated to the location of any of the other five. So, by the Orbit-Stabilizer Theorem, the rotation group has order $4 \cdot 6 = 24$. By Theorem 29.2, G is one of Z_{24}, D_{12}, or S_4. But each of the first two groups have exactly two elements of order 4, while G has more than two. So, G is isomorphic to S_4. ☐

The group of rotations of a tetrahedron (the *tetrahedral group*) is isomorphic to A_4; the group of rotations of a cube or an octahedron (the *octahedral group*) is isomorphic to S_4; the group of rotations of a dodecahedron or an icosahedron (the *icosahedral group*) is isomorphic to A_5. (Coxeter [1, pp. 271–273] specifies which portions of the polyhedra are being permuted in each case.) These five solids are illustrated in Figure 29.5.

That these five solids are the only possible regular solids (that is, all faces are congruent and all solid angles at the vertices are equal) was one of the great discoveries of the ancient Greeks. Euclid discussed them at length in his book *Elements*. Plato theorized that all matter was made up of minute particles of earth, air, fire, and water. He believed that earth particles were cubes, air particles were octahedra, fire particles were tetrahedra, and water particles were icosahedra. Pythagoras and his followers mystically associated the dodecahedron with the cosmos. They believed that understanding of the dodecahedron was too dangerous for ordinary people and tried to restrict this knowledge to their own cult. Over 2,000 years later, the astronomer Johannes Kepler tried, in vain, to model the planetary system after the way in which the five regular solids can be inscribed in one another. It was his conjecture that the orbits of the planets were circumscribed by these solids. (See Figure 29.6.)

A complete list of the finite symmetry groups in $\mathbf{R}^3$ (not just the groups of rotations) is given in [1, p. 413]. We mention in passing that, with the exception of the tetrahedron, the symmetry groups of the Platonic solids are just direct products of the rotation subgroup and the normal subgroup consisting of the

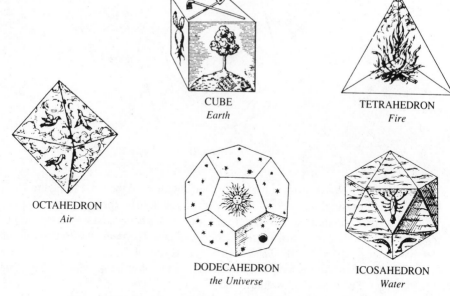

CUBE
Earth

TETRAHEDRON
Fire

OCTAHEDRON
Air

DODECAHEDRON
the Universe

ICOSAHEDRON
Water

Figure 29.5 The five regular solids as depicted by Johannes Kepler in *Harmonices Mundi, Book II* (1619).

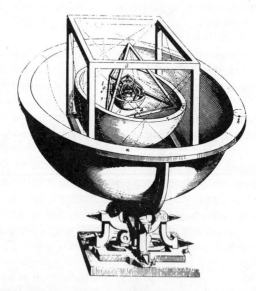

Figure 29.6 Kepler's cosmic mystery pictures the spheres of the six planets nested in the five perfect solids of Pythagoras and Plato. The outermost solid is the cube.

identity and inversion in the center of the solid. (An *inversion* in a point Q is the isometry that fixes Q and carries every other point P to the point P' such that Q bisects the line segment from P to P'. For example, in $\mathbf{R}^3$, the transformation $(x, y, z) \rightarrow (-x, -y, -z)$ is an inversion in the origin.)

EXERCISES

Perhaps the most valuable results of all education is the ability to make yourself do the thing you have to do, when it ought to be done, whether you like it or not.

Thomas Henry Huxley, *Technical Education*

1. Show that a function from $\mathbf{R}^n$ to $\mathbf{R}^n$ that preserves distance (see page 373) is a one-to-one function.
2. Show that the translations in $\mathbf{R}^n$ form a group.
3. Exhibit a plane figure whose plane symmetry group is Z_5.
4. Show that the group of rotations in $\mathbf{R}^3$ of a 3-prism (that is, a prism with equilateral ends as in the following figure) is isomorphic to D_3.

5. What is the order of the (entire) symmetry group in $\mathbf{R}^3$ of a 3-prism?
6. What is the order of the symmetry group in $\mathbf{R}^3$ of a 4-prism (a box with square ends that is not a cube)?
7. What is the order of the symmetry group in $\mathbf{R}^3$ of an n-prism?
8. Show that the symmetry group in $\mathbf{R}^3$ of a box of dimensions $2'' \times 3'' \times 4''$ is isomorphic to $Z_2 \oplus Z_2 \oplus Z_2$.
9. Describe the symmetry group of a line segment viewed as
   a. a subset of $\mathbf{R}^1$
   b. a subset of $\mathbf{R}^2$
   c. a subset of $\mathbf{R}^3$
   (This exercise is referred to in this chapter.)
10. Determine the group of rotations for each of the solids shown in exercise 48 of Chapter 9.
11. Exactly how many elements of order 4 does the group of Example 1 have?
12. Why is inversion not listed as one of the four kinds of isometries in $\mathbf{R}^2$?
13. In $\mathbf{R}^2$, inversion through a point is the same as a rotation. Explain why inversion through a point in $\mathbf{R}^3$ cannot be duplicated by a rotation in $\mathbf{R}^3$.

14. Reflection in a line L in $\mathbf{R}^3$ is the isometry that takes each point Q to the point Q' with the property that L is a perpendicular bisector of the line segment joining Q and Q'. Describe a rotation that has this same effect.

15. In $\mathbf{R}^2$, a rotation fixes a point; in $\mathbf{R}^3$, a rotation fixes a line. In $\mathbf{R}^4$, what does a rotation fix? Generalize these observations to $\mathbf{R}^n$.

16. Show that an isometry of a plane preserves angles.

17. Show that an isometry of a plane is completely determined by the image of three noncolinear points.

18. Suppose that an isometry of a plane leaves three noncolinear points fixed. Which isometry is it?

19. Suppose that an isometry of a plane fixes exactly one point. What type of isometry must it be?

20. Suppose A and B are rotations of 180° about the points a and b, respectively. What is A followed by B? How is the composite motion related to the points a and b?

REFERENCE

1. H. S. M. Coxeter, *Introduction to Geometry*, 2nd ed., New York: Wiley, 1969.

SUGGESTED READINGS*

J. Rosen, *Symmetry Discovered*, Cambridge: Cambridge University Press, 1975.

This excellent book is written for first- or second-year college students. It includes sections on group theory, spatial symmetry, temporal symmetry, color symmetry, and chapters on symmetry in nature and the uses of symmetry in science.

Doris Schattschneider, "The Taxicab Group," *The American Mathematical Monthly*, 91(1984): 423–428.

The *taxicab metric* is the function d_t, defined on $\mathbf{R}^2$ by $d_t(a, b) = |a_1 - b_1| + |a_2 - b_2|$ for any pair of points $a = (a_1, a_2)$ and $b = (b_1, b_2)$. A *taxicab isometry* is a mapping from $\mathbf{R}^2$ to $\mathbf{R}^2$ that preserves the taxicab metric (that is, $d_t(a, b) = d_t(a\phi, b\phi)$). In this note it is shown that the group of taxicab isometries is the semidirect product of the dihedral group of order 8 and the group of all translations of the plane. (G is the *semidirect product* of A and B if $G = AB$, B is normal in G, and $A \cap B = \{e\}$.)

SUGGESTED FILMS

Dihedral Kaleidoscopes with H. S. M. Coxeter, International Film Bureau $13\frac{1}{2}$ minutes, color.

See Chapter 2 for a description of this film.

---

*See also the suggested readings from Chapter 1.

Symmetries of the Cube, H. S. M. Coxeter and W. O. J. Moser, International Film Bureau, 13½ minutes, color.

In this beautiful film, the symmetry properties of the cube and the octahedron are explored. At one point in the film, a cube is cut along its nine planes of symmetry to yield forty-eight congruent tetrahedra. One of the tetrahedra is then placed in an octahedral kaleidoscope made from three mirrors. The image produced is the entire cube, thereby proving the symmetry group of a cube has order 48 and is generated by three reflections. The film also shows how the cube and the octahedron are related and why they have the same symmetry group. The use of stop-action photography and ultraviolet light to illuminate the models yields particularly dramatic effects.

Frieze Groups and Crystallographic Groups

30

Symmetry, considered as a law of regular composition of structural objects, is similar to harmony. More precisely, symmetry is one of its components, while the other component is dissymmetry. In our opinion the whole esthetics of scientific and artistic creativity lies in the ability to feel this where others fail to perceive it.

A. V. Shubnikov and V. A. Koptsik, *Symmetry in Science and Art*

The Frieze Groups

In this chapter, we discuss an interesting collection of infinite symmetry groups that arises from periodic designs in a plane. There are two types of such groups. The *discrete frieze groups* are the plane symmetry groups of patterns whose subgroup of translations is isomorphic to Z. These kinds of designs are the ones used as decorative strips [4] and for patterns on jewelry, as illustrated in Figure 30.1. In mathematics, familiar examples include the graphs of $y = \sin x$, $y = \tan x$, $y = |\sin x|$, and $|y| = \sin x$.

In previous chapters, it was our custom to view two isomorphic groups as the same group since we could not distinguish between them algebraically. In the case of the frieze groups, we will soon see that, although some of them are isomorphic as groups (i.e., algebraically the same), geometrically they are quite different. To emphasize this difference, we will treat them separately. In each of the following cases, the given pattern extends infinitely far in both directions.

Pattern I (Figure 30.2) consists of translations only. Letting x denote a translation to the right of one unit (that is, the distance between two consecutive R's), we may write the symmetry group of pattern I as

$$F_1 = \{x^n \mid n \in Z\}.$$

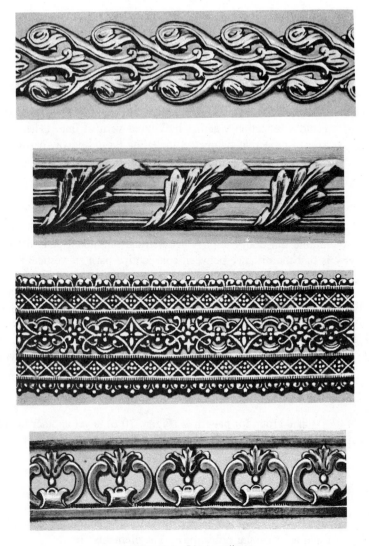

Figure 30.1   Frieze patterns.

R   R   R   R

Figure 30.2   Pattern I.

The group for pattern II (Figure 30.3), like that of pattern I, is infinitely cyclic. Letting x denote a glide-reflection, we may write the symmetry group of pattern II as

$$F_2 = \{x^n \mid n \in Z\}.$$

Notice that the translation subgroup is just $\langle x^2 \rangle$.

The symmetry group of pattern III (Figure 30.4) is generated by a translation x and a reflection y across the dotted vertical line. (There are infinitely many axes of reflective symmetry, including those midway between consecutive pairs of opposite facing R's. Any one will do.) The entire group is

$$F_3 = \{x^n y^m \mid n \in Z, m = 0 \text{ or } 1\}.$$

Note that the pair of elements xy and y have order 2, they generate F_3, and their product $(xy)y = x$ has infinite order. Thus, by Theorem 28.5, F_3 is the infinite dihedral group. A geometrical fact about pattern III worth mentioning is that the distance between consecutive paris of vertical reflection axes is half the length of the smallest translation vector.

In pattern IV (Figure 30.5), the symmetry group is generated by a translation x and a rotation y of $180°$ about a point p midway between consecutive R's (such a rotation is often called a *half-turn*). This group, like F_3, is also infinitely dihedral. (Another rotation point lies between a top and bottom R. As in pattern III, the distance between consecutive points of rotational symmetry is half the length of the smallest translation vector.)

$$F_4 = \{x^n y^m \mid n \in Z, m = 0 \text{ or } 1\}.$$

The group for pattern V (Figure 30.6) is yet another infinite dihedral group generated by a glide-reflection x and a rotation y of $180°$ about the point p. Notice that pattern V has vertical reflection symmetry, but it is just xy. The rotation points are midway between the vertical reflection axes.

$$F_5 = \{x^n y^m \mid n \in Z, m = 0 \text{ or } 1\}.$$

The symmetry group of pattern VI (Figure 30.7) is generated by a translation x and a horizontal reflection y. The elements are

$$F_6 = \{x^n y^m \mid n \in Z, m = 0 \text{ or } 1\}.$$

Note that since x and y commute, F_6 is not infinite dihedral. In fact, F_6 is isomorphic to $Z \oplus Z_2$. Pattern VI is left invariant under a glide-reflection also, but in this case the glide-reflection is considered trivial since it is the product of x and y. (Conversely, a glide-reflection is nontrivial if its translation component and reflection component are not elements of the symmetry group.)

The symmetry group of pattern VII (Figure 30.8) is generated by a translation x, a horizontal reflection y, and a vertical reflection z. It is isomorphic to the

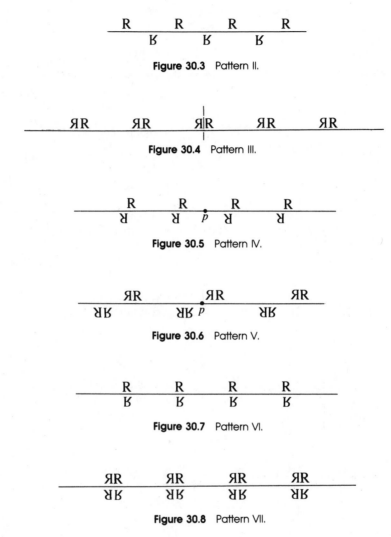

Figure 30.3 Pattern II.

Figure 30.4 Pattern III.

Figure 30.5 Pattern IV.

Figure 30.6 Pattern V.

Figure 30.7 Pattern VI.

Figure 30.8 Pattern VII.

direct product of the infinite dihedral group and Z_2. The product of y and z is a 180° rotation.

$$F_7 = \{x^n y^m z^k \mid n \in Z,\ m = 0 \text{ or } 1,\ k = 0 \text{ or } 1\}.$$

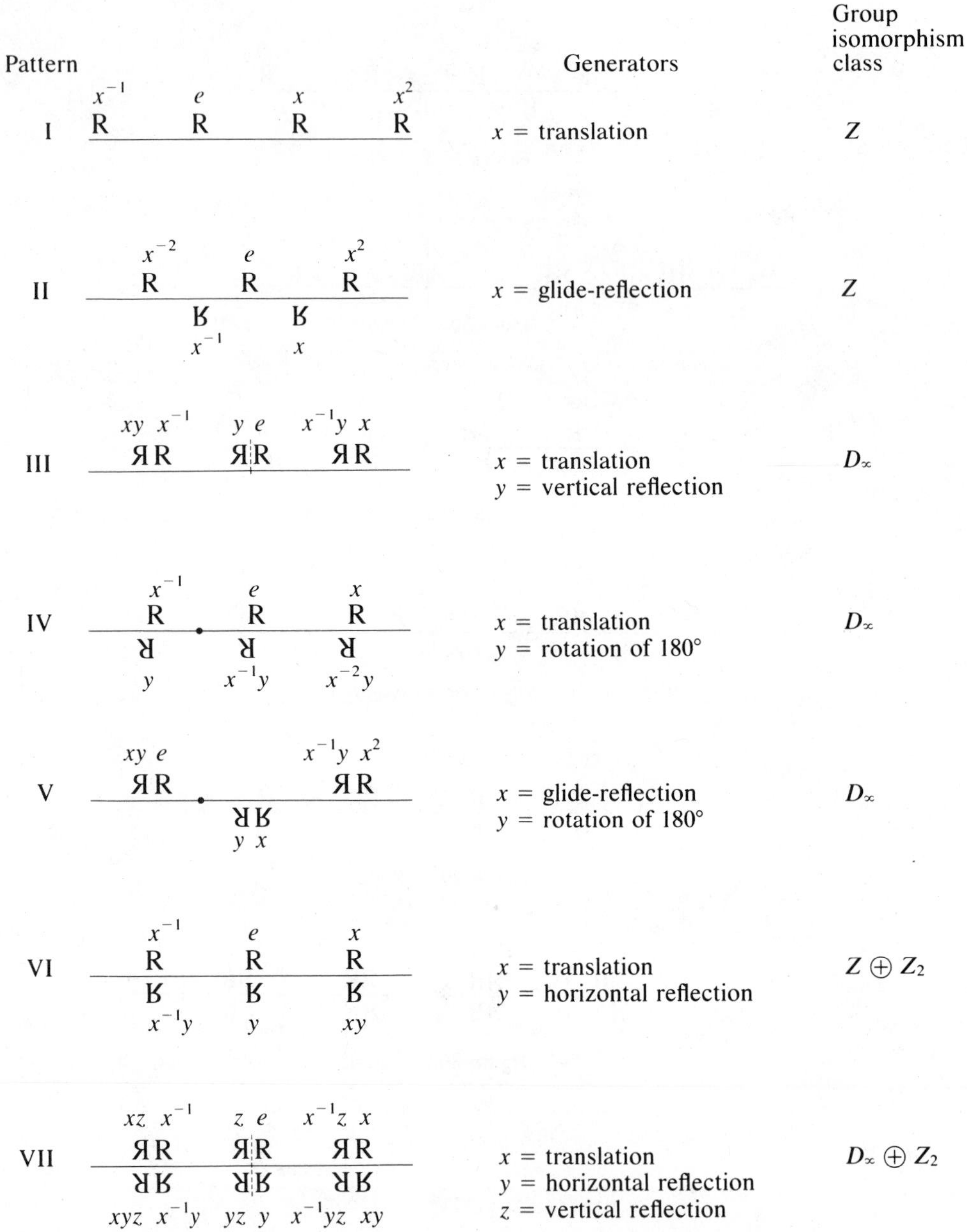

	Pattern	Generators	Group isomorphism class
I	x^{-1} $\quad$ e $\quad$ x $\quad$ x^2 R $\quad$ R $\quad$ R $\quad$ R	x = translation	Z
II	x^{-2} $\quad$ e $\quad$ x^2 R $\quad$ R $\quad$ R $\quad$ Я $\quad$ Я $\quad$ x^{-1} $\quad$ x	x = glide-reflection	Z
III	$xy\ x^{-1}$ $\quad$ $y\ e$ $\quad$ $x^{-1}y\ x$ ЯR $\quad$ ЯR $\quad$ ЯR	x = translation y = vertical reflection	D_∞
IV	x^{-1} $\quad$ e $\quad$ x R $\quad$ R $\quad$ R Я $\quad$ Я $\quad$ Я y $\quad$ $x^{-1}y$ $\quad$ $x^{-2}y$	x = translation y = rotation of 180°	D_∞
V	$xy\ e$ $\quad\quad$ $x^{-1}y\ x^2$ ЯR $\quad\quad$ ЯR $\quad$ ЯЯ $\quad$ $y\ x$	x = glide-reflection y = rotation of 180°	D_∞
VI	x^{-1} $\quad$ e $\quad$ x R $\quad$ R $\quad$ R Я $\quad$ Я $\quad$ Я $x^{-1}y$ $\quad$ y $\quad$ xy	x = translation y = horizontal reflection	$Z \oplus Z_2$
VII	$xz\ x^{-1}$ $\quad$ $z\ e$ $\quad$ $x^{-1}z\ x$ ЯR $\quad$ ЯR $\quad$ ЯR ЯЯ $\quad$ ЯЯ $\quad$ ЯЯ $xyz\ x^{-1}y$ $\quad$ $yz\ y$ $\quad$ $x^{-1}yz\ xy$	x = translation y = horizontal reflection z = vertical reflection	$D_\infty \oplus Z_2$

Figure 30.9 The seven frieze patterns and their groups of symmetries.

Table 30.1   Recognition Chart for Frieze Patterns[a]

Type	R_{180} Rotation	Horizontal Reflection	Vertical Reflection	Nontrivial Glide-Reflection
I	no	no	no	no
II	no	no	no	yes
III	no	no	yes	no
IV	yes	no	no	no
V	yes	no	yes	yes
VI	no	yes	no	no
VII	yes	yes	yes	no

[a]The distance between consecutive pairs of reflection axes and consecutive rotation points is half the length of the smallest translation vector. In patterns V and VII, the rotation points are on the vertical axes of reflection or midway between adjacent vertical axes of reflection. In pattern V, each vertical reflection is the product of a rotation and a glide-reflection. In pattern VII, each rotation is the product of horizontal and vertical reflections.

The preceding discussion is summarized in Figure 30.9. Table 30.1 provides an identification algorithm for the frieze patterns.

In describing the seven frieze groups, we have not explicitly said how multiplication is done algebraically. However, each group element corresponds to some isometry, so multiplication is the same as function composition. Thus, we can always use the geometry to determine the product of any particular string of elements.

For example, we know that every element of F_7 can be written in the form $x^n y^m z^k$. So, just for fun, let's determine the appropriate values for n, m, and k for the element $g = (x^{-1} yz)(xz)$. We may do this simply by looking at the effect that g has on pattern VII. For convenience, we will pick out a particular R in the pattern and trace the action of g one step at a time. To distinguish this R, we enclose it in a shaded box. Also, we draw the axis of the vertical reflection z as a dotted line segment. See Figure 30.10.

Now comparing the starting position of the shaded R with its final position, we see that $x^{-1}yzxz = x^{-2} y$. Exercise 7 suggests how one may arrive at the same result through purely algebraic manipulation.

The Crystallographic Groups

The seven frieze groups catalog all symmetry groups that leave a design invariant under all multiples of just one translation. However, there are seventeen additional kinds of discrete plane symmetry groups that arise from infinitely repeating designs in a plane. These groups are the symmetry groups of plane patterns whose subgroup of translations is isomorphic to $Z \oplus Z$. Consequently, the patterns are invariant under linear combinations of two linearly independent translations. These seventeen groups were first studied by nineteenth-century

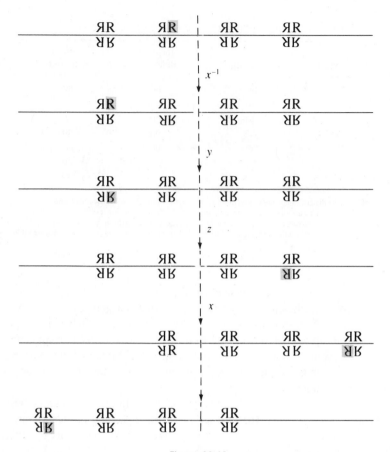

Figure 30.10

crystallographers and are often called the *plane crystallographic* groups. Another term occasionally used for these groups is *wallpaper groups*. A group-theoretic proof that there are exactly seventeen types of wallpaper patterns is given in [8].

Our approach to the crystallographic groups will be geometric. It is adapted from the excellent article by Schattschneider [7]. Our goal is to enable the reader to determine which of the seventeen plane symmetry groups corresponds to a given periodic pattern. We begin with some examples.

The simplest of the seventeen crystallographic groups contains translations only. In Figure 30.11, we give an illustration of a representative pattern for this group (imagine the pattern repeated to fill the entire plane). The crystallographic notation for it is $p1$. (This notation is explained in [7].)

The symmetry group of the pattern in Figure 30.12 contains translations and glide-reflections. This group has no rotational or reflective symmetry. The crystallographic notation for it is *pg*.

Figure 30.11   *Study of Regular Division of the Plane with Fish and Birds,*
1938. Escher graphic with symmetry group p1. The
arrows are translation vectors.

Figure 30.13 has translational symmetry and threefold rotational symmetry
(that is, the figure can be rotated 120° about certain points and be brought into
coincidence with itself). The notation for this group is *p*3.

Representative patterns for all seventeen plane crystallographic groups, together with their notation, are given in Figures 30.14 and 30.15. Figure 30.16
illustrates the seventeen classes of symmetry patterns generated by a triangle
motif.

Identification of Plane Periodic Patterns

To decide which of the seventeen classes any particular plane periodic pattern
belongs to, we may use the algorithm given in Table 30.2 (reproduced from [7,
p. 443]). This is done by determining the highest order of rotational symmetry,
and whether or not the pattern has reflection symmetry or nontrivial glide-

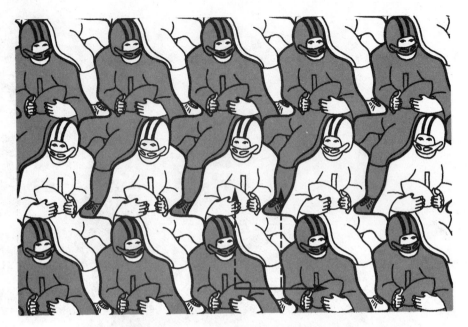

Figure 30.12 Escher-like tessellation by J. L. Teeters with symmetry group *pg* (disregard shading). The solid arrow is a translation vector. The dashed arrows are glide-reflection vectors.

reflection symmetry. These three pieces of information will narrow the list of candidates to at most two. The final test, if necessary, is to look for the distinguishing property listed in Table 30.2.

For example, consider the two patterns in Figure 30.17 generated by a hockey stick motif. Both patterns have a highest order of rotational symmetry of 3; both have reflectional and nontrivial glide-reflectional symmetry. Now, according to Table 30.2, these patterns must be of type *p3m1* or *p31m*. But notice that the pattern on the left has all its threefold centers of rotation on the reflection axis, while the pattern on the right does not. Thus, the left pattern is *p3m1*, and the right pattern is *p31m*.

Table 30.2 also contains two other features that are often useful. A *lattice of points* of a pattern is a set of images of any particular point acted on by the translation group of the pattern. The possible lattices for periodic patterns in a plane, together with lattice units, are shown in Figure 30.18. A *generating region* (or *fundamental region*) of a periodic pattern is the smallest portion of the lattice unit whose images under the full symmetry group of the pattern cover the plane. Examples of generating regions for the patterns represented in Figures 30.11, 30.12, and 30.13 are given in Figure 30.19. In Figure 30.19, the portion of the lattice unit with vertical bars is the generating region. The only symmetry pattern

Figure 30.13   *Study of Regular Division of the Plane with Human Figures,*
1938. Escher graphic with symmetry *p*3 (disregard shad-
ing). The inserted arrows are translation vectors.

in which the lattice unit and the generating unit coincide is the *p*1 pattern
illustrated in Figure 30.11. Table 30.2 tells which proportion of the lattice unit
constitutes the generating region of each plane-periodic plane pattern.

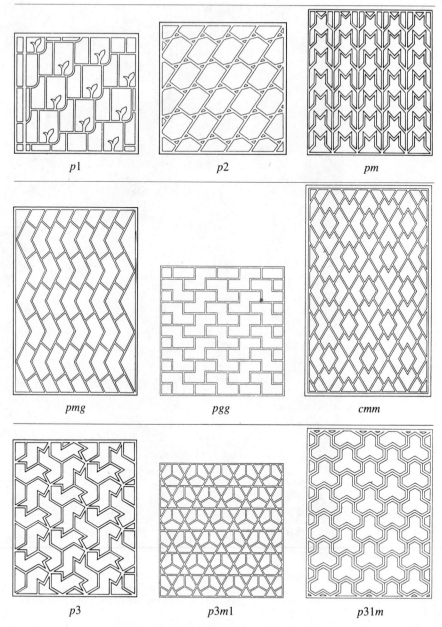

Figure 30.14 The plane symmetry groups.*

*In Figures 30.14 and 30.15, all designs except *pm, p3,* and *pg* are found in [2]. The designs for *p3* and *pg* are based on elements of Chinese lattice designs found in [2]; the design for *pm* is based on a weaving pattern from the Sandwich Islands, found in [4].

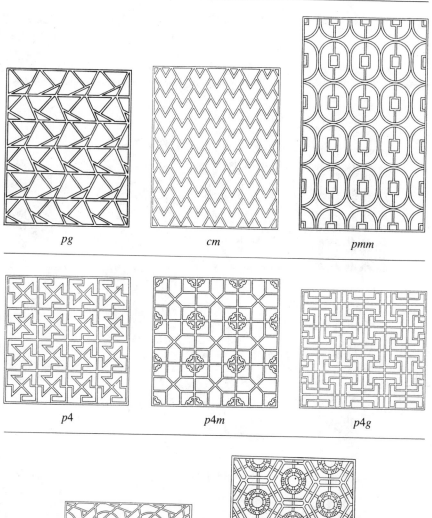

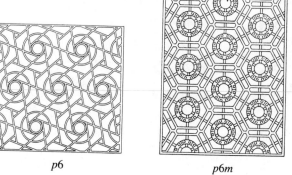

Figure 30.15 The plane symmetry groups.

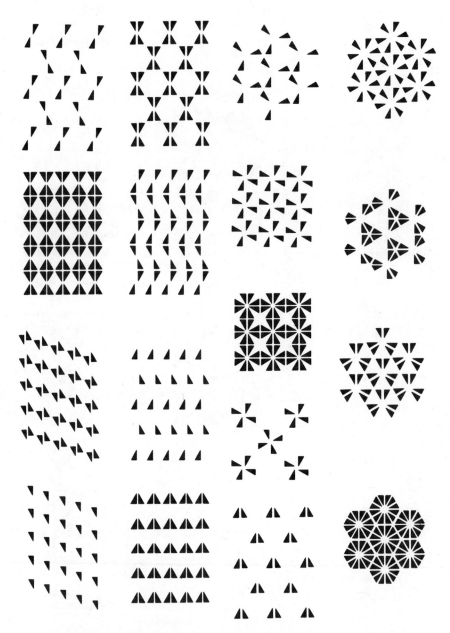

Figure 30.16 The seventeen plane periodic patterns formed with a triangle motif.

Table 30.2 Identification Chart for Plane Periodic Patterns[a]

Type	Lattice	Highest Order of Rotation	Reflections	Nontrivial Glide-Reflections	Generating Region	Helpful Distinguishing Properties
p1	parallelogram	1	no	no	1 unit	
p2	parallelogram	2	no	no	1/2 unit	
pm	rectangular	1	yes	no	1/2 unit	
pg	rectangular	1	no	yes	1/2 unit	
cm	rhombic	1	yes	yes	1/2 unit	
pmm	rectangular	2	yes	no	1/4 unit	
pmg	rectangular	2	yes	yes	1/4 unit	parallel reflection axes
pgg	rectangular	2	no	yes	1/4 unit	
cmm	rhombic	2	yes	yes	1/4 unit	perpendicular reflection axes
p4	square	4	no	no	1/4 unit	
p4m	square	4	yes	yes	1/8 unit	fourfold centers on reflection axes
p4g	square	4	yes	yes	1/8 unit	fourfold centers not on reflection axes
p3	hexagonal	3	no	no	1/3 unit	
p3m1	hexagonal	3	yes	yes	1/6 unit	all threefold centers on reflection axes
p31m	hexagonal	3	yes	yes	1/6 unit	not all threefold centers on reflections axes
p6	hexagonal	6	no	no	1/6 unit	
p6m	hexagonal	6	yes	yes	1/12 unit	

[a]A rotation through an angle of 360°/n is said to have order n. A glide-reflection is nontrivial if its component translation and reflection are not symmetries of the pattern.

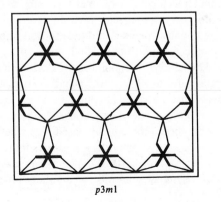

p3m1

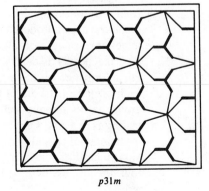

p31m

Figure 30.17 Patterns generated by a hockey stick motif.

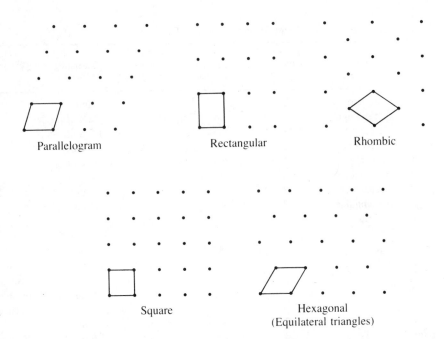

Parallelogram Rectangular Rhombic

Square Hexagonal
(Equilateral triangles)

Figure 30.18 Possible lattices for periodic plane patterns.

Notice that Table 30.2 reveals that the only possible n-fold rotational symmetries are when $n = 1, 2, 3, 4,$ and 6. This fact is commonly called the *crystallographic restriction*. The first proof of this was given by the Englishman W. Barlow. The information in Table 30.2 can also be used in reverse to create patterns with a specific symmetry group. The patterns in Figure 30.17 were made this way. Generators and relations for the crystallographic groups are given in [2, p. 136].

In sharp contrast to the situation for finite symmetry groups, the transition from two-dimensional crystallographic groups to three-dimensional crystallographic groups introduces a great many more possibilities. Indeed, there are 230 three-dimensional crystallographic groups (often called *space groups*). These were independently determined by Fedorov, Schonflies, and Barlow in the 1890s. For information on these groups refer to [9] or [10]. David Hilbert, one of the leading mathematicians of this century, focused attention on the crystallographic groups in his famous lecture in 1900 at the International Congress of Mathematicians at Paris. One of twenty-three problems he posed was whether or not the number of crystallographic groups in n-dimensions is always finite. This was answered affirmatively by L. Biebebach in 1910. We mention in passing that in four dimensions, there are 4,783 symmetry groups for infinitely repeating patterns.

Figure 30.19 A lattice unit and generating region for the pattern in Figures 30.11, 30.12, and 30.13. Generating regions are shaded with vertical bars.

Perhaps it is fitting to conclude this chapter by recounting an episode in the history of science where an understanding of the symmetry group of an object was crucial to a great discovery. In the early 1950s, a handful of scientists were attempting to learn the structure of the DNA molecule—the basic genetic material. One of these was a graduate student named Francis Crick; another was an x-ray crystallographer Rosalind Franklin. On one occasion, Crick was shown one of Franklin's research reports and an x-ray diffraction photograph of DNA.

At this point, we let Horace Judson [5, pp. 165–166], our source, continue the story.

> Crick saw in Franklin's words and numbers something just as important, indeed eventually just as visualizable. There was drama, too: Crick's insight began with an extraordinary coincidence. Crystallographers distinguish 230 different space groups, of which the face-centered monoclinic cell with its curious properties of symmetry is only one—though in biological substances a fairly common one. The principal experimental subject of Crick's dissertation, however, was the X-ray diffraction of the crystals of a protein that was of exactly the same space group as DNA. So Crick saw at once the symmetry that neither Franklin nor Wilkins had comprehended, that Perutz, for that matter, hadn't noticed, that had escaped the theoretical crystallographer in Wilkins' lab, Alexander Stokes—namely, that the molecule of DNA, rotated a half turn, came back to congruence with itself. The structure was dyadic, one half matching the other half in reverse.

This was a crucial fact. Shortly thereafter, Watson and Crick built an accurate model of DNA. In 1962, Watson, Crick, and Wilkins received the Nobel Prize in medicine and physiology for their discovery. The opinion has been expressed that, had Franklin correctly recognized the symmetry of the DNA molecule, she might have been the one to unravel the mystery and receive the the Nobel prize [5, p. 172].

EXERCISES

You can see a lot just by looking.
Yogi Berra

1. Show that the frieze group, F_6, is isomorphic to $Z \oplus Z_2$.
2. How many nonisomorphic frieze groups are there?
3. In the frieze group F_7, write x^2yzxz in the form $x^ny^mz^k$.
4. In the frieze group F_7, write $x^{-3}zxyz$ in the form $x^ny^mz^k$.
5. In the frieze group F_7, show that $yz = zy$ and $xy = yx$.
6. In the frieze group F_7, show that $zxz = x^{-1}$.
7. Use the results of exercises 5 and 6 to do exercises 3 and 4 through symbol manipulation only (i.e., without referring to the pattern). (This exercise is referred to in this chapter.)
8. Prove that in F_7, the cyclic subgroup generated by x is a normal subgroup.
9. Quote a previous result that tells why the subgroups $\langle x, y \rangle$ and $\langle x, z \rangle$ must be normal in F_7.

10. Look up the word *frieze* in an ordinary dictionary. Explain why the frieze groups are appropriately named.

11. Determine which of the seven frieze groups is the symmetry group of each of the following patterns.

a.

b.

c.

d.

e.

f.

12. Determine the frieze group corresponding to each of the following patterns:
   a. $y = \sin x$
   b. $y = |\sin x|$
   c. $|y| = \sin x$
   d. $y = \tan x$
   e. $y = \csc x$

13. Determine the symmetry group of the tessellation of the plane exemplified by the brickwork shown.

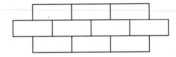

14. Determine the plane symmetry group for each of the patterns in Figure 30.16.

15. Determine which of the seventeen crystallographic groups is the symmetry group of each of the following patterns.

a.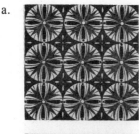

b.

c.

d.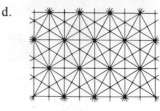

16. Look up Escher's "Flying Fish" print *Depth*, 1955 [6, p. 21]. Describe the isomorphism class of the three-dimensional symmetry group of the pattern depicted in the print.

17. Look up Escher's print *Cubic Space Division*, 1952 [6, p. 117]. Is the symmetry group of the pattern depicted the same as that of the "Flying Fish" print? Notice that the symmetry group of the structure depicted in this print contains subgroups isomorphic to $Z \oplus Z \oplus Z$ and S_4. Explain why the entire group is not the direct product of these two subgroups.

18. Determine the plane symmetry group of the Escher print *Study of Regular Division of the Plane with Birds*, 1955 [6, p. 125].

19. Determine the plane symmetry group of the Escher print *Study of Regular Division of the Plane with Angels and Devils*, 1941 [6, p. 141].

20. Determine the plane symmetry group of the Escher print [6, p. 77, no. 84].

21. Determine which of the frieze groups is the symmetry group of each of the following patterns.
   a. $\cdots$ D D D D $\cdots$
   b. $\cdots$ V Λ V Λ $\cdots$
   c. $\cdots$ L L L L $\cdots$
   d. $\cdots$ V V V V $\cdots$
   e. $\cdots$ N N N N $\cdots$
   f. $\cdots$ H H H H $\cdots$
   g. $\cdots$ L ⅃ L ⅃ $\cdots$

22. In the following figure is a point labeled 1. Let α be the translation of the plane that carries the point labeled 1 to the point labeled α and let β be the translation of the plane that carries the point labeled 1 to the point labeled β. The image of 1 under the composition of α and β is labeled $\alpha\beta$. In the corresponding fashion, label the remaining points in the figure in the form $\alpha^x \beta^y$.

REFERENCES

1. H. S. M. Coxeter and W. O. J. Moser, *Generators and Relations for Discrete Groups*, 4th ed., Berlin: Springer-Verlag, 1980.

2. Daniel S. Dye, *A Grammar of Chinese Lattice*, Harvard-Yenching Institute Monograph Series, vol. VI, Cambridge, Mass.: Harvard University Press, 1937. (Reprinted as *Chinese Lattice Designs*, New York: Dover, 1974.)

3. Bruno Ernst, *The Magic Mirror of M. C. Escher*, New York: Random House, 1976. (Paperback edition, New York: Ballantine Books, 1977.)

4. Owen Jones, *The Grammar of Ornament*, New York: Van Nostrand Reinhold, 1972. (Reproduction of the same title, first published in 1856 and reprinted in 1910 and 1928.)

5. Horace Freeland Judson, *The Eighth Day of Creation*, New York: Simon and Schuster, 1979.

6. J. L. Locher, editor, *The World of M. C. Escher*. (Paperback edition, New York: Harry N. Abrams, 1971.)

7. D. Schattschneider, "The Plane Symmetry Groups: Their Recognition and Notation," *American Mathematical Monthly* 85 (1978): 439–450.

8. R. L. E. Schwarzenberger, "The 17 Plane Symmetry Groups" *Mathematical Gazette* 58 (1974): 123–131.

9. A. V. Shubnikov and V. A. Koptski, *Symmetry in Science and Art*, New York: Plenum Press, 1974.

10. H. Weyl, *Symmetry*, Princeton: Princeton University Press, 1952.

SUGGESTED READINGS

M. Gardner, "The Eerie Mathematical Art of Maurits C. Escher," *Scientific American*, April (1966): 110–121.

The author analyzes many of Escher's most famous works.

R. Hughes, "*n*-Dimensional Reality," *Time*, April 17 (1972): 64.

A brief review of Escher's work is given, just one month after his death.

C. MacGillavry, *Fantasy and Symmetry—The Periodic Drawings of M. C. Escher*, New York: Harry N. Abrams, 1976.

This is a collection of Escher's periodic drawings together with a mathematical discussion of each one.

B. Rose, and R. Stafford, "An Elementary Course in Mathematical Symmetry," *American Mathematical Monthly* 88 (1981): 59–64.

> This article contains identification algorithms for the frieze groups and the wallpaper groups.

D. Schattschneider and W. Walker, *M. C. Escher Kaleidocycles,* New York: Ballantine Books, 1977.

> This delightful book contains precut and creased material for seventeen three-dimensional models adorned with Escher's graphics. The book also has a discussion of the geometric solids used for the models.

SUGGESTED FILMS

Adventures in Perception, Film Productie, 22 min., color. Available on free loan from Royal Netherlands Embassy, 4200 Linnean Avenue, N.W., Washington, D.C. 20008.

> This excellent film features some fifty of Escher's works with Escher himself discussing some of them.

Maurits Escher: Painter of Fantasies, A Coronet Film, $26\frac{1}{4}$ min., color, 1970.

> This award-winning film features Escher talking informally about his work. On several occasions, he mentions mathematics and at one point he constructs a Möbius strip. The film has pleasant background music and presents a close-up view of numerous prints that can serve as starting points for mathematical discussions.

M. C. Escher

I never got a pass mark in math. The funny thing is I seem to latch on to mathematical theories without realizing what is happening. No indeed, I was a pretty poor pupil at school. And just imagine—mathematicians now use my prints to illustrate their books. Fancy me consorting with all these learned folk, as though I were their long-lost brother. I guess they are quite unaware of the fact that I'm ignorant about the whole thing.

M. C. Escher

M. C. ESCHER was born on June 17, 1898, in the Netherlands. His artistic work prior to 1937 was dominated by the representation of visible reality, such as landscapes and buildings. Gradually, he became less and less interested in the visible world and became increasingly absorbed in an inventive approach to space. He studied the abstract space-filing patterns used in the Moorish mosaic in the Alhambra in Spain. He also studied the mathematician George Polya's paper on the seventeen plane crystallographic groups. Instead of the geometrical motifs used by the Moors and Polya, Escher preferred to use animals, plants, or people in his space-filling prints.

Escher was fond of incorporating various mathematical ideas into his works. Among these are infinity, Möbius bands, stellations, deformations, reflections, Platonic solids, spirals, and the hyperbolic plane. This latter idea was suggested to Escher by a figure in a paper by the geometer H. S. M. Coxeter.

Although Escher originals are now quite expensive, it was not until 1951 that he derived a significant portion of his income from his prints. Today, Escher is widely known and appreciated as a graphic artist. His graphics have appeared on postage stamps, bank notes, a candy box (in the shape of an icosahedron!), note cards, T-shirts, jigsaw puzzles, record album covers, and covers of dozens of scientific publications. His prints have been used to illustrate ideas in hundreds of scientific works. Despite this popularity among scientists, Escher has never been held in high esteem in traditional art circles. Escher died on March 27, 1973, in Holland.

Cayley Digraphs
of Groups

31

The important thing in Science is not so much to obtain new facts as to discover new ways of thinking about them.

Sir William Lawrence Bragg, *Beyond Reductionism*

Motivation

In this chapter, we introduce a graphical representation of a group given by a set of generators and relations. The idea was originated by Cayley in 1878. Although this topic is not usually covered in an abstract algebra book, we include it for four reasons: it provides a method to visualize a group (in fact, the word *graph* is Latin for "picture"); it connects two important branches of modern mathematics—groups and graphs; it gives a review of some of our old friends—cyclic groups, dihedral groups, direct products, generators and relations; and most importantly, it is fun!

Intuitively, a directed graph (or digraph) is a finite set of points, called *vertices,* and a set of arrows, called *arcs,* connecting some of the vertices. Although there is a rich and important general theory of directed graphs with many applications (see [2]), we are interested only in those that arise from groups.

The Cayley Digraph of a Group

DEFINITION   Cayley Digraph of a Group

Let G be a finite group and S a set of generators for G. We define a digraph $\text{Cay}(S:G)$, called the *Cayley digraph of G with generating set S* as follows.

1. Each element of G is a vertex of $\text{Cay}(S:G)$.
2. For x and y in G, there is an arc from x to y if and only if $xs = y$ for some $s \in S$.

To tell from the digraph which particular generator connects two vertices, Cayley proposed that each generator be assigned a color, and that the arrow joining x to xs be colored with the color assigned to s. He called the resulting figure the *color graph of the group*. This terminology is still widely used. Rather than use colors to distinguish the different generators, we will use solid arrows, broken arrows, and dotted arrows. In general, if there is an arc from x to y, there need not be an arc from y to x. An arrow emanating from x and pointing to y indicates that there is an arc from x to y.

Following are numerous examples of Cayley digraphs. Note that there are several ways to draw the digraph of a group given by a particular generating set. However, it is not the appearance of the graph that is relevant but the manner in which the vertices are connected. These connections are uniquely determined by the generating set. Thus, distances between vertices and angles formed by the arcs have no significance. (In the digraphs below, a headless arrow joining two vertices x and y indicates that there is an arc from x to y and an arc from y to x. For example, this occurs when a generator has order 2.)

Example 1 $Z_6 = \langle 1 \rangle.$

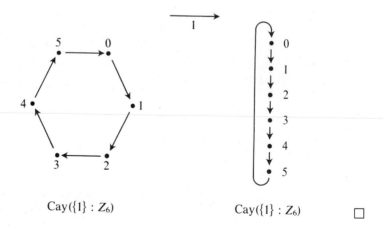

Cay($\{1\}$: Z_6)   Cay($\{1\}$: Z_6)   □

Example 2 $Z_3 \oplus Z_2 = \langle(1, 0), (0, 1)\rangle$.

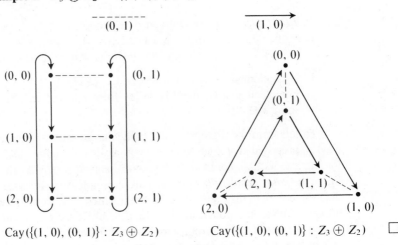

Cay($\{(1, 0), (0, 1)\} : Z_3 \oplus Z_2$)   Cay($\{(1, 0), (0, 1)\} : Z_3 \oplus Z_2$)   □

Example 3 $D_4 = \langle R_{90}, H\rangle$.

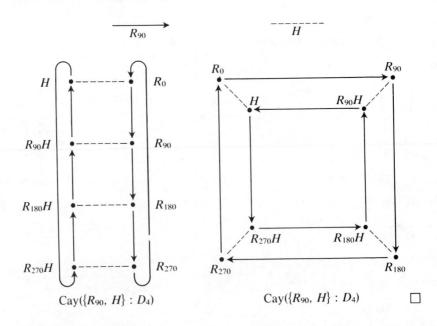

Cay($\{R_{90}, H\} : D_4$)   Cay($\{R_{90}, H\} : D_4$)   □

Example 4 $S_3 = \langle(12), (123)\rangle$.

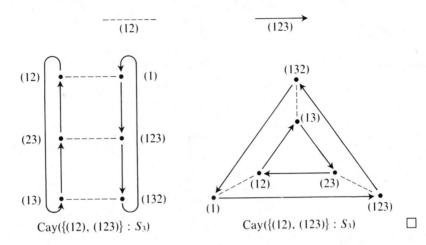

Cay({(12), (123)} : S_3)   Cay({(12), (123)} : S_3)

Example 5 $S_3 = \langle(12), (13)\rangle$.

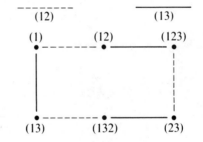

Example 6 $A_4 = \langle(12)(34), (123)\rangle$.

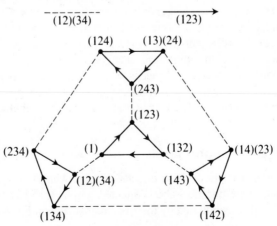

Example 7 $Q_4 = \langle a, b \mid a^4 = e, a^2 = b^2, b^{-1}ab = a^3 \rangle.$

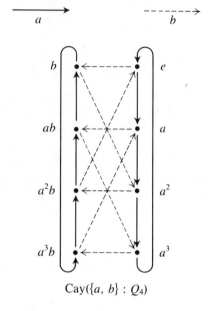

Cay($\{a, b\} : Q_4$)

□

Example 8 $D_\infty = \langle a, b \mid a^2 = b^2 = e \rangle.$

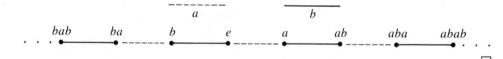

□

The Cayley digraph provides a quick and easy way to determine the value of any product of the generators and their inverses. Consider, for example, the product $ab^3ab^{-1}a^{-1}b^{-3}$ from the group given in Example 7. To reduce this to one of the eight elements used to label the vertices, we need only begin at the vertex e and follow the arcs from each vertex to the next as specified in the given product. Of course, a^{-1} means traverse the a arc in reverse. (Observations such as $b^{-3} = b$ also help.) Tracing the product through, we obtain ab. Similarly, one can verify or discover relations among the generators.

Hamiltonian Circuits and Paths

Now that we have these directed graphs, what is it that we care to know about them? One question about directed graphs that has been the object of much research was initiated by the Irish mathematician Sir William Hamilton in 1859,

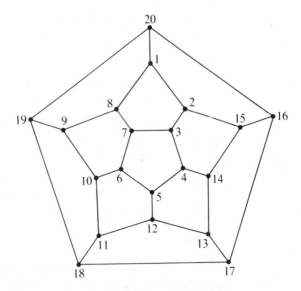

Figure 31.1 Around the world.

when he invented a puzzle called "Around the World." His idea was to label the twenty vertices of a regular dodecahedron with the names of famous cities. One solves this puzzle by starting at any particular city (vertex) and travelling "around the world," moving along the arcs in such a way that each other city is visited exactly once before returning to the original starting point. One solution to this puzzle is given in Figure 31.1, where the vertices are visited in the order indicated.

Obviously, this idea can be applied to any digraph; that is, one starts at some vertex and attempts to traverse the digraph by moving along arcs in such a way that each vertex is visited exactly once before returning to the starting vertex. (To go from x to y, there must be an arc from x to y.) Such a sequence of arcs is called a *Hamiltonian circuit* in the digraph. A sequence of arcs that

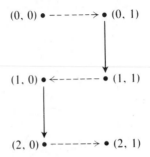

Figure 31.2 Hamiltonian path in $\text{Cay}(\{(1, 0), (0, 1)\}: Z_3 \oplus Z_2)$ from $(0,0)$ to $(2,1)$.

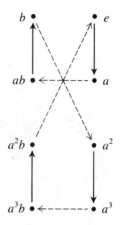

Figure 31.3 Hamiltonian circuit in Cay({a, b}: Q₄).

passes through each vertex exactly once without returning to the starting point is called a *Hamiltonian path*. In the remainder of this chapter, we concern ourselves with the existence of Hamiltonian circuits and paths in Cayley digraphs.

Figures 31.2 and 31.3 show a Hamiltonian path for the digraph given in Example 2 and a Hamiltonian circuit for the digraph given in Example 7, respectively.

Is there a Hamiltonian circuit in

$$\text{Cay}(\{(1, 0), (0, 1)\}: Z_3 \oplus Z_2)?$$

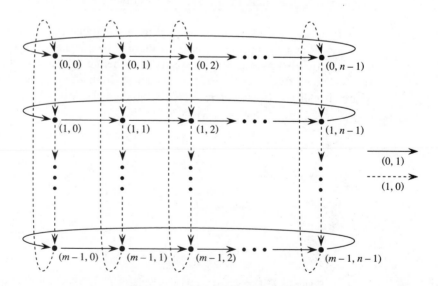

Figure 31.4 Cay({(1, 0), (0, 1)}: $Z_m \oplus Z_n$).

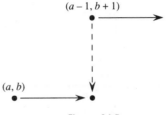

Figure 31.5

More generally, let us investigate the existence of Hamiltonian circuits in

$$\text{Cay}(\{(1, 0), (0, 1)\}: Z_m \oplus Z_n)$$

where m and n are relatively prime and both greater than 1. Visualize the Cayley digraph as a rectangular grid coordinatized with $Z_m \oplus Z_n$ as in Figure 31.4. Suppose there is a Hamiltonian circuit in the digraph and (a, b) is some vertex where the circuit exits horizontally. (Clearly, such a vertex exists). Then the circuit must exit $(a - 1, b + 1)$ horizontally also, for otherwise the circuit passes through $(a, b + 1)$ twice—see Figure 31.5. Repeating this argument again and again, we see that the circuit exits horizontally from each of the vertices $(a, b), (a - 1, b + 1), (a - 2, b + 2), \ldots$, which is just the coset $(a, b) + \langle(-1, 1)\rangle$. But when m and n are relatively prime, $\langle(-1, 1)\rangle$ is the entire group. Obviously there cannot be a Hamiltonian circuit consisting entirely of horizontal moves. Let us record what we have just proved.

Theorem 31.1 Cay$(Z_m \oplus Z_n)$ *is not Hamiltonian when* gcd$(m, n) = 1$. Cay$(\{(1, 0), (0, 1)\}: Z_m \oplus Z_n)$ *does not have a Hamiltonian circuit when m and n are relatively prime and greater than* 1.

What about when m and n are not relatively prime? In general the answer is somewhat complicated (see [7]) but the following special case is easy to prove.

Theorem 31.2 Cay$(Z_m \oplus Z_n)$ *is Hamiltonian when* $n \mid m$. Cay$(\{(1, 0), (0, 1)\}: Z_m \oplus Z_n)$ *has a Hamiltonian circuit when n divides m*.

Proof. Say, $m = kn$. Then we may think of $Z_m \oplus Z_n$ as k blocks of size $n \times n$. (See Figure 31.6 for an example.) Start at $(0, 0)$ and cover the vertices of the top block as follows. Use the generator $(0, 1)$ to move horizontally across the first row to the end. Then use the generator $(1, 0)$ to move vertically to the point below, and cover the remaining points in the second row by moving horizontally. Keep this process up until arriving at the point $(n - 1, 0)$—the lower left-hand corner of the first block. Next, move vertically to the second block and repeat the process used in the first block. Keep this up until the bottom block is covered. Complete the circuit by moving vertically back to $(0, 0)$. ∎

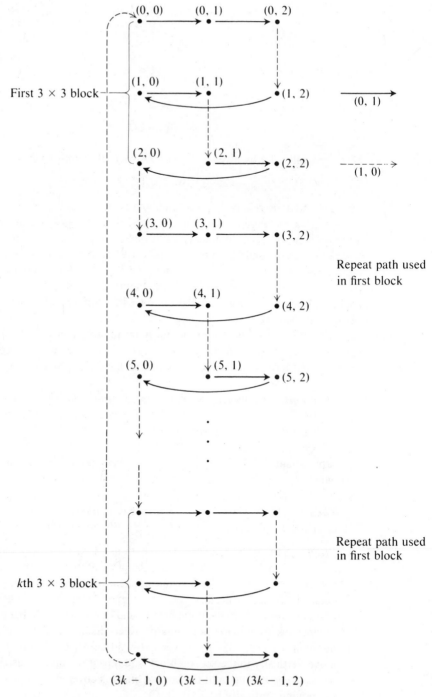

Figure 31.6 Cay({(1, 0), (0, 1)}: $Z_{3k} \oplus Z_3$).

Notice that the circuit given in the proof of Theorem 31.2 is easy to visualize but somewhat cumbersome to describe in words. A much more convenient way to describe a Hamiltonian path or circuit is to specify the starting vertex and the sequence of generators in the order that they are to be applied. In Example 5, for instance, we may start at (1) and alternate the generators (12) and (13) until we return to (1). In Example 3, we may start at R_0 and successively apply R, R, R, H, R, R, R, H. When k is a positive integer and a, b, . . . , c is a sequence of group elements, we use $k * (a, b, . . . , c)$ to denote the concatenation of k copies of the sequence $(a, b, . . . , c)$. Thus, $2 * (R, R, R, H)$ and $2 * (3 * R, H)$ both mean R, R, R, H, R, R, R, H. With this notation, we may conveniently denote the Hamiltonian circuit given in Theorem 31.2 as

$$m * ((n - 1) * (0, 1), (1, 0)).$$

We leave it as an exercise to show that if x_1, x_2, . . . , x_n is a sequence of generators determining a Hamiltonian circuit starting at some vertex, then the same sequence determines a Hamiltonian circuit for any starting vertex.

From Theorem 31.1, we know there are some Cayley digraphs of Abelian groups that do not have any Hamiltonian circuit. But Theorem 31.3 (first proved by Holsztýnski and Nathanson [5]) shows that each of these Cayley digraphs does have a Hamiltonian path. There are some Cayley digraphs for *non-Abelian* groups that do not even have a Hamiltonian path, but we will not discuss them here. (See [5] for more information.)

Theorem 31.3   *Abelian Groups Have Hamiltonian Paths*
Let G be a finite Abelian group, and let S be any (nonempty*) generating set for G. Then Cay(S:G) has a Hamiltonian path.

Proof.   We induct on $|S|$. If $|S| = 1$, say, $S = \{a\}$, then the digraph is just a circle labeled with e, a, a^2, . . . , a^{m-1} where $|a| = m$. Obviously, there is a Hamiltonian path for this case. Now assume $|S| > 1$. Choose some $s \in S$. Let $T = S - \{s\}$, i.e., T is S with s removed, and put $H = \langle T \rangle$. (Notice that H may be equal to G.)

Because $|T| < |S|$ and H is a finite Abelian group, the induction hypothesis guarantees that there is a Hamiltonian path $(a_1, a_2, . . . , a_k)$ in Cay(T:H). We will show that

$(a_1, a_2, . . . , a_k, s, a_1, a_2, . . . , a_k, s, . . . , a_1,$
$\qquad\qquad\qquad\qquad a_2, . . . , a_k, s, a_1, a_2, . . . , a_k)$

where a_1, a_2, . . . , a_k occurs $|G|/|H|$ times and s occurs $|G|/|H| - 1$ times, is a Hamiltonian path in Cay(S:G).

---

*If S is the empty set, it is customary to define $\langle S \rangle$ to be the identity group. We prefer to ignore this trivial case.

Because $S = T \cup \{s\}$ and T generates H, the coset Hs generates the factor group G/H. (Since G is Abelian, this group exists.) Hence, the cosets of H are H, Hs, Hs^2, . . . , Hs^n, where $n = |G|/|H| - 1$. Starting from the identity element of G, the path given by $(a_1, a_2, \ldots, a_k)$ visits each element of H exactly once (because $(a_1, a_2, \ldots, a_k)$ is a Hamiltonian path in $\text{Cay}(T{:}H)$). The generator s then moves us to some element of the coset Hs. Starting from here, the path $(a_1, a_2, \ldots, a_k)$ visits each element of Hs exactly once. Then, s moves us to the coset Hs^2, and we visit each element of this coset exactly once. Continuing this process, we successively move to Hs^3, Hs^4, . . . , Hs^n, visiting each vertex in each of these cosets exactly once. Because each vertex of $\text{Cay}(S{:}G)$ is in exactly one coset Hs^i, this implies that we visit each vertex of $\text{Cay}(S{:}G)$ exactly once, Thus, we have a Hamiltonian path. ∎

We next look at three generated Cayley digraphs.

Example 9   Let

$$D_3 = \langle r, f \mid r^3 = f^2 = e, \, rf = fr^2 \rangle.$$

Then a Hamiltonian circuit in

$$\text{Cay}(\{(r, 0), (f, 0), (e, 1)\}{:} D_3 \oplus Z_6)$$

is given in Figure 31.7.

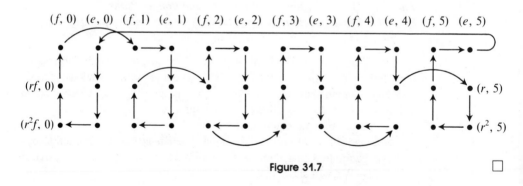

Figure 31.7   □

Although it is not easy to prove, it is true that

$$\text{Cay}(\{(r, 0), (f, 0), (e, 1)\}{:} D_n \oplus Z_m)$$

has a Hamiltonian circuit for all n and m. (See [9].) Example 10 shows the circuit for this digraph when m is even.

Example 10   Let

$$D_n = \langle r, f \mid r^n = f^2 = e, \, rf = fr^{-1} \rangle.$$

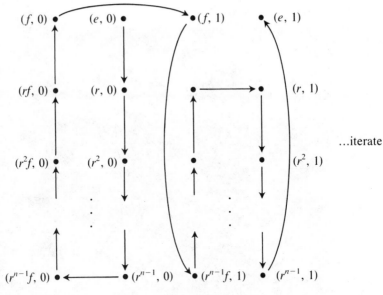

...iterate

Figure 31.8

Then a Hamiltonian circuit in

$$\text{Cay}(\{(r, 0), (f, 0), (e, 1)\}: D_n \oplus Z_m)$$

with m even is traced in Figure 31.8. The sequence of generators that traces the circuits is

$$m * [(n - 1) * (r, 0), (f, 0), (n - 1) * (r, 0), (e, 1)]. \qquad \square$$

Some Applications

Hamiltonian paths and circuits in Cayley digraphs have arisen in a variety of group theory contexts. A Hamiltonian path in a Cayley digraph of a group is simply an ordered listing of the group elements without repetition. The vertices of the digraph are the group elements and the arcs of the path are generators of the group. In 1948, R. A. Rankin used these ideas (although not the terminology) to prove that certain bell-ringing exercises could not be done by the traditional methods employed by bell ringers. (See [3, Chap. 24] for the group-theoretical aspects of bell ringing). In 1981, Hamiltonian paths in Cayley digraphs were used in an algorithm for creating computer graphics of Escher-type repeating patterns in the hyperbolic plane [4]. The program can produce repeating hyperbolic patterns in color from among five infinite classes of symmetry groups. The program has now been improved so that the user may choose from many kinds of color symmetry. Two Escher drawings and their computer counterparts are given in Figures 31.9–31.12.

Figure 31.9 M. C. Escher's *Circle Limit I* [6]

Figure 31.10   A computer duplication of the pattern of M. C. Escher's
Circle Limit I [6]. The program used a Hamiltonian path
in a Cayley digraph of the underlying symmetry group.

Figure 31.11 M. C. Escher's *Circle Limit IV* [6]

Figure 31.12 A computer duplication of the pattern of M. C. Escher's
Circle Limit IV [6]. The program used a Hamiltonian path
in a Cayley digraph of the underlying symmetry group.

In this chapter, we have shown how one may construct a directed graph from a group. It is also possible to associate a group—called the *automorphism group*—with every directed graph. (See [8, pp. 16–20] or [1, pp. 104–107] for details.) In fact, several of the twenty-six sporadic simple groups were first constructed in this way.

EXERCISES

It is the function of creative men to perceive the relations between thoughts, or things, or forms of expression that may seen utterly different, and to be able to combine them into some new forms—the power to connect the seemingly unconnected.

William Plomer

1. Find a Hamiltonian circuit in the digraph given in Example 7 different form the one in Figure 31.3.

2. Find a Hamiltonian circuit in

$$\text{Cay}(\{(a, 0), (b, 0), (e, 1)\}: Q_4 \oplus Z_2).$$

3. Find a Hamiltonian circuit in

$$\text{Cay}(\{(a, 0), (b, 0), (e, 1)\}: Q_4 \oplus Z_m).$$

where m is even.

4. Write the sequence of generators for each of the circuits found in exercises 1, 2, and 3.

5. Use the Cayley digraph in Example 7 to evaluate the product $a^3ba^{-1}ba^3b^{-1}$.

6. Let x and y be two vertices of a Cayley digraph. Explain why two paths from x to y in the digraph yield a group relation.

7. Use the Cayley digraph in Example 7 to verify the relation $aba^{-1}b^{-1}a^{-1}b^{-1} = a^2ba^3$.

8. Identify the following Cayley digraph of a familiar group.

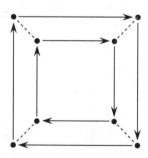

9. Let $D_4 = \langle r, f \mid r^4 = e = f^2, rf = fr^{-1} \rangle$. Verify that

$$6 * [3 * (r, 0), (f, 0), 3 * (r, 0), (e, 1)]$$

is a Hamiltonian circuit in

$$\text{Cay}(\{(r, 0), (f, 0), (e, 1)\}: D_4 \oplus Z_6).$$

10. Draw a picture of $\text{Cay}(\{2, 5\}: Z_8)$.

11. If $s_1, s_2, \ldots, s_n$ is a sequence of generators that determines a Hamiltonian circuit beginning at some vertex, explain why the same sequence determines a Hamiltonian circuit beginning at any point.

12. Show that the Cayley digraph given in Example 7 has a Hamiltonian path from e to a.

13. Show that there is no Hamiltonian path in

$$\text{Cay}(\{(1, 0), (0, 1)\}: Z_3 \oplus Z_2)$$

from $(0, 0)$ to $(2, 0)$.

14. Draw $\text{Cay}(\{2, 3\}: Z_6)$. Is there a Hamiltonian circuit in this digraph?

15. a. Let G be a group of order n generated by a set S. Show that a sequence $s_1, s_2, \ldots, s_{n-1}$ from S is a Hamiltonian path in $\text{Cay}(S:G)$ if and only if for all i and j with $1 \leq i \leq j < n$, we have $s_i s_{i+1} \cdots s_j \neq e$.

    b. Show that the sequence in part a is a Hamiltonian circuit if and only if $s_1 s_2 \cdots s_n = e$, and that whenever $1 \leq i \leq j < n$, we have $s_i s_{i+1} \cdots s_j \neq e$.

16. Let $D_4 = \langle a, b \mid a^2 = b^2 = (ab)^4 = e \rangle$. Draw $\text{Cay}(\{a, b\}: D_4)$. Why is it reasonable to say that this digraph is undirected?

17. Let D_n be as in Example 10. Show that $2 * [(n - 1) * r, f]$ is a Hamiltonian circuit in $\text{Cay}(\{r, f\}: D_n)$.

18. Let $Q_8 = \langle a, b \mid a^8 = e, a^4 = b^2, b^{-1}ab = a^{-1} \rangle$. Find a Hamiltonian circuit in $\text{Cay}(\{a, b\}: Q_8)$.

19. Let Q_8 be as in exercise 18. Find a Hamiltonian circuit in

$$\text{Cay}(\{(a, 0), (b, 0), (e, 1)\}: Q_8 \oplus Z_5).$$

20. Prove that the Cayley digraph given in Example 6 does not have a Hamiltonian circuit. Does it have a Hamiltonian path?

21. Find a Hamiltonian circuit in

$$\text{Cay}(\{(R_{90}, 0), (H, 0), (R_0, 1)\}: D_4 \oplus Z_3).$$

Does this circuit generalize to the case $D_{n+1} \oplus Z_n$ for all $n \geq 3$?

22. Let Q_8 be as in exercise 18. Find a Hamiltonian circuit in

$$\text{Cay}(\{(a, 0), (b, 0), (e, 1)\}: Q_8 \oplus Z_m) \text{ for all even } m.$$

23. Find a Hamiltonian circuit in

$$\text{Cay}(\{(a,\ 0),\ (b,\ 0),\ (e,\ 1)\}\colon Q_4 \oplus Z_3).$$

24. Find a Hamiltonian circuit in

$$\text{Cay}(\{(a,\ 0),\ (b,\ 0),\ (e,\ 1)\}\colon Q_4 \oplus Z_m) \text{ for all odd } m \geq 3.$$

25. Write the sequence of generators that describes the Hamiltonian circuit in Example 9.

26. Let D_n be as in Example 10. Find a Hamiltonian circuit in

$$\text{Cay}(\{(r,\ 0),\ (f,\ 0),\ (e,\ 1)\}\colon D_4 \oplus Z_5).$$

Does your circuit generalize to the case $D_n \oplus Z_{n+1}$ for all $n \geq 4$?

27. Prove that $\text{Cay}(\{(0,\ 1),\ (1,\ 1)\}\colon Z_m \oplus Z_n)$ has a Hamiltonian circuit for all m and n greater than 1.

28. Suppose a Hamiltonian circuit exists for $\text{Cay}(\{(1,\ 0),\ (0,\ 1)\}\colon Z_m \oplus Z_n)$ and the circuit exits from vertex $(a,\ b)$ vertically. Show that the circuit exits from every member of the coset $(a,\ b) + \langle(1,\ -1)\rangle$ vertically.

29. Let $D_2 = \langle r,\ f \mid r^2 = f^2 = e,\ rf = fr^{-1}\rangle$. Find a Hamiltonian circuit in $\text{Cay}(\{(a,\ 0),\ (b,\ 0),\ (e,\ 1)\}\colon D_2 \oplus Z_3)$.

30. Let Q_8 be as in exercise 18. Find a Hamiltonian circuit in $\text{Cay}(\{(a,\ 0),\ (b,\ 0),\ (e,\ 1)\}\colon Q_8 \oplus Z_3)$.

31. A finite group is called *Hamiltonian* if all of its subgroups are normal. (One non-Abelian example is Q_4.) Show that Theorem 31.3 can be generalized to include all Hamiltonian groups.

32. (Factor Group Lemma) Let S be a generating set for a group G, N a cyclic normal subgroup of G, and

$$\bar{S} = \{sN \mid s \in S\}.$$

If $(a_1 N,\ \ldots,\ a_r N)$ is a Hamiltonian circuit in $\text{Cay}\ (\bar{S}\colon G/N)$ and the product $a_1 \ldots a_r$ generates N, prove that

$$|N| * (a_1,\ \ldots,\ a_r)$$

is a Hamiltonian circuit in $\text{Cay}(S\colon G)$.

REFERENCES

1. Sabra S. Anderson, *Graph Theory and Finite Combinatorics*, Chicago: Markham, 1970.
2. Claude Berge, *Graphs and Hypergraphs*, Amsterdam: North-Holland, 1973.
3. F. J. Budden, *The Fascination of Groups*, Cambridge: Cambridge University Press, 1972.

4. Douglas Dunham, John Lindgren, and David Witte, "Creating Repeating Hyperbolic Patterns," *Computer Graphics* 15 (1981): 215–223.
5. W. Holsztyński and R. F. E. Strube, "Paths and Circuits in Finite Groups," *Discrete Mathematics* 22 (1978): 263–272.
6. J. L. Locher, ed., *The World of M. C. Escher,* New York: Harry N. Abrams, 1971.
7. W. T. Trotter, Jr., and P. Erdös, "When the Cartesian Product of Directed Cycles Is Hamiltonian," *Journal of Graph Theory* 2 (1978): 137–142.
8. Arthur T. White, *Graphs, Groups and Surfaces,* New York: Elsevier Science, 1984.
9. David Witte, Gail Letzter, and Joseph A. Gallian, "On Hamiltonian Circuits in Cartesian Products of Cayley Digraphs," *Discrete Mathematics* 43 (1983): 297–307.

SUGGESTED READINGS

Frank Budden, "Cayley Graphs for Some Well-Known Groups," *The Mathematical Gazette* 69 (1985): 271–278.

This article contains the Cayley graphs of A_4, Q_4, and S_4 using a variety of generators and relations.

E. L. Burrows and M. J. Clark, "Pictures of Point Groups," *Journal of Chemical Education* 51 (1974): 87–90.

Chemistry students may be interested in reading this article. It gives a comprehensive collection of the Cayley digraphs of the important point groups.

Douglas Dunham, John Lindgren, and David Witte, "Creating Repeating Hyperbolic Patterns," *Computer Graphics* 15 (1981): 215–223.

In this beautifully illustrated paper, a process for creating repeating patterns of the hyperbolic plane is described. The paper is a blend of group theory, geometry, and art.

Joseph A. Gallian and David Witte, "Hamiltonian Checkerboards," *Mathematics Magazine* 57 (1984): 291–294.

This paper gives some additional examples of Hamiltonian circuits in Cayley digraphs.

Paul Hoffman, "The Man Who Loves Only Numbers," *The Atlantic Monthly* 260 (1987): 60–74.

A charming portrait of Paul Erdös, the most prolific and most eccentric mathematician in the world.

A. T. White, "Ringing the Cosets," *The American Mathematical Monthly* 94 (1987): 721–746.

This article analyzes the practice of bell ringing by way of Cayley digraphs.

David Witte and Joseph A. Gallian, "A Survey: Hamiltonian Cycles in Cayley Graphs," *Discrete Mathematics* 51 (1984): 293–304.

This paper surveys the results, techniques, applications, and open problems in the field.

William Rowan Hamilton

After Isaac Newton, the greatest mathematician of the English-speaking peoples is William Rowan Hamilton.

Sir Edmund Whittaker, *Scientific American*

ÉIRE 29

Quaternions discovery by **Hamilton** 1843

This stamp featuring the quaternions was issued in 1983.

WILLIAM ROWAN HAMILTON was born on August 3, 1805, in Dublin, Ireland. Although Hamilton did not attend attend school before entering college, he was widely recognized as a child prodigy. At three, he was skilled at reading and arithmetic. At five, he read and translated Latin, Greek and Hebrew; at fourteen, he had mastered fourteen languages, including Arabic, Sanskrit, Hindustani, Malay, and Bengali.

Hamilton's undergraduate career at Trinity College in Dublin was brilliant. At the age of twenty-one, he wrote a paper entitled *A Theory of Systems of Rays,* introducing techniques that have become indispensable in physics and have created the field of mathematical optics. This work was so extraordinary that a year later, when the chair of professor of astronomy at Trinity—the holder of which is given the title of Royal Astronomer of Ireland—became vacant, Hamilton was unanimously elected to the position, over many distinguished candidates, in spite of the fact that he was still an undergraduate and had not applied for the appointment!

In 1833 Hamilton provided the first modern treatment of complex numbers. In 1843, he made what he considered his greatest discovery—the algebra of quaternions. The quaternions represent a natural generalization of the complex numbers with three numbers $i, j,$ and k whose squares are -1. With these, rotations in three and four dimensions can be algebraically treated. Of greater significance, however, is the fact that the quaternions are noncommutative under multiplication. This was the first ring to be discovered in which the commutative property does not hold. After ten years of fruitless thought, the essential idea for the quaternions suddenly came to him. Many years later, in a letter to his son, Hamilton described the circumstances of his discovery:

But on the 16th day of the same month [October 1843]—which happened to be a Monday and a Council day of the Royal Irish Academy—I was walking to attend and preside, and

424

your mother was walking with me, along the Royal Canal, to which she had perhaps been driven; and although she talked with me now and then, yet an under-current of thought was going on in my mind, which gave at last a result, whereof it is not too much to say that I felt at once the importance. An electric circuit seemed to close; and a spark flashed forth, the herald (as I foresaw immediately) of many long years to come of definitely directed thought and work, by myself if spared, and at all events on the part of others, if I should ever be allowed to live long enough distinctly to communicate the discovery. I pulled out on the spot a pocket-book, which still exists, and made an entry there and then. Nor could I resist the impulse—unphilosophical as it may have been—to cut with a knife on a stone of Brougham Bridge, as we passed it, the fundamental formula with the symbols i, j, k;

$$i^2 = j^2 = k^2 = ijk = -1,$$

which contains the solution of the Problem, but of course as an inscription, has long since mouldered away.

Today Hamilton's name is attached to a number of concepts, such as the Hamiltonian function, which represents the total energy in a physical system; the Hamilton-Jacobi differential equations; and the Cayley-Hamilton Theorem from linear algebra. He also coined the terms *vector, scalar,* and *tensor.*

In his later years, Hamilton was plagued by alcoholism. He died on September 2, 1865, at the age of sixty-one.

Paul Erdös

Paul Erdös is a socially helpless Hungarian who has thought about more mathematical problems than anyone else in history.

The Atlantic Monthly

PAUL ERDÖS (pronounced AIR-dish) is one of the best known and highly respected mathematicians of this century. Unlike most of his contemporaries who concentrate on theory building, Erdös focuses on problem-solving and problem-posing. Like Euler, whose solutions to special problems pointed the way to much of the mathematical theory we have today, the problems and methods of solution of Erdös have helped pioneer new theories, such as combinatorial and probabilistic number theory, combinatorial geometry, probabilistic and transfinite combinatorics, and graph theory.

Erdös was born on March 26, 1913, in Hungary. Both of his parents were high-school mathematics teachers. They provided his early training and encouraged the development of his mathematical talent. Erdös was a mathematical prodigy. At the age of four, he told his mother that if you take 250 away for 100, you have 150 below 0. His first research paper, published when he was eighteen, gave a new proof that there must exist a prime number between n and $2n$. Erdös, a Jew, left Hungary in 1934 at the age of twenty-one because of the rapid rise of anti-Semitism in Europe. Ever since then, he has been travelling. Erdös has no family, no property, no fixed address. He travels from place to place, never staying more than a month, giving lectures for small honoraria and staying with mathematicians. All that he owns he carries with him in a medium-sized suitcase, frequently visiting as many as fifteen places in a month. He discusses his record for travel as follows. "There was a Saturday meeting in Winnipeg on number theory and computing. On Saturday evening we had a dinner in a Hungarian restaurant, a farewell dinner for the speaker. Then on Sunday morning I flew to Toronto. I was met at the airport, and we went to Waterloo to a picnic. In the evening I was taken back to Toronto, and I flew to London where I lectured at 11 o'clock at Imperial College." His motto is "Another roof, another proof." Although in his seventies, he puts in nineteen-hour days doing mathematics.

One might say that Erdös has lived a sheltered life. He buttered his first piece of bread at age twenty-one. He has never cooked anything or even boiled water. He has never driven a car.

One of Erdös's customs is to offer cash prizes for the solutions to unsolved problems. These awards range from $5 to $10,000, depending on how difficult he judges them to be.

Erdös usually dispenses with common greetings and small talk. Typically, he will greet someone with, "Consider the following problem." He sends off over a 1,000 letters a year, all to fellow mathematicians and exclusively about mathematics.

Erdös has written over 1,000 research papers. It is believed that the previous record was held by Cayley, at 927. He has coauthored papers with over 250 people. These people are said to have Erdös number 1. People who do not have Erdös number 1, but who have written a paper with someone who does, are said to have Erdös number 2, and so on inductively. Of course, these numbers keep changing. The highest known Erdös number in 1987 was 7.

Erdös's papers cover a broad range of topics, but the majority are in number theory, combinatorics, and graph theory. In 1951, he received the American Mathematical Society's Cole Prize for number theory. In 1984, he received the prestigious $50,000 Wolf Prize. He kept $750, established a scholarship fund, and gave the rest away.

Among his friends, Erdös is admired for his great generosity, his extraordinary concern for human rights, and his keen interest in helping promising young mathematicians.

Introduction to Algebraic Coding Theory

32

Damn it, if the machine can detect an error, why can't it locate the position of the error and correct it?

Richard W. Hamming

Motivation

One of the most interesting and important applications of finite fields has been the development of algebraic coding theory. This theory originated in the late 1940s, and was created in response to practical communication problems. (Algebraic coding has nothing to do with secret codes.)

To motivate this theory, imagine that we wish to transmit one of two possible signals to a spacecraft approaching Mars. If video pictures reveal the conditions of the proposed landing site and they look unfavorable, we will command the craft to orbit the planet; otherwise, we will command the craft to land. The signal for orbiting will be a 0, and the signal for landing will be 1. But it is possible that some sort of interference (called *noise*) could cause an incorrect message to be received. To decrease the chance of this happening, redundancy is built into the transmission process. For example, if we wish the craft to orbit Mars, we could send five 0's. The craft's onboard computer is programmed to take any five-digit message received and decode the result by majority rule. So, if 00000 is sent and 10001 is received, the computer decides that 0 was the intended message. Notice that, for the computer to make the wrong decision, at least three errors must occur during transmission. If we assume that errors occur independently, it is less likely that three errors occur than two or fewer

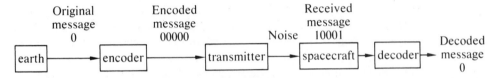

Figure 32.1 Encoding and decoding by fivefold repetition.

errors. For this reason, this decision process is frequently called the *maximum-likelihood decoding* procedure. Our particular situation is illustrated in Figure 32.1. The general coding procedure is illustrated in Figure 32.2.

In practice, the means of transmission are telephone, telegraph, radiowave, microwave, or even a magnetic tape or disc. The noise might be human error, cross talk, lightning, thermal noise, impulse noise, or deterioration of a tape or disc. Throughout this chapter, we assume that errors in transmission occur independently. Different methods are needed when this is not the case.

Now, let's consider a more complicated situation. This time, assume we wish to send a sequence of 0's and 1's of length 500. Further, suppose that the probability that an error is made in the transmission of any particular digit is .01. If we send this message directly without any redundancy, the probability that it will be received error-free is $(.99)^{500}$, or approximately .0066.

On the other hand, if we adopt a threefold repetition scheme by sending each digit three times and decoding each block of three digits received by majority rule, we can do much better. For example, the sequence 1011 is encoded as 111000111111. If the received message is 011000001110, the decoded message is 1001. Now, what is the probability that our 500-digit message is error-free? Well, if a 1, say, is sent, it is decoded as a 0 only if the block received is 001, 010, 100, or 000. The probability that this occurs is

$$(.01)(.01)(.99) + (.01)(.99)(.01) + (.99)(.01)(.01) + (.01)(.01)(.01)$$
$$= (.01)^2[3(.99) + .01]$$
$$= .000298 < .0003.$$

Thus, the probability that any particular digit in the sequence is decoded correctly is greater than .9997, and it follows that the probability of decoding the entire 500-digit message correctly is greater than $(.9997)^{500}$, or approximately .86, a dramatic improvement over .0066.

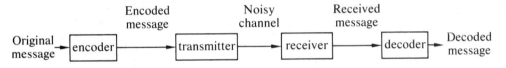

Figure 32.2 General encoding-decoding.

This example illustrates the three basic features of a code. One has a set of messages, a method of encoding these messages, and a method of decoding the received messages. The encoding procedure builds some redundancy into the original messages; the decoding procedure corrects or detects certain prescribed errors. Repetition codes have the advantage of simplicity of encoding and decoding, but they are too inefficient. In a fivefold repetition code, 80% of all transmitted information is redundant. The goal of coding theory is to devise message encoding and decoding methods that are reliable, efficient, and reasonably easy to implement.

Before plunging into the formal theory, it is instructive to look at a sophisticated example.

Example 1 The Hamming (7, 4) Code

This time, our message set consists of all possible 4-tuples of 0's and 1's (i.e., we wish to send a sequence of 0's and 1's of length 4). Encoding will be done by viewing these messages as 1×4 matrices with entries from Z_2 and multiplying each of the 16 messages on the right by the matrix

$$G = \begin{bmatrix} 1 & 0 & 0 & 0 & 1 & 1 & 0 \\ 0 & 1 & 0 & 0 & 1 & 0 & 1 \\ 0 & 0 & 1 & 0 & 1 & 1 & 1 \\ 0 & 0 & 0 & 1 & 0 & 1 & 1 \end{bmatrix}.$$

(All arithmetic is done modulo 2.) The resulting 7-tuples are called *code words*. (See Table 32.1.)

Table 32.1

Message	Encoder G	Code Word
0000	→	0000000
0001	→	0001011
0010	→	0010111
0100	→	0100101
1000	→	1000110
1100	→	1100011
1010	→	1010001
1001	→	1001101
0110	→	0110010
0101	→	0101110
0011	→	0011100
1110	→	1110100
1101	→	1101100
1011	→	1011010
0111	→	0111001
1111	→	1111111

Notice that the first four digits of each code word constitute just the original message corresponding to the code word. The last three digits of the code word constitute the redundancy features. For this code, we use the *nearest-neighbor* decoding method (which, in the case that the errors occur independently, is the same as the maximum-likelihood decoding). For any received word v, we assume the word sent is the code word v', which differs from v in the fewest number of positions. If the choice of v' is not unique, we can decide not to decode or arbitrarily choose one of the code words closest to v. (The first option is usually selected when retransmission is practical.) Once we have decoded the received word, we can obtain the intended message by deleting the last three digits of v'. For instance, suppose 1000 were the intended message. It would be encoded and transmitted as $u = 1000110$. If the received word is $v = 1100110$ (an error in the second position), it would still be decoded as u, since v and u differ in only one position, while v and any other code word would differ in at least three positions. Similarly, the intended message 1111 would be encoded as 1111111. If, instead of this, the word 0111111 is received, our decoding procedure still gives us the intended message 1111. $\square$

The code in Example 1 is one of an infinite class of important codes discovered by Richard Hamming in 1948. The Hamming codes are the most widely used codes.

The Hamming (7, 4) encoding scheme can be conveniently illustrated with the use of a Venn diagram (with three overlapping circles). The four message digits are placed in the four overlapping regions (starting at the top region and then going from left to right) while the remaining regions are assigned 0 or 1, so that the total number of 1's in each circle is even. See Figure 32.3.

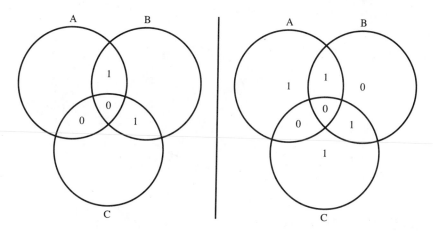

Figure 32.3 Venn diagram of the message 1001 and the encoded message 1001101.

Consider the Venn diagram of the received word 0001101:

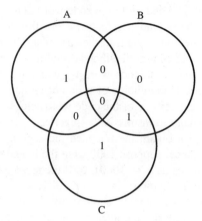

How may we detect and correct an error? Well, observe that both circles *A* and *B* have an odd number of 1's. This tells us something is wrong. At the same time, we note that circle *C* has an even number of 1's. Thus, the portion of the diagram in both *A* and *B* but not *C* is the source of the error. See Figure 32.4.

Quite often, codes are utilized to detect errors rather than correct them. This is especially appropriate when it is easy to retransmit a message. If a received word is not a code word, we have detected an error. For example, computers are designed to use a parity check for numbers. Inside the computer, each number is represented as a string of 0's and 1's. If there is an even number of 1's in this representation, a 0 is attached to the string; if there is an odd number of 1's in the representation, a 1 is attached to the string. Thus, each number stored in the computer memory has an even number of 1's. Now, when the computer reads a number from memory, it performs a parity check. If the read number

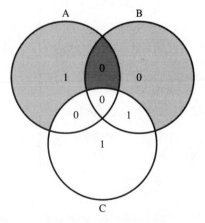

Figure 32.4 Circles *A* and *B* but not *C* have wrong parity.

has an odd number of 1's, the computer will know an error has been made, and it will reread the number. Note that an even number of errors will not be detected by a parity check.

The methods of error detection introduced in Chapters 0 and 2 are based on the same principle. An extra character is appended to a string of numbers so that a particular condition is satisfied. If we find that such a string does not satisfy that condition, we know an error has occurred.

Linear Codes

We now formalize some of the ideas introduced in the preceding discussion.

DEFINITION Linear Code
An (n, k) *linear code* over a finite field F is a k-dimensional subspace V of the vector space

$$F^n = \underbrace{F \oplus F \oplus \cdots \oplus F}_{n \text{ copies}}$$

over F. The members of V are called the *code words*. The ratio k/n is called the *information rate* of the code. When F is Z_2, the code is called *binary*.

One should think of an (n, k) linear code over F as a set of n-tuples from F, where each n-tuple is comprised of two parts: the message part, consisting of k digits; and the redundancy part, consisting of the remaining $n - k$ digits. Note that an (n, k) linear code over a finite field F of order q has q^k code words, since every member of the code is uniquely expressible as a linear combination of the k basis vectors with coefficients from F. The set of q^k code words is closed under addition and scalar multiplication by members of F. Also, since errors in transmission may occur in any of the n positions, there are q^n possible vectors that can be received. Where there is no possibility of confusion, it is customary to denote an n-tuple $(a_1, a_2, \ldots, a_n)$ more simply as $a_1 a_2 \cdots a_n$, as we did in Example 1.

Example 2 The set

$$\{0000000, 0010111, 0101011, 1001101,$$
$$1100110, 1011010, 0111100, 1110001\}$$

is a $(7, 3)$ binary code. This code has information rate 3/7. □

Example 3 The set $\{0000, 0101, 1010, 1111\}$ is a $(4, 2)$ binary code. □

Although binary codes are by far the most important ones, other codes are occasionally used.

Example 4 The set

$$\{0000, 0121, 0212, 1022, 1110, 1201, 2011, 2102, 2220\}$$

is a (4, 2) linear code over Z_3. A linear code over Z_3 is called a *ternary code*.

□

To facilitate our discussion of the error-correcting and error-detecting capability of a code, we introduce the following terminology.

DEFINITION   Hamming Distance, Hamming Weight
The *Hamming distance* between two vectors of a vector space is the number of components in which they differ. The *Hamming weight* of a vector is the number of nonzero components of the vector.

We will use $d(u, v)$ to denote the Hamming distance between the vectors u and v, and wt(u) for the Hamming weight of the vector u.

Example 5   Let $s = 0010111$, $t = 0101011$, $u = 1001101$, and $v = 1101101$. Then, $d(s, t) = 4$, $d(s, u) = 4$, $d(s, v) = 5$, $d(u, v) = 1$; wt(s) = 4, wt(t) = 4, wt(u) = 4, and wt(v) = 5. □

For linear codes, the Hamming distance and Hamming weight have the following important properties.

Theorem 32.1   *Properties of Hamming Distance and Hamming Weight*
For any vectors u, v, and w of a linear code, $d(u, v) \leq d(u, w) + d(w, v)$ and $d(u, v) = \mathrm{wt}(u - v)$.

Proof.   Exercise.   ∎

With the preceding definitions and Theorem 32.1, we can now explain why the codes given in Examples 1, 2, and 4 will correct any single error, but why the code in Example 3 will not.

Theorem 32.2   *Correcting Capability of a Linear Code*
If the Hamming weight of every nonzero code word in a linear code is at least $2t + 1$, then the code can correct any t or fewer errors. Furthermore, the same code can detect any 2t or fewer errors.

Proof.   We will use nearest-neighbor decoding; that is, for any received vector v, we will assume that the corresponding code word sent is a code word v' such that the Hamming distance $d(v, v')$ is a minimum. (If there is more than one such v', we decide arbitrarily.) Now, suppose a transmitted code word u is received as the vector v and at most t errors were made in transmission. Then, by definition, $d(v, u) \leq t$. If w is any code word other than u, then $w - u$ is a nonzero code word. Thus, by assumption,

$$2t + 1 \leqslant \text{wt}(w - u) = d(w, u) \leqslant d(w, v) + d(v, u) \leqslant d(w, v) + t,$$

and it follows that $t + 1 \leqslant d(w, v)$. So, the code word closest to the received vector v is u and, therefore, v is correctly decoded as u.

To show that the code can detect $2t$ errors, we suppose a transmitted code word u is received as the vector v and at least one error, but no more than $2t$ errors, was made in transmission. Because only code words are transmitted, an error will be detected whenever a received word is not a code word. But u cannot be a code word, since $d(v, u) \leqslant 2t$, while we know that the minimum distance between distinct code words is at least $2t + 1$. ∎

Example 6 The Hamming (7, 4) code given in Example 1 will correct any single error and detect any pair of errors. For, by inspection, the minimum Hamming weight of any nonzero code word is $3 = 2 \cdot 1 + 1$. □

It is natural to wonder how the matrix G used to produce the Hamming code in Example 1 was chosen. Better yet, in general, how can one find a function G that carries a subspace V of F^k to a subspace of F^n in such a way that for any k-tuple v in V, the image vector vG will agree with v in the first k components and build in some redundancy in the last $n - k$ components? Such a function is a $k \times n$ matrix of the form

$$\begin{bmatrix} 1 & 0 & \cdots & 0 & a_{11} & \cdots & a_{1n-k} \\ 0 & 1 & \cdots & 0 & \vdots & & \vdots \\ \vdots & \vdots & & \vdots & \vdots & & \vdots \\ \vdots & \vdots & & \vdots & \vdots & & \vdots \\ 0 & 0 & \cdots & 1 & a_{k1} & \cdots & a_{kn-k} \end{bmatrix}$$

where the a_{ij}'s belong to F. A matrix of this form is called the *standard generator matrix* (or *standard encoding matrix*) for the resulting code.

Any $k \times n$ matrix whose rows are linearly independent will transform the subspace of F^k to a k-dimensional subspace of F^n that could be used to build redundancy, but using the standard generating matrix has the advantage that the original message constitutes the first k components of the transformed vectors. An (n, k) linear code in which the k information digits occur at the beginning of each code word is called a *systematic code*. Schematically, we have

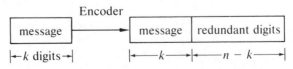

Notice that, by definition, a standard generating matrix produces a systematic code.

Example 7 From the set of message

$$\{000, 001, 010, 100, 110, 101, 011, 111\},$$

we may construct a (6, 3) linear code over Z_2 with the standard generating matrix

$$G = \begin{bmatrix} 1 & 0 & 0 & 1 & 1 & 0 \\ 0 & 1 & 0 & 1 & 0 & 1 \\ 0 & 0 & 1 & 1 & 1 & 1 \end{bmatrix}.$$

The resulting code words are given in Table 32.2.

Table 32.2

Message	Encoder G	Code Word
000	$\rightarrow$	000000
001	$\rightarrow$	001111
010	$\rightarrow$	010101
100	$\rightarrow$	100110
110	$\rightarrow$	110011
101	$\rightarrow$	101001
011	$\rightarrow$	011010
111	$\rightarrow$	111100

Since the minimum weight of any nonzero code word is 3, this code will correct any single error and detect any double error.   ☐

Example 8 Here we take a set of messages as

$$\{00, 01, 02, 10, 11, 12, 20, 21, 22\},$$

and we construct a (4, 2) linear code over Z_3 with the standard generator matrix

$$G = \begin{bmatrix} 1 & 0 & 2 & 1 \\ 0 & 1 & 2 & 2 \end{bmatrix}.$$

The resulting code words are given in Table 32.3.

Table 32.3

Message	Encoder G	Code Word
00	$\rightarrow$	0000
01	$\rightarrow$	0122
02	$\rightarrow$	0211
10	$\rightarrow$	1021
11	$\rightarrow$	1110
12	$\rightarrow$	1202
20	$\rightarrow$	2012
21	$\rightarrow$	2101
22	$\rightarrow$	2220

This code will correct any single error and detect any double error. □

Parity-Check Matrix Decoding

Now that we have a convenient method for encoding messages, we need a convenient method for decoding the received messages. Unfortunately, this is not as easy to do; however, in the case where at most one error per code word occurred, there is a fairly simple method for decoding.

To describe this method, suppose that V is a systematic linear code over the field F given by the standard generating matrix $G = [I_k \mid A]$, where I_k represents the $k \times k$ identity matrix and A is the $k \times (n - k)$ matrix obtained from G by deleting the first k columns of G. Then, the $n \times (n - k)$ matrix

$$H = \begin{bmatrix} -A \\ \hline I_{n-k} \end{bmatrix},$$

where $-A$ is the negative of A and I_{n-k} is the $(n - k) \times (n - k)$ identity matrix, is called the *parity-check matrix* for V. (In the literature, the transpose of H is called the parity-check matrix, but H is much more convenient for our purposes.) The decoding procedure is:

1. For any received word w, compute wH.
2. If wH is the zero vector, assume no error was made.
3. If for some nonzero element $s \in F$, we have wH is s times the ith row of H, assume the word was $w - 0 \cdots s \cdots 0$, where s occurs in the ith component.
3'. When the code is binary, (3) reduces to: If wH is the ith row of H, assume an error was made in the ith component of w.
4. If wH does not fit into either of categories (2) or (3), at least two errors occurred in transmission and we do not decode.

Example 9 Consider the Hamming (7, 4) code given in Example 1. The generating matrix is

$$G = \begin{bmatrix} 1 & 0 & 0 & 0 & 1 & 1 & 0 \\ 0 & 1 & 0 & 0 & 1 & 0 & 1 \\ 0 & 0 & 1 & 0 & 1 & 1 & 1 \\ 0 & 0 & 0 & 1 & 0 & 1 & 1 \end{bmatrix}$$

and the corresponding parity-check matrix is

$$H = \begin{bmatrix} 1 & 1 & 0 \\ 1 & 0 & 1 \\ 1 & 1 & 1 \\ 0 & 1 & 1 \\ 1 & 0 & 0 \\ 0 & 1 & 0 \\ 0 & 0 & 1 \end{bmatrix}.$$

Now, if the received vector is $v = 0000110$, we find $vH = 110$. Since this is the first row of H, we assume an error was made in the first position of v. Thus, the transmitted code word is assumed to be 1000110, and the corresponding message was 1000. Similarly, if $w = 0111111$ is the received word, then $wH = 110$ and we assume an error was made in the first position. So, we assume 1111111 was sent and 1111 was the intended message. If the encoded message 1001100 were received as $z = 1001010$, we find $zH = 111$. Since this matches the third row of H, we would decode z as 1011010 and incorrectly assume that the message 1011 was intended. This illustrates the major disadvantage of parity-check matrix decoding. It cannot detect all double errors. On the other hand, since the minimum weight of any nonzero code word is 3, we know that nearest-neighbor decoding will detect every double error. $\qquad\square$

Notice that when only one error was made in transmission, the parity-check decoding procedure gave us the originally intended message. We will soon see under what conditions this is true, but first we need an important fact relating a code given by a generator matrix and its parity-check matrix.

Lemma *Orthogonality Relation*
Let C be an (n, k) linear code over F with generator matrix G and parity-check matrix H. Then, for any vector v in F^n, we have $vH = 0$ (the zero vector) if and only if v belongs to C.

Proof. First note that since H has rank $n - k$, we may think of H as a linear transformation from F^n onto F^{n-k}. Therefore, it follows from the dimension theorem for linear transformations that $n = n - k + \dim \text{Ker } H$ so that Ker H has dimension k. (Alternatively, one can use a group theory argument to show that $|\text{Ker } H| = |F|^k$.) Then, since the dimension of C is also k, it suffices to show that $C \subseteq \text{Ker } H$. To do this, let $G = [I_k \mid A]$ so that $H = [-A/I_{n-k}]$. Then,

$$GH = [I_k \mid A] \begin{bmatrix} -A \\ \hline I_{n-k} \end{bmatrix} = -A + A = [0] \qquad \text{(the zero matrix).}$$

Now, by definition, any vector v in C has the form mG, where m is a message vector. Thus, $vH = (mG)H = m\,[0] = 0$ (the zero vector). $\qquad\blacksquare$

By definition, the parity-check matrix method correctly decodes any received word in which no error has been made. But it will do more.

Theorem 32.3 *Parity-Check Matrix Decoding*
Parity-check matrix decoding will correct any single error if and only if the rows of the parity-check matrix are nonzero and no one row is a scalar multiple of any other.

Proof. For simplicity's sake, we prove only the binary case. In this special situation, the condition on the rows is that they are nonzero and distinct.

So, let H be the parity-check matrix, and let's assume that this condition holds for the rows. Suppose the transmitted code word w was received with only one error, and this error occurred in the ith position. Denoting the vector that has a 1 in the ith position and 0's elsewhere by e_i, we may write the received word as $w + e_i$. Now, using the Orthogonality Lemma, we obtain

$$(w + e_i)H = wH + e_iH = 0 + e_iH = e_iH.$$

But this last vector is precisely the ith row of H. Thus, if there is exactly one error in transmission, we can use the rows of the parity-check matrix to identify the location of the error, provided that these rows are distinct. (If two rows, say, the ith and jth, are the same, we would know that the error occurred in either the ith position or the jth position, but we would not know in which.)

Conversely, suppose the parity-check matrix method correctly decodes all received words in which at most one error has been made in transmission. If the ith row of the parity-check matrix H were the zero vector and if the code word $u = 0 \cdots 0$ were received as e_i, we would find $e_iH = 0 \cdots 0$ and we would erroneously assume that the vector e_i was sent. Thus, no row of H is the zero vector. Now, suppose that the ith row of H and the jth row of H are equal. Then, if some code word w is transmitted and the received word is $w + e_i$ (that is, a single error in the ith position), we find

$$(w + e_i)H = wH + e_iH = i\text{th row of } H = j\text{th row of } H.$$

Thus, our decoding procedure incorrectly tells us to assume errors were made in both the ith and jth positions. ∎

Coset Decoding

There is another convenient method for decoding that utilizes the fact that an (n, k) linear code C over a finite field F is a subgroup of the additive group of $V = F^n$. The method was devised by David Slepian in 1956 and is called *coset decoding* (or *standard decoding*). To use this method, we proceed by constructing a table, called a *standard array*. The first row of the table is the set C of code words beginning in column one with the identity $0 \cdots 0$. To form additional rows of the table, choose an element v of V not listed in the table thus far. Among all the elements of the coset $v + C$, choose one of minimum weight, say, v'. Complete the next row of the table by placing under the column headed by the code word c the vector $v' + c$. Continue this process until all the vectors in V are listed in the table. (Note that an (n, k) linear code over Z_p will have $|V:C| = p^{n-k}$ rows.) The words in the first column are called the *coset leaders*. The decoding procedure is simply to decode any received word w as the code word at the head of the column containing w.

Table 32.4 A Standard Array for a (6,3) Linear Code

Coset Leaders	Words						
000000	100110	010101	001011	110011	101101	011110	111000
100000	000110	110101	101011	010011	001101	111110	011000
010000	110110	000101	011011	100011	111101	001110	101000
001000	101110	011101	000011	111011	100101	010110	110000
000100	100010	010001	001111	110111	101001	011010	111100
000010	100100	010111	001001	110001	101111	011100	111010
000001	100111	010100	001010	110010	101100	011111	111001
100001	000111	110100	101010	010010	001100	111111	011001

Example 10 Consider the (6, 3) binary linear code

$$C = \{000000, 100110, 010101, 001011, 110011, 101101, 011110, 111000\}.$$

The first row of the standard array is just the elements of C. Obviously, 100000 is not in C and has minimum weight among the elements of $100000 + C$, so it can be used to lead the second row. Table 32.4 is the completed table.

If the word 101001 is received, it is decoded as 101101, since 101001 lies in the column headed by 101101. Similarly, the received word 011001 is decoded as 111000. □

Recall that the first method of decoding that we introduced was the nearest-neighbor method; that is, any received word w is decoded as the code word c such that $d(w, c)$ is a minimum (ties are decided arbitrarily). The next result shows that coset decoding is the same as nearest-neighbor decoding.

Theorem 32.4 *Coset Decoding Is Nearest-Neighbor Decoding*
In coset decoding, a received word w is decoded as the code word c such that $d(w, c)$ is a minimum.

Proof. Let C be a linear code and let w be any received word. Suppose v is the coset leader for the coset $w + C$. Then, $w + C = v + C$, so $w = v + c$ for some c in C. Thus, using coset decoding, w is decoded as c. Now, if c' is any code word, then $w - c' \in w + C = v + C$ so that $\text{wt}(w - c') \geq \text{wt}(v)$, since the coset leader v was chosen as a vector of minimum weight among the members of $v + C$. Therefore,

$$d(w, c') = \text{wt}(w - c') \geq \text{wt}(v) = \text{wt}(w - c) = d(w, c).$$

So, using nearest-neighbor decoding, w is also decoded as c. ■

When we know a parity-check matrix for a linear code, coset decoding can be considerably simplified.

DEFINITION Syndrome
If an (n, k) linear code over F has parity-check matrix H, then for any
vector u in F^n, the vector uH is called the *syndrome** of u.

The importance of syndromes stems from the following property.

Theorem 32.5 *Same Coset—Same Syndrome*
Let C be an (n, k) linear code over F with a parity-check matrix H. Then,
two vectors of F^n are in the same coset of C if and only if they have the
same syndrome.

Proof. Two vectors u and v are in the same coset of C if and only if $u - v$
is in C. So, by the Orthogonality Lemma, u and v are in the same coset if
and only if $0 = (u - v)H = uH - vH$. ∎

We may now use syndromes for decoding any received word w:

1. Calculate wH, the syndrome of w.
2. Find the coset leader v so that $wH = vH$.
3. Assume the vector sent is $w - v$.

With this method, we can decode any received word with a table that has
only two rows—one row of coset leaders and another with the corresponding
syndromes.

Example 11 Consider the code given in Example 10. The parity-check matrix
for this code is

$$H = \begin{bmatrix} 1 & 1 & 0 \\ 1 & 0 & 1 \\ 0 & 1 & 1 \\ 1 & 0 & 0 \\ 0 & 1 & 0 \\ 0 & 0 & 1 \end{bmatrix}.$$

The list of coset leaders and corresponding syndromes is

Coset leaders	000000	100000	010000	001000	000100	000010	000001	100001
Syndromes	000	110	101	011	100	010	001	111

So, to decode the received word $v = 101001$, we compute $vH = 100$. Since
the coset leader 000100 has 100 as its syndrome, we assume $v - 000100 = 101101$ was sent. If the received word is $w = 011001$, we compute $wH = 111$,

---

*This term was coined by D. Hagelbarger in 1959.

and assume $w - 100001 = 111000$ was sent. Notice that these answers are in agreement with those obtained by using the standard-array method of Example 10.   □

The term syndrome is a descriptive one. In medicine, it is used to designate a collection of symptoms that typify a disorder. In coset decoding, the syndrome typifies an error pattern.

In this chapter, we have presented algebraic coding theory in its simplest form. A more sophisticated treatment would make substantially more use of group theory, ring theory, and especially finite-field theory. For example, Gorenstein (see Chapter 27 for a biography) and Zierler, in 1961, made use of the fact that the multiplicative subgroup of a finite field is cyclic. They associated each digit of certain codes with a field element in such a way that an algebraic equation could be derived with its zeros determining the locations of the errors.

In some instances, two error-correcting codes are employed. The European Space Agency space probe Giotto, which encountered Halley's Comet, had two error-correcting codes built into its electronics. One code checked for independently occurring errors and another checked for bursts of errors. Giotto achieved an error detection rate of .999999.

Although algebraic coding theory has been in existence only since 1950, there is already extensive literature on the subject. Indeed, the book by MacWilliams and Sloane [1] lists 1,478 entries in the bibliography!

EXERCISES

You know it ain't easy, you know how hard it can be.
John Lennon and Paul McCartney, *The Ballad of John and Yoko*

1. Find the Hamming weight of each code word in Example 1.
2. Find the Hamming distance between the following pairs of vectors: {1101, 0111}, {0220, 1122}, {11101, 00111}.
3. Referring to Example 1, use the nearest-neighbor method to decode the received words 0000110 and 1110100.
4. For any vector space V and any u, v, w in V, prove that the Hamming distance has the following properties:
   a. $d(u, v) = wt(u - v)$
   b. $d(u, v) = d(v, u)$ (*symmetry*)
   c. $d(u, v) = 0$ if and only if $u = v$
   d. $d(u, w) \leq d(u, v) + d(v, w)$ (*triangle inequality*)
   e. $d(u, v) = d(u + w, v + w)$ (*translation invariance*)

5. Determine the (6, 3) binary linear code with generator matrix

$$G = \begin{bmatrix} 1 & 0 & 0 & 0 & 1 & 1 \\ 0 & 1 & 0 & 1 & 0 & 1 \\ 0 & 0 & 1 & 1 & 1 & 0 \end{bmatrix}.$$

6. Show that for binary vectors, $\text{wt}(u + v) \geq \text{wt}(u) - \text{wt}(v)$ and equality occurs if and only if the ith component of u is 1 whenever the ith component of v is 1.

7. Let C be a binary linear code. Show that every member of C has even weight or exactly half of the members of C have even weight. (Compare with exercise 19 of Chapter 5.)

8. Let C be an (n, k) linear code. For each i with $1 \leq i \leq n$, let $C_i = \{v \in C \mid \text{the } i\text{th component of } v \text{ is } 0\}$. Show that C_i is a subcode of C. (A *subcode* of a code is a subset of the code that is itself a code.)

9. Let C be a binary linear code. Show that the code words of even weight form a subcode of C.

10. Let

$$C = \{0000000, 1110100, 0111010, 0011101,$$
$$1001110, 0100111, 1010011, 1101001\}$$

What is the error-correcting capability of C? What is the error-detecting capability of C?

11. Suppose the parity-check matrix of a binary linear code is

$$H = \begin{bmatrix} 1 & 0 \\ 0 & 1 \\ 1 & 1 \\ 1 & 0 \\ 0 & 1 \end{bmatrix}.$$

Can the code correct any single error?

12. Use the generator matrix

$$G = \begin{bmatrix} 1 & 0 & 1 & 1 \\ 0 & 1 & 2 & 1 \end{bmatrix}$$

to construct a (4, 2) ternary linear code. What is the parity-check matrix for this code? What is the error-correcting capability of this code? What is the error-detecting capability of this code? Use parity-check decoding to decode the received word 1201.

13. Find all code words of the (7, 4) binary linear code whose generator matrix is

$$G = \begin{bmatrix} 1 & 0 & 0 & 0 & 1 & 1 & 1 \\ 0 & 1 & 0 & 0 & 1 & 0 & 1 \\ 0 & 0 & 1 & 0 & 1 & 1 & 0 \\ 0 & 0 & 0 & 1 & 0 & 1 & 1 \end{bmatrix}.$$

Find the parity-check matrix of this code. Will this code correct any single error?

14. Show that in a binary linear code, either all the code words begin with 0, or exactly half begin with 0. What about the other components?

15. Suppose a code word v is received as the received as the vector u. Show that coset decoding will decode u as the code word v if and only if $u - v$ is a coset leader.

16. Consider the binary linear code

$C = \{00000, 10011, 01010, 11001, 00101, 10110, 01111, 11100\}.$

Construct the standard array for C. Use nearest-neighbor decoding to decode 11101. If the received word 11101 has exactly one error, can we determine the intended code word?

17. Construct a (6, 3) binary linear code with generator matrix

$$G = \begin{bmatrix} 1 & 0 & 0 & 1 & 1 & 0 \\ 0 & 1 & 0 & 0 & 1 & 1 \\ 0 & 0 & 1 & 1 & 0 & 1 \end{bmatrix}.$$

Decode each of the received words,

$$001001, \ 011000, \ 000110, \ 100001,$$

by the following methods:
a. nearest-neighbor method
b. parity-check matrix method
c. coset decoding using the standard array
d. coset decoding using the syndrome method

18. Suppose the minimum weight of any nonzero code word in a linear code is 6. How many errors can the code correct? How many errors can the code detect?

19. Using the code and the parity-check matrix given in Example 9, show that parity-check matrix decoding cannot detect any multiple errors (i.e., two or more errors).

20. Suppose the last row of the standard array for a binary linear code is

10000 00011 11010 01001 10101 00110 11111 01100.

Complete the array.

21. How many code words are there in a (6, 4) linear ternary code? How many possible received words are there for this code?

22. If the parity-check matrix for a binary linear code is

$$H = \begin{bmatrix} 1 & 1 & 0 \\ 0 & 1 & 1 \\ 1 & 0 & 1 \\ 1 & 0 & 0 \\ 0 & 1 & 0 \\ 0 & 0 & 1 \end{bmatrix},$$

will the code correct any single error? Why?

23. Suppose that the parity-check matrix for a ternary code is

$$\begin{bmatrix} 2 & 1 \\ 2 & 2 \\ 1 & 2 \\ 1 & 0 \\ 0 & 1 \end{bmatrix}$$

Can the code correct all single errors? Give a reason for your answer.

24. Prove that for nearest-neighbor decoding the converse of Theorem 32.2 is true.

25. Can a (6, 3) binary linear code be double-error-correcting using the nearest-neighbor method?

26. Prove that there is no 2×5 standard generating matrix G that will produce a (5, 2) linear code over Z_3 capable of detecting all possible triple errors.

27. Why can't the nearest-neighbor method with a (4, 2) binary linear code correct all single errors?

28. Suppose that one row of a standard array for a binary code is

000100   110000   011110   111101   101010   001001   100111   010011.

Determine the row that contains 100001.

29. Use the field $F = Z_2[x]/\langle x^2 + x + 1 \rangle$ to construct a (5, 2) linear code that will correct any single error.

30. Find the standard generator matrix for a (4, 2) linear code over Z_3 that encodes 20 as 2012 and 11 as 1100. Determine the entire code and the parity-check matrix for the code.

31. Assume C is an (n, k) binary linear code and, for each position $i = 1, 2, \ldots, n$, the code C has at least one vector with a 1 in the ith position. Show that the average weight of a code word is $n/2$.

32. Let C be an (n, k) linear code over F such that the minimum weight of any nonzero code word is $2t + 1$. Show that not every vector of weight $t + 1$ in F^n can occur as a coset leader.

33. Let C be an (n, k) binary linear code over F. If $v \in F^n$ but $v \notin C$, show that $C \cup (v + C)$ is a linear code.

REFERENCE

1. F. J. MacWilliams and N. J. A. Sloane, *The Theory of Error-Correcting Codes,* Parts I, II, Amsterdam: North-Holland, 1977.

SUGGESTED READINGS

Norman Levinson, "Coding Theory: A Counterexample to G. H. Hardy's Conception of Applied Mathematics," *American Mathematical Monthly* 77 (1970): 249–258.

The eminent mathematician G. H. Hardy insisted that "real" mathematics was almost wholly useless. In this article, the author argues that coding theory refutes Hardy's notion. Levinson uses the finite field of order 16 to construct a linear code that can correct any three errors.

R. J. McEliece, "The Reliability of Computer Memories," *Scientific American* 252(1) (1985): 88–95.

This well-written article discusses why and how error-correcting codes are employed in computer memories.

W. W. Peterson, "Error-Correcting Codes," *Scientific American* 206 (2) (1962): 96–108.

This article gives a lucid discussion of the Hamming (15, 11) binary code.

T. M. Thompson, *From Error-Correcting Codes Through Sphere Packing to Simple Groups,* Washington, D. C.: The Mathematical Association of America, 1983.

Chapter 1 of this book gives a fascinating historical account of the origins of error-correcting codes.

Richard W. Hamming

For introduction of error-correcting codes, pioneering work in operating systems and programming languages, and the advancement of numerical computation.

Citation for the Piore Award, 1979

RICHARD W. HAMMING was born in Chicago, Illinois, on February 11, 1915. He graduated from the University of Chicago with B.S. degree in mathematics. In 1939, he received an M.A. degree in mathematics from the University of Nebraska, and in 1942, a Ph.D. in mathematics from the University of Illinois.

During the latter part of World War II, Hamming was at Los Alamos, where he was involved in computing atomic-bomb designs on large-scale computers. In 1946, he joined Bell Telephone Laboratories, where he worked in mathematics, computing, engineering, and science.

Richard Hamming was one of the first users of early electronic computers. His patch wiring for the IBM CPC became widely used. His work on the IBM 650 in 1956 led to the development of a programming language that was the precursor of modern, high-level languages. It first demonstrated many of the format conversions for numbers, overflow, and fault conventions that are used in today's high-level languages.

In 1950, Hamming published his famous paper in error-detecting and error-correcting codes. This work started a branch of information theory. The Hamming codes are used in many modern computers. Hamming's work in the field of numerical analysis has also been of fundamental importance. The Hamming window for smoothing data prior to Fourier analysis is widely used today.

Hamming retired from Bell Laboratories in 1976 and took up teaching at the Naval Postgraduate School. Thus far, he has written seventy-five research papers and a half dozen books, and he is the recipient of numerous prestigious awards.

An Introduction to Galois Theory

33

Galois theory is a showpiece of mathematical unification, bringing together several different branches of the subject and creating a powerful machine for the study of problems of considerable historical and mathematical importance.

Ian Stewart, *Galois Theory*

Fundamental Theorem of Galois Theory

The Fundamental Theorem of Galois Theory is one of the most elegant theorems in mathematics. Look at Figures 33.1 and 33.2. Figure 33.1 pictures the lattice of subgroups of the group of automorphisms of $Q(\sqrt[4]{2}, i)$. The integer along an upward lattice line from H_1 to H_2 is the index of H_1 in H_2. Figure 33.2 shows the lattice of subfields of $Q(\sqrt[4]{2}, i)$. The integer along an upward line from K_1 to K_2 is the degree of K_2 over K_1. Notice that the lattice in Figure 33.2 is the lattice of Figure 33.1 turned upside down. This is only one of many relationships between these two lattices. Under suitable conditions, the Fundamental Theorem of Galois Theory relates, in a multitude of ways, the lattice of subfields of an algebraic extension E of a field F to the subgroup structure of the group of automorphisms of E that send each element of F to itself. Historically, this relationship was discovered in the process of attempting to solve a polynomial equation $f(x) = 0$ by radicals.

Before we can give a precise statement of the Fundamental Theorem of Galois Theory, we need some terminology and notation.

DEFINITIONS Automorphism, Group Fixing F, Fixed Field of H

Let E be an extension field of the field F. An *automorphism of E* is a ring isomorphism from E onto E. The *automorphism group of E fixing F, $G(E/F)$,*

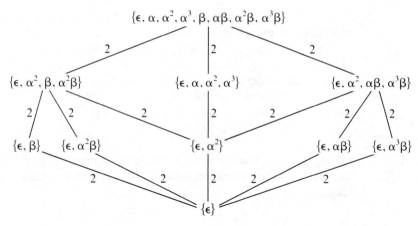

Figure 33.1 Lattice of subgroups of the group of field automorphisms of $Q(\sqrt[4]{2}, i)$, where $\alpha : i \to i$ and $\sqrt[4]{2} \to -i\sqrt[4]{2}$, $\beta : i \to -i$ and $\sqrt[4]{2} \to \sqrt[4]{2}$.

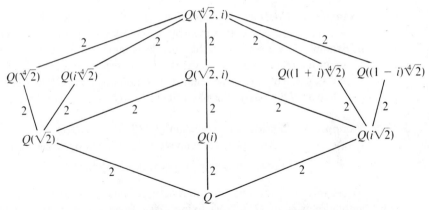

Figure 33.2 Lattice of subfields of $Q(\sqrt[4]{2}, i)$.

is the set of all automorphisms of E that take every element of F to itself. If H is a subgroup of $G(E/F)$, the set

$$E_H = \{x \in E \mid x\phi = x \text{ for all } \phi \in H\}$$

is called the *fixed field of H*.

As was the case for groups, it is easy to show that the set of automorphisms of E forms a group under composition. We leave as exercises the verifications that the automorphism group of E fixing F is a subgroup of the automorphism group of E and, for any subgroup H of $G(E/F)$, the fixed field E_H of H is a subfield of E. The group $G(E/F)$ is sometimes called the *Galois group of E over*

F. Be careful not to misinterpret $G(E/F)$ as something having to do with factor rings or factor groups. It does not.

The following examples will help you assimilate these definitions. In each example, we simply indicate how the automorphisms are defined. We leave as exercises the verifications that the mappings are indeed automorphisms.

Example 1   Consider the extension $Q(\sqrt{2})$ of Q. Since

$$Q(\sqrt{2}) = \{a + b\sqrt{2} \mid a, b \in Q\}$$

and any automorphism of a field containing Q must act as the identity on Q (exercise 1), an automorphism ϕ of $Q(\sqrt{2})$ is completely determined by $\sqrt{2}\phi$. Thus,

$$2 = 2\phi = (\sqrt{2}\sqrt{2})\phi = (\sqrt{2}\phi)^2$$

and, therefore, $\sqrt{2}\phi = \pm\sqrt{2}$. This proves that the group $G(Q(\sqrt{2})/Q)$ has two elements, the identity mapping and the mapping that sends $a + b\sqrt{2}$ to $a - b\sqrt{2}$.   □

Example 2   Consider the extension $Q(\sqrt[3]{2})$ of Q. An automorphism ϕ of $Q(\sqrt[3]{2})$ is completely determined by $\sqrt[3]{2}\phi$. By an argument analogous to that in Example 1, we see that $\sqrt[3]{2}\phi$ must be a cube root of 2. Since $Q(\sqrt[3]{2})$ is a subset of the real numbers and $\sqrt[3]{2}$ is the only real cube root of 2, we must have $\sqrt[3]{2}\phi = \sqrt[3]{2}$. Thus, ϕ is the identity automorphism and $G(Q(\sqrt[3]{2})/Q)$ has only one element. Obviously, the fixed field of $G(Q(\sqrt[3]{2})/Q)$ is $Q(\sqrt[3]{2})$.   □

Example 3   Consider the extension $Q(\sqrt[4]{2}, i)$ over $Q(i)$. Any automorphism ϕ of $Q(\sqrt[4]{2}, i)$ fixing $Q(i)$ is completely determined by $\sqrt[4]{2}\phi$. Since

$$2 = 2\phi = (\sqrt[4]{2})^4\phi = (\sqrt[4]{2}\phi)^4,$$

we see that $\sqrt[4]{2}\phi$ must be a fourth root of 2. Thus, there are at most four possible automorphisms of $Q(\sqrt[4]{2}, i)$ fixing $Q(i)$. If we define ϕ so that $i\phi = i$ and

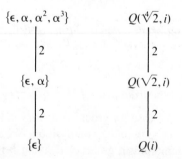

Figure 33.3   Lattice of subgroups of $G(Q(\sqrt[4]{2}, i)/Q(i))$ and lattice of subfields of $Q(\sqrt[4]{2}, i)$ containing $Q(i)$.

$\sqrt[4]{2}\phi = i\sqrt[4]{2}$, then $\phi \in G(Q(\sqrt[4]{2}, i)/Q(i))$ and ϕ has order 4. Thus, $G(Q(\sqrt[4]{2}, i)/Q(i))$ is a cyclic group of order 4. The fixed field of $\{\varepsilon, \phi^2\}$ (where ε is the identity automorphism) is $Q(\sqrt{2}, i)$. The lattice of subgroups of $G(Q(\sqrt[4]{2}, i)/Q(i))$ and the lattice of subfields between $Q(\sqrt[4]{2}, i)$ and $Q(i)$ are given in Figure 33.3. As in Figures 33.1 and 33.2, the integers along the lines in the group lattice represent the index of a subgroup in the group above it, and the integers along the lines of the field lattice represent the degree of the extension of a field over the field below it. ☐

Example 4 Consider the extension $Q(\sqrt{3}, \sqrt{5})$ of Q. Since

$$Q(\sqrt{3}, \sqrt{5}) = \{a + b\sqrt{3} + c\sqrt{5} + d\sqrt{3}\sqrt{5} \mid a, b, c, d \in Q\},$$

any automorphism ϕ of $Q\sqrt{3}, \sqrt{5})$ is completely determined by the two values $\sqrt{3}\phi$ and $\sqrt{5}\phi$. This time there are four automorphisms:

ε	α	β	$\alpha\beta$
$\sqrt{3} \to \sqrt{3}$	$\sqrt{3} \to -\sqrt{3}$	$\sqrt{3} \to \sqrt{3}$	$\sqrt{3} \to -\sqrt{3}$
$\sqrt{5} \to \sqrt{5}$	$\sqrt{5} \to \sqrt{5}$	$\sqrt{5} \to -\sqrt{5}$	$\sqrt{5} \to -\sqrt{5}$

Obviously, $G(Q(\sqrt{3}, \sqrt{5})/Q)$ is isomorphic to $Z_2 \oplus Z_2$. The fixed field of $\{\varepsilon, \alpha\}$ is $Q(\sqrt{5})$, the fixed field of $\{\varepsilon, \beta\}$ is $Q(\sqrt{3})$, the fixed field of $\{\varepsilon, \alpha\beta\}$ is $Q(\sqrt{3}\sqrt{5})$. The lattice of subgroups of $G(Q(\sqrt{3}, \sqrt{5})/Q)$ and the lattice of subfields of $Q(\sqrt{3}, \sqrt{5})$ are given in Figure 33.4. ☐

Example 5 is a bit more complicated than our previous examples. In particular, the automorphism group is non-Abelian.

Example 5 Direct calculations show that $\omega = -\frac{1}{2} + i\sqrt{3}/2$ satisfies the equations $\omega^3 = 1$ and $\omega^2 + \omega + 1 = 0$. Now, consider the extension $Q(\omega, \sqrt[3]{2})$

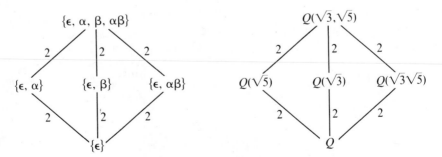

Figure 33.4 Lattice of subgroups of $G(Q(\sqrt{3}, \sqrt{5})/Q)$ and lattice of subfields of $Q(\sqrt{3}, \sqrt{5})$.

of Q. We may describe the automorphisms of $Q(\omega, \sqrt[3]{2})$ by specifying how they act on ω and $\sqrt[3]{2}$. There are six in all:

ε	α	β	β^2	$\alpha\beta$	$\alpha\beta^2$
$\omega \rightarrow \omega$	$\omega \rightarrow \omega^2$	$\omega \rightarrow \omega$	$\omega \rightarrow \omega$	$\omega \rightarrow \omega^2$	$\omega \rightarrow \omega^2$
$\sqrt[3]{2} \rightarrow \sqrt[3]{2}$	$\sqrt[3]{2} \rightarrow \sqrt[3]{2}$	$\sqrt[3]{2} \rightarrow \omega\sqrt[3]{2}$	$\sqrt[3]{2} \rightarrow \omega^2\sqrt[3]{2}$	$\sqrt[3]{2} \rightarrow \omega\sqrt[3]{2}$	$\sqrt[3]{2} \rightarrow \omega^2\sqrt[3]{2}$

Since $\alpha\beta \neq \beta\alpha$, we know $G(Q(\omega, \sqrt[3]{2})/Q)$ is isomorphic to S_3. (See Example 4 of Chapter 9.) The lattices of subgroups and subfields are given in Figure 33.5.

The lattices in Figure 33.5 have been arranged so that the field occupying the same position as some group is the fixed field of that group. For instance, $Q(\sqrt[3]{2}\omega)$ is the fixed field of $\{\varepsilon, \alpha\beta^2\}$. □

The preceding examples show that, in certain cases, there is an intimate connection between the lattice of subfields between E and F and the lattice of subgroups of $G(E/F)$. In general, if E is an extension of F, and we let $\mathscr{F}$ be the lattice of subfields of E containing F and let $\mathscr{G}$ be the lattice of subgroups of $G(E/F)$, then for each K in $\mathscr{F}$, the group $G(E/K)$ is in $\mathscr{G}$ and, for each H in $\mathscr{G}$, the field E_H is in $\mathscr{F}$. Thus, we may define a mapping $g:\mathscr{F} \rightarrow \mathscr{G}$ by $Kg = G(E/K)$

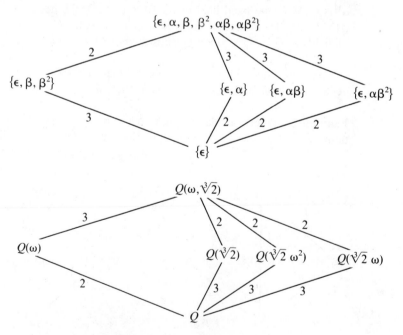

Figure 33.5 Lattice of subgroups of $G(Q(\omega, \sqrt[3]{2})/Q)$ and lattice of subfields of $Q(\omega, \sqrt[3]{2})$, where $\omega = -\frac{1}{2} + i\sqrt{3}/2$.

and a mapping $f:\mathcal{G} \to \mathcal{F}$ by $Hf = E_H$. It is easy to show that if K and L belong to $\mathcal{F}$ and $K \subseteq L$, then $Kg \supseteq Lg$. Similarly, if G and H belong to $\mathcal{G}$ and $G \subseteq H$, then $Gf \supseteq Hf$. Thus, f and g are inclusion-reversing mappings between $\mathcal{F}$ and $\mathcal{G}$. We leave it as an exercise to show that for any K in $\mathcal{F}$, we have $Kgf \supseteq K$ and, for any G in $\mathcal{G}$, we have $Gfg \supseteq G$. When E is an arbitrary extension of F, these inclusions may be strict. (See Example 2.) However, when E is a suitably chosen extension of F, the Fundamental Theorem of Galois Theory, Theorem 33.1, says that f and g are inverses of each other so that the inclusions are equalities. In particular, f and g are inclusion-reversing isomorphisms between the lattices $\mathcal{F}$ and $\mathcal{G}$. A stronger result than that given in Theorem 33.1 is true, but our theorem illustrates the fundamental principles involved. The student is referred to [1, p. 455] for additional details and proofs.

Theorem 33.1 *Fundamental Theorem of Galois Theory*
Let F be a field of characteristic 0 or a finite field. If E is the splitting field over F for some polynomial in $F[x]$, then the mapping from the set of subfields of E containing F to the set of subgroups of $G(E/F)$ given by $K \to G(E/K)$ is a one-to-one correspondence. Furthermore, for any subfield K of E containing F,

1. *$[E:K] = |G(E/K)|$ and $[K:F] = |G(E/F)|/|G(E/K)|$. (The index of $G(E/K)$ in $G(E/F)$ equals the degree of K over F.)*
2. *If K is the splitting field of some polynomial in $F[x]$, then $G(E/K)$ is a normal subgroup of $G(E/F)$ and $G(K/F)$ is isomorphic to $G(E/F)/G(E/K)$.*
3. *$K = E_{G(E/K)}$. (The fixed field of $G(E/K)$ is K.)*
4. *If H is a subgroup of $G(E/F)$, then $H = G(E/E_H)$. (The automorphism group of E fixing E_H is H.)*

Generally speaking, it is much easier to determine a lattice of subgroups than a lattice of subfields. For example, it is usually quite difficult to determine, directly, how many subfields a given field has, and it is often difficult to decide whether or not two field extensions are the same. The corresponding questions about groups are much more tractable. Hence, the Fundamental Theorem of Galois Theory can be a great labor-saving device. Here is an illustration.

Example 6 Let $\omega = \cos(360°/7) + i \sin(360°/7)$, so that $\omega^7 = 1$, and consider the field $Q(\omega)$. How many subfields does it have and what are they? First, observe that $Q(\omega)$ is the splitting field of $x^7 - 1$ over Q, so that we may apply the Fundamental Theorem of Galois Theory. A simple calculation shows that the automorphism ϕ that sends ω to ω^3 has order 6. Thus,

$$[Q(\omega):G] = |G(Q(\omega)/Q)| \geq 6.$$

Also, since

$$x^7 - 1 = (x - 1)(x^6 + x^5 + x^4 + x^3 + x^2 + x + 1)$$

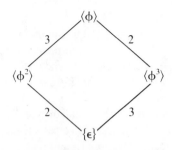

Figure 33.6 Lattice of subgroups of $G(Q(\omega)/Q)$, where $\omega = \cos(360°/7) + i\sin(360°/7)$.

and ω is a zero of $x^7 - 1$, we see that

$$|G(Q(\omega)/Q)| = [Q(\omega):Q] \leq 6.$$

Thus, $G(Q(\omega)/Q)$ is a cyclic group of order 6. So, the lattice of subgroups of $G(Q(\omega)/Q)$ is trivial to compute. See Figure 33.6.

This means that $Q(\omega)$ contains exactly two proper extensions of Q, one of degree 3 and one of degree 2. Noting that $\omega + \omega^6$ is fixed by ϕ^3, it follows that $Q \subsetneq Q(\omega + \omega^6) \subseteq Q(\omega)_{\langle\phi^3\rangle}$. Since $[Q(\omega)_{\langle\phi^3\rangle}:Q] = 3$ and $[Q(\omega + \omega^6):Q]$ divides $[Q(\omega)_{\langle\phi^3\rangle}:Q]$, we see that $Q(\omega + \omega^6) = Q(\omega)_{\langle\phi^3\rangle}$. A similar argument shows that $Q(\omega^3 + \omega^5 + \omega^6)$ is the fixed field of ϕ^2. Thus, we have found all subfields of $Q(\omega)$.   □

Solvability of Polynomials by Radicals

For Galois, the elegant correspondence between groups and fields given by Theorem 33.1 was only a means to an end. The goal Galois sought was the solution to a problem that had stymied mathematicians for centuries. Methods for solving linear and quadratic equations were known thousands of years ago (the quadratic formula). In the sixteenth century, Italian mathematicians developed formulas for solving any third- or fourth-degree equation. Their formulas involved only the operations of addition, subtraction, multiplication, division, and extraction of roots (radicals). For example, the equation

$$x^3 + bx + c = 0$$

has the three solutions

$$A + B, \qquad -(A + B)/2 + (A - B)\sqrt{-3}/2,$$
$$-(A + B)/2 - (A - B)\sqrt{-3}/2,$$

where

$$A = \sqrt[3]{\frac{-c}{2} + \sqrt{\frac{b^3}{27} + \frac{c^2}{4}}} \qquad \text{and} \qquad B = \sqrt[3]{\frac{-c}{2} - \sqrt{\frac{b^2}{27} + \frac{c^2}{4}}}.$$

The formulas for the general cubic $x^3 + ax^2 + bx + c = 0$ and the general quartic (fourth-degree polynomial) are even more complicated, but nevertheless can be given in terms of radicals of rational expressions of the coefficients. (See [2, 392–394].)

Both Abel and Galois proved that there is no general solution of a fifth-degree equation by radicals. In particular, there is no "quintic formula." Before discussing Galois's method, which provided a group-theoretic criterion for the solution of an equation by radicals and led to the modern-day Galois Theory, we need a few definitions.

DEFINITION Solvable by Radicals

Let F be a field, and let $f(x) \in F[x]$. We say $f(x)$ is *solvable by radicals over F* if $f(x)$ splits in some extension $F(a_1, a_2, \ldots , a_n)$ of F and there exist positive integers $k_1, \ldots , k_n$ such that $a_1^{k_1} \in F$ and $a_i^{k_i} \in F(a_1, \ldots , a_{i-1})$ for $i = 2, \ldots , n$.

So, a polynomial in $F[x]$ is solvable by radicals if we can obtain all of its zeros by adjoining nth roots (for various n) to F. In other words, each zero of the polynomial can be written as an expression (usually a messy one) involving elements of F combined by the operations of addition, subtraction, multiplication, division, and extraction of roots.

Thus, the problem of solving a polynomial equation for its zeros can be transferred to a problem about field extensions. At the same time, we can use the Fundamental Theorem of Galois Theory to transfer a problem about field extensions to a problem about groups. This is exactly how Galois showed that there are fifth-degree polynomials that cannot be solved by radicals, and this is exactly how we will do it. Before giving an example of such a polynomial, we need some additional group theory.

DEFINITION Solvable Group

We say a group G is *solvable* if G has a series of subgroups

$$\{e\} = H_0 \subset H_1 \subset H_2 \subset \cdots \subset H_k = G,$$

where, for each $0 \leq i < k$, H_i is normal in H_{i+1} and H_{i+1}/H_i is Abelian.

Obviously, Abelian groups are solvable. So are the dihedral groups and any group whose order has the form p^n, where p is a prime. In a certain sense, solvable groups are almost Abelian. On the other hand, it follows directly from the definitions that any non-Abelian simple group is not solvable. In particular, A_5 is not solvable. It follows from exercise 24 of Chapter 27 that S_5 is not solvable.

Theorem 33.2 *Factor Group of a Solvable Group Is Solvable.*
A factor group of a solvable group is solvable.

Proof. Suppose G has a series of subgroups

$$\{e\} = H_0 \subset H_1 \subset H_2 \subset \cdots \subset H_k = G,$$

where, for each $0 \le i < k$, H_i is normal in H_{i+1} and H_{i+1}/H_i is Abelian. If N is any normal subgroup of G, then

$$\{e\} = H_0 N/N \subset H_1 N/N \subset H_2 N/N \subset \cdots \subset H_k N/N = G/N$$

is the requisite series of subgroups that guarantees G/N is solvable. (See exercise 26.) ∎

We now are able to make the critical connection between solvability of polynomials by radicals and solvable groups.

Theorem 33.3 *Solvable by Radicals Implies a Solvable Group*
Let F be a field of characteristic 0 and let $f(x) \in F[x]$. Let E be the splitting field for $f(x)$ over F. If $f(x)$ is solvable by radicals over F, then the Galois group $G(E/F)$ is solvable.

Proof. The idea behind this proof is to construct a sequence of field extensions

$$F = L_0 \subset L_1 \subset L_2 \subset \cdots \subset L_t = L$$

such that L contains E and the group $G(L/F)$ is solvable. We then obtain the desired conclusion from Theorem 33.2 by observing that $G(E/F)$ is a factor group of $G(L/F)$.

To do this, suppose $f(x)$ splits in $F(a_1, a_2, \ldots, a_r)$, where $a_1^{n_1} \in F$ and $a_i^{n_i} \in F(a_1, \ldots, a_{i-1})$ for $i = 2, \ldots, r$. The first thing we want to do is to construct an extension of F that contains all of the roots of unity that we will need. To do this, let n be the largest n_i for $i = 1, \ldots, r$. For each $j = 3, 4, \ldots, n$, let α_j be a primitive jth root of unity. (See Example 2 of Chapter 18.) We construct a sequence of field extensions of F as follows.

$$F_0 = F, \qquad F_1 = F_0(\alpha_3), \qquad F_2 = F_1(\alpha_4), \ldots, \qquad F_{n-2} = F_{n-3}(\alpha_n).$$

Then F_{i+1} is a splitting field of $x^{i+3} - 1$ over F_i, and $G(F_{i+1}/F_i)$ is Abelian. (See exercise 15.) Let K_0 denote F_{n-2}, and observe that K_0 contains all jth roots of unity for $j \le n$. We next construct another sequence of field extensions as follows. Let

$$K_1 = K_0(a_1), \qquad K_2 = K_1(a_2) \qquad \ldots, \qquad K_r = K_{r-1}(a_r).$$

For each $i = 1, \ldots, r$, let $b_i = a_i^{n_i}$. Then, since K_i contains all of the n_ith roots of unity, it follows that K_i is the splitting field of $x^{n_i} - b_i$ over K_{i-1}. We further claim that $G(K_i/K_{i-1})$ is Abelian. To see this, observe that any automorphism in $G(K_i/K_{i-1})$ is completely determined by its action on a_i. Also, since a_i is a zero of $x^{n_i} - b_i$, we know that any element of

$G(K_i/K_{i-1})$ sends a_i to another zero of $x^{n_i} - b_i$. Since the zeros of $x^{n_i} - b_i$ are a_i, $\alpha_{n_i}a_i$, $\alpha_{n_i}^2 a_i$, and $\alpha_{n_i}^{n_i-1}a_i$, any element of $G(K_i/K_{i-1})$ sends a_i to $\alpha_{n_i}^j a_i$ for some j. Let ϕ and σ be two elements of $G(K_i/K_{i-1})$. Then

$$a_i\phi = \alpha_{n_i}^j a_i \quad \text{and} \quad a_i\sigma = \alpha_{n_i}^k a_i \quad \text{for some } j \text{ and } k.$$

Thus,

$$a_i\phi\sigma = (\alpha_{n_i}^j a_i)\sigma = \alpha_{n_i}^j \alpha_{n_i}^k a_i = \alpha_{n_i}^{j+k}a_i,$$

while

$$a_i\sigma\phi = (\alpha_{n_i}^k a_i)\phi = \alpha_{n_i}^k \alpha_{n_i}^j a_i = a_{n_i}^{k+j}a_i$$

so that $\phi\sigma$ and $\sigma\phi$ agree on a_i and K_{i-1}. This shows that $\phi\sigma = \sigma\phi$, and, therefore, $G(K_i/K_{i-1})$ is Abelian.

At this point, we have now constructed a sequence of field extensions $F = L_0 \subset L_1 \subset \cdots \subset L_t = L$ such that L contains E and $G(L_{i+1}/L_i)$ is Abelian. Since the Fundamental Theorem of Galois Theory tells us that $G(L_{i+1}/L_i)$ is isomorphic to $G(L/L_i)/G(L/L_{i+1})$, the series of subgroups of $G(L/F)$,

$$\{e\} = G(L/L_t) \subset G(L/L_{t-1}) \subset \cdots \subset G(L/L_0) = G(L/F)$$

demonstrates that $G(L/F)$ is solvable. Finally, because $G(E/F)$ is isomorphic to $G(L/F)/G(L/E)$, Theorem 33.2 guarantees that $G(E/F)$ is solvable. ■

It is worth remarking that the converse of Theorem 33.3 is true also; that is, if E is the splitting field of a polynomial $f(x)$ over a field F of characteristic 0 and $G(E/F)$ is solvable, then $f(x)$ is solvable by radicals over F.

One of the major unsolved problems in algebra, first posed by Emmy Noether, is determining which finite groups can occur as Galois groups over Q. Many people suspect that the answer is "all of them." It is known that every solvable group is a Galois group. John Thompson has recently proved that certain kinds of simple groups, including the Monster, are Galois groups. The suggested reading by Ian Stewart provides more information on this topic.

Insolvability of a Quintic

We will finish our introduction to Galois theory by explicitly exhibiting a polynomial that has integer coefficients and that is not solvable by radicals over Q.

Consider $g(x) = 3x^5 - 15x + 5$. By Eisenstein's Criterion, $g(x)$ is irreducible over Q. Since $g(x)$ is continuous and $g(-2) = -61$ and $g(-1) = 17$, we know that $g(x)$ has a real zero between -2 and -1. A similar analysis shows that $g(x)$ also has real zeros between 0 and 1, and 1 and 2.

Each of these real zeros has multiplicity 1. (See Theorem 22.6.) Furthermore,

$g(x)$ has no more than three real zeros, because Rolle's Theorem from calculus guarantees that between each pair of real zeros of $g(x)$, there must be a zero of $g'(x) = 15x^4 - 15$. So, for $g(x)$ to have four real zeros, $g'(x)$ would have to have three real zeros and it does not. Thus, the other two zeros of $g(x)$ are nonreal complex numbers, say, $a + bi$ and $a - bi$. (See exercise 48 of Chapter 17.)

Now, let's denote the five zeros of $g(x)$ by a_1, a_2, a_3, a_4, a_5. Since any automorphism of $K = Q(a_1, a_2, a_3, a_4, a_5)$ is completely determined by its action on the a's and must permute the a's, we know that $G(K/Q)$ is isomorphic to a subgroup of S_5, the symmetric group on five symbols. Since a_1 is a zero of an irreducible polynomial of degree 5 over Q, we know that $[Q(a_1):Q] = 5$, and, therefore, 5 divides $[K:Q]$. Thus, the Fundamental Theorem of Galois Theory tells us that 5 also divides $G(K/Q)$. So, by Cauchy's Theorem, we may conclude that $G(K/Q)$ has an element of order 5. Since the only elements in S_5 of order 5 are the 5-cycles, we know that $G(K/Q)$ contains a 5-cycle. The mapping from **C** to **C**, sending $a + bi$ to $a - bi$, is also an element of $G(K/Q)$. Since this mapping fixes the three real zeros and interchanges the two complex zeros of $g(x)$, we know that $G(K/Q)$ contains a 2-cycle. But, the only subgroup of S_5 that contains both a 5-cycle and a 2-cycle is S_5. (See exercise 27 of Chapter 27.) So, $G(K/Q)$ is isomorphic to S_5. Finally, since S_5 is not solvable (see exercise 21), we have succeeded in exhibiting a fifth-degree polynomial that is not solvable by radicals.

EXERCISES

Seeing much, suffering much, and studying much are the three pillars of learning.

Benjamin Disraeli

1. Let E be an extension field of Q. Show that any automorphism of E acts as the identity on Q. (This exercise is referred to in this chapter.)

2. Let E be a field extension of a field F. Show that the automorphism group of E fixing F is indeed a group. (This exercise is referred to in this chapter.)

3. Let E be a field extension of a field F and let H be a subgroup of $G(E/F)$. Show that the fixed field of H is indeed a field. (This exercise is referred to in this chapter.)

4. Referring to Example 6, show that the automorphism ϕ has order 6. Show that $\omega + \omega^6$ is fixed by ϕ^3 and $\omega^3 + \omega^5 + \omega^6$ is fixed by ϕ^2.

5. Let $f(x) \in F[x]$ and let the zeros of $f(x)$ be $a_1, a_2, \ldots, a_n$. If $K = F(a_1, a_2, \ldots, a_n)$, show that $G(K/F)$ is isomorphic to a group of permutations of the a_i's . (When K is the splitting field of $f(x)$ over F, the group $G(K/F)$ is called the *Galois group* of $f(x)$.)

6. Show that the Galois group of a polynomial of degree n has order dividing $n!$.

7. Let E be the splitting field of $x^4 + 1$ over Q. Find $G(E/Q)$. Find all subfields of E. Find the automorphisms of E that have fixed fields $Q(\sqrt{2})$, $Q(\sqrt{-2})$, and $Q(i)$. Is there an automorphism of E whose fixed field is Q?

8. Determine the group of field automorphisms of GF(4).

9. Let $E = Q(\sqrt{2}, \sqrt{5})$. What is the order of the group $G(E/Q)$? What is the order of $G(Q(\sqrt{10})/Q)$?

10. Given that the automorphism group of $Q(\sqrt{2}, \sqrt{5}, \sqrt{7})$ is isomorphic to $Z_2 \oplus Z_2 \oplus Z_2$, determine the number of subfields of $Q(\sqrt{2}, \sqrt{5}, \sqrt{7})$ that have degree 4 over Q.

11. Suppose F is a field of characteristic 0 and E is the splitting field for some polynomial over F. If $G(E/F)$ is isomorphic to A_4, show that there is no subfield K of E such that $[K:F] = 2$.

12. Show that the Galois group of $x^3 - 3$ over Q is isomorphic to S_3.

13. Suppose K is the splitting field of some polynomial over a field F of characteristic 0. If $[K:F] = p^2q$, where p and q are distinct primes, show that K has subfields L_1, L_2, and L_3 such that $[L_1:F] = pq$, $[L_2:F] = p^2$, and $[L_3:F] = q$.

14. Suppose E is the splitting field of some polynomial over a field F of characteristic 0. If $G(E/F)$ is isomorphic to D_6, draw the subfield lattice for the fields between E and F.

15. Let F be a field of characteristic 0. If K is the splitting field of $x^n - 1$ over F, prove that $G(K/F)$ is Abelian. (This exercise is referred to in the proof of Theorem 33.3.)

16. Suppose E is the splitting field of some polynomial over a field F of characteristic 0. If $[E:F]$ is finite, show that there is only a finite number of fields between E and F.

17. Suppose E is the splitting field of some polynomial over a field F of characteristic 0. If $G(E/F)$ is an Abelian group of order 10, draw the subfield lattice for the fields between E and F.

18. Let ω be a nonreal complex number such that $\omega^5 = 1$. If ϕ is the automorphism of $Q(\omega)$ that carries ω to ω^4, find the fixed field of $\langle \phi \rangle$.

19. Determine the isomorphism class of the group $G(\text{GF}(64)/\text{GF}(2))$.

20. Determine the isomorphism class of the group $G(\text{GF}(729)/\text{GF}(9))$.

21. Show that S_5 is not solvable.

22. Show that the dihedral groups are solvable.

23. Show that a group of order p^n, where p is prime, is solvable.

24. Show that S_n is solvable when $n \leq 4$.

25. Show that a subgroup of a solvable group is solvable.

26. Complete the proof of Theorem 33.2 by showing that the series of groups given satisfies the definition for solvability.

REFERENCES

1. J. B. Fraleigh, *A First Course in Abstract Algebra,* 4th ed., Reading, Mass.: Addison-Wesley, 1989.
2. Samuel M. Selby, *Standard Mathematical Tables,* Cleveland: The Chemical Rubber Company, 1965.

SUGGESTED READINGS

Lisl Gaal, *Classical Galois Theory with Examples,* Chicago: Markham, 1971.

This book has a large number of examples pertaining to Galois theory worked out in great detail.

D. G. Mead, "The Missing Fields," *The American Mathematical Monthly* 94 (1987): 12–13.

This article uses Galois theory to show that, for any positive integer n, there is an extension K of Q with $[K:Q] = n$, and yet there is no field properly between K and Q.

Tony Rothman, "The Short Life of Évariste Galois," *Scientific American,* April (1982): 136–149.

This article gives an elementary discussion of Galois's proof that the general fifth-degree equation cannot be solved by radicals. The article also goes into detail about Galois's controversial life and death. In this regard, Rothman refutes several accounts given by other Galois biographers.

Ian Stewart, "The Duelist and the Monster," *Nature* 317 (1985): 12–13.

This nontechnical article discusses recent work of John Thompson pertaining to the question "which groups can occur as Galois groups."

Philip Hall

He was pre-eminent as a group theorist and made many fundamental discoveries; the conspicuous growth of interest in group theory in this century owes much to him.

J. E. Roseblade

PHILIP HALL was born on April 11, 1904, in London. Abandoned by his father shortly after birth, Hall was raised by his mother, a dressmaker. He demonstrated academic prowess early by winning a scholarship to Christ's Hospital, where he had several outstanding mathematics teachers. At Christ's Hospital, Hall won a medal for the best English essay, the gold medal in mathematics, and a scholarship to King's College, Cambridge.

Although abstract algebra was a field neglected at King's College, Hall studied Burnside's book *Theory of Groups* and some of Burnside's later papers. After graduating in 1925, he stayed on at King's College for further study and was elected to a Fellowship in 1927. That same year Hall discovered a major "Sylow-like" theorem about solvable groups: If a solvable group has order mn where m and n are relatively prime, then every subgroup whose order divides m is contained in a group of order m and all subgroups of order m are conjugate. Over the next three decades, Hall developed a general theory of finite solvable groups that had a profound influence on John Thompson's spectacular achievements of the 1960s. In the 1930s, Hall also developed a general theory of groups of prime-power order that has become a foundation of modern finite group theory. In addition to his fundamental contributions to finite groups, Hall wrote many seminal papers on infinite groups.

Among the concepts that have Hall's name attached to them are Hall subgroup, Hall algebra, Hall-Littlewood polynomials, Hall divisors, the marriage theorem from graph theory, and the Hall commutator collecting process. Beyond his own discoveries, Hall had an enormous influence on algebra through his research students. No fewer than one dozen have become eminent mathematicians in their own right.

Hall had a deep love of poetry, flowers, and country walks. He enjoyed music and art. He died on December 30, 1982.

An Introduction to Boolean Algebras

34

In recent years much interest has arisen in what are known as "Boolean algebras." The principal reason for this great interest is that many applications of this discipline have been found in connection with various systems of automation.

John T. Moore, *Elements of Abstract Algebra*

Motivation

To be worthy of study, an abstract mathematical system should have several concrete realizations. Groups, for instance, arise in connection with symmetries, matrices with nonzero determinants, and permutations. Rings encompass the integers, polynomials, and matrices. Fields generalize familiar systems such as the real numbers, the integers modulo a prime, and polynomial rings modulo a maximal ideal. In this chapter, we introduce another system, called a Boolean algebra, that has three major concrete models—the algebra of sets, the algebra of electrical circuits, and the algebra of logic. We will briefly discuss these three models.

Informally, a Boolean algebra is a set of objects with two binary operations and one unary operation that satisfy certain conditions. Examples 1–3 motivate the formal definition of Boolean algebra.

Example 1 Algebra of Sets
Let X be a finite nonempty set and let $P(X)$ denote the set of all subsets X. For any pair A, B in $P(X)$ we let

$$A \cap B = \{x \mid x \in A \text{ and } x \in B\} \quad \text{(intersection)}$$
$$A \cup B = \{x \mid x \in A \text{ or } x \in B\} \quad \text{(union)}$$
$$A' = \{x \in X \mid x \notin A\} \quad \text{(complement)}$$

The student should check that the operations $\cap$ and $\cup$ are commutative, associative, and each is distributive over the other; that is,

$$A \cup B = B \cup A \qquad \text{and} \qquad A \cap B = B \cap A;$$
$$A \cup (B \cup C) = (A \cup B) \cup C \qquad \text{and} \qquad A \cap (B \cap C) = (A \cap B) \cap C;$$
$$A \cup (B \cap C) = (A \cup B) \cap (A \cup C) \qquad \text{and}$$
$$A \cap (B \cup C) = (A \cap B) \cup (A \cap C).$$

Also note that for any A in $P(X)$ we have

$$A \cup \emptyset = A, \qquad A \cap X = A,$$

and

$$A \cup A' = X, \qquad A \cap A' = \emptyset. \qquad \square$$

Example 2   Algebra of Switching Functions

For any positive integer n, a function from the Cartesian product of n copies of $\{0, 1\}$ to $\{0, 1\}$ is called a *switching function on n variables*. (Later in this chapter, we will see how an electrical circuit with n switches defines such a function.) If f and g are switching functions on n variables, then so are the functions defined by

$$(f + g)(x) = \max\{f(x), g(x)\}$$

and

$$(f \cdot g)(x) = \min\{f(x), g(x)\}.$$

Furthermore, for any switching function f, we may define a new switching function f' by

$$f'(x) = \begin{cases} 0 & \text{if } f(x) = 1, \\ 1 & \text{if } f(x) = 0. \end{cases}$$

Again, we may observe that switching function addition and multiplication are commutative, associative, and distributive. Moreover, the constant functions $z = 0$ and $u = 1$ have the properties that

$$z + f = f, \qquad u \cdot f = f,$$

and

$$f + f' = u, \qquad f \cdot f' = z.$$

Table 34.1 shows all switching functions on two variables.

Table 34.1   All Switching Functions on Two Variables

	f_0	f_1	f_2	f_3	f_4	f_5	f_6	f_7	f_8	f_9	f_{10}	f_{11}	f_{12}	f_{13}	f_{14}	f_{15}
(0,0)	0	1	0	1	0	1	0	1	0	1	0	1	0	1	0	1
(1,0)	0	0	1	1	0	0	1	1	0	0	1	1	0	0	1	1
(0,1)	0	0	0	0	1	1	1	1	0	0	0	0	1	1	1	1
(1,1)	0	0	0	0	0	0	0	0	1	1	1	1	1	1	1	1

We may compute the sum of any pair of these functions by adding their corresponding columns. The resulting column is the column of the sum. For instance, $f_3 + f_5 = f_7$ (keep in mind that $1 + 1 = 1$).

Similarly, $f_3 \cdot f_5 = f_1$ and $f_3' = f_{12}$. ☐

Example 3 Divisors of 105

Consider the set S of all positive divisors of 105. Specifically,

$$S = \{1, 3, 5, 7, 15, 21, 35, 105\}.$$

For any pair a, b in S, define $a \vee b = \text{lcm}(a,b)$ and $a \wedge b = \gcd(a, b)$. Finally, for any a in S, let $a' = 105/a$. Then the operations $\vee$ and $\wedge$ are commutative, associative, and distributive. Furthermore for any a in S

$$a \vee 1 = a, \qquad a \wedge 105 = a,$$

and

$$a \vee a' = 105, \qquad a \wedge a' = 1.$$ ☐

Examples 1–3 are illustrations of Boolean algebras.

Definition and Properties

DEFINITION Boolean Algebra

A *Boolean algebra* is a set B with two binary operations $\vee$ (read "or") and $\wedge$ (read "and") and a unary operation $'$ (read "complement") that satisfy the following axioms (for all a, b, c in B).

1. Commutativity

$$a \vee b = b \vee a, \qquad a \wedge b = b \wedge a.$$

2. Associativity

$$a \vee (b \vee c) = (a \vee b) \vee c, \qquad a \wedge (b \wedge c) = (a \wedge b) \wedge c.$$

3. Distributivity

$$a \wedge (b \vee c) = (a \wedge b) \vee (a \wedge c),$$
$$a \vee (b \wedge c) = (a \vee b) \wedge (a \vee c).$$

4. Existence of zero and unity. There are elements 0 and 1 in B such that for all a in B,

$$a \vee 0 = a \quad \text{and} \quad a \wedge 1 = a.$$

5. Complementation

$$a \vee a' = 1, \qquad a \wedge a' = 0.$$

It is worth mentioning that a Boolean algebra is almost a commutative ring with unity. The only ring property that a Boolean algebra lacks is the existence of inverses for one of the operations. On the other hand, a commutative ring with unity is usually not a Boolean algebra because addition does not distribute over multiplication; that is,

$$a + bc \neq (a + b)(a + c).$$

Before stating some basic properties of Boolean algebras, we look at one nonexample.

Example 4   Let $X = \{1, 2, 4, 7, 14, 28\}$—the divisors of 28. For any a, b in X, define

$$a \vee b = \text{lcm}(a, b) \qquad \text{and} \qquad a \wedge b = \text{gcd}(a, b).$$

The operations $\wedge$ and $\vee$ satisfy axioms 1–4 of the definition of Boolean algebra with 1 as the zero and 28 as the unity. However, axiom 5 cannot be satisfied, no matter how we define a'. For consider the possibilities for $14'$. The only x's that yield $14 \vee x = \text{lcm}(14, x) = 28$ are 4 and 28. So, $14' = 4$ or $14' = 28$. But $14 \wedge 4 = \text{gcd}(14, 4) = 2$ and $14 \wedge 28 = \text{gcd}(14, 28) = 14$. Since neither 2 nor 14 is the zero element, there is no complement for 14. $\square$

Theorem 34.1 collects many elementary properties of a Boolean algebra. The proofs are left as exercises.

Theorem 34.1   *Properties of Boolean Algebras*
Let B be Boolean algebra and let a, b belong to B. Then

1. $a \wedge 0 = 0$.
2. $a \vee 1 = 1$.
3. *(Absorption laws)*

$$a \wedge (a \vee b) = a, \qquad a \vee (a \wedge b) = a.$$

4. *(Idempotent laws)*

$$a \wedge a = a, \qquad a \vee a = a.$$

5. *If $a \vee b = 1$ and $a \wedge b = 0$, then $b = a'$.*
6. *(DeMorgan's Laws)*

$$(a \wedge b)' = a' \vee b', \qquad (a \vee b)' = a' \wedge b'.$$

7. *(Involution)*

$$(a')' = a.$$

8. *The zero and unity elements are unique.*

The perceptive student may have noticed that there is a symmetry in properties 1–4 and 6 of Theorem 34.1. In particular, in each case, one may obtain the right-hand equation from the left-hand equation by interchanging $\wedge$ for $\vee$ and 0 for 1. Indeed, the axioms for a Boolean algebra share this property. Consequently, the *duality principle* holds for Boolean algebras; that is, for any proposition that is true, there is a corresponding proposition obtained by making these interchanges that is also true.

Example 5 The dual of $a \wedge b = 0$ is $a \vee b = 1$; the dual of $a \wedge (b \vee c) = (a \wedge b) \vee c$ is $a \vee (b \wedge c) = (a \vee b) \wedge c$. ☐

The Algebra of Electric Circuits

We next give a brief indication of how Boolean algebra is used in electrical engineering. On-off electrical switches, diodes, magnetic dipoles, and transistors are examples of two-state devices called *switches*. In electrical networks, these two states may be "current flows" (on) or "current does not flow" (off), magnetized or not magnetized, high potential or low potential, closed (current flows) or open (current does not flow). Abstractly, we represent the "on" state by 1 and the "off" state by 0. We use letters $a, b, c, \ldots$ to denote the states of switches that could be on or off so that these variables can take on the values 0 and 1. (See Figure 34.1.)

Two switches, a and b, connecting two terminals are said to be connected in *series* if current will flow between the two terminals only when both a and b are on. This situation is diagrammed in Figure 34.2, and it is denoted by $a \cdot b$ (or just ab).

Switches a and b connecting two terminals are said to be connected in *parallel* if current will flow between the two terminals when either a or b is on. This is depicted in Figure 34.3 and is denoted by $a + b$.

Following our convention of using 0 for off and 1 for on, the diagrams in Figures 34.2 and 34.3 and the notation defined previously suggest that we may define two binary operations $+$ and $\cdot$ on $\{0, 1\}$, as in Table 34.2.

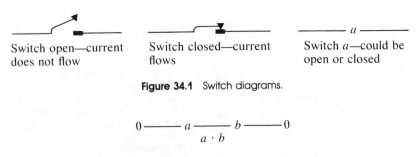

Switch open—current does not flow

Switch closed—current flows

Switch a—could be open or closed

Figure 34.1 Switch diagrams.

$$0 \underline{\hspace{1.5cm}} a \underline{\hspace{1.5cm}} b \underline{\hspace{1.5cm}} 0$$
$$a \cdot b$$

Figure 34.2 Switches a and b connected in series.

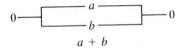

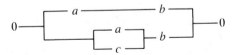

Figure 34.3 Switches a and b connected in parallel.

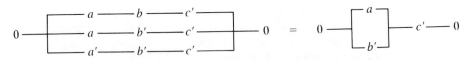

Figure 34.4 The diagram for $ab + (a + c)b$.

Figure 34.5 $abc' + ab'c' + a'b'c' = (a + b')c'$.

With these operations, we can now combine switches in series and parallel to make more complicated devices called *series-parallel switching circuits*. One such example is given in Figure 34.4.

It is also convenient to define a unary operation on circuits. For any circuit C made up of switches a, b, c, . . . , the circuit C' is made up of the same switches, but for any choice of states for a, b, c, . . . , C' has the opposite state as C. So, for example, if $a + b$ is on, $(a + b)'$ is off. We say two circuits, C_1 and C_2, made up of switches a, b, c, . . . , are equivalent (and write $C_1 = C_2$) if, for any states for a, b, c, . . . , current flows through C_1 if and only if current flows through C_2. Figure 34.5 shows two equivalent circuits.

It should not come as a surprise that the set of all equivalence classes of circuits (made up of a finite number of switches) with the binary operations $+$ and $\cdot$ and the unary operation $'$ is a Boolean algebra. Of course, we may determine whether or not current will flow through a circuit composed of switches a, b, c, . . . , by simply replacing each variable representing an off switch with 0 and each variable representing an on switch with 1 and using the operations

$$0 + 0 = 0, \qquad 1 + 0 = 1, \qquad 1 + 1 = 1,$$
$$0 \cdot 0 = 0, \qquad 1 \cdot 0 = 0, \qquad 1 \cdot 1 = 1.$$

Table 34.3 gives an example.

Table 34.2 Operation Table for $+$ and $\cdot$

a	b	$a + b$	ab
0 (off)	0 (off)	0 (off)	0 (off)
1 (on)	0 (off)	1 (on)	0 (off)
0 (off)	1 (on)	1 (on)	0 (off)
1 (on)	1 (on)	1 (on)	1 (on)

Table 34.3   Values for the Circuit $ab + (a + c)b'$

a	b	c	ab	a + c	b'	(a + c)b'	ab + (a + c)b'
0	0	0	0	0	1	0	0
1	0	0	0	1	1	1	1
0	1	0	0	0	0	0	0
0	0	1	0	1	1	1	1
1	1	0	1	1	0	0	1
1	0	1	0	1	1	1	1
0	1	1	0	1	0	0	0
1	1	1	1	1	0	0	1

In computer design, switches are symbolically represented by so-called gates, as shown in Figure 34.6. The gate representations for the circuits in Figures 34.4 and 34.5 are shown in Figures 34.7 and 34.8.

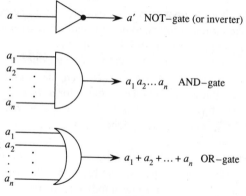

Figure 34.6

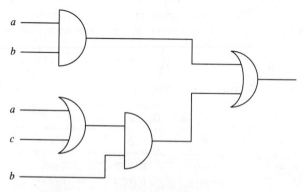

Figure 34.7   Gate diagram for $ab + (a + c)b$.

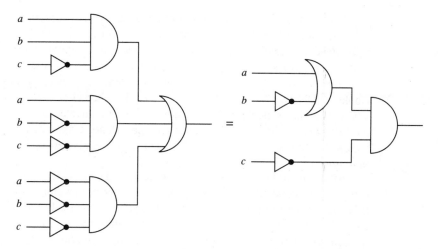

Figure 34.8 Gate diagram for $abc' + ab'c' + a'b'c' = (a + b')c'$

The Algebra of Logic

Just as the set of all equivalence classes of circuits forms a Boolean algebra, so too does the set of all equivalence classes of propositions. For our purposes, a *proposition* is a declarative statement that we can decide (in theory, at least) is either true or false. For example, the proposition "John F. Kennedy was assassinated" is true, while the proposition "Franklin D. Roosevelt was assassinated" is false. Propositions can be combined to form new propositions by using the logical connectives "and" and "or." In general, if A and B are propositions, mathematicians often use $A \wedge B$ to denote the proposition "A and B," and $A \vee B$ to denote the proposition "A or B." For instance, if we use A to denote the proposition "John F. Kennedy was assassinated" and B to denote the proposition "Franklin D. Roosevelt was assassinated," then $A \wedge B$ represents the proposition "John F. Kennedy was assassinated and Franklin D. Roosevelt was assassinated," which is false. On the other hand, the symbolism $A \vee B$ represents the proposition "John F. Kennedy was assassinated or Franklin D. Roosevelt was assassinated," which is true.

We define $A \wedge B$ to be true if and only if both A and B are true, and $A \vee B$ to be true if and only if either A or B (or both) is true. The latter operation is sometimes called the *inclusive or*. (Another term for it is *and/or*). The *exclusive or* connecting two propositions means that one of the propositions is true and one is false. For any proposition A, there is another proposition (often denoted by A' and called the *negation* of A) that asserts that proposition A is false. For example, if A denotes the proposition "John F. Kennedy was assassinated," then A' is the proposition "It is false that John F. Kennedy was assassinated," or more naturally, "John F. Kennedy was not assassinated." In general, proposition A' is true only when proposition A is false and vice versa.

If A and B are propositions formed by combining propositions P_1, P_2, P_3, . . . , and P_1', P_2', P_3', . . . , with the connectives $\wedge$ and $\vee$, we say A is *equivalent* to B (and write $A = B$) if, for every assignment of truth or falsity to the propositions P_1, P_2, P_3, . . . , A is true if and only if B is true. With this definition of equivalence, the set of all equivalence classes of propositions is a Boolean algebra. The zero element is the equivalence class containing $P \wedge P'$, where P is any statement. (Any element in this equivalence class is called a *contradiction*.) The unity element is the equivalence class containing $P \vee P'$, where P is any statement. (Any statement in this equivalence class is called a *tautology*.)

As a consequence of this, we can decide mechanically whether or not a complicated proposition made by combining other propositions is true or false, if we know the truth or falsity of the propositions involved. This can be done easily by assigning the letter T to each proposition that is true and the letter F to each proposition that is false. We then use the operations

$$F \wedge F = F, \qquad F \vee F = F, \qquad F \wedge T = F,$$
$$F \vee T = T, \qquad T \wedge T = T, \qquad T \vee T = T,$$

and the axioms for a Boolean algebra on the set $\{T, F\}$ to reduce the resulting expression to a single T or a single F. If it reduces to T, the proposition under consideration is true; otherwise, it is false. Table 34.4 gives several examples. Such a table is often called a *truth table*.

Table 34.4   Truth Table for Various Propositions

P	Q	R	$P \vee Q$	$P \wedge Q$	$(P \wedge Q) \vee [(P \vee R) \wedge Q]$	$(P \vee Q') \wedge R'$
F	F	F	F	F	F	T
F	F	T	F	F	F	F
F	T	F	T	F	F	F
F	T	T	T	F	T	F
T	F	F	T	F	F	T
T	F	T	T	F	F	F
T	T	F	T	T	T	T
T	T	T	T	T	T	F

Finite Boolean Algebras

Our last theorem provides an important arithmetical criterion that every finite Boolean algebra must satisfy. Notice that, with this result, we may instantly reject the set in Example 4 as a Boolean algebra, no matter how $\vee$ and $\wedge$ are defined.

> **Theorem 34.2** $|B| = 2^n$.
> *A finite Boolean algebra has 2^n elements for some integer n.*

Proof. Let B be any finite Boolean algebra with more than one element. For any pair of elements a and b in B, we define

$$a + b = (a \vee b) \wedge (a \wedge b)'.$$

(This operation is called the *symmetric difference* of a and b. See exercise 22.) One may tediously verify that B is in fact an Abelian group under $+$. Since $a + a = 0$ for all a, we see that every nonzero element of B has order 2. It now follows from Cauchy's Theorem that the only prime divisor of $|B|$ is 2. Thus, $|B| = 2^n$ for some positive integer n. ∎

Although it appears that there are many diverse models for Boolean algebras (e.g., algebra of sets, algebra of switching functions, algebra of circuits, algebra of logic), it can be proved (see [1, p. 378]) that every finite Boolean algebra is isomorphic to an algebra of sets. This is analogous to the fact that every group is isomorphic to a group of permutations. (See Cayley's Theorem in Chapter 6.)

EXERCISES

The road to wisdom?—Well it's plain and simple to express:
Err
and err
and err again
but less
and less
and less.

Piet Hein, "The Road to Wisdom," Grooks

1. Prove that the following identities hold in a Boolean algebra.
    a. $a \vee (a' \wedge b) = a \vee b$
    b. $(a \wedge b) \vee (a \wedge b') \vee (a' \wedge b') = a \vee b'$
    c. $(a \vee b) \wedge (a \vee b') \wedge (a' \vee b') = a \wedge b'$

2. Verify that the set and operations given in Example 3 satisfy axiom 4 for a Boolean algebra.

3. Show that the set of divisors of 36 cannot be a Boolean algebra.

4. Prove Theorem 34.1.

5. Referring to Example 2, calculate the following:
    a. $f_2 + f_8, f_2 \cdot f_8$
    b. $f_7, + f_{14}, f_7 \cdot f_{14}$
    c. f_7', f_{14}'

6. Verify that the following two circuits are equivalent.

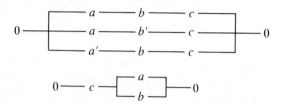

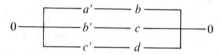

7. Write an algebraic expression for the following circuit.

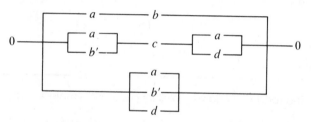

8. Simplify the expression $([(a' \wedge b') \vee c] \wedge (a \vee c))'$.

9. Write an expression that represents the following circuit.

10. Draw the gate representation for the circuit in Exercise 7.

11. Draw the gate representation for the circuit in Exercise 9.

12. In a Boolean algebra containing a and b, prove that $a = b$ if and only if $(a \wedge b') \vee (a' \wedge b) = 0$.

13. Let $S = \{1, 3, 7, 21\}$. For x, y in S, define $x \vee y = \text{lcm}(x, y)$ and $x \wedge y = \gcd(x, y)$. Define $x' = 21/x$. Show that S with these operations satisfies axiom 4 for a Boolean Algebra. (In fact, all the axioms are satisfied.)

14. (Poretzky's Law) Let B be a Boolean algebra, and let a belong to B. Prove that $a = 0$ if and only if $b = (a \wedge b') \vee (a' \wedge b)$ for each b in B.

15. Show, by example, that in a Boolean algebra the cancellation law does not hold; that is, $a \vee c = b \vee c$ does not always imply $a = b$.

16. For any a, b, c from a Boolean algebra, prove
   a. $a \vee c = b \vee c$ and $a \vee c' = b \vee c'$ imply $a = b$.
   b. $a \vee c = b \vee c$ and $a \wedge c = b \wedge c$ imply $a = b$.

17. How would you define an isomorphism for a Boolean algebra?

18. Suppose ϕ is an isomorphism from a Boolean algebra A onto a Boolean algebra B. Prove that ϕ carries the unity of A to the unity of B. Also prove that $a'\phi = (a\phi)'$ for every a in A follows from the other conditions.

19. Draw a circuit diagram that will be closed if and only if exactly one of a, b, and c is closed.

20. Draw a circuit diagram that will be closed if and only if at most one of a, b, and c is closed.

21. Draw a circuit and a gate diagram that represent the expression $(a + b)c + ab'$.

22. Draw a Venn diagram for the symmetric difference of sets A and B. (See the proof of Theorem 34.2 for the definition.)

23. Verify Table 34.4.

REFERENCE

1. Harold S. Stone, *Discrete Mathematical Structures and Their Applications*, Chicago: Science Research Associates, 1973.

SUGGESTED READINGS

F. Hohn, "Some Mathematical Aspects of Switching," *The American Mathematical Monthly* 62 (1955): 75–90.

In this article, the author discusses the Boolean algebra of switching circuits in greater detail than we have done.

F. Hohn, *Applied Boolean Algebra: An Elementary Introduction*, New York: MacMillan, 1960.

This brief monograph covers circuits, propositional logic, and the algebra of sets.

Claude E. Shannon

Probably no single work in this century has more profoundly altered man's understanding of communication than C. E. Shannon's "A mathematical theory of communication" first published in 1948. There resulted theorems of great power, elegance, generality and beauty. They have shed much understanding on the elusive true nature of the communication process and have delineated its interest limitations.

David Slepian, *Key Papers in the Development of Information Theory*

CLAUDE E. SHANNON was born on April 30, 1916, in Petoskey, Michigan, and grew up in nearby Gaylord. He received a B.S. degree in electrical engineering from the University of Michigan in 1936. In 1940, he received an M.S. degree in electrical engineering and a Ph.D. degree in mathematics, simultaneously, from the Massachusetts Institute of Technology. In his master's thesis, he showed that Boolean algebra provides an excellent tool for analyzing electrical switching circuits. After spending 1941 at the Institute for Advanced Study at Princeton, Shannon went to Bell Laboratories as a research mathematician. There he made major contributions to coding theory, cryptography, computing circuit design, and information theory, a field he founded. In 1957, Shannon became a faculty member at MIT and served as a consultant to Bell Laboratories until 1972.

Shannon's theory defined information. It explained crucial relationships among the elements of a communications system—signal power, bandwidth (the frequency range of an information channel), and noise (static). The computer term *bit* was coined by Shannon.

Among the numerous prestigious awards Shannon has received are the National Medal of Science and honorary degrees from Princeton and Yale. He is a member of the National Academy of Sciences, and has written more than fifty technical articles and co-authored two books. His hobbies include chess, playing clarinet, and listening to music. He also builds electronic gadgets.

SUPPLEMENTARY EXERCISES FOR CHAPTERS 26–34

Nothing worthwhile comes easily.... Work, continuous work and hard work, is the only way to accomplish results that last.

Hamilton Holt

1. Let $G = \langle x, y \mid x = (xy)^3, y = (xy)^4 \rangle$. What familiar group is G isomorphic to?

2. Let $G = \langle z \mid z^6 = 1 \rangle$ and $H = \langle x, y \mid x^2 = y^3 = 1, xy = yx \rangle$. Show that G and H are isomorphic.

3. Show that a group of order $315 = 3^3 \cdot 5 \cdot 7$ has a subgroup of order 45.

4. Let G be a group of order p^2q, where p and q are primes, $q < p$, and q does not divide $p^2 - 1$. Show that G is Abelian.

5. Let H denote a Sylow 7-subgroup of a group G and K a Sylow 5-subgroup of G. Assume $|H| = 49$, $|K| = 5$, and K is a subgroup of $N(H)$. Show that H is a subgroup of $N(K)$.

6. Determine all groups of order 30.

7. Suppose K is a normal Sylow p-subgroup of H and H is a normal subgroup of G. Prove that K is a normal subgroup of G. (Compare this with exercise 39 of Chapter 10.)

8. Prove that A_5, the group of even permutations on five objects, cannot have a subgroup of order 15.

9. Let H and K be subgroups of G. Prove that HK is a subgroup of G if $H \leqslant N(K)$.

10. Suppose H is a subgroup of a finite group G and H contains $N(P)$, where P is some Sylow p-subgroup of G. Prove that $N(H) = H$.

11. Prove that a simple group G of order 168 cannot contain an element of order 21.

12. Prove that the only group of order 561 is Z_{561}.

13. Prove that the center of a non-Abelian group of order 105 has order 5.

14. Let n be an odd integer at least 3. Prove that every Sylow subgroup of D_n is cyclic.

15. Let G be the digraph obtained from $\text{Cay}(\{(1, 0), (0, 1)\}: Z_3 \oplus Z_5)$ by deleting the vertex $(0, 0)$. (Also, delete each arc to or from $(0, 0)$.) Prove that G has a Hamiltonian circuit.

16. Prove that the digraph obtained from $\text{Cay}(\{(1, 0), (0, 1)\}: Z_4 \oplus Z_7)$ by deleting the vertex $(0, 0)$ has a Hamiltonian circuit.

17. Let G be a finite group generated by a and b. Let $s_1, s_2, \ldots, s_n$ be the arcs of a Hamiltonian circuit in the digraph $\text{Cay}(\{a, b\}: G)$. We say the vertex

$s_1 s_2 \cdots s_i$ *travels by* a if $s_{i+1} = a$. Show that if a vertex x travels by a, then every vertex in $x \langle ab^{-1} \rangle$ travels by a.

18. Recall, the dot product $u \cdot v$ of two vectors $u = (u_1, u_2, \ldots, u_n)$ and $v = (v_1, v_2, \ldots, v_n)$ from F^n is

$$u_1 v_1 + u_2 v_2 + \cdots + u_n v_n$$

(where the addition and multiplication is that of F). Let C be an (n, k) linear code. Show that

$$C^\perp = \{v \in F^n \mid v \cdot u = 0 \text{ for all } u \in C\}$$

is an $(n, n - k)$ linear code. This code called the *dual* of C.

19. Find the dual of each of the following binary codes:
   a. $\{00, 11\}$
   b. $\{000, 011, 101, 110\}$
   c. $\{0000, 1111\}$
   d. $\{0000, 1100, 0011, 1111\}$

20. Let C be a binary linear code such that $C \subseteq C^\perp$. Show that wt(v) is even for all v in C.

21. Let C be an (n, k) binary linear code over F. If $v \in F^n$, but $v \notin C^\perp$, show that $v \cdot u = 0$ for exactly half of the elements u in C.

22. Suppose C is an (n, k) binary linear code and the vector $11 \cdots 1 \in C^\perp$. Show that wt(v) is even for every v in C.

23. Suppose C is an (n, k) binary linear code and $C = C^\perp$. (Such a code is called *self-dual*.) Prove that n is even. Prove that $11 \cdots 1$ is a code word.

24. If G is a finite solvable group, show that there exist subgroups

$$\{e\} = H_0 \triangleleft H_1 \triangleleft H_2 \triangleleft \cdots \triangleleft H_n = G$$

such that H_{i+1}/H_i has prime order.

25. If a group G has a normal subgroup N such that both N and G/N are solvable, show that G is solvable.

26. Show that the polynomial $x^5 - 6x + 3$ over Q is not solvable by radicals.

27. Let a, b, c belong to a Boolean algebra. If $a \vee b \vee c = a \wedge b \wedge c$, prove that $a = b = c$.

The End.

Title of song by John Lennon and Paul McCartney, *Abbey Road*, side 2, October 1969

Selected Answers

Many of the proofs given below are merely sketches. In these cases, the student should supply the complete proof.

CHAPTER 0

To make headway, improve your head. B. C. Forbes

1. $\{1, 3, 5, 7\}$; $\{1, 5, 7, 11\}$; $\{1, 3, 7, 9, 11, 13, 17, 19\}$; $\{1, 2, 3, 4, 6, 7, 8, 9, 11, 12, 13, 14, 16, 17, 18, 19, 21, 22, 23, 24\}$

3. 1; 0; 4; 5

5. 1942, June 18; 1953, December 13.

7. By using 0 as an exponent if necessary, we may write $a = p_1^{m_1} \cdots p_k^{m_k}$ and $b = p_1^{n_1} \cdots p_k^{n_k}$ where the p's are distinct primes and the m's and n's are nonnegative. Then $\text{lcm}(a, b) = p_1^{s_1} \cdots p_k^{s_k}$ where $s_i = \max(m_i, n_i)$ and $\gcd(a, b) = p_1^{t_1} \cdots p_k^{t_k}$ where $t_i = \min(m_i, n_i)$. Then $\text{lcm}(a, b) \cdot \gcd(a, b) = p_1^{m_1+n_1} \cdots p_k^{m_k+n_k} = ab$.

9. $7(5n + 3) - 5(7n + 4) = 1$.

11. Use the "GCD is a linear combination" theorem.

13. No. There is no smallest positive rational number.

15. Any common divisor of a and b would also divide $at + bs = 1$.

17. $\gcd(34, 126) = 2$; $2 = 26 \cdot 34 - 7 \cdot 126$

19. Use proof by contradiction.

21. Let S be a set with $n + 1$ elements and pick some a in S. By induction, S has 2^n subsets that do not contain a. But there is a one-to-one correspondence between the subsets of S that do not contain a and those that do. So there are $2 \cdot 2^n = 2^{n+1}$ subsets in all.

23. Say $p_1 p_2 \ldots p_r = q_1 q_2 \ldots q_s$ where the p's and q's are positive primes. By the Generalized Euclid's Lemma, p_1 divides some q_i, say q_1 (we may relabel the q's if necessary). Then $p_1 = q_1$ and $p_2 \ldots p_r = q_2 \ldots q_s$. Repeating this argument at each step we obtain $p_2 = q_2, \ldots, p_r = q_r$ and $r = s$.

25. By the Second Principle of Mathematical Induction, $f_n = f_{n-1} + f_{n-2} < 2^{n-1} + 2^{n-2} = 2^{n-2}(2 + 1) < 2^n$.

27. The statement is true for any divisor of $8^4 - 4 = 4092$.

29. Observe that the number with the decimal representation $a_9 a_8 \ldots a_1 a_0$ is $a_9 \cdot 10^9 + a_8 \cdot 10^8 + \cdots + a_1 \cdot 10 + a_0$. Then use exercise 12 and the fact that $a_i 10^i \bmod 9 = a_i$ to deduce that the check digit is $(a_9 + a_8 + \cdots + a_i + a_0) \bmod 9$.

31. For the case that the check digit is not involved see the answer to exercise 29. If a transposition involving the check digit $c = a_1 + a_2 + \cdots + a_{10}$ goes undetected then $a_{10} = a_1 + a_2 + \cdots + a_9 + c$. Substitution yields $2(a_1 + \cdots + a_9) = 0 \bmod 9$. Therefore, $10(a_1 + \cdots + a_9) = a_1 + \cdots + a_9 = 0 \bmod 9$. It follows that $c = a_{10}$. In this case the transposition does not yield an error.

33. Say the number is $a_8 a_7 \ldots a_1 a_0 = a_8 \cdot 10^8 + a_7 \cdot 10^7 + \cdots + a_1 \cdot 10 + a_0$. Then the error is undetected if and only if $a_i 10^i - a_i' 10^i = 0 \bmod 7$. Multiplying both sides by 5^i and noting that $50 = 1 \bmod 7$ we obtain $a_i - a_i' = 0 \bmod 7$.

35. One need only verify the equation for $n = 0, 1, 2, 3, 4, 5$. Alternatively, observe that $n^3 - n = n(n-1)(n+1)$.

37. Observe that $1^2 = 3^2 = 5^2 = 7^2 = 1 \bmod 8$. Alternatively, note that $(2k+1)^2 - 1 = 4k^2 + 4k = 4k(k+1)$.

39. $a^2 + b^2 = a^2 + b^2$ gives $(a, b)R(a, b)$; since $a^2 + b^2 = c^2 + d^2$ implies $c^2 + d^2 = a^2 + b^2$, R is symmetric; if $a^2 + b^2 = c^2 + d^2$ and $c^2 + d^2 = e^2 + f^2$ then $a^2 + b^2 = c^2 + f^2$ and R is transitive. The equivalence classes are circles in the plane centered at $(0, 0)$.

41. No. $(1, 0) \in R$ and $(0, -1) \in R$, but $(1, -1) \notin R$.

43. Consider the set of integers with aRb if $|a - b| \leq 1$.

45. Let $S = \{1, 2, 3\}$ and $R = \{(1, 1), (2, 2), (1, 2), (2, 1)\}$.

CHAPTER 1

It requires a very unusual mind to make an analysis of the obvious.   Alfred North Whitehead

1. Three rotations: $0°$, $120°$, $240°$, and three reflections across lines from vertices to midpoints of opposite sides

3. No

5. D_n has n rotations of the form $k(360°/n)$ where $k = 0, \ldots, n - 1$. D_n has n reflections. When n is odd the axes of reflection are the lines from the vertices to the midpoints of the opposite sides. When n is even, half of the axes of reflection are obtained by joining opposite vertices; the other half, by joining midpoints of opposite sides.

7. A rotation followed by a rotation either fixes every point (and so is the identity) or fixes only the center of rotation. However, a reflection fixes a line.

9. Observe that $1 \cdot 1 = 1$; $1(-1) = -1$; $(-1)1 = -1$; $(-1)(-1) = 1$. These relationships also hold when 1 is replaced by "rotation" and -1 is replaced by "reflection".

11. $HD = DV$.

13. R_0, R_{180}, H, V.

15. See answer for exercise 13.

17. In each case the group is D_6.

19. Let L denote the line segment joining the centers of the two ends of a cigarette and L' any perpendicular bisector of L. A cigarette has rotational symmetry about L of every degree and $180°$ rotational symmetry about L'. A cigarette has reflective symmetry about all planes containing L and the plane containing L' that is perpendicular to L. A cigar has the rotational and reflective symmetries involving L but not those involving L'.

21. D_{11}.

23. D_{28}.

CHAPTER 2

It's easy! John Lennon and Paul McCartney, *All You Need Is Love*, single

1. Does not contain the identity; closure fails.

3. Suppose $\begin{bmatrix} a & b \\ c & d \end{bmatrix}$ is the inverse. Then, $\begin{bmatrix} 2 & 2 \\ 1 & 1 \end{bmatrix}\begin{bmatrix} a & b \\ c & d \end{bmatrix} = \begin{bmatrix} 1 & 0 \\ 0 & 1 \end{bmatrix}$ so that $2a + 2c = 1$ and $a + c = 0$.

5. Apply the definition.

7. (i) $2a + 3b$; (ii) $-2a + 2(-b + c)$; (iii) $-3(a + 2b) + 2c = 0$.

9.

	0	1	2	3
0	0	1	2	3
1	1	2	3	0
2	2	3	0	1
3	3	0	1	2

11. Under modulo 4, 2 does not have an inverse. Under modulo 5, each element has an inverse.

15. Use exercise 5 to verify closure. It also follows from exercise 5 that $(\det A^{-1}) = (\det A)^{-1}$. Thus $\det A^{-1} = \pm 1$ whenever $\det A = \pm 1$.

17. 29.

19. $(ab)^n$ need not equal $a^n b^n$ in a non-Abelian group.

21. Use exercise 20.

23. The identity is 25.

25. $\{1, 3, 5, 9, 13, 15, 19, 23, 25, 27, 39, 45\}$

27.

	1	5	7	11
1	1	5	7	11
5	5	1	11	7
7	7	11	1	5
11	11	7	5	1

29. Use exercise 28.

31. aca^{-1}

33. If $x^3 = e$ and $x \neq e$, then $(x^{-1})^3 = e$ and $x \neq x^{-1}$. So, nonidentity solutions come in pairs. If $x^2 \neq e$, then $x^{-1} \neq x$ and $(x^{-1})^2 \neq e$. So solutions to $x^2 \neq e$ come in pairs.

35. $x\phi_g\phi_h = (g^{-1}xg)\phi_h = h^{-1}(g^{-1}xg)h = (gh)^{-1}xgh = x\phi_{gh}$

37. Use associativity. For example, $4 \cdot 5 = 4 \cdot (4 \cdot 1) = (4 \cdot 4) \cdot 1 = 0 \cdot 1 = 1$.

39. $(R_{36}F)^{-1} = F^{-1}R_{36}^{-1} = FR_{324}$

41. If n is not prime, the set is not closed under multiplication modulo n. If n is prime the set is closed and for every r in the set there are integers s and t such that $1 = rs + nt = rs \bmod n$.

43. 4

47. a. 3; b. dot product is not 0; the correct number cannot be determined; yes. c. no; the dot product is 0. d. Use Theorem 2.4.

49. 2; Say a_i' is substituted for a_i $(a_i' \neq a_i)$. If i is even, argue as in the proof of Theorem 2.4. Otherwise consider cases: $0 \leqslant a_i$, $a_i' < 5$; $5 \leqslant a_i$, $a_i' < 9$; $0 \leqslant a_i < 5$ and $5 \leqslant a_i < 9$; $0 \leqslant a_i' < 5$ and $5 \leqslant a_i < 9$. In every case a_i' and a_i contribute different amounts to the dot product. Theorem 2.4 does not apply because of the r term. All transposition errors except $09 \leftrightarrow 90$ are detected.

CHAPTER 3

The brain is as strong as its weakest think.   Eleanor Doan

1. $|Z_{12}| = 12; |U(10)| = 4; |U(12)| = 4; |U(20)| = 8; |D_4| = 8$
   In $Z_{12}, |0| = 1; |1| = |5| = |7| = |11| = 12; |2| = |10| = 6; |3| = |9| = 4; |4| = |8| = 3;$
   $|6| = 2.$
   In $U(10), |1| = 1; |3| = |7| = 4; |9| = 2.$
   In $U(20), |1| = 1; |3| = |7| = |13| = |17| = 4; |9| = |11| = |19| = 2.$
   In $D_4, |R_0| = 1; |R_{90}| = |R_{270}| = 4; |R_{180}| = |H| = |V| = |D| = |D'| = 2.$
   In each case notice that the order of the element divides the order of the group.
3. In $Q, |0| = 1$ and all other elements have infinite order. In $Q^*, |1| = 1, |-1| = 2,$ and all other elements have infinite order.
5. Each is the inverse of the other.
7. $U(14) = \{1, 3, 5, 9, 11, 13\}$
   $\langle 3 \rangle = \{3, 3^2, 3^3, 3^4, 3^5, 3^6\} = \{3, 9, 13, 11, 5, 1\} = U(14);$
   $U(14) \neq \langle 11 \rangle$
9. By brute force show that $k^4 = 1$ for all k.
11. For any integer $n \geq 3$, D_n contains elements a and b of order 2 while $|ab| = n$. In general, there is no relationship between $|a|, |b|,$ and $|ab|$.
13. Write $k = sm$. If $x \equiv 1 \bmod k$, then $x - 1 = tk = tsm$ so $x \equiv 1 \bmod m$.
15. If $x \in Z(G)$, then $x \in C(a)$ for all a, so $x \in \bigcap_{a \in G} C(a)$. If $x \in \bigcap_{a \in G} C(a)$, then $xa = ax$ for all a in G so $x \in Z(G)$.
17. a. $C(5) = G; C(7) = \{1, 3, 5, 7\}$
    b. $Z(G) = \{1, 5\}$
    c. $|2| = 2; |3| = 4.$ They divide the order of the group.
19. Mimic the proof of Theorem 3.5.
21. Suppose $x^n = e$ and $y^n = e$. Then $(xy^{-1})^n = x^n(y^{-1})^n = x^n(y^n)^{-1} = ee^{-1} = e.$ For the second part, use D_4.
25. 2
27. Note that $\begin{bmatrix} 1 & 1 \\ 0 & 1 \end{bmatrix}^n = \begin{bmatrix} 1 & n \\ 0 & 1 \end{bmatrix}.$
29. For any positive integer n, a rotation of $360°/n$ has order n. A rotation of $\sqrt{2}°$ has infinite order.
31. $\langle R_0 \rangle, \langle R_{90} \rangle, \langle R_{180} \rangle, \langle D \rangle, \langle D' \rangle, \langle H \rangle, \langle V \rangle. \langle R_{270} \rangle$ is not on the list since $\langle R_{90} \rangle = \langle R_{270} \rangle. \{R_0, R_{180}, D, D'\}$ and $\{R_0, R_{180}, H, V\}$ are not cyclic.
33. Certainly, $(a^{n/k})^k = a^n = e.$ If $(a^{n/k})^t = e$ for some positive $t < k$, then $a^{nt/k} = e$ and $nt/k < n$, a contradiction.
35. No. $7 \in H$ but $7 \cdot 7 \notin H$.
39. $|\langle 3 \rangle| = 4$
41. Let $\begin{bmatrix} a & b \\ c & d \end{bmatrix}$ and $\begin{bmatrix} a' & b' \\ c' & d' \end{bmatrix}$ belong to H. It suffices to show that $a - a' + b - b' + c - c' + d - d' = 0$. This follows from $a + b + c + d = 0 = a' + b' + c' + d'$. If 0 is replaced by 1, H is not a subgroup.
43. If 2^a and $2^b \in K$, then $2^a(2^b)^{-1} = 2^{a-b} \in K$ since $a - b \in H$.
45. $\begin{bmatrix} 2 & 0 \\ 0 & 2 \end{bmatrix}^{-1} = \begin{bmatrix} 1/2 & 0 \\ 0 & 1/2 \end{bmatrix}$ is not in H.
47. If $a + bi$ and $c + di \in H$, then $(a + bi)(c + di)^{-1} = (ac + bd) + (bc - ad)i$ and $(ac + bd)^2 + (bc - ad)^2 = 1$ so that H is a subgroup. H is the unit circle in the complex plane.

CHAPTER 4

There will be an answer, let it be.   John Lennon and Paul McCartney, *Let It Be*, single

1. For Z_6, generators are 1 and 5; for Z_8, generators are 1, 3, 5, and 7; for Z_{20}, generators are 1, 3, 7, 9, 11, 13, 17, and 19.

3. $\langle 20 \rangle = \{20, 10, 0\}$
   $\langle 10 \rangle = \{10, 20, 0\}$

5. $\langle 3 \rangle = \{3, 9, 7, 1\}$
   $\langle 7 \rangle = \{7, 9, 3, 1\}$

7. $U(8)$ or D_3

9. Six subgroups; generators are the divisors of 20.
   Six subgroups; generators are a^k where k is a divisor of 20.

11. Certainly, $a^{-1} \in \langle a \rangle$. So $\langle a^{-1} \rangle \subseteq \langle a \rangle$. So, by symmetry, $\langle a \rangle \subseteq \langle a^{-1} \rangle$ as well.

13. Let $k = \mathrm{lcm}(m, n) \bmod 24$. Then $\langle a^m \rangle \cap \langle a^n \rangle = \langle a^k \rangle$.

15. $|g|$ divides 12 is equivalent to $g^{12} = e$. So, if $a^{12} = e$ and $b^{12} = e$, then $(ab^{-1})^{12} = a^{12} (b^{12})^{-1} = ee^{-1} = e$. The general result is given in exercise 21 of Chapter 3.

17. $\langle 1 \rangle$, $\langle 7 \rangle$, $\langle 11 \rangle$, $\langle 17 \rangle$, $\langle 19 \rangle$, $\langle 29 \rangle$.

19. a. $|a|$ divides 12. b. $|a|$ divides m. c. By Theorem 4.3, $|a| = 1, 2, 3, 4, 6, 8, 12,$ or 24. If $|a| = 2$, then $a^8 = (a^2)^4 = e^4 = e$. A similar argument eliminates all other possibilities except 24.

21. Yes, by Theorem 4.3. The subgroups of Z are of the form $\{0, \pm n, \pm 2n, \pm 3n, \ldots\}$ where n is any integer.

23. Suppose $|a^t| = s$. Clearly, $(a^t)^{n/\gcd(n,t)} = e$ so that s divides $n/\gcd(n, t)$. Also, since $a^{ts} = e$ we know n divides ts, say $nk = ts$. Then $[n/\gcd(n, t)]k = [t/\gcd(n, t)]s$. Since $n/\gcd(n, t)$ and $t/\gcd(n, t)$ are relatively prime, it follows from Euclid's Lemma (Chapter 0) that $n/\gcd(n, t)$ divides s. Thus $s = n/\gcd(n, t)$.

25. Two: $2a$ and a^{-1}

27. $\{1000000, 3000000, 5000000, 7000000\}$
   By Theorem 4.3, $\langle 1000000 \rangle$ is the unique subgroup of order 8 and only those on the list are generators.

29. Let $G = \{a_1, a_2, \ldots, a_k\}$. Now let $|a_i| = n_i$. Consider $n = n_1 n_2 \ldots n_k$.

31. Mimic exercise 30.

33. Mimic exercise 32.

35. Suppose a and b are relatively prime positive integers and $\langle a/b \rangle = Q^+$. Then there is some positive integer n such that $(a/b)^n = 2$. Clearly $n \neq 0, 1,$ or -1. If $n > 1$, $a^n = 2b^n$ so that 2 divides a. But then 2 divides b as well. A similar contradiction occurs if $n < -1$.

37.

	7	35	49	77
7	49	77	7	35
35	77	49	35	7
49	7	35	49	77
77	35	7	77	49

The identity is 49. The group is not cyclic.

39. Under the given conditions, ab has order $\mathrm{lcm}(m, n)$. For the second part look at D_3.

41. An infinite cyclic group does not have an element of prime order. A finite cyclic group can have only one subgroup for each divisor of its order. A subgroup of order p has exactly $p - 1$ elements of order p. Another element of order p would give another subgroup of order p.

43. $4, 3 \cdot 4, 7 \cdot 4, 9 \cdot 4$
45. 1 of order 1; 33 of order 2; 2 of order 3; 10 of order 11; 20 of order 33
47. 1, 2, 10, 20
49. If $|a| = 2$ and $|b| = 2$ and a and b commute, then $\{e, a, b, ab\}$ is a subgroup. The subgroup is not cyclic.
51. Use exercise 14 of Chapter 3 and Theorem 4.3.
53. In a cyclic group there are at most n solutions to the equation $x^n = e$.
55. First observe that 1 and $n - 1$ are generators. We must find another. If $n = p$ or $2p$, where p is prime, then 3 is a generator. Otherwise, apply Bertrand's Postulate to the largest prime divisor of n.
57. Use exercise 23.
59. Observe that $\langle a \rangle \cap H \subseteq \langle a \rangle$ so that $\langle a \rangle \cap H$ has the form $\langle a^k \rangle$ where k divides n. Since $a^k \in H$, $k = n$.

SUPPLEMENTARY EXERCISES FOR CHAPTERS 1–4

I have learned throughout my life as a composer chiefly through my mistakes and pursuits of false assumptions, not by my exposure to founts of wisdom and knowledge.
Igor Stravinsky, "Contingencies," *Themes and Episodes*

1. a. Let $x^{-1}h_1x$ and $x^{-1}h_2x$ belong to $x^{-1}Hx$. Then $(x^{-1}h_1x)(x^{-1}h_2x)^{-1} = x^{-1}h_1h_2^{-1}x \in x^{-1}Hx$ also.
    b. Let $\langle h \rangle = H$. Then $\langle x^{-1}hx \rangle = x^{-1}Hx$.
    c. $(x^{-1}h_1x)(x^{-1}h_2x) = x^{-1}h_1h_2x = x^{-1}h_2h_1x = (x^{-1}h_2x)(x^{-1}h_1x)$
3. Suppose $\mathrm{cl}(a) \cap \mathrm{cl}(b) \neq \emptyset$. Say $x^{-1}ax = y^{-1}by$. Then $(xy^{-1})^{-1}a(xy^{-1}) = b$. Thus, for any $u^{-1}bu$ in $\mathrm{cl}(b)$, we have $u^{-1}bu = (xy^{-1}u)^{-1}a(xy^{-1}u) \in \mathrm{cl}(a)$. This shows that $\mathrm{cl}(b) \subseteq \mathrm{cl}(a)$. By symmetry, $\mathrm{cl}(a) \subseteq \mathrm{cl}(b)$. $a = e^{-1}ae \in \mathrm{cl}(a)$ so the union of the conjugacy classes is G.
5. If both ab and ba have infinite order, we are done. Suppose $|ab| = k$. Then

$$\underbrace{(ab)(ab) \ldots (ab)}_{k \text{ factors}} = e.$$

Thus, $b[\underbrace{(ab)(ab) \ldots (ab)}_{k \text{ factors}}]a = bea = ba$.

So $(ba)^{k+1} = ba$, and $(ba)^k = e$. This proves $|ba| \leq |ab|$. By symmetry, $|ab| \leq |ba|$.
7. By exercise 6, for every x in G, $|x^{-1}ax| = |a|$ so that $x^{-1}ax = a$ or $ax = xa$.
9. 1 of order 1, 15 of order 2, 8 of order 15, 4 of order 5, 2 of order 3.
11. Let $|G| = 5$. Let $a \neq e$ belong to G. If $|a| = 5$ we are done. If $|a| = 3$, then $\{e, a, a^2\}$ is a subgroup of G. Let b be either of the remaining two elements of G. Then the set $\{e, a, a^2, b, ab, a^2b\}$ consists of six different elements, a contradiction. Thus, $|a| \neq 3$. Similarly, $|a| \neq 4$. We may now assume that every nonidentity element of G has order 2. Pick $a \neq e$ and $b \neq e$ in G with $a \neq b$. Then $\{e, a, b, ab\}$ is a subgroup of G. Let c be the remaining element of G. Then $\{e, a, b, ab, c, ac, bc, abc\}$ is a set of eight distinct elements of G, a contradiction. It now follows that if $a \in G$, and $a \neq e$, then $|a| = 5$.
13. $a^n(b^n)^{-1} = (ab^{-1})^n$, so G^n is a subgroup. For the non-Abelian group try D_3.
15. Suppose $G = H \cup K$. Pick $h \in H$ with $h \notin K$. Pick $k \in K$, but $k \notin H$. Then, $hk \in G$, but $hk \notin H$ and $hk \notin K$. $U(8)$ is the union of the three subgroups.
17. If $|a| = p^k$ and $|b| = p^r$ with $k \leq r$, say, then $|ab^{-1}|$ divides p^r.
19. Note that $ba^2 = ab$ and $a^3 = b^2 = e$ imply $ba = a^2b$. Thus, every member of the group can

be written in the form a^ib^j. Therefore, the group is $\{e,\ a,\ a^2,\ b,\ ab,\ a^2b\}$. D_3 satisfies these conditions.

21. $xy = yx$ if and only if $xyx^{-1}y^{-1} = e$. But, $(xy)x^{-1}y^{-1} = x^{-1}(xy)y^{-1} = ee = e$.

23. Let $x \in N(g^{-1}Hg)$. Then $x^{-1}(g^{-1}Hg)x = g^{-1}Hg$. Thus $(gxg^{-1})^{-1}Hgxg^{-1} = H$. This means that $gxg^{-1} \in N(H)$. So $x \in g^{-1}N(H)g$. Reverse the argument to show $g^{-1}N(H)g \subseteq N(g^{-1}Hg)$.

CHAPTER 5

You cannot have the success without the failures.   H. G. Hasler, *The Observer*

1. a. 2 b. 3 c. 5
3. a. 3 b. 3 c. 6 d. 12
5. 12
7. $|(123)(45678)| = 15$
9. a. even b. odd c. even d. odd e. even
11. even; odd
13. An even number of two cycles followed by an even number of two cycles gives an even number of two cycles in all. So the finite subgroup test is verified.
15. even
17.

$$\alpha^{-1} = \begin{bmatrix} 1 & 2 & 3 & 4 & 5 & 6 \\ 2 & 1 & 3 & 5 & 4 & 6 \end{bmatrix}$$

$$\alpha\beta = \begin{bmatrix} 1 & 2 & 3 & 4 & 5 & 6 \\ 1 & 6 & 2 & 3 & 4 & 5 \end{bmatrix}$$

$$\beta\alpha = \begin{bmatrix} 1 & 2 & 3 & 4 & 5 & 6 \\ 6 & 2 & 1 & 5 & 3 & 4 \end{bmatrix}$$

19. Suppose H contains at least one odd permutation, say, σ. Let A be the set of even permutations in H, and B be the set of odd permutations in H. Then $\sigma A \subseteq B$ and $|\sigma A| = |A|$, so $|A| \leq |B|$. Also, $\sigma B \subseteq A$ and $|\sigma B| = |B|$, so $|B| \leq |A|$.
21. No. The identity is even.
23. $C(\alpha_3) = \{\alpha_1, \alpha_2, \alpha_3, \alpha_4\}$, $C(\alpha_{12}) = \{\alpha_1, \alpha_7, \alpha_{12}\}$
25. $(123)(321) = (1)$, $(1478)(8741) = (1)$
27. Let α, $\beta \in \text{stab}(a)$. Then $a(\alpha\beta) = (a\alpha)\beta = a\beta = a$. Also, $a\alpha = a$ implies $a = a\alpha^{-1}$.
29. Let $\alpha = (123)$ and $\beta = (145)$.
33. Say $\alpha = a_1a_2 \ldots a_n$ and $\beta = b_1 \ldots b_m$ where the a's and b's are *cycles*. Then $\alpha\beta^{-1} = a_1a_2 \ldots a_nb_m^{-1} \ldots b_1^{-1}$ is a finite number of cycles.
35. Hint: $(12)(13) = (123)$ and $(12)(34)$ and $(132)(324)$
37. 4
39. 2; Adapt proof of Theorem 2.4; $09 \leftrightarrow 90$ is undetected; the methods are essentially the same.

CHAPTER 6

Think and you won't sink.   B. C. Forbes, *Epigrams*

1. Try $n \rightarrow 2n$.
3. $(xy)\phi = \sqrt{xy} = \sqrt{x}\ \sqrt{y} = x\phi y\phi$
5. Try $1 \rightarrow 1$, $3 \rightarrow 5$, $5 \rightarrow 7$, $7 \rightarrow 11$.

7. D_{12} has elements of order 12 and S_4 does not.
9. The mapping $h \rightarrow x^{-1}hx$ is an isomorphism.
11. Let $G = Z_6$ and $H = Z_3$.
13. $a^{-1}xa = a^{-1}ya$ implies $x = y$, so ϕ_a is one-to-one. If $b \in G$, then $aba^{-1}\phi_a = b$, so ϕ_a is onto; $xy\phi_a = a^{-1}xya = a^{-1}xaa^{-1}ya = x\phi_a y\phi_a$ so ϕ_a is operation-preserving.
15. Let $\alpha \in \text{Aut}(G)$. We show α^{-1} is operation-preserving: $(xy)\alpha^{-1} = (x\alpha^{-1})(y\alpha^{-1})$ if and only if $(xy)\alpha^{-1}\alpha = [(x\alpha^{-1})(y\alpha^{-1})]\alpha$, that is, if and only if $xy = (x\alpha^{-1})\alpha(y\alpha^{-1})\alpha = xy$. So α^{-1} is operation-preserving. That $\text{Inn}(G)$ is a group follows from the equation $\phi_g\phi_h = \phi_{gh}$.
17. Use the fact that $x^4 = 1$ for all x in $U(16)$. In general, $x \rightarrow x^n$ is an automorphism of $U(16)$ when n is odd.
19. If $\alpha \in S_8$, write

$$\alpha = \begin{bmatrix} 1 & 2 & \cdots & 7 & 8 \\ 1\alpha & 2\alpha & & 7\alpha & 8 \end{bmatrix}$$

Then map α to

$$\begin{bmatrix} 1 & 2 & \cdots & 7 \\ 1\alpha & 2\alpha & & 7\alpha \end{bmatrix}$$

21. $\{\phi_e, \phi_a, \phi_b, \phi_{ba}\}$
23. We show ϕ preserves addition and multiplication.

$$[(a + bi) + (c + di)]\phi = [(a + c) + (b + d)i]\phi = (a + c) - (b + d)i$$

Also,

$$(a + bi)\phi + (c + di)\phi = (a - bi) + (c - di) = (a + c) - (b + d)i$$

Now multiplication:

$$[(a + bi)(c + di)]\phi = [(ac - bd) + (bc + ad)i]\phi = (ac - bd) - (bc + ad)i$$

while

$$(a + bi)\phi(c + di)\phi = (a - bi)(c - di) = (ac - bd) - (bc + ad)i$$

25. Show that Q is not cyclic.
27. Try $a + bi \rightarrow \begin{bmatrix} a & -b \\ b & a \end{bmatrix}$.
29. Yes, by Cayley's Theorem.
31. $\log_{10}(xy) = \log_{10}x + \log_{10}y$. The marks on a slide rule represent logarithmic lengths. Manipulating the slide rule corresponds to adding (or subtracting) these lengths. The given equation shows this corresponds to multiplying (or dividing) the marks.
33. $\phi_g = \phi_h$ implies $g^{-1}xg = h^{-1}xh$ for all x. This implies $(gh^{-1})^{-1}xgh^{-1} = x$ and therefore $gh^{-1} \in Z(G)$.
35. The elements of D_n are permutations on the vertices of a regular n-gon.

CHAPTER 7

There is always a right and a wrong way, and the wrong way always seems the more reasonable.
George Moore, *The Bending of the Bough*

1. $|(0, 0)| = 1$
   $|(0, 2)| = |(1, 2)| = |(1, 0)| = 2$
   $|(0, 1)| = |(0, 3)| = |(1, 1)| = |(1, 3)| = 4$

3. Every nonidentity element in the group has order 2. Each of these generates a subgroup of order 2.
5. Suppose $\langle (a, b) \rangle = Z \oplus Z$. Then there exist integers n and m such that $(na, mb) = (1, 1)$. This implies that $a = \pm 1$ and $b = \pm 1$. Thus $(2, 3) \notin \langle (a, b) \rangle$.
7. Try $(g_1, g_2) \rightarrow (g_2, g_1)$.
9. Yes, by Theorem 7.2.
11. Each of Z_8, Z_4, $Z_{8000000}$, and $Z_{4000000}$ has a unique subgroup of order 4. If $|(a, b)| = 4$, then a and b both belong to the unique subgroup of order 4. So the number of choices for a and b (actually 12 in all) is the same in either group.
13. Try $(a, b) \rightarrow a + bi$.
17. 3
19. Map $\begin{bmatrix} a & b \\ c & d \end{bmatrix}$ to (a, b, c, d). Let $\mathbf{R}^k$ denote $\underbrace{\mathbf{R} \oplus \mathbf{R} \oplus \cdots \oplus \mathbf{R}}_{k \text{ factors}}$. Then the group of $n \times n$ matrices under addition is isomorphic to $\mathbf{R}^{n^2}$.
21. $(g, g)(h, h)^{-1} = (gh^{-1}, gh^{-1})$
    When $G = \mathbf{R}$, $G \oplus G$ is the plane and H is the line $y = x$.
23. $\langle (3, 0) \rangle$, $\langle (3, 1) \rangle$, $\langle (3, 2) \rangle$, $\langle (0, 1) \rangle$
25. Try $g \rightarrow (g, e)$.
27. Try $3^m 6^m \rightarrow (m, n)$.
29. D_{24} has elements of order 24 while $D_3 \oplus D_4$ does not.
31. Use the observation that $(h, k)^n = (h^n, k^n)$.
33. 12.

CHAPTER 8

With every mistake we must surely be learning.   George Harrison, *The Beatles*, (The White Album)

1. Certainly, every nonzero real number is of the form $\pm r$ where r is a positive real. Real numbers commute and $\mathbf{R}^+ \cap \{1, -1\} = \{1\}$.
3. See Theorem 8.2.
5. Yes, it is the direct product. No, it does not.
7. $U(165) \approx U(11) \oplus U(15) \approx U(5) \oplus U(33) \approx U(3) \oplus U(55)$
9. First observe that every proper subgroup of D_4 is Abelian. Now use Theorem 8.1 and exercise 2 of Chapter 7.
11. Hint: $U(27) \approx Z_{18}$
13. Mimic the analysis for elements of order 12 in $U(720)$ at the end of this chapter.
15. If $G = HK$, no. If $G = H \times K$, then $|hk| = \text{lcm}(|h|, |k|)$.
17. 60
19. They are both isomorphic to $Z_{10} \oplus Z_4$.
21. That $U(n)^2$ is a subgroup follows from exercise 13 of supplementary exercises for Chapters 1–4. $1^2 = (n - 1)^2$ shows that it is a proper subgroup.
23. For the first question note that $\langle 3 \rangle \cap \langle 6 \rangle = \{1\}$ and $\langle 10 \rangle \cap \langle 3 \rangle \langle 6 \rangle = \{1\}$. For the second question observe that $12 = 3^{-1} 6^2$.
25. Consider $U(49)$.
27. Consider $U(65)$.
29. $HK = \langle 2 \rangle$; $HK = \langle \text{gcd}(a, b) \rangle$.
31. $Z_2 \oplus Z_4$.

33. Observe that
$$\begin{bmatrix} 1 & a & b \\ 0 & 1 & c \\ 0 & 0 & 1 \end{bmatrix} = \begin{bmatrix} 1 & a & 0 \\ 0 & 1 & 0 \\ 0 & 0 & 1 \end{bmatrix} \begin{bmatrix} 1 & 0 & b-ac \\ 0 & 1 & 0 \\ 0 & 0 & 1 \end{bmatrix} \begin{bmatrix} 1 & 0 & 0 \\ 0 & 1 & c \\ 0 & 0 & 1 \end{bmatrix}.$$

No, elements of H do not commute with elements of L.

SUPPLEMENTARY EXERCISES FOR CHAPTERS 5–8

For those who keep trying failure is temporary.   Frank Tyger

1. Consider the finite and infinite cases separately. In the finite case, note that $|H| = |H\phi|$. Now use Theorem 4.3. For the infinite case, use exercise 2 of Chapter 6.
3. Observe $(x^{-1}y^{-1}xy)\phi = (x\phi)^{-1}(y\phi)^{-1}x\phi y\phi$ so ϕ carries commutators to commutators.
5. All nonidentity elements of G and H have order 3. $G \neq H$.
7. Certainly the set HK has $|H|\,|K|$ symbols. However, not all symbols need represent distinct group elements. That is, we may have $hk = h'k'$ although $h \neq h'$ and $k \neq k'$. We must determine the extent to which this happens. For every t in $H \cap K$, $hk = (ht)(t^{-1}k)$ so each group element in HK is represented by at least $|H \cap K|$ products in HK. But $hk = h'k'$ implies $t = h^{-1}h' = k(k')^{-1} \in H \cap K$ so that $h' = ht$ and $k' = t^{-1}k$. Thus each element in HK is represented by exactly $|H \cap K|$ products. So, $|HK| = |H||K|/|H \cap K|$.
9. $U(n)$ where $n = 4, 8, 3, 6, 12, 24$
11. Hint:
$$3 + 2i = \sqrt{13}\left(\frac{3}{\sqrt{13}} + \frac{2}{\sqrt{13}}i\right)$$

13. Suppose $\phi:Q \to \mathbf{R}$ is an isomorphism. Let $1\phi = x_0$. Show that $(a/b)\phi = (a/b)x_0$ for all integers a, b ($b \neq 0$). If x_0 is rational, the image of ϕ contains only rationals. If x_0 is irrational, the image of ϕ contains only irrationals and 0. In either case, the image of ϕ is not $\mathbf{R}$.
15. In Q, equation $2x = a$ has a solution for all a. The corresponding equation $x^2 = b$ in Q^+ does not have a solution for all b.
17. We must show that $(h_1k_1)(h_2k_2) = (h_2k_2)(h_1k_1)$. But since H and K are Abelian, h's commute with h's and k's commute with k's. Also, since G is an internal direct product, h's and k's commute.
19. Look at Example 3 of Chapter 8.
21. See exercise 1 of Chapter 6.
23. If $|x| = pm$ where p is prime, then $|x^m| = p$.
25. Count elements of order 2.
27. Count elements of order 2.
29. If $|\text{Inn}(G)| = 1$, the only inner automorphism of G is the identity. Thus for any x and a in G, $x = x\phi_a = a^{-1}xa$ so that $ax = xa$. Conversely, if G is Abelian ϕ_a is the identity.

CHAPTER 9

Failure is the path of least persistence.   Author unknown.

1. $H = \{\alpha_1, \alpha_2, \alpha_3, \alpha_4\}$, $\alpha_5H = \{\alpha_5, \alpha_6, \alpha_7, \alpha_8\}$, $\alpha_9H = \{\alpha_9, \alpha_{10}, \alpha_{11}, \alpha_{12}\}$
3. $H, 1 + H, 2 + H$
5. a. yes b. yes c. no
7. $8/2 = 4$ so there are four cosets. Let $H = \{1, 11\}$. The cosets are $H, 7H, 13H, 19H$.

9. Since $|a^4| = 15$ there are two cosets: $\langle a^4 \rangle$ and $a\langle a^4 \rangle$.

11. The correspondence $ah \to bh$ is one-to-one and onto.

13. Say the points in H lie on the line $y = mx$. Then $(a, b) + H = \{(a + x, b + mx) \mid x \in R\}$. This set is the line $y - b = m(x - a)$.

15. The subgroup is the solution set of the system

$$3x + 2y - 3z = 0$$
$$5x + y + 4z = 0.$$

17. 1,2,3,4,5,6,10,12,15,20,30,60.

19. Use Lagrange's Theorem and one of its corollaries.

21. By exercise 20 we have $5^6 = 1 \bmod 7$. So $5^{15} = 5^6 \cdot 5^6 \cdot 5^2 \cdot 5 = 1 \cdot 1 \cdot 4 \cdot 5 = 6 \bmod 7$.

23. By Corollary 1 of Theorem 9.1 we know that any nonidentity element of a non-Abelian group G of order 10 has order 2 or 5. By exercise 38 of Chapter 2, there must be an element of order 5. Call it a. Then for any $b \notin \langle a \rangle$ we have $G = \langle a \rangle \cup b\langle a \rangle$. Cancellation shows that $b^2 \notin b\langle a \rangle$. Now show $b^2 = e$. A non-Abelian group of order $2p$ has p elements of order 2. An Abelian group of order $2p$ has one element of order 2.

25. Use the coset representatives $(0, 1)$, $(1, 1,)$, $(2, 1)$, $(3, 1)$.

27. Let H be the subgroup of order p and K be the subgroup of order q. Then $H \cup K$ has $p + q - 1 < pq$ elements. Let a be any element in G that is not in $H \cup K$. By Lagrange's Theorem, $|a| = p$, q, or pq. But $|a| \neq p$ for if so, then $\langle a \rangle = H$. Similarly, $|a| \neq q$.

29. 1, 3, 11, 33. If $|x| = 33$, then $|x^{11}| = 3$. Elements of order 11 occur in multiples of 10.

31. Certainly, $a \in \mathrm{orb}_G(a)$. Now suppose $c \in \mathrm{orb}_G(a) \cap \mathrm{orb}_G(b)$. Then $c = a\alpha$ and $c = b\beta$ for some α and β and therefore $a\alpha\beta^{-1} = b$. So, if $x \in \mathrm{orb}_G(b)$, then $x = b\gamma = a\alpha\beta^{-1}\gamma$ for some γ. This proves $\mathrm{orb}_G(b) \subseteq \mathrm{orb}_G(a)$. By symmetry, $\mathrm{orb}_G(a) \subseteq \mathrm{orb}_G(b)$.

33. a. $\mathrm{stab}_G(1) = \{(1), (24)(56)\}$; $\mathrm{orb}_G(1) = \{1,2,3,4\}$
    b. $\mathrm{stab}_G(3) = \{(1), (24)(56)$; $\mathrm{orb}_G(3) = \{3,4,1,2\}$
    c. $\mathrm{stab}_G(5) = \{(1), (12)(34), (13)(24), (14)(23)\}$; $\mathrm{orb}_G(5) = \{5,6\}$

35. Let F_1, F_2, F_3, F_4, F_5 be the five reflections in D_5. The subgroup lattice of D_5 is

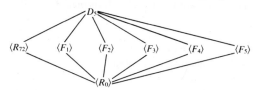

37. 2520.

39. If both H and K were Abelian, then G would also be. So, one of H or K must be non-Abelian. Since the classification of groups of order $\leqslant 7$ shows that the smallest non-Abelian group has order 6, G must have order at least 12.

41. It is the set of all permutations that carry face 2 to face 1.

43. $aH = bH$ if and only if $\det(a) = \pm\det(b)$.

45. 50

47. The order of the symmetry group would have to be $6 \cdot 20 = 120$.

CHAPTER 10

There's a mighty big difference between good, sound reasons and reasons that sound good.
Burton Hillis

1. If $z \in Z(G)$, $xz = zx$ so $xZ(G) = Z(G)x$.

3. $\alpha^{-1}A_n\alpha$ is even for all α in S_n.

5. Recall that if A and B are matrices, then $\det(A^{-1}BA) = (\det A)^{-1}\det B \det A$.

7. Let $x \in G$. If $x \in H$, then $xH = H = Hx$. If $x \notin H$, then xH is the set of elements in G, not in H. But Hx is also the elements in G, not in H.

9. $G/H \approx Z_4$
   $G/K \approx Z_2 \oplus Z_2$

11. 6

13. 2

15. $H = \{0 + \langle 20\rangle, \ 4 + \langle 20\rangle, \ 8 + \langle 20\rangle, \ 12 + \langle 20\rangle, \ 16 + \langle 20\rangle\}$. $G/H = \{0 + \langle 20\rangle + H, \ 1 + \langle 20\rangle + H, \ 2 + \langle 20\rangle + H, \ 3 + \langle 20\rangle + H\}$.

17. $40/10 = 4$

19. Z_4

21. ∞; no, $(6, 3) + \langle(4, 2)\rangle$ has order 2

23. Z_8

25. yes; no

27. Since $\det(AB^{-1}) = \det A(\det B)^{-1}$, H is a subgroup. Since $\det(C^{-1}AC) = (\det C)^{-1}\det A \det C = \det A$, H is normal.

29. Take $G = Z_6$, $H = \{0, 3\}$, $a = 1$, $b = 4$.

31. $aHbH = abH = baH = bHaH$.

33. Because $x^{-1}Hx$ and H are both subgroups of the same order.

35. Use the "G/Z Theorem."

37. If $H \triangleleft G$, then $(xN)^{-1}hNxN = x^{-1}hxN \in H/N$ so $H/N \triangleleft G/N$. Now assume $H/N \triangleleft G/N$. Then $(xN)^{-1}hNxN = x^{-1}hxN \in H/N$. Thus $x^{-1}hxN = h'N$ for some $h' \in H$. So, $x^{-1}hx = h'n \in H$.

39. Use exercise 7 and observe $VK \neq KV$.

41. $x^{-1}(H \cap N)x = x^{-1}Hx \cap x^{-1}Nx = H \cap N$

43. Observe that $n^{-1}m^{-1}nm = (n^{-1}m^{-1}n)m \in M$ and $n^{-1}m^{-1}nm = n^{-1}(m^{-1}nm) \in N$.

45. $\gcd(|x|, |G/H|) = 1$ implies $\gcd(|xH|, |G/H|) = 1$. But $|xH|$ divides $|G/H|$. Thus $|xH| = 1$ and therefore $xH = H$.

47. Note that G/H is a group and use Corollary 3 of Theorem 9.1.

49. Use exercise 43.

51. Use Theorems 10.4 and 10.3.

53. Say $|gH| = n$. Then $|g| = nt$ (by exercise 34) and $|g'| = n$. For the second part consider $Z/\langle k\rangle$.

55. It is not a group table. No, because $\mathcal{H}$ is not normal in D_4.

57. no

59. By exercise 58, A_5 would have an element of the form $(ab)(cd)$ that commutes with every element of A_5. Try (abc).

61. Use Theorem 10.4.

CHAPTER 11

Sixty minutes of thinking of any kind is bound to lead to confusion and unhappiness.   James Thurber

1. Observe that $(xy)^r = x^r y^r$.

3. Observe that $\det(AB) = (\det A)(\det B)$.

5. Observe that $(f + g)' = f' + g'$.

7. $[(g, h)(g', h')]\phi = (gg', hh')\phi = gg' = (g, h)\phi(g', h')\phi$

9. $[2/3 + 2/3] \neq [2/3] + [2/3]$

11. $(a, b) \to b$ is a homomorphism from $A \oplus B$ onto B with kernel $A \oplus \{e\}$.

13. 3, 13, 23.
15. Use the First Isomorphism Theorem.
17. Use the First Isomorphism Theorem.
19. 4 onto; 10 to.
21. For each k with $0 \leqslant k \leqslant n - 1$ the mapping $1 \to k$ determines a homomorphism.
23. Use properties 10 and 12 of Theorem 11.1.
25. $7\phi^{-1} = 7$ Ker $\phi = \{7, 17\}$
27. 11 Ker ϕ
29. $((a, b) + (c, d))\phi = (a + c, b + d)\phi = (a + c) - (b + d) = a - b + c - d = (a, b)\phi + (c, d)\phi$. Ker $\phi = \{(a, a) \mid a \in Z\}$. $3\phi^{-1} = \{(a + 3, a) \mid a \in Z\}$.
31. $(xy)\phi = (xy)^4 = x^4y^4 = x\phi y\phi$. Ker $\phi = \{\pm 1, \pm i\}$.
33. Show the mapping from K to KN/N given by $k \to kN$ is an onto homomorphism with kernel $K \cap N$.
35. For each divisor d of k there is a unique subgroup of Z_k of order d and this subgroup is generated by $\phi(d)$ elements. A homomorphism from Z_n to a subgroup of Z_k must carry 1 to a generator of the subgroup. Furthermore, the order of the image of 1 must divide n so we need only consider those divisors d of k that also divide n.
37. D_4, $\{e\}$, Z_2, $Z_2 \oplus Z_2$
39. It is divisible by 10.
41. It is infinite.
43. Let ϕ be the natural homomorphism from G onto G/N. Let $\overline{H}$ be a subgroup of G/N and let $H = \overline{H}\phi^{-1}$. Then H is a subgroup of G and $H/N = H\phi = (\overline{H}\phi^{-1})\phi = \overline{H}$.
45. The mapping $g \to \phi_g$ is a homomorphism with kernel $Z(G)$.
47. $(f + g)(3) = f(3) + g(3)$. The kernel is the set of elements in $Z[x]$ whose graph passes through the point $(3, 0)$.
49. Use exercise 41 of Chapter 10 and exercise 33 to prove the first assertion. To verify $G/H \cap K$ is not cyclic, observe that it has two subgroups of order 2.

CHAPTER 12

Think before you think! Stanislaw J. Lec, *Unkempt Thoughts*

1. $n = 4$
   Z_4, $Z_2 \oplus Z_2$
3. $n = 36$
   $Z_9 \oplus Z_4$, $Z_3 \oplus Z_3 \oplus Z_4$, $Z_9 \oplus Z_2 \oplus Z_2$, $Z_3 \oplus Z_3 \oplus Z_2 \oplus Z_2$
5. The only Abelian groups of order 45 are Z_{45} and $Z_3 \oplus Z_3 \oplus Z_5$. In the first group $|3| = 15$, in the second one $|(1, 1, 1)| = 15$. $Z_3 \oplus Z_3 \oplus Z_5$ does not have an element of order 9.
7. $Z_9 \oplus Z_3 \oplus Z_4$; $Z_9 \oplus Z_3 \oplus Z_2 \oplus Z_2$
9. $Z_4 \oplus Z_2 \oplus Z_3 \oplus Z_5$
11. $360 = 8 \cdot 9 \cdot 5$
    $Z_8 \oplus Z_9 \oplus Z_5$
    $Z_4 \oplus Z_2 \oplus Z_9 \oplus Z_5$
    $Z_2 \oplus Z_2 \oplus Z_2 \oplus Z_9 \oplus Z_5$
    $Z_8 \oplus Z_3 \oplus Z_3 \oplus Z_5$
    $Z_4 \oplus Z_2 \oplus Z_3 \oplus Z_3 \oplus Z_5$
    $Z_2 \oplus Z_2 \oplus Z_2 \oplus Z_3 \oplus Z_3 \oplus Z_5$
13. $Z_2 \oplus Z_2$

15. a. 1 b. 1 c. 1 d. 1 e. 1 f. There is a unique Abelian group of order n if and only if n is not divisible by the square of any prime.

17. $Z_2 \oplus Z_2$

19. $Z_3 \oplus Z_3$

21. n is square-free (no prime factor of n occurs more than once).

23. Among the first 11 elements in the table there are 9 elements of order 4. None of the other isomorphism classes has this many.

25. $Z_4 \oplus Z_2 \oplus Z_2$; One internal direct product is $\langle 7 \rangle \times \langle 101 \rangle \times \langle 199 \rangle$.

27. 3; 6; 12.

29. $Z_4 \oplus Z_4$.

31. Use the Fundamental Theorem, Theorem 7.1, and Theorem 4.3.

33. If $|C| = p^n$, use the Fundamental Theorem and Theorem 8.1. If every element has order a power of p use the corollary to the Fundamental Theorem.

35. By the Fundamental Theorem of Finite Abelian Groups it suffices to show that every group of the form $Z_{p_1^{n_1}} \oplus Z_{p_2^{n_2}} \cdots \oplus Z_{p_k^{n_k}}$ is a subgroup of a U-group. Consider first a group of the form $Z_{p_1^{n_1}} \oplus Z_{p_2^{n_2}}$ (p_1 and p_2 need not be distinct). By Dirichlet's Theorem, for some s and t there are distinct primes q and r such that $q = tp_1^{n_1} + 1$ and $r = sp_2^{n_2} + 1$. Then $U(qr) = U(q) \oplus U(r) \approx Z_{tp_1^{n_1}} \oplus Z_{sp_2^{n_2}}$ and this latter group contains a subgroup isomorphic to $Z_{p_1^{n_1}} \oplus Z_{p_2^{n_2}}$. The general case follows in the same way.

SUPPLEMENTARY EXERCISES FOR CHAPTERS 9–12

It took me so long to find out.   John Lennon and Paul McCartney, *Day Tripper*, single

1. Say $aH = Hb$. Then $a = hb$ for some h in H. Then $Ha = Hhb = Hb = aH$.

3. Observe that $x^{-1}N = yN$ so that $N = xyN$.

5. Say $aH = bK$. Then $H = a^{-1}bK$ and $K = b^{-1}aH$. Thus $a^{-1}b \in H$ and $(a^{-1}b)^{-1} = b^{-1}a \in H$. So, $K = b^{-1}aH = H$. For the example try A_4.

7. $aH = bH$ implies $a^{-1}b \in H$. So $(a^{-1}b)^{-1} = b^{-1}a \in H$. Thus, $Hb^{-1}a = H$ or $Hb^{-1} = Ha^{-1}$. These steps are reversible.

9. Suppose diag(G) is normal. Then $(e, a)^{-1}(b, b)(e, a) = (b, a^{-1}ba) \in$ diag (G). Thus $b = a^{-1}ba$. If G is Abelian, $(g, h)^{-1}(b, b)(g, h) = (g^{-1}bg, h^{-1}bh) = (b, b)$. When $G = R$, diag(G) is the line $y = x$. The index of diag(G) is $|G|$.

11. Let $\alpha \in \text{Aut}(G)$ and $\phi_a \in \text{Inn}(G)$. Then $x\alpha^{-1}\phi_a\alpha = [a^{-1}(xa^{-1})a]\alpha = a^{-1}\alpha xa\alpha = (a\alpha)^{-1}x\,(a\alpha) = x\phi_{a\alpha}$

13. $R^{\#}$ (See Example 2 of Chapter 11.)

15. a.

$$Z(H) = \left\{ \begin{bmatrix} 1 & 0 & b \\ 0 & 1 & 0 \\ 0 & 0 & 1 \end{bmatrix} \,\middle|\, b \in Q \right\}$$

The mapping

$$\begin{bmatrix} 1 & 0 & b \\ 0 & 1 & 0 \\ 0 & 0 & 1 \end{bmatrix} \to b$$

is an isomorphism.

b. The mapping

$$\begin{bmatrix} 1 & a & b \\ 0 & 1 & c \\ 0 & 0 & 1 \end{bmatrix} \rightarrow (a, c)$$

is a homomorphism with $Z(H)$ as the kernel.
c. The proofs are valid with **R** and Z_p.

17. $b(a/b + Z) = a + Z = Z$
19. Use exercise 6 of the supplemental exercises for Chapters 1–4. For the example, try $G = D_4$.
21. $Z_2 \oplus Z_2$ has 3 subgroups of order 2.
    $Z_3 \oplus Z_3$ has 4 subgroups of order 3.
    $Z_p \oplus Z_p$ has $p + 1$ subgroups of order p.
    It suffices to count elements of order p and divide by $p - 1$. Thus there are $(p^2 - 1)/(p - 1)$ $= p + 1$ subgroups of order p.
23. Use exercise 43 of Chapter 10.
25. The mapping $g \rightarrow g^n$ is a homomorphism from G onto G^n with kernel G_n.
27. Use Theorem 9.2 and exercise 7 of Chapter 10.
29. The number is m in all cases.
31. For each x with $|x| > 2$, both x and x^{-1} are factors in the product and therefore cancel out. This reduces the problem to groups of the form $Z_2 \oplus Z_2 \oplus \cdots \oplus Z_2$ of order at least 4. Now appeal to exercise 30. If the subgroup of order 2^n is cyclic, the product is the unique element of order 2.

CHAPTER 13

Though the proportion of those who think be extremely small, yet every individual flatters himself that he is one of the number. C. C. Colton, Lacon

1. For any $n > 1$, the ring $M_2(Z_n)$ of 2×2 matrices with entries from Z_n is a finite noncommutative ring. The set $M_2(2Z)$ of 2×2 matrices with even integer entries is an infinite noncommutative ring that does not have a unity.
3. The only property that is not immediate is closure under multiplication:

$$(a + b \sqrt{2})(c + d \sqrt{2}) = (ac + 2bd) + (ad + bc) \sqrt{2}$$

5. The proof given for a group applies to a ring as well.
7. In Z_p nonzero elements have multiplicative inverses. Use them.
9. Note that $(na)(ma) = (nm)a^2 = (mn)a^2 = (ma)(na)$.
11. $(x_1, \ldots, x_n)(a_1, \ldots, a_n) = (x_1, \ldots, x_n)$ for all x_i in R_i if and only if $x_i a_i = x_i$ for all x_i in R_i and $i = 1, \ldots, n$.
13. $U(Z_n) = U(n)$, the set of integers less than or equal to n and relatively prime to n.
15. $f(x) = 1$ and $g(x) = -1$.
17. If a is a unit, then $b = a(a^{-1}b)$.
19. Consider $a^{-1} - a^{-2}b$.
21. Try the ring $M_2(Z)$.
23. The equation has nontrivial solutions if and only if $p = 1 \bmod 4$.

CHAPTER 14

Think for yourself.   Title of a song by George Harrison, *Rubber Soul*

1. The ring need not process a unity.
3. Part 3: $0 = 0(-b) = (a + (-a))(-b) = a(-b) + (-a)(-b) = -(ab) + (-a)(-b)$.
    So, $ab = (-a)(-b)$.
    Part 4: $a(b - c) = a(b + (-c)) = ab + a(-c) = ab + (-(ac)) = ab - ac$.
    Part 5: Use Part 2.
    Part 6: Use Part 3.
5. Hint: Z is a cyclic group under addition, and every subgroup of a cyclic group is cyclic.
9. If a and b belong to the intersection, then they belong to each member of the intersection. Thus $a - b$ and ab belong to each member of the intersection. So, $a - b$ and ab belong to the intersection.
11. Let a, b belong to the center. Then $(a - b)x = ax - bx = xa - xb = x(a - b)$. Also, $(ab)x = a(bx) = a(xb) = (ax)b = (xa)b = x(ab)$.
13. Every subgroup of Z_n is closed under multiplication.
15. $ara - asa = a(r - s)a$. $(ara)(asa) = ara^2sa = arsa$. $a1a = a^2 = 1$ so $1 \in S$.
17. The subring test is satisfied.
19. Look at $(1, 0, 1)$ and $(0, 1, 1)$.
21. $2Z \cup 3Z$ contains 2 and 3, but not $2 + 3$.
23. $\{m/2^n \mid m \in Z, n \in Z^+\}$
25. $(a + b)(a - b) = a^2 + ba - ab - b^2 = a^2 - b^2$ if and only if $ba - ab = 0$.
27. Say $n = 2m$. Then $-a = (-a)^n = (-a)^{2m} = [(-a^2]^m = (a^2)^m = a^n = a$.

CHAPTER 15

Work now or wince later.   B. C. Forbes, *Epigrams*

3. Let $ab = 0$ and $a \neq 0$. Then $ab = a \cdot 0$ so $b = 0$.
5. Let $k \in Z_n$. If $\gcd(k, n) = 1$, then k is a unit. If $\gcd(k, n) = d > 1$, write $k = sd$. Then $k(n/d) = sd(n/d) = sn = 0 \bmod n$.
7. Let $s \in R$, $s \neq 0$. Consider the set $S = \{sr \mid r \in R\}$. If $S = R$, then $sr = 1$ (the unity) for some r. If $S \neq R$, then there are distinct r_1 and r_2 such that $sr_1 = sr_2$. In this case, $s(r_1 - r_2) = 0$. To see what happens when the "finite" condition is dropped consider Z.
9. $(a_1 + b_1\sqrt{d}) - (a_2 + b_2\sqrt{d}) = (a_1 + a_2) - (b_1 + b_2)\sqrt{d}$; $(a_1 + b_1\sqrt{d})(a_2 + b_2\sqrt{d}) = (a_1a_2 + b_1b_2d) + (a_1b_2 + a_2b_1)\sqrt{d}$. Thus the set is a ring. Since $Z[\sqrt{d}]$ is a subring of the ring of real numbers, it has no zero-divisors.
11. The even integers.
15. Suppose $a \neq 0$ while $a^n = 0$ (where we take n to be as small as possible). Then $0 = a \cdot 0 = a^n = a \cdot a^{n-1}$, so by cancellation, $a^{n-1} = 0$.
17. If $a^2 = a$ and $b^2 = b$, then $(ab)^2 = a^2b^2 = ab$.
19. $Z_4 \oplus Z_4$.
21. $a^2 = a$ implies $a(a - 1) = 0$. So if a is a unit, $a - 1 = 0$ and $a = 1$.
23. See Theorems 3.1 and 14.3.
25. Note that $ab = 1$ implies $aba = a$. Thus $0 = aba - a = a(ba - 1)$. So, $ba - 1 = 0$.
27. A subdomain of an integral domain D is a subset of D that is an integral domain under the operations of D. To show that P is a subdomain show that it is a subring and contains 1. Every

subdomain contains 1 and is closed under addition and subtraction so every subdomain contains P. $|P| = $ char D.

29. No, $(1, 0) \cdot (0, 1) = (0, 0)$.
31. b. Use the fact that there exist integers s and t such that $1 = sn + tm$, but remember that you cannot use negative exponents in a ring.
35. Z_8
37. Let $S = \{a_1, a_2, \ldots, a_n\}$ be the *nonzero* elements of the ring. First show that $S = \{a_1a_1, a_1a_2, \ldots, a_1a_n\}$. Thus, $a_1 = a_1a_i$ for some i. Then a_i is the unity, for if a_k is any element of S, we have $a_1a_k = a_1a_ia_k$ so that $a_1(a_k - a_ia_k) = 0$.
39. Say $|x| = n$ and $|y| = m$ with $n < m$. Consider $n(xy) = (nx)y = x(ny)$.
41. Use exercise 39.
43. Try Z_2.
45. $n \begin{bmatrix} a & b \\ c & d \end{bmatrix} = \begin{bmatrix} 0 & 0 \\ 0 & 0 \end{bmatrix}$ for all members of $M_2(R)$ if and only if $na = 0$ for all a in R.
47. Use exercise 46.
49. a. 2 b. 2, 3 c. 2, 3, 6, 11 d. 2, 3, 9, 10
51. 2
53. See Example 9.
55. Use exercise 23 and part a of exercise 42.
57. Choose $a \neq 0$ and $a \neq 1$ and consider the image of $1 + a$.
59. $\phi(x) = \phi(x \cdot 1) = \phi(x) \cdot \phi(1)$ so $\phi(1) = 1$. Also, $1 = \phi(1) = \phi(xx^{-1}) = \phi(x)\phi(x^{-1})$.

CHAPTER 16

Not one student in a thousand breaks down from overwork.   William Allan Neilson

1. $\sqrt{2} \cdot 1$ is not rational.
3. Let $a + bi$, $c + di \in S$. Then $(a + bi) - (c + di) = a - c + (b - d)i$ and $b - d$ is even. Also $(a + bi)(c + di) = ac - bd + (ad + cb)i$ and $ad + cb$ is even. Finally, $(1 + 2i)(1 + i) = -1 + 3i \notin S$.
5. $ar_1 - ar_2 = a(r_1 - r_2)$; $(ar_1)r = a(r_1r)$
   $4R = \langle 8 \rangle$
7. Mimic exercise 9 of Chapter 14.
9. a. $a = 1$ b. $a = 3$ c. $a = \gcd(m, n)$
11. a. $a = 12$
    b. $a = 48$. To see this, note that every element of $\langle 6 \rangle \langle 8 \rangle$ has the form $6t_18k_1 + 6t_28k_2 + \cdots + 6t_n8k_n = 48s \in \langle 48 \rangle$. So, $\langle 6 \rangle \langle 8 \rangle \subseteq \langle 48 \rangle$. Also, since $48 \in \langle 6 \rangle \langle 8 \rangle$, we have $\langle 48 \rangle \subseteq \langle 6 \rangle \langle 8 \rangle$.
13. By exercise 12, we have $AB \subseteq A \cap B$. So, let $x \in A \cap B$. To show that $x \in AB$, start by writing $1 = a + b$ where $a \in A$, $b \in B$.
15. Let $u \in I$ be a unit and let $r \in R$. Then $r = r(u^{-1}u) = (ru^{-1})u \in I$.
17. Use the observation that every member of R can be written in the form
$$\begin{bmatrix} 2q_1 + r_1 & 2q_2 + r_2 \\ 2q_3 + r_3 & 2q_4 + r_4 \end{bmatrix}.$$
19. $(br_1 + a_1) - (br_2 + a_2) = b(r_1 - r_2) + (a_1 - a_2) \in B$; $r'(br + a) + b(r'r) + r'a \in B$
21. $(b + A)(c + A) = bc + A = cb + A = (c + A)(b + A)$. If 1 is the unity of R, then $1 + A$ is the unity of R/A.

23. Use Theorems 16.4 and 16.3.
25. Suppose $f(x) + A \neq A$. Then $f(x) + A = f(0) + A$ and $f(0) \neq 0$. Thus

$$(f(x) + A)^{-1} = \frac{1}{f(0)} + A$$

27. Since $(3 + i)(3 - i) = 10$, $10 + \langle 3 + i \rangle = 0 + \langle 3 + i \rangle$. Also, $i + \langle 3 + i \rangle = -3 + \langle 3 + i \rangle = 7 + \langle 3 + i \rangle$. So, $Z[i]/\langle 3 + i \rangle = \{k + \langle 3 + i \rangle | k = 0, 1, \ldots, 9\}$.
29. Use Theorems 16.3 and 16.4.
31. Let $f, g \in I$ and $h \in Z[x]$. Then $(f - g)(0) = f(0) - g(0)$ and $(f \cdot h)(0) = f(0)h(0)$ so that I is an ideal. Also, if $h(0)k(0)$ is even then one of $h(0)$ or $k(0)$ is even. Yes, it is maximal.
33. $3x + 1 + I$
35. Let $a, b \in I_p$. Say $|a| = p^n$ and $|b| = p^m$. Then $p^{n+m}(a - b) = 0$ so $|a - b|$ divides p^{n+m}. Also, $p^n(ra) = r(p^na) = 0$ so $|ra|$ divides p^n.
37. Say $b, c \in \text{Ann}(A)$. Then $(b - c)a = ba - ca = 0 - 0 = 0$. Also, $(rb)a = (ba)r = 0 \cdot r = 0$.
39. a. $\langle 3 \rangle$ b. $\langle 3 \rangle$ c. $\langle 3 \rangle$
41. Suppose $(x + \sqrt{\langle 0 \rangle})^n = 0 + \sqrt{\langle 0 \rangle}$. We must show that $x \in \sqrt{\langle 0 \rangle}$. We know that $x^n + \sqrt{\langle 0 \rangle} = 0 + \sqrt{\langle 0 \rangle}$ so that $x^n \in \sqrt{\langle 0 \rangle}$. Then, for some m, $(x^n)^m = 0$ and $x \in \sqrt{\langle 0 \rangle}$.
43. The set $Z_2[x]/\langle x^2 + x + 1 \rangle$ has only four elements and each of the nonzero ones has a multiplicative inverse. For example,

$$(x + \langle x^2 + x + 1 \rangle)(x + 1 + \langle x^2 + x + 1 \rangle) = 1 + \langle x^2 + x + 1 \rangle$$

45. $x + 2 + \langle x^2 + x + 1 \rangle$ is not zero but its square is.
47. If f and $g \in A$, then $(f - g)(0) = f(0) - g(0)$ is even and $(f \cdot g)(0) = f(0) \cdot g(0)$ is even. $f(x) = \sqrt{2} \in R$ and $g(x) = 2 \in A$ but $f(x)g(x) \notin A$.
49. Hint: Any ideal of R/I has the form A/I where A is an ideal of R.

SUPPLEMENTARY EXERCISES FOR CHAPTERS 13–16

If at first you don't succeed, try, try, again. Then quit. There's no use being a damn fool about it.
W. C. Fields

1. In Z_{10} they are 0, 1, 5, 6.
3. We must show that $a^n = 0$ implies $a = 0$. First show for the case when n is a power of 2. If n is not a power of 2, say 13 for example, note that $a^{13} = 0$ implies $a^{16} = 0$.
5. Suppose $A \not\subseteq C$ and $B \not\subseteq C$. Pick $a \in A$ and $b \in B$ so that $a, b \notin C$. But $ab \in C$ and C is prime.
7. Suppose I is an ideal and $a \in I$ with $a \neq 0$. Then $1 = a^{-1}a \in I$ so $I = F$.
9. Observe that $Z + Z/A = \{(b, 0) | b = 0, 1, \ldots, p - 1\} \approx Z_p$, a field.
11. Suppose $a_1, a_2 \in A$ but $a_1 \notin B$ and $a_2 \notin C$. Use $a_1 + a_2$ to derive a contradiction.
13. Clearly $\langle a \rangle$ contains the right-hand side. Now show that the right-hand side contains a and is an ideal.
15. Since A is an ideal, $ab \in A$. Since B is an ideal, $ab \in B$. So $ab \in A \cap B = \{0\}$.
17. 6
19. Use exercise 4.
21. Consider $x^2 + 1 + \langle x^4 + x^2 \rangle$.
23. Consider Z_8.

25. Say char $R = p$ (remember p must be prime). Then char $R/A =$ the additive order of $1 + A$. But $|1 + A|$ divides $|1| = p$.
27. Use Theorems 15.2, 16.3, and 16.4.
29. $\begin{bmatrix} 1 & 0 \\ 0 & 0 \end{bmatrix} \begin{bmatrix} 1 & 1 \\ 1 & 1 \end{bmatrix} = \begin{bmatrix} 1 & 1 \\ 0 & 0 \end{bmatrix}$ which is not in the set.
31. $Z[i]/A$ has two elements. (From this it follows that A is maximal. See Theorem 16.4.)

CHAPTER 17

A sign in a Pentagon office reads: If at first you don't succeed, forget it.

1. In Z_5, $1 = 6$. But in Z_{10}, $5 \neq 30$.
3. a. No. Suppose $2 \rightarrow a$. Now consider $2 + 2$ and $2 \cdot 2$.
   b. no
7. Multiplication is not preserved.
9. yes
11. The set of all polynomials passing through the point $(1, 0)$.
13. The group A/B is cyclic of order 4. The ring A/B has no unity.
15. The zero map and the identity map.
17. If $2 + 8k = x^3$ for some k, then $2 = x^3$ mod 8 has a solution. But direct substitution of 0, 1, 2, 3, 4, 5, 6, 7 shows that there is no solution.
19. Say $m = a_k a_{k-1} \ldots a_1 a_0$ and $n = b_k b_{k-1} \ldots b_1 b_0$. Then $m - n = (a_k - b_k)10^k + (a_{k-1} - b_{k-1})10^{k-1} + \cdots + (a_1 - b_1)10 + (a_0 - b_0)$. Now use the test for divisibility by 9.
21. Use the appropriate divisibility tests.
23. Mimic Example 8.
25. Observe that $2 \cdot 10^{75} + 2 = 1$ mod 3 and $10^{100} + 1 = 2 = -1$ mod 3.
27. This follows directly from Theorem 15.3 and Theorem 11.1, part 7.
29. No. The kernel must be an ideal.
31. a. Suppose $ab \in \phi^{-1}(A)$. Then $a\phi b\phi \in A$ so that $a \in \phi^{-1}(A)$ or $b \in \phi^{-1}(A)$.
    b. Consider the natural homomorphism from R to S/A. Then use Theorems 17.3 and 16.4.
33. c. Use $(r, s) \rightarrow (s, r)$.
35. Observe $x^4 = 1$ has two solutions in **R** but four in C.
37. Use exercises 30 and 36.
39. To check that multiplication is operation preserving, observe that $xy \rightarrow a(xy) = a^2xy = axay$.
41. First note that any field containing Z and i must contain $Q[i]$. Then prove $(a + bi)/(c + di) \in Q[i]$.
43. The subfield of E is $\{ab^{-1} \mid a, b \in D, b \neq 0\}$.
45. The set of even integers is a subring of the rationals.
47. Try $ab^{-1} \rightarrow a/b$.
49. Say 1_R is the unity of R and 1_S is the unity of S. Pick a $\in R$ such that $a\phi \neq 0$. Then $1_S(a\phi) = (1_R a)\phi = (1_R \phi)(a\phi)$. Now cancel. For the example, consider the mapping from Z to $2Z$ that sends n to $2n$.
51. Certainly, the unity 1 is contained in every subfield. So, if a field has characteristic p, the subfield $\{0, 1, \ldots, p - 1\}$ is contained in every subfield. If a field has characteristic 0, then $\{(n1)(m1)^{-1} \mid n, m \in Z, m \neq 0\}$ is a subfield contained in every subfield. This subfield is isomorphic to Q (map $(n1)(m1)^{-1}$ to n/m).

CHAPTER 18

You know my methods, apply them! Sherlock Holmes, *The Hound of the Baskervilles*

1. $f + g = 3x^4 + 2x^3 + 2x + 2$
   $f \cdot g = 2x^7 + 3x^6 + x^5 + 2x^4 + 3x^2 + 2x + 2$
3. Let $f(x) = x^4 + x$ and $g(x) = x^2 + x$. Then $f(0) = 0 = g(0); f(1) = 2 = g(1);$
   $f(2) = 0 = g(2)$.
5. Use Corollary 1 of Theorem 18.2.
7. Since R is isomorphic to the subring of constant polynomials, char $R \leq$ char $R[x]$. On the other hand, char $R = c$ implies

$$c(a_nx^n + \cdots + a_0) = (ca_n)x^n + \cdots + (ca_0) = 0$$

so that char $R[x] \leq$ char R.
9. Use exercise 8 and observe that $a_n\phi x^n + \cdots + a_0\phi = b_n\phi x^n + \cdots + b_0\phi$ if and only if $a_i\phi = b_i\phi$ for all i.
11. quotient $2x^2 + 2x + 1$; remainder 2
13. It is its own inverse.
15. no (See exercise 16.)
17. Observe that $Z[x]/\langle x \rangle$ is isomorphic to Z. Now use Theorems 16.3 and 16.4.
19. Use Corollary 3 of Theorem 18.2.
21. If $f(x) \neq g(x)$, then deg $[f(x) - g(x)] <$ deg $p(x)$. But the minimum degree of any member of $\langle p(x) \rangle$ is deg $p(x)$.
23. Start with $(x - 1/2)(x + 1/3)$ and clear fractions.
25. See exercise 2.
27. By the Corollary to Theorem 18.3, $I = \langle x - 1 \rangle$.
29. Use the Factor Theorem.
31. For any a in $U(p)$, $a^{p-1} = 1$ so every member of $U(p)$ is a zero of $x^{p-1} - 1$. Now use the Factor Theorem and a degree argument.
33. Use exercise 31.
35. Observe that, modulo 101, $(50!)^2 = (50!)(-1)(-2) \cdots (-50) = (50!)(100)(99) \cdots (51) = 100!$ and use exercise 32.
37. Take $R = Z$ and $I = \langle 2 \rangle$.
39. Mimic Example 3.
41. Write $f(x) = (x - a)g(x)$. Use the product rule to compute $f'(x)$.

CHAPTER 19

Experience enables you to recognize a mistake when you make it again. Franklin P. Jones

1. $f(x)$ factors over D as $ah(x)$ where a is not a unit in D.
3. Apply Einstein's Criterion.
5. a. If $f(x) = g(x) h(x)$, then $af(x) = ag(x) h(x)$.
   b. If $f(x) = g(x) h(x)$, then $f(ax) = g(ax) h(ax)$.
   c. If $f(x) = g(x) h(x)$, then $f(x + a) = g(x + a)h(x + a)$.
7. Find an irreducible polynomial $p(x)$ of degree 2 over Z_5. Then $Z_5[x]/\langle p(x) \rangle$ is a field of order 25.
9. Note that -1 is a root. No, since 4 is not a prime.
11. Use direct calculations to show that it has no zeros.

13. $(x + 3)(x + 5)(x + 6)$
15. a. Consider the number of distinct expressions of the form $(x - c)(x - d)$.
    b. Reduce the problem to (a).
17. Use exercise 16.
19. $x^n + p$ where p is prime is irreducible over Q.
21. $x^2 + 1$, $x^2 + x + 2$, $x^2 + 2x + 2$.
23. 1 has multiplicity 1, 3 has multiplicity 2.
25. We know $a_n(r/s)^n + a_{n-1}(r/s)^{n-1} + \cdots + a_0 = 0$. So $a_n r^n + a_{n-1}sr^{n-1} + \cdots + s^n a_0 = 0$. This shows that $s \mid a_n r^n$ and $r \mid s^n a_0$. Now use Euclid's Lemma and the fact that r and s are relatively prime.
27. Use exercise 3a and clear fractions.
29. If there is an a in Z_p such that $a^2 = -1$, then $x^4 + 1 = (x^2 + a)(x^2 - a)$.
    If there is an a in Z_p such that $a^2 = 2$, then $x^4 + 1 = (x^2 + ax + 1)(x^2 - ax + 1)$.
    If there is an a in Z_p such that $a^2 = -2$, then $x^4 + 1 = (x^2 + ax - 1)(x^2 - ax - 1)$. Now show that one of these three cases must occur.
31. Since $(f + g)(a) = f(a) + g(a)$ and $(f \cdot g)(a) = f(a)g(a)$ the mapping is a homomorphism. Clearly, $p(x)$ belongs to the kernel. By Theorem 19.5, $\langle p(x) \rangle$ is a maximal idea so $\langle p(x) \rangle$ = kernel.
33. Use the Corollary to Theorem 19.4 and exercise 5b with $a = -1$.
35. The analysis is identical except $0 \le q, r, t, u \le n$. Now just as when $n = 2$, we have $q = r = t = 1$. But this time $0 \le u \le n$. However, when $u > 2$, $P(x) = x(x + 1)(x^2 + x + 1) \cdot (x^2 - x + 1)^u$ has $(-u + 2)x^{2u+3}$ as one of its terms. Since the coefficient of x^{2u+3} represents the number of dice with the label $2u + 3$, the coefficient cannot be negative. Thus, $u \le 2$, as before.
37. Although the probabilities of rolling any particular sum is the same with either pair of dice, the probability of rolling doubles is different (1/6 with ordinary dice, 1/9 with Sicherman dice). Thus the probability of going to jail is different. Other probabilities are also affected. For example, if in jail one cannot land on Virginia by rolling a pair of 2's with Sicherman dice but one is twice as likely to land on St. James with a pair of 3's with the Sicherman dice as with ordinary dice.

CHAPTER 20

He thinks things through very carefully before going off half-cocked.
General Carl Spaatz, in *Presidents Who Have Known Me*, George E. Allen

1. Say r is irreducible and u is a unit. If $ru = ab$ where a and b are not units, then $r = a(bu^{-1})$ where a and bu^{-1} are not units.
3. Clearly, $\langle ab \rangle \subseteq \langle b \rangle$. If $\langle ab \rangle = \langle b \rangle$, then $b = rab$ so that $1 = ra$ and a is a unit.
5. Say $x = a + bi$ and $y = c + di$. Then

$$xy = (ac - bd) + (bc + ad)i$$

So

$$d(xy) = (ac - bd)^2 + (bc + ad)^2 = (ac)^2 + (bd)^2 + (bc)^2 + (ad)^2$$

On the other hand,

$$d(x)d(y) = (a^2 + b^2)(c^2 + d^2) = a^2c^2 + b^2d^2 + b^2c^2 + a^2d^2$$

7. Suppose $a = bu$ where u is a unit. Then $d(b) \le d(bu) = d(a)$. Also, $d(a) \le d(au^{-1}) = d(b)$.

9. $3 \cdot 7$ and $(1 + 2\sqrt{-5})(1 - 2\sqrt{-5})$. Mimic Example 7 to show that these are irreducible.

11. Mimic Example 7 and observe $10 = 2 \cdot 5$ and $10 = (2 - \sqrt{-6})(2 + \sqrt{-6})$. A PID is a UFD.

13. Suppose $3 = \alpha\beta$ where α, $\beta \in Z[i]$ and neither is a unit. Then $9 = d(3) = d(\alpha)d(\beta)$ so that $d(\alpha) = 3$. But there are no integers such that $a^2 + b^2 = 3$. Observe that $2 = -i(1 + i)^2$ and $5 = (1 + 2i)(1 - 2i)$.

15. Say $a = bu$ where u is a unit. Then $ra = rbu = (ru)b \in \langle b \rangle$ so that $\langle a \rangle \subseteq \langle b \rangle$. By symmetry, $\langle b \rangle \subseteq \langle a \rangle$. If $\langle a \rangle = \langle b \rangle$, then $a = bu$ and $b = av$. Thus $a = avu$ and $vu = 1$.

17. Use exercise 14 with $d = -1$. 5 and $1 + 2i$; 13 and $3 + 2i$; 17 and $4 + i$.

19. Mimic Example 1.

21. Use the fact that x is a unit if and only if $N(x) = 1$.

23. See Example 2.

25. ± 1, $\pm i$

27. $(-1 + \sqrt{5})(1 + \sqrt{5}) = 4 = 2 \cdot 2$. Now use exercise 20.

29. Use exercise 28 and Theorem 16.4.

31. Suppose R satisfies the ascending chain condition and there is an ideal I of R that is not finitely generated. Then pick $a_1 \in I$. Since I is not finitely generated, $\langle a_1 \rangle$ is a proper subset of I so we may choose $a_2 \in I$ but $a_2 \notin \langle a_1 \rangle$. As before $\langle a_1, a_2 \rangle$ is proper so we may choose $a_3 \in I$ but $a_3 \notin \langle a_1, a_2 \rangle$. Continuing in this fashion, we obtain a chain of infinite length $\langle a_1 \rangle \subset \langle a_1, a_2 \rangle \subset \langle a_1, a_2, a_3 \rangle \subset \cdots$. Now suppose every ideal of R is finitely generated and there is a chain $I_1 \subseteq I_2 \subseteq I_3 \subseteq \cdots$. Let $I = \cup I_i$. Then $I = \langle a_1, a_2, \ldots, a_n \rangle$. Since $I = \cup I_i$ each a_i belongs to some member of the union, say $I_{i'}$. Letting $k = \max \{i' \mid i = 1, \ldots, n\}$, we see that all $a_i \in I_k$. Thus $I \subseteq I_k$ and the chain has length at most k.

33. Say $I = \langle a + bi \rangle$. Then $a^2 + b^2 + I = (a + bi)(a - bi) + I = I$ and $a^2 + b^2 \in I$. For any $c, d \in Z$, let $c = q_1(a^2 + b^2) + r_1$ and $d = q_2(a^2 + b^2) + r_2$ where $0 \le r_1, r_2 < a^2 + b^2$. Then $c + di + I = r_1 + r_2i + I$.

SUPPLEMENTARY EXERCISES FOR CHAPTERS 17–20

Yes I get by with a little help from my friends.

John Lennon and Paul McCartney, *With a little help from my friends*, Sgt. Pepper's Lonely Hearts Club Band

1. Use Theorem 17.3, supplementary exercise 8 for Chapters 13–16, Theorem 16.4 and Example 10 of Chapter 16.

3. To show the isomorphism, use the First Isomorphism Theorem.

5. Use the First Isomorphism Theorem.

7. Consider the obvious homomorphism from $Z[x]$ onto $Z_2[x]$. Then use the First Isomorphism Theorem and Theorem 16.3.

9. As in exercise 13 of Chapter 6 the mapping is onto, 1-1, and preserves multiplication. Also $a^{-1}(x + y)a = a^{-1}xa + a^{-1}ya$ so that it preserves addition as well.

11. $Z[i]/\langle 2 + i \rangle = \{0 + \langle 2 + i \rangle, 1 + \langle 2 + i \rangle, 2 + \langle 2 + i \rangle, 3 + \langle 2 + i \rangle, 4 + \langle 2 + i \rangle\}$ Note that

$$5 + \langle 2 + i \rangle = (2 + i)(2 - i) + \langle 2 + i \rangle = 0 + \langle 2 + i \rangle$$

13. Observe that $(3 + 2\sqrt{2})(3 - 2\sqrt{2}) = 1$.

15. We are given $(k + 1)^2 = k + 1 \bmod n$. So, $k^2 + 2k + 1 = k + 1 \bmod n$ or $k^2 = -k = n - k$ $\bmod n$. Also, $(n - k)^2 = n^2 - 2nk + k^2 = k^2 \bmod n$ so $(n - k)^2 = n - k \bmod n$.

17. Suppose $0 \le a < p^k$ and $a^2 = a \bmod p^k$. Then $p^k \mid a(a - 1)$ and since a and $a - 1$ are relatively prime $p^k \mid a$ or $p^k \mid a - 1$. Thus, $a = 0$ or $a - 1 = 0$.
19. Use the Mod 2 Irreducibility Test.
21. Use Theorem 16.4.
23. Use Theorem 16.4.
25. Say a/b, $c/d \in R$. Then $(ad - bc)/bd$ and $ac/bd \in R$ by Euclid's Lemma. The field of quotients is Q.
27. $Z[i]/\langle 3 \rangle$ is a field and $Z_3 \oplus Z_3$ is not.

CHAPTER 21

When I was young I observed that nine out of every ten things I did were failures, so I did ten times more work.   George Bernard Shaw

1. $\mathbf{R}^n$ has basis $(1, 0, \ldots, 0), (0, 1, 0, \ldots, 0), \ldots, (0, 0, \ldots, 1)$.
   $M_2(Q)$ has basis $\begin{bmatrix} 1 & 0 \\ 0 & 0 \end{bmatrix}, \begin{bmatrix} 0 & 1 \\ 0 & 0 \end{bmatrix}, \begin{bmatrix} 0 & 0 \\ 1 & 0 \end{bmatrix}, \begin{bmatrix} 0 & 0 \\ 0 & 1 \end{bmatrix}$
   $Z_p[x]$ has basis $1, x, x^2, \ldots$
   $\mathbf{C}$ has basis $1, i$
3. $(a_2 x^2 + a_1 x + a_0) - (a_2' x^2 + a_1' x + a_0') = (a_2 - a_2')x^2 + (a_1 - a_1')x + (a_0 - a_0')$ and $a(a_2 x^2 + a_1 x + a_0) = aa_2 x^2 + aa_1 x + aa_0$. A basis is $\{1, x, x^2\}$. Yes.
5. linearly dependence since $-3(2, -1,0) - (1, 2, 5) + (7, -1, 5) = (0, 0, 0)$
7. Suppose $au + b(u + v) + c(u + v + w) = 0$. Then $(a + b + c)u + (b + c)v + cw = 0$. Since $\{u, v, w\}$ are linearly independent we obtain $c = 0$, $b + c = 0$, and $a + b + c = 0$. So, $a = b = c = 0$.
9. If the set is linearly independent, it is a basis. If not, then delete one of the vectors that is a linear combination of the others (see exercise 8). This new set still spans V. Repeat this process until we obtain a linearly independent subset. This subset will still span V since we only deleted vectors that are linear combinations of the remaining ones.
11. Let u_1, u_2, u_3 be a basis for U and w_1, w_2, w_3 be a basis for W. Use the fact that $u_1, u_2, u_3, w_1, w_2, w_3$ are linearly dependent over F. In general, if $\dim U + \dim W > \dim V$, then $U \cap W \ne \{0\}$.
13. no
15. yes; 2
17. $\begin{bmatrix} a & a + b \\ a + b & b \end{bmatrix} - \begin{bmatrix} a' & a' + b' \\ a' + b' & b' \end{bmatrix} = \begin{bmatrix} a - a' & a + b - a' - b' \\ a + b - a' - b' & b - b' \end{bmatrix}$
   and $c \begin{bmatrix} a & a + b \\ a + b & b \end{bmatrix} = \begin{bmatrix} ac & ac + bc \\ ac + bc & bc \end{bmatrix}$
19. If v and $v' \in U \cap W$ and a is a scalar, then $v + v' \in U$, $v + v' \in W$, $av \in U$, and $av \in W$. So $U \cap W$ is a subspace. (See exercise 11.) If $u_1 + w_1$ and $u_2 + w_2 \in U + W$, then $(u_1 + w_1) - (u_2 + w_2) = (u_1 - u_2) + (w_1 - w_2) \in U + W$ and $a(u_1 + w_1) = au_1 + aw_1 \in U + W$.
21. p^n
23. 2
25. Yes, because Z_7 is a field and, therefore, $1/2$, $-2/3$, and $-1/6$ exist in Z_7. Specifically, $1/2 = 4$, $-2/3 = 4$, $-1/6 = 1$.

27. If V and W are vector spaces over F, then the mapping must preserve addition and scalar multiplication. That is, $T{:}V \to W$ must satisfy $(u + v)T = uT + vT$ for all vectors u and v in V, and $a(uT) = (au)T$ for all vectors u in V and scalars a in F.

29. Suppose v and u belong to the kernel and a is a scalar. Then $(v + u)T = vT + uT = 0 + 0 = 0$ and $(av)T = a(vT) = a \cdot 0 = 0$.

31. Let $\{v_1, v_2, \ldots, v_n\}$ be a basis for V. Map $a_1v_1 + a_2v_2 + \cdots + a_nv_n$ to $(a_1, a_2, \ldots, a_n)$.

CHAPTER 22

Well here's another clue for you all.   John Lennon and Paul McCartney, *Glass Onion*

1. Compare with exercise 24 of the supplementary exercises for Chapters 13–16.
3. $Q(\sqrt{-3})$
5. $Q(\sqrt{-3})$
7. Note that $x = \sqrt{1 + \sqrt{5}}$ implies $x^4 - 2x^2 - 4 = 0$.
9. $a^5 = a^2 + a + 1$; $a^{-2} = a^2 + a + 1$; $a^{100} = a^2$
11. The set of all expressions of the form

$$(a_n\pi^n + a_{n-1}\pi^{n-1} + \cdots + a_0)/(b_m\pi^m + b_{m-1}\pi^{m-1} + \cdots + b_0)$$

where $b_m \neq 0$.

13. Hint: Note that any isomorphism must act as the identity on the rationals.
15. Hint: Use exercise 42 of Chapter 15.
17. $a = 4/3$, $b = 2/3$, $c = 5/6$
19. Use the fact that $1 + i = -(4 - i) + 5$ and $4 - i = 5 - (1 + i)$.
21. If the zeros of $f(x)$ are $a_1, a_2, \ldots, a_n$ then the zeros of $f(x + a)$ are $a_1 - a, a_2 - a, \ldots, a_n - a$. Now use exercise 20.
23. Q and $Q(\sqrt{2})$

CHAPTER 23

Work is the greatest thing in the world, so we should always save some of it for tomorrow.
Don Herald

1. It follows from Theorem 23.1 that if $p(x)$ and $q(x)$ are both monic irreducible polynomials in $F[x]$ with $p(a) = q(a) = 0$, then deg $p(x) =$ deg $q(x)$. If $p(x) \neq q(x)$, then $(p - q)(a) = p(a) - q(a) = 0$ and $\deg(p(x) - q(x)) < \deg p(x)$, contradicting Theorem 23.1. To prove Theorem 23.3, use the Division Algorithm (Theorem 18.2).
3. If $f(x) \in F[x]$ does not split in E, then it has a nonlinear factor $q(x)$ which is irreducible over E. But then $E[x]/\langle q(x)\rangle$ is a proper algebraic extension of E.
5. Use exercise 4.
7. Suppose $Q(\sqrt{a}) = Q(\sqrt{b})$. If $\sqrt{b} \in Q$, then $\sqrt{a} \in Q$ and we may take $c = \sqrt{a}/\sqrt{b}$. If $\sqrt{b} \notin Q$, then $\sqrt{a} \notin Q$. Write $\sqrt{a} = r + s\sqrt{b}$. It follows that $r = 0$ and $a = bs^2$. The other half follows from exercise 20 of Chapter 22.
9. Observe that $[F(a){:}F]$ must divide $[E{:}F]$.
11. Note that $[F(a, b){:}F]$ is divisible by both $m = [F(a){:}F]$ and $n = [F(b){:}F]$.
13. Let $g(x) \in F(a^3)[x]$ be the minimal polynomial for a over $F(a^3)$. Since a is a zero of $x^3 - a^3 = (x - a)(x^2 + ax + a^2)$, $g(x) = x - a$ or $g(x) = x^3 - a^3$.

15. Suppose $E_1 \cap E_2 \neq F$. Then $[E_1{:}E_1 \cap E_2] [E_1 \cap E_2{:}F] = [E_1{:}F]$ implies $[E_1{:}E_1 \cap E_2] = 1$ so that $E_1 = E_1 \cap E_2$. Similarly, $E_2 = E_1 \cap E_2$.

17. E must be an algebraic extension of **R** so that $E \subseteq$ **C**. But then $[C{:}E][E{:}R] = [C{:}R] = 2$.

19. Let a be a zero of $p(x)$ in some extension of F. First note $[E(a){:}E] \leq [F(a){:}F] = \deg p(x)$. Then observe that $[E(a){:}F(a)][F(a){:}F] = [E(a){:}E][E{:}F]$. This implies $\deg p(x)$ divides $[E(a){:}E]$ so that $\deg p(x) = [E(a){:}E]$.

21. Hint: if $\alpha + \beta$ and $\alpha\beta$ are algebraic, then so is $\sqrt{(\alpha + \beta)^2 - 4\alpha\beta}$.

23. $\sqrt{b^2 - 4ac}$

25. Use the Factor Theorem.

27. Say a is a generator of F^*. Then $F = Z_p(a)$ and it suffices to show a is algebraic over Z_p. If $a \in Z_p$, we are done. Otherwise, $1 + a = a^k$ for some $k \neq 0$. If $k > 0$, we are done. If $k < 0$, then $a^{-k} + a^{1-k} = 1$ and we are done.

CHAPTER 24

Tell me tell me tell me come on tell me the answer.   John Lennon and Paul McCartney, *Helter Skelter*

1. $[GF(729){:}GF(9)] = 3$
   $[GF(64){:}GF(8)] = 2$

3. Use Theorem 24.2

5. Let a be a zero of $f(x)$ in some extension and write $f(x) = (x - a) g(x)$. Then $|Z_2(a){:}Z_2| = 3$. If $f(x)$ splits in $Z_2(a)$, we are done. If not, let b be a zero of $g(x)$ is some extension of $Z_2(a)$. Then $|Z_2(a,b){:}Z_2| = 2$. Thus $|Z_2(a, b){:}Z_2| = 6$ and $Z_2(a, b)$ is the splitting field of $f(x)$. (See exercise 25 of Chapter 23.)

7. Use the fact that $|GF(8){:}GF(2)| = 3$ and $GF(8)$ is the splitting field of $x^8 - x$.

9. Direct calculations show that given $x^3 + 2x + 1 = 0$, we have $x^2 \neq 1$ and $x^{13} \neq 1$.

11. Note that the group has prime order.

13. Find a cubic irreducible polynomial $p(x)$ over Z_3, then $Z_3[x]/\langle p(x)\rangle$ is a field of order 27.

15. Use long division.

17. Theorem 24.3 reduces the problem to constructing the subgroup lattices for Z_{18} and Z_{30}.

19. identical

21. Consider $g(x) = x^2 - a$. Note that $|GF(p)[x]/\langle g(x)\rangle| = p^2$ so that $g(x)$ has a zero in $GF(p^2)$. Now use Theorem 24.3.

23. Use exercise 22.

25. Since F^* is a cyclic group of order 124, it has a unique subgroup of order 2.

27. Use exercise 42 of Chapter 15

CHAPTER 25

Why, sometimes I've believed as many as six impossible things before breakfast.   Lewis Carroll

1.

5. Use $\sin^2 \theta + \cos^2 \theta = 1$.

7. Use $\cos 2\theta = 2 \cos^2 \theta - 1$.

9. Observe that $\sin 15° \in Q(\sqrt{2}, \sqrt{3})$.

11. Solving two linear equations with coefficients from F involves only the operations of F.

13. Use Theorem 19.1 and exercise 25 of Chapter 19.

15. Try $\theta = 90°$.

17. If so, then an angle of $40°$ is constructible. Now use exercise 10.

19. No, since $[Q(\sqrt[3]{3}):Q] = 3$.

21. No, since $[Q(\sqrt[3]{\pi}):Q]$ is infinite.

SUPPLEMENTARY EXERCISES FOR CHAPTERS 21–25

The things taught in colleges and schools are not an education, but the means of education.
Ralph Waldo Emerson, *Journals*

1. Use Theorem 22.5.

3. Suppose b is one solution of $x^n = a$. Since F^* is a cyclic group of order $q - 1$, it has a cyclic subgroup of order n, say $\langle c \rangle$. Then each member of $\langle c \rangle$ is a solution to the equation $x^n = 1$. It follows that $b\langle c \rangle$ is the solution set of $x^n = a$.

5. $(5a^2 + 2)/a = 5a + 2a^{-1}$. Now observe that since $a^2 + a + 1 = 0$ we know $a(-a - 1) = 1$ so that $a^{-1} = -a - 1$. Thus $(5a^2 + 2)/a = -2 + 3a$.

7. 5

9. Since $F(a) = F(a^{-1})$, we have degree of $a = [F(a):F] = [F(a^{-1}):F] = $ degree of a^{-1}.

11. If ab is a zero of $c_n x^n + \cdots + c_1 x + c_0 \in F[x]$, then a is a zero of $c_n b^n x^n + \cdots + c_1 bx + c_0 \in F(b)[x]$.

13. Every element of $F(a)$ can be written in the form $f(a)/g(a)$ where $f(x), g(x) \in F[x]$. If $f(a)/g(a)$ is algebraic and not in F, then there is some $h(x) \in F[x]$ such that $h(f(a)/g(a)) = 0$. By clearing fractions and collecting like powers of a, we obtain a polynomial in a with coefficients from F equal to 0. But then a would be algebraic over F.

CHAPTER 26

All wish to possess knowledge, but few, comparatively speaking, are willing to pay the price.
Juvenal

1. $a = e^{-1}ae$; $c^{-1}ac = b$ implies $a = cbc^{-1}$; $a = x^{-1}bx$ and $b = y^{-1}cy$ imply $a = x^{-1}y^{-1}cyx = (yx)^{-1}cyx$.

3. $cl(a) = \{a\}$ if and only if for all x in G, $x^{-1}ax = a$. This is equivalent to $a \in Z(G)$.

5. See Example 3 of Chapter 5.

7. Use exercise 7 of Supplementary Exercises for Chapters 5–8.

9. Use exercise 40 of Chapter 10 and exercise 7 of the Supplementary Exercises for Chapter 5–8.

11. 8

13. 15

15. By exercise 14, G has seven subgroups of order 3.

17. 10; $\langle(123)\rangle, \langle(234)\rangle, \langle(134)\rangle, \langle(345)\rangle, \langle(245)\rangle$.

19. If $p \nmid q - 1$ and $q \nmid p^2 - 1$, then a group of order $p^2 q$ is Abelian.

21. 21

23. By Sylow, $n_7 = 1$ or 15 and $n_5 = 1$ or 21. Counting elements reveals that at least one of these must be 1. Then the product of the Sylow 7-subgroup and the Sylow 5-subgroup is a subgroup of order 35.

25. By Sylow, $n_{17} = 1$ or 35. Assume $n_{17} = 35$. Then the union of the Sylow 17-subgroups has 561 elements. By Sylow, $n_5 = 1$. Thus we may form a cyclic subgroup of order 85 (exercise 40 of Chapter 10 and Theorem 26.7). But then there are 64 elements of order 85. This gives too many elements.

27. Use the "G/Z Theorem" (10.3).

29. Let H be the Sylow 3-subgroup and suppose the Sylow 5-subgroups are not normal. By Sylow, there must be six Sylow 5-subgroups, call them $K_1, \ldots, K_6$. These subgroups have 24 elements of order 5. Also, the cyclic subgroups $HK_1, \ldots, HK_6$ each have eight generators. Thus there are 48 elements of order 15.

31. Mimic the proof of Sylow's First Theorem.

33. Use exercise 32.

35. Automorphisms preserve order.

37. Use the fact that HK is a cyclic group of order 15.

39. Normality of H implies $cl(h) \subseteq H$ for h in H. Now observe that $h \in cl(h)$. This is true only when H is normal.

41. The mapping from H to $x^{-1}Hx$ given by $h \to x^{-1}hx$ is an isomorphism.

43. $|C(a)|/|G|$.

45. $Pr(D_4) = 5/8$, $Pr(S_3) = 1/2$, $Pr(A_4) = 1/3$

CHAPTER 27

Learn to reason forward and backward on both sides of a question.   Thomas Blandi

1. Use the 2-odd Test.

3. Use the Index Theorem.

5. Suppose G is a simple group of order 525. Let L_7 be a Sylow 7-subgroup of G. It follows from Sylow's theorems that $|N(L_7)| = 35$. Let L be a subgroup of $N(L_7)$ of order 5. Since $N(L_7)$ is cyclic (Theorem 26.7), $N(L) \geq N(L_7)$ so that 35 divides $|N(L)|$. But L is contained in a Sylow 5-subgroup (Theorem 26.4) which is Abelian (see the corollary of Theorem 26.2). Thus 25 divides $|N(L)|$ as well. It follows that 175 divides $|N(L)|$. The Index Theorem now yields a contradiction.

7. $n_{11} = 12$. Use the N/C Theorem (Example 13 of Chapter 11) to show that there is an element of order 22, but A_{12} has no element of order 22.

9. A_{12} has no element of order 33.

11. If we can find a pair of distinct Sylow 2-subgroups A and B so that $|A \cap B| = 8$, then $N(A \cap B) \geq AB$ so that $N(A \cap B) = G$. Now let H and K be any distinct pair of Sylow 2-subgroups. Then $16 \cdot 16/|H \cap K| = |HK| \leq 112$ (supplementary exercise 7 for Chapters 5–8) so that $|H \cap K|$ is at least 4. If $|H \cap K| = 4$, then $N(H \cap K)$ picks up at least eight elements from H and at least eight from K (see exercise 32 of Chapter 26). Thus $|N(H \cap K)| \geq 16$ and is divisible by 8. So $|N(H \cap K)| = 16, 56,$ or 112. The last two cases give us a normal subgroup. If $|N(H \cap K)| = 16$, then $N(H \cap K)$ and H have at least eight elements in common and so do $N(H \cap K)$ and K. This guarantees the A and B.

15. Use the Index Theorem.

17. The Sylow Test for Nonsimplicity forces $n_5 = 6$ and $n_3 = 4$ or 10. The Index Theorem rules out $n_3 = 4$. Now appeal to Sylow's Third Theorem (26.5).

19. Let α be as in the proof of the Generalized Cayley Theorem. Then Ker $\alpha \leqslant H$ and $|G/\text{Ker } \alpha|$ divides $|G{:}H|!$ Now show $|\text{Ker } \alpha| = |H|$.

21. $n_5 = 6$ and $n_3 = 10$ or 40. If there are two Sylow 2-subgroups L_2 and $L_2{'}$ whose intersection has order 4, show that $N(L_2 \cap L_2{'})$ has index at most 5. Now use the Embedding Theorem. If $n_3 = 40$, the union of all the Sylow subgroups has more than 120 elements. If $n_3 = 10$, use the N/C Theorem to show that there is an element of order 6. But A_6 has no element of order 6.

23. Mimic the proof that A_5 is simple.

25. By direct computation, show that PSL(2, 7) has more than four Sylow 3-subgroups, more than one Sylow 7-subgroup, and more than one Sylow 2-subgroup. Hint: Observe that $\begin{bmatrix} 2 & 1 \\ 2 & 3 \end{bmatrix}$ has order 3. Now use conjugation to find four other subgroups of order 3; observe that $\left|\begin{bmatrix} 5 & 5 \\ 1 & 4 \end{bmatrix}\right| = 7$ and use conjugation to find another subgroup of order 7; observe that $\left|\begin{bmatrix} 5 & 1 \\ 3 & 5 \end{bmatrix}\right| = 8$ and use conjugation to find another subgroup of order 8. Now, argue as we did to show that A_5 is simple. In the cases that the supposed normal subgroup N has order 2 or 4, show that in G/N, the Sylow 7-subgroup is normal. But then, G has a normal subgroup of order 14 or 28, which were already ruled out.

27. Mimic exercise 26.

29. Let $p = [G{:}H]$ and $q = [G{:}K]$. It suffices to show that $p = q$. But if $p < q$, say, then $q \nmid p!$ so $|G| \nmid p!$ This contradicts the Index Theorem.

CHAPTER 28

The dictionary is the only place where success comes before work.   Arthur Brisbane

1. Let a be any reflection in D_n and let $b = aR_{360/n}$. Then $aZ(D_n)$ and $bZ(D_n)$ have order 2 and generate $D_n/Z(D_n)$. Now use Theorem 28.5 and the fact that $|D_n/Z(D_n)| = n = |D_{n/2}|$.

3. Since $b = b^{-1}$, we have $bab = a^2$. Then $a = a^6 = (bab)^3 = ba^3b$ so that $ba = a^3b$. Thus $a^3b = a^2b$ and $a = e$.

5. Let F be the free group on $\{a_1, a_2, \ldots, a_n\}$. Let N be the smallest normal group containing $\{w_1, w_2, \ldots, w_t\}$ and M be the smallest normal subgroup containing $\{w_1, w_2, \ldots, w_t, w_{t+1}, \ldots, w_{t+k}\}$. Then $F/N \approx G$ and $F/M \approx \bar{G}$. The mapping from F/N to F/M given by $aN \rightarrow aM$ induces a homomorphism from G onto $\bar{G}$.

   To prove the corollary observe that the theorem shows that K is a homomorphic image of G so that $|K| \leqslant |G|$.

7. Clearly, a and ab belong to $\langle a, b \rangle$, so $\langle a, ab \rangle \subseteq \langle a, b \rangle$. Now show that a and b belong to $\langle a, ab \rangle$.

9. Use Theorem 28.5.

11. Since $x^2 = y^2 = e$, we have $(xy)^{-1} = y^{-1}x^{-1} = yx$. Also $xy = z^{-1}yz$ so that $(xy)^{-1} = (z^{-1}yz)^{-1} = z^{-1}y^{-1}z = z^{-1}yz = xy$.

13. a. $b^6a^3 = a$ b. b^7a

15. Center is $\langle x^2 \rangle$. $|xy| = 8$.

17. Use the fact that the mapping from G onto G/N by $x \rightarrow xN$ is a homomorphism.

19. This is equivalent to showing that every left coset of $\langle b \rangle$ has the form $a^i\langle b \rangle$. But $\langle b \rangle$ absorbs all powers of b and commutes with all powers of a. So, if w is a word in a and b, then $w\langle b \rangle = a^i\langle b \rangle$ where i is the sum of all the exponents of a.

21. 6; The given relations imply that $a^2 = e$. G is isomorphic to Z_6.
23. 1 and n have order 2 and are generators.

CHAPTER 29

If at first you don't succeed—that makes you about average. Bradenton, *Florida Herald*

1. If T is a distance-preserving function and the distance between points a and b is positive, then the distance between aT and bT is positive.
3.

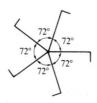

5. 12
7. $4n$
9. a. Z_2
   b. $Z_2 \oplus Z_2$
   c. $G \oplus Z_2$ where G is isomorphic to the plane symmetry group of a circle.
11. 6
13. An inversion in $\mathbf{R}^3$ leaves only a single point fixed while a rotation leaves a line fixed.
15. In $\mathbf{R}^4$, a plane is fixed. In $\mathbf{R}^n$, a subspace of dimension $n - 2$ is fixed.
17. Let T be an isometry, p, q, and r the three noncolinear points, and s any other point in the plane. Then the quadrilateral determined by pT, qT, rT, and sT is congruent to the one formed by p, q, r, and s. Thus sT is uniquely determined by pT, qT, and rT.
19. a rotation

CHAPTER 30

The thing that counts is not what we know but the ability to use what we know. Leo L. Spears

1. Try $x^n y^m \to (n, m)$.
3. xy
5. Use Figure 30.9.
7. $x^2 yzxz = x^2 yx^{-1} = x^2 x^{-1} y = xy$
   $x^{-3} zxzy = x^{-3} x^{-1} y = x^{-4} y$
9. A subgroup of index 2 is normal.
11. a. V b. I c. II d. VI e. VII f. III
13. *cmm*
15. a. *p4m* b. *p3* c. *p31m* d. *p6m*
17. No. This print has S_4 as a subgroup but "Flying Fish" does not. The elements of $Z \oplus Z \oplus Z$ do not commute with the elements of S_4.
19. *p4g* (ignoring shading)
21. a. VI b. V c. I d. III e. IV f. VII g. IV

CHAPTER 31

Oh when you were young, did you question all the answers.   Graham Nash, *Wasted on the Way*, single

1. $4 * (b, a)$
3. $(m/2) * \{3 * [(a, 0), (b, 0)], (a, 0), (e, 1), 3 * (a, 0), (b, 0), 3 * (a, 0), (e, 1)\}$
5. $a^3 b$
7. Both yield paths from e to $a^3 b$.
11. Say we start at x. Then we know the vertices $x, xs_1, xs_1 s_2, \ldots, xs_1 s_2 \ldots s_{n-1}$ are distinct and $x = xs_1 s_2 \ldots s_n$. So if we apply the same sequence beginning at y, then cancellation shows that $y, ys_1, ys_1 s_2, \ldots, ys_1 s_2 \ldots s_{n-1}$ are distinct and $y = ys_1 s_2 \ldots s_n$.
13. If there were a Hamiltonian path from $(0, 0)$ to $(2, 0)$, there would be a Hamiltonian circuit in the digraph since $(2, 0) + (1, 0) = (0, 0)$.
15. a. If $s_1, s_2, \ldots, s_{n-1}$ traces a Hamiltonian path and $s_i s_{i+1} \ldots s_j = e$, then the vertex $s_1 s_2 \ldots s_{i-1}$ appears twice. Conversely, if $s_i s_{i+1} \ldots s_j \neq e$, then the sequence $e, s_1, s_1 s_2, \ldots, s_1 s_2 \ldots s_{n-1}$ yields the n vertices (otherwise, cancellation gives a contradiction).
    b. This is immediate from (1).
17. The sequence traces the digraph in a clockwise fashion.
19. Abbreviate $(a, 0)$, $(b, 0)$, and $(e, 1)$ by a, b and 1 respectively. A circuit is $4 * (4 * 1, a)$, $3 * a, b, 7 * a, 1, b, 3 * a, b, 6 * a, 1, a, b, 3 * a, b, 5 * a, 1, a, a, b, 3 * a, b, 4 * a, 1, 3 * a, b, 3 * a, b, 3 * a, b$.
21. Abbreviate $(R_{90}, 0)$, $(H, 0)$, and $(R_0, 1)$ by R, H, and 1 respectively. A circuit is $3 * (R, 1, 1)$, $H, 2 * (1, R, R), R, 1, R, R, 1, H, 1, 1$.
23. Abbreviate $(a, 0)$, $(b, 0)$, and $(e, 1)$ by a, b, and 1 respectively. A circuit is $2 * (1, 1, a), a, b, 3 * a, 1, b, b, a, b, b, 1, 3, * a, b, a, a$.
25. Abbreviate $(r, 0)$, $(f, 0)$ and $(e, 1)$ by r, f, and 1 respectively. Then the sequence is $r, r, f, r, r, 1, f, r, r, f, r, 1, r, f, r, r, f, 1, r, r, f, r, r, 1, f, r, r, f, r, 1, r, f, r, r, f, 1$.
27. $m * ((n - 1) * (0, 1), (1, 1))$
29. Abbreviate $(r, 0)$, $(f, 0)$ and $(e, 1)$ by r, f and 1 respectively. A circuit is $1, r, 1, 1, f, r, 1, r, 1, r, f, 1$.
31. In the proof of Theorem 31.3, we used the hypothesis that G is Abelian in two places: We needed H to satisfy the induction hypothesis, and we needed to form the factor group G/H. Now, if we assume only that G is Hamiltonian, then H also is Hamiltonian and G/H exists.

CHAPTER 32

We must view with profound respect the infinite capacity of the human mind to resist the introduction of useful knowledge.   Thomas R. Lounsbury

1. $\text{wt}(0001011) = 3$; $\text{wt}(0010111) = 4$; $\text{wt}(0100101) = 3$.
3. 1000110; 1110100.
5. 000000, 100011, 010101, 001110, 110110, 101101, 011011, 111000
7. Let v be any vector. If u is a vector of weight 1, then $\text{wt}(v)$ and $\text{wt}(v + u)$ have opposite parity. Since any vector u of odd weight is the sum of an odd number of vectors of weight 1, it follows that $\text{wt}(v)$ and $\text{wt}(v + u)$ have opposite parity. Now, mimic the proof of exercise 19 of Chapter 5.
9. Observe that a vector has even weight if and only if it can be written as a sum of an even number of vectors of weight 1.
11. No, by Theorem 32.3.

13. 0000000, 1000111, 0100101, 0010110, 0001011, 1100010, 1010001, 1001100, 0110011, 0101110, 0011101, 1110100, 1101001, 1011010, 0111000, 1111111

$$H = \begin{bmatrix} 1 & 1 & 1 \\ 1 & 0 & 1 \\ 1 & 1 & 0 \\ 0 & 1 & 1 \\ 1 & 0 & 0 \\ 0 & 1 & 0 \\ 0 & 0 & 1 \end{bmatrix}$$

yes

15. Suppose u is decoded as v and x is the coset leader of the row containing u. Coset decoding means v is at the head of the column containing u. So, $x + v = u$ and $x = u - v$. Now suppose $u - v$ is a coset leader and u is decoded as y. Then y is at the head of the column containing u. Since v is a code word, $u = u - v + v$ is in the row containing $u - v$. Thus $u - v + y = u$ and $y = v$.

17. 000000, 100110, 010011, 001101, 110101, 101011, 011110, 111000

$$H = \begin{bmatrix} 1 & 1 & 0 \\ 0 & 1 & 1 \\ 1 & 0 & 1 \\ 1 & 0 & 0 \\ 0 & 1 & 0 \\ 0 & 0 & 1 \end{bmatrix}$$

001001 is decoded as 001101 by all four methods.
011000 is decoded as 111000 by all four methods.
000110 is decoded as 100110 by all four methods.
There are three code words whose distance from 100001 is 2, so there is no way to correct the errors; parity-check matrix decoding does not decide 10001; standard-array and syndrome methods decode 100001 as 000000, 110101, or 101011, depending upon which vectors are used as coset leaders.

19. For any received word w, there are only eight possibilities for wH But each of these eight possibilities satisfies condition 2 or condition 3' on page 437 so decoding assumes no error was made or one error was made.

21. There are 3^4 code words and 3^6 possible received words.

23. No; row 3 is twice row 1.

25. No. For if so, nonzero code words would be all words with weight at least 5. But this set is not closed under addition.

27. Use exercise 24 together with the fact that the set of codewords is closed under addition.

29. Abbreviating the coset $a + \langle x^2 + x + 1 \rangle$ with a, the following generating matrix will produce the desired code:

$$\begin{bmatrix} 1 & 0 & 1 & 1 & x \\ 0 & 1 & x & x+1 & x+1 \end{bmatrix}$$

31. Use exercise 14.

33. Let $c, c' \in C$. Then, $c + (v + c') = v + c + c' \in v + C$ and $(v + c) + (v + c') = c + c' \in C$, so the set $C \cup (v + C)$ is closed under addition.

CHAPTER 33

Wisdom rises upon the ruins of folly. Thomas Fuller, *Gnomologia*

1. Note that $1\phi = 1$. Thus $n\phi = n$. Also, $1 = 1\phi = (nn^{-1})\phi = (n\phi)(n^{-1}\phi) = n(n^{-1}\phi)$ so that $1/n = n^{-1}\phi$.

3. If a and b are fixed by elements of H, so are $a + b$, $a - b$, $a \cdot b$, and a/b.

5. It suffices to show that each member of $G(K/F)$ defines a permutation on the a_i's. Let $\alpha \in G(K/F)$ and write

$$f(x) = c_n x^n + c_{n-1}x^{n-1} + \cdots + c_0 = c_n(x - a_1)(x - a_2) \cdots (x - a_n)$$

   Then $f(x) = f(x)\alpha = c_n(x - a_1\alpha)(x - a_2\alpha) \cdots (x - a_n\alpha)$.
   Thus $f(a_i) = 0$ implies $a_i = a_j\alpha$ for some j so that α permutes the a_i's.

7. By exercise 6 of Chapter 19, the splitting field is $Q(\sqrt{2}, i)$. Since $[Q(\sqrt{2}, i):Q] = 4$, $|G(E/Q)| = 4$. It follows that $G(E/Q) = \{\varepsilon, \alpha, \beta, \alpha\beta\}$ where $\sqrt{2}\,\alpha = -\sqrt{2}$ and $i\alpha = i$, $\sqrt{2}\,\beta = \sqrt{2}$, and $i\beta = -i$ and the proper subfields of E are Q, $Q(\sqrt{2})$, $Q(\sqrt{-2})$, and $Q(i)$. β has fixed field $Q(\sqrt{2})$, α has fixed field $Q(i)$, and $\alpha\beta$ has fixed field $Q(\sqrt{-2})$. No automorphism of E has fixed field Q.

9. $|G(E/Q)| = |E:Q| = 4$
   $|G(Q(\sqrt{10})/Q)| = [Q(\sqrt{10}):Q] = 2$

11. Recall A_4 has no subgroup of order 6. (See Example 13 of Chapter 10.)

13. Use Sylow's Theorem.

15. Let $\omega = \cos(360°/n) + i\sin(360°/n)$. Then $Q(\omega)$ is the splitting field. Every α in $G(K/F)$ has the form $\omega\alpha = \omega^i$. Since such elements commute the group is Abelian.

17. Use the lattice of Z_{10}.

19. Z_6 (Be sure you know why the group is cyclic.)

21. See exercise 24 of Chapter 27.

23. Use exercise 31 of Chapter 26.

25. Let $\{e\} = H_0 \subset H_1 \subset \cdots \subset H_n = G$ be the series that shows that G is solvable. Then $H_0 \cap H \subset H_1 \cap H \subset \cdots \subset H_n \cap H$ shows that H is solvable.

CHAPTER 34

Won't you please help me John Lennon and Paul McCartney, *Help*

1. a. $a \vee (a' \wedge b) = (a \vee a') \wedge (a \vee b) = 1 \wedge (a \vee b) = a \vee b$

    b. $(a \wedge b) \vee (a \wedge b') \vee (a' \wedge b') = [a \wedge (b \vee b')] \vee (a' \wedge b')$
    $$= (a \wedge 1) \vee (a' \wedge b')$$
    $$= a \vee (a' \wedge b')$$
    $$= (a \vee a') \wedge (a \vee b')$$
    $$= 1 \wedge (a \vee b') = a \vee b'$$

    c. This is the dual to (b).

3. Use Theorem 34.2.

5. $f_2 + f_8 = f_{10}; f_2 \cdot f_8 = f_0$
   $f_7 + f_{14} = f_{15}; f_7 \cdot f_{14} = f_6$
   $f_7' = f_8, f_{14}' = f_1$

7. $a'b + b'c + c'd$

9. $ab + (a + b')c(a + d) + a + b' + d$

11.

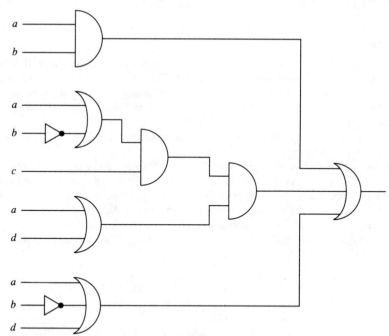

13.   This follows directly from the fact that x and $21/x$ are relatively prime.

15.   Let $a = \{1\}$, $b = \{2\}$, $c = \{1, 2\}$, and $\vee$ be set union, then $a \vee c = b \vee c$ but $a \neq b$.

17.   Suppose $(B_1, +, \cdot, ')$ and $(B_2, \vee, \wedge, -)$ are Boolean algebras. An isomorphism ϕ from B_1 to B_2 is a one-to-one, onto mapping from B_1 to B_2 such that $(a + b)\phi = a\phi \vee b\phi$; $(a \cdot b)\phi = a\phi \wedge b\phi$; $a'\phi = \overline{a\phi}$.

19.

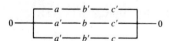

21.

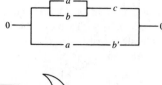

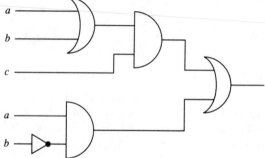

SUPPLEMENTARY EXERCISES FOR CHAPTERS 26–34

Give me your answer.   John Lennon and Paul McCartney, *When I'm sixty-four*, Sgt. Pepper's Lonely Hearts Club Band

1. Z_6
3. Let $|G| = 315$ and let H be a Sylow 3-subgroup and K a Sylow 5-subgroup. If $H \lhd G$, then $HK = 45$. If H is not normal, then by Sylow's Third Theorem, $|G/N(H)| = 7$, so that $|N(H)| = 45$.
5. Observe that $K \subseteq N(H)$ implies HK is a group of order 245. Now, use the previous exercise.
7. Note that $g^{-1}Kg \subseteq g^{-1}Hg = H$. Now, use the corollary to Sylow's Third Theorem.
9. Use the same proof as for exercise 40 of Chapter 10.
11. Since $n_7 = 8$, we know by the Embedding Theorem (Chapter 27) that $G \leqslant A_8$. But A_8 does not have an element of order 21.
13. Let G be a non-Abelian group of order 105. By Theorem 10.3, $G/Z(G)$ is not cyclic. So $|Z(G)| \neq 3, 7, 15, 21,$ or 35. This leaves only 1 or 5 for $|Z(G)|$. Let H, K, and L be Sylow 3-, Sylow 5- and Sylow 7-subgroups of G, respectively. Now, counting shows that $K \lhd G$ or $L \lhd G$. Thus, $|KL| = 35$ and KL is a cyclic subgroup of G. But, KL has 24 elements of order 35 (since $|U(Z_{35})| = 24$). Thus, a counting argument shows that $K \lhd G$ and $L \lhd G$. Now, $|HK| = 15$ and HK is a cyclic subgroup of G. Thus, $HK \subseteq C(K)$ and $KL \subseteq C(K)$. This means that 105 divides $|C(K)|$. So $K \subseteq Z(G)$.
15.

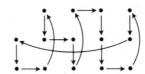

17. It suffices to show that x travels by a implies xab^{-1} travels by a (for we may successively replace x by xab^{-1}). If xab^{-1} traveled by b then the vertex xa would appear twice in the circuit.
19. a. $\{00, 11\}$
    b. $\{000, 111\}$
    c. $\{0000, 1100, 1010, 1001, 0101, 0110, 0011, 1111\}$
    d. $\{0000, 1100, 0011, 1111\}$
21. The mapping $T_v:F^n \rightarrow \{0, 1\}$ given by $uT_v = u \cdot v$ is an onto homomorphism. Therefore, $|F^n|/\text{Ker } T_v| = 2$.
23. It follows from exercise 18 that if C is an (n, k) linear code, then $C^\perp$ is an $(n, n - k)$ linear code. Thus, in this problem, $k = n - k$. To prove the second claim, use exercise 21, the definition of $C^\perp$, and the hypothesis is that $C^\perp = C$.
25. Let the two normal series with Abelian factors be

$$\{e\} = N_0 \lhd N_1 \lhd \cdots \lhd N_t = N$$

and

$$N/N = H_0/n \lhd H_1/N \lhd \cdots \lhd H_s/N = G/N$$

Then

$$\{e\} = N_0 \lhd N_1, \lhd \cdots \lhd N_t \lhd H_1 \lhd \cdots \lhd H_s = G$$

27. $a \lor b\lor c = a \land b \land c$ implies that $a' \land [a \lor (b \lor c)] = a' \land a \land b \land c = 0$. Thus, $(a' \land a) \lor [a' \land (b \lor c)] = 0$ so that $a' \land (b \lor c) = 0$. Then $a \lor [a' \land (b \lor c)] = a \lor 0 = a$. This implies that $(a \lor a') \land (a \lor b \lor c) = a$. It follows that $a \lor b \lor c = a$. By symmetry, $a \lor b \lor c = b$ and $a \lor b \lor c = c$.

Notations

SET THEORY

$\bigcap_{i \in I} S_i$ intersection of sets S_i, $i \in I$

$\bigcup_{i \in I} S_i$ union of sets S_i, $i \in I$

$[a]$ $\{x \in S \mid x \sim a\}$, equivalence class of S containing a

SPECIAL SETS

Z integers, additive group of integers, ring of integers

Q rational numbers, field of rational numbers

Q^+ multiplicative group of positive rational numbers

$F^{\#}$ set of nonzero elements of F

$\mathbf{R}$ real numbers, field of real numbers

$\mathbf{R}^+$ multiplicative group of positive real numbers

$\mathbf{C}$ complex numbers

FUNCTIONS AND ARITHMETIC

f^{-1} the inverse of the function f

$g \circ f$, $\alpha\beta$ composite function

$\phi(a)$, $a\phi$ image of a under ϕ

$\phi \colon A \to B$ mapping of A to B

$t \mid s$ t divides s

$t \nmid s$ t does not divide s

$\gcd(m, n)$ greatest common divisor of integers m and n

$\operatorname{lcm}(s, t)$ least common multiple of the integers s and t

ALGEBRAIC SYSTEMS

(The number after each item indicates where the notation is defined.)

D_4 group of symmetries of a square, dihedral group of order eight, 25

D_n	dihedral group of order $2n$, 26
e	identity element, 34
Z_n	cyclic group $\{0, 1, \ldots, n - 1\}$ under addition modulo n, 35
det A	the determinant of A, 36
$GL(2, F)$	2×2 matrices of nonzero determinant with coefficients from the field F (the general linear group), 36
$U(n)$	group of units modulo n (that is, the set of integers less than n and relatively prime to n under multiplication modulo n), 36
$\mathbf{R}^n$	$\{(a_1, a_2, \ldots, a_n) \mid a_1, a_2, \ldots, a_n \in \mathbf{R}\}$, 37
$SL(2, F)$	group of 2×2 matrices over F with determinant 1, 38
$a^{-1}, -a$	multiplicative inverse of a, additive inverse of a, 41
$|G|$	order of the group G, 54
$|g|$	order of the element g, 54
$H \leq G$	subgroup inclusion, 53
$H < G$	subgroup $H \neq G$, 53
$\langle a \rangle$	$\{a^n \mid n \in Z\}$, cyclic group generated by a, 57
$Z(G)$	$\{x \in G \mid xy = yx \text{ for all } y \in G\}$, the center of G, 58
$C(a)$	$\{g \in G \mid ga = ag\}$, the centralizer of a in G, 60
$C(H)$	$\{x \in G \mid xh = hx \text{ for all } h \in H\}$, the centralizer of H, 62
$\phi(n)$	Euler phi function of n, 71
$N(H)$	$\{x \in G \mid x^{-1}Hx = H\} = \{x \in G \mid Hx = xH\}$, the normalizer of H in G, 80
G^n	$\{g^n \mid g \in G\}$, 81
S_n	group of one-to-one functions from $\{1, 2, \ldots, n\}$ to itself, 85
A_n	alternating group of degree n, 91
$G \approx H$	G and H are isomorphic, 100
ϕ_a	mapping given by $x\phi_a = a^{-1}xa$ for all x, 106
Aut(G)	group of automorphisms of the group G, 106
Inn(G)	group of inner automorphisms of G, 106
$G_1 \oplus G_2 \oplus \cdots \oplus G_n$	external direct product of groups $G_1, G_2, \ldots, G_n$, 113
HK	$\{hk \mid h \in H, k \in K\}$, 119
$H \times K$	internal direct product of H and K, 119
$H_1 \times H_2 \times \cdots \times H_n$	internal direct product of $H_1, \ldots, H_n$, 121
$U_k(n)$	$\{x \in U(n) \mid x = 1 \bmod k\}$, 122
G'	commutator subgroup, 128
$\oplus_{i=1}^n G_i$	$G_1 \oplus G_2 \oplus \cdots \oplus G_n$, 129
xH	$\{xh \mid h \in H\}$, 130
$|G{:}H|$	the index of H in G, 133
stab$_G(a)$	$\{\alpha \in G \mid a\alpha = a\}$, the stabilizer of a under the permutation group G, 135

$\text{orb}_G(i)$	$\{i\alpha \mid \alpha \in G\}$, the orbit of i under the permutation group G, 135
$H \lhd G$	H is a normal subgroup of G, 145
G/N; R/A	factor group; factor ring, 146, 211
$\text{Ker } \phi$	kernel of the homomorphism ϕ, 159
$g\phi^{-1}$	inverse image of g under ϕ, 161
$K\phi^{-1}$	inverse image of K under ϕ, 161
$Z[x]$	ring of polynomials with integer coefficients, 188
$M_2(Z)$	ring of all 2×2 matrices with integer entries, 188
$R_1 \oplus R_2 \oplus \cdots \oplus R_n$	direct sum of rings, 189
$U(R)$	group of units of the ring R, 190
nZ	ring of multiples of n, 194
$Z[i]$	ring of Gaussian integers, 194
$Z_n[i]$	ring of Gaussian integers modulo n, 197
$\langle a \rangle$	principal ideal generated by a, 211
$A + B$	sum of ideals A and B, 216
AB	product of ideals A and B, 216
$\text{Ann}(A)$	annihilator of A, 217
$\sqrt{A}$	nil radical of A, 217
$F(x)$	field of quotients of $F[x]$, 228
$R[x]$	ring of polynomials over R, 233
$\deg f(x)$	degree of the polynomial $f(x)$, 235
$g(x) \mid f(x)$	$g(x)$ divides $f(x)$, 238
$\Phi_p(x)$	pth cyclotomic polynomial, 246
$M_2(Q)$	ring of 2×2 matrices over Q, 280
$\langle v_1, v_2, \ldots, v_n \rangle$	subspace spanned by $v_1, v_2, \ldots, v_n$, 281
$F(a_1, a_2, \ldots, a_n)$	extension of F by $a_1, a_2, \ldots, a_n$, 289
$f'(x)$	the derivative of $f(x)$, 293
$[E:F]$	index of E over F, 300
$GF(p^n)$	Galois field of order p^n, 310
$GF(p^n)^*$	nonzero elements of $GF(p^n)$, 311
$\text{cl}(a)$	$\{x^{-1}ax \mid x \in G\}$, the conjugacy class of a, 327
$Pr(G)$	probability that two elements from G commute, 329
n_p	the number of Sylow p-subgroups of a group, 332
$PSL(n, q)$	$SL(n, q)/Z(SL(n, q))$, the projective special linear group, 350
$W(S)$	set of all words from S, 359
$\langle x, y, \ldots, z \mid w_1 = w_2 = \cdots = w_t \rangle$	group with generators $x, y, \ldots, z$ and relations $w_1 = w_2 = \cdots = w_t$, 362
D_∞	infinite dihedral group, 365
$\text{Cay}(S{:}G)$	Cayley digraph of the group G with generating set S, 404
$k * (a, b, \ldots, c)$	concatenation of k copies of $(a, b, \ldots, c)$, 413
(n, k)	linear code, k dimensional subspace of F^n, 433
F^n	$F \times F \times \cdots \times F$, direct product of n copies of the field F, 433
$d(u, v)$	Hamming distance between vectors u and v, 434

$\text{wt}(u)$	the number of nonzero components of the vector u (the Hamming weight of u), 434
$G(E/F)$	the automorphism group of E fixing F, 448
E_H	fixed field of H, 449
$a \wedge b$	a and b, 464
$a \vee b$	a or b, 464
a'	complement of a, 465
$a \cdot b$	switches a and b connected in series, 466
$a + b$	switches a and b connected in parallel, 466
$A \wedge B$	A and B (where A and B are propositions), 469
$A \vee B$	A or B (where A and B are propositions), 469
A'	negation of A (where A is a proposition), 470
$C^{\perp}$	dual code of a code C, 476

Index of Mathematicians

(Biographies appear on pages in boldface).

Index of Terms